Human Molecular Genetics

Human Molecular Genetics

Edited by Mark Williams

AMERICAN
MEDICAL PUBLISHERS
www.americanmedicalpublishers.com

American Medical Publishers,
41 Flatbush Avenue,
1st Floor, New York,
NY 11217, USA

Visit us on the World Wide Web at:
www.americanmedicalpublishers.com

ISBN: 978-1-63927-254-9

Cataloging-in-Publication Data

Human molecular genetics / edited by Mark Williams.
     p. cm.
Includes bibliographical references and index.
ISBN 978-1-63927-254-9
1. Human molecular genetics. 2. Medical genetics. 3. Human genetics.
4. Molecular genetics. I. Williams, Mark.
RB155.5 .H86 2022
616.042--dc23

Table of Contents

Preface

The biological field in which the functions and structure of human genes are studied at the molecular level is referred to as human molecular genetics. It uses various tools from molecular biology and genetics. The molecular basis of biological activity between biomolecules in the diverse systems of a cell is studied under molecular biology. Genetics deals with the study of genes, heredity and genetic variation in organisms. Human molecular genetics finds its application in the study of developmental biology as well as in the treatment of genetic diseases. There are diverse techniques that are used within this field such as forward genetics, reverse genetics and DNA replication. This book attempts to understand the multiple branches that fall under human molecular genetics and how such concepts have practical applications. The topics covered herein deal with the core aspects of this field. This book includes contributions of experts and scientists which will provide innovative insights into this field.

This book has been the outcome of endless efforts put in by authors and researchers on various issues and topics within the field. The book is a comprehensive collection of significant researches that are addressed in a variety of chapters. It will surely enhance the knowledge of the field among readers across the globe.

It gives us an immense pleasure to thank our researchers and authors for their efforts to submit their piece of writing before the deadlines. Finally in the end, I would like to thank my family and colleagues who have been a great source of inspiration and support.

Editor

The human protein disulfide isomerase gene family

James J Galligan[1] and Dennis R Petersen[2*]

Abstract

Enzyme-mediated disulfide bond formation is a highly conserved process affecting over one-third of all eukaryotic proteins. The enzymes primarily responsible for facilitating thiol-disulfide exchange are members of an expanding family of proteins known as protein disulfide isomerases (PDIs). These proteins are part of a larger superfamily of proteins known as the thioredoxin protein family (TRX). As members of the PDI family of proteins, all proteins contain a TRX-like structural domain and are predominantly expressed in the endoplasmic reticulum. Subcellular localization and the presence of a TRX domain, however, comprise the short list of distinguishing features required for gene family classification. To date, the *PDI* gene family contains 21 members, varying in domain composition, molecular weight, tissue expression, and cellular processing. Given their vital role in protein-folding, loss of PDI activity has been associated with the pathogenesis of numerous disease states, most commonly related to the unfolded protein response (UPR). Over the past decade, UPR has become a very attractive therapeutic target for multiple pathologies including Alzheimer disease, Parkinson disease, alcoholic and non-alcoholic liver disease, and type-2 diabetes. Understanding the mechanisms of protein-folding, specifically thiol-disulfide exchange, may lead to development of a novel class of therapeutics that would help alleviate a wide range of diseases by targeting the UPR.

Keywords: Disulfide bond, Thioredoxin, Calsequestrin, UPR, Unfolded protein response, ER stress

Introduction

Increasing evidence supports an important role for misfolded proteins in the pathogenesis of numerous diseases including diabetes, Alzheimer disease, Parkinson disease, and both alcoholic and non-alcoholic liver disease [1]. Accumulation of misfolded proteins (or erred protein load) is generally caused by either decreased disposal of erred protein or a decrease in the correct folding of proteins [1]. Disulfide bond formation represents a fundamentally important post-translational modification that is a critical step in the folding of nascent peptides in the endoplasmic reticulum (ER) [2]. These covalent linkages are formed between the side-chains of cysteine residues and represent a key rate-limiting step in protein maturation [3]. The enzymatic formation, breakage, and subsequent rearrangement of cysteine linkages are crucial to protein structure and function and primarily mediated

by members of the protein disulfide isomerase (PDI) family [4]. All genes in the *PDI* family are part of a superfamily referred to as the thioredoxin (TRX) superfamily, which also includes the glutaredoxins, TRXs, ferroredoxins, and peroxidoxins [5].

The *PDI* gene family currently comprises 21 genes, varying in size, expression, localization, and enzymatic function. Although it is implied that all members of the PDI family possess the ability to rearrange disulfide bonds, only a subset is considered orthologous and able to carry out these reactions, with the other members being paralogous and linked to the family through evolution, not function [4]. While these proteins may be functionally different, the unifying feature of all PDI family members is the presence of a TRX-like domain [2]. These may be present as either a catalytically active **a** or **a'** domain (the presence of a CXXC motif) or a catalytically inactive **b** or **b'** domain (for a more detailed review on the precise role of these domains, see the work of Ellgaard et al.) [2,4]. Extensive research has assessed the roles of these domains and revealed the **b'** domain to be

* Correspondence: Dennis.Petersen@UCDenver.edu
[2]Molecular Toxicology and Environmental Health Sciences Program, Department of Pharmacy and Pharmaceutical Sciences, University of Colorado Anschutz Medical Campus, Aurora, CO 80045, USA
Full list of author information is available at the end of the article

the primary peptide- or protein-binding domain [4]. Previous literature has highlighted the features of a number of *PDI* family members; however, with an increasing amount of cDNA and EST sequence information deposited in the NCBI database, a composite review is required to further characterize and compare all 21 current members of the *PDI* gene family (Table 1).

Domain composition of the PDI family proteins

Proteins in the PDI family are largely expressed in the ER, although few family members have been detected in other subcellular compartments [6]. Due to their localization, the presence of a short NH_2-terminal signal peptide exists in all members of the family. These peptides are typically 15–30 amino acids (a.a.) in length and are cleaved upon entry into the ER [7]; this has led to some sequence discrepancy among multiple PDI proteins. As indicated, the common thread between all members of the PDI proteins is the presence of at least one TRX-like domain, whether it is catalytically active (**a**) or inactive (**b**) [2]. The active site of the **a**-type domains also varies greatly, with the "classical" motif being comprised of Cys-Gly-His-Cys. The cysteine

residues in these active sites are considered redox active, undergoing active shuffling of disulfide bonds [2]. The surrounding a.a. largely play a role in the regulation of the pKa of the cysteines, dictating the local redox potential and thus regulating the catalytic ability of these cysteines to actively oxidize or reduce disulfide bonds (for a more comprehensive review on the redox potential of PDI, see the work of Hatahet et al.) [3]. Extensive biochemical and biophysical experimentation has taken place analyzing TRX–like domain containing proteins; however, a complete crystal structure is currently not available for most family members. Another common characteristic of the PDI family of proteins is the presence of a COOH-terminal ER retention sequence comprised of a Lys-Asp-Glu-Leu-like sequence [2]. Whereas these sequences may differ greatly in a.a. composition, only four PDI proteins do not contain this sequence. Figure 1 shows the domain composition of the 21 proteins in the PDI family.

Evolutionary divergence of the *PDI* gene family

As mentioned, all genes encompassed in the *PDI* family belong to the *TRX* superfamily of genes [5]. The unifying

Table 1 Human *PDI* genes as listed in the Human Gene Nomenclature Committee (HGNC) database

Gene name	Other aliases	Protein name	Chromosomal location	Amino acids
AGR2	XAG-2, HAG-2, AG2, PDIA17	Anterior gradient protein 2 homolog	7p21.3	175
AGR3	HAG3, hAG-3, BCMP11, PDIA18	Anterior gradient protein 3 homolog	7p21.1	166
CASQ1	PDIB1	Calsequestrin-1	1q21	396
CASQ2	PDIB2	Calsequestrin-2	1p13.3-p11	399
DNAJC10	MTHr, ERdj5	DnaJ (Hsp40) homolog, subfamily C, member 10	2q32.1	793
ERP27	FLJ32115, ERp27, PDIA8	Endoplasmic reticulum resident protein 27	12p12.3	273
ERP29	ERp28, ERp31, ERp29, PDI-DB, PDIA9	Endoplasmic reticulum resident protein 29	12q24.13	261
ERP44	KIAA0573, PDIA10	Endoplasmic reticulum resident protein 44	9q22.33	406
P4HB	DIA1, PROHB, DSI, GIT, PDI, PO4HB, P4Hb	Protein disulfide-isomerase	17q25	508
PDIA2	PDA2, PDIp	Protein disulfide-isomerase A2	16p13.3	525
PDIA3	P58, ERp61, ERp57, ERp60, GRP57, PI-PLC, HsT17083	Protein disulfide-isomerase A3	15q15	505
PDIA4	ERP70, ERP72	Protein disulfide-isomerase A4	7q35	645
PDIA5	PDIR, FLJ30401	Protein disulfide-isomerase A5	3q21.1	519
PDIA6	P5, ERp5	Protein disulfide-isomerase A6	2p25.1	440
PDILT	PDIA7, ERp65	Protein disulfide-isomerase-like protein of the testis	16p12.3	584
TMX1	TMX, PDIA11	Thioredoxin-related transmembrane protein 1	14q22.1	280
TMX2	PDIA12	Thioredoxin-related transmembrane protein 2	11cen-q22.3	296
TMX3	FLJ20793, KIAA1830, PDIA13	Protein disulfide-isomerase TMX3	18q22	454
TMX4	DJ971N18.2, KIAA1162, PDIA14	Thioredoxin-related transmembrane protein 4	20p12	349
TXNDC5	MGC3178, FLJ21353, FLJ90810, EndoPDI, Hcc-2, ERp46, PDIA15	Thioredoxin domain-containing protein 5	6p24.3	432
TXNDC12	TLP19, ERP18, ERP19, hAG-1, AGR1, PDIA16	Thioredoxin domain-containing protein 12	1p32.3	172

DNAJC10 has been included following identification of this gene in the *PDI* gene family. Synonyms of these genes have also been corrected.

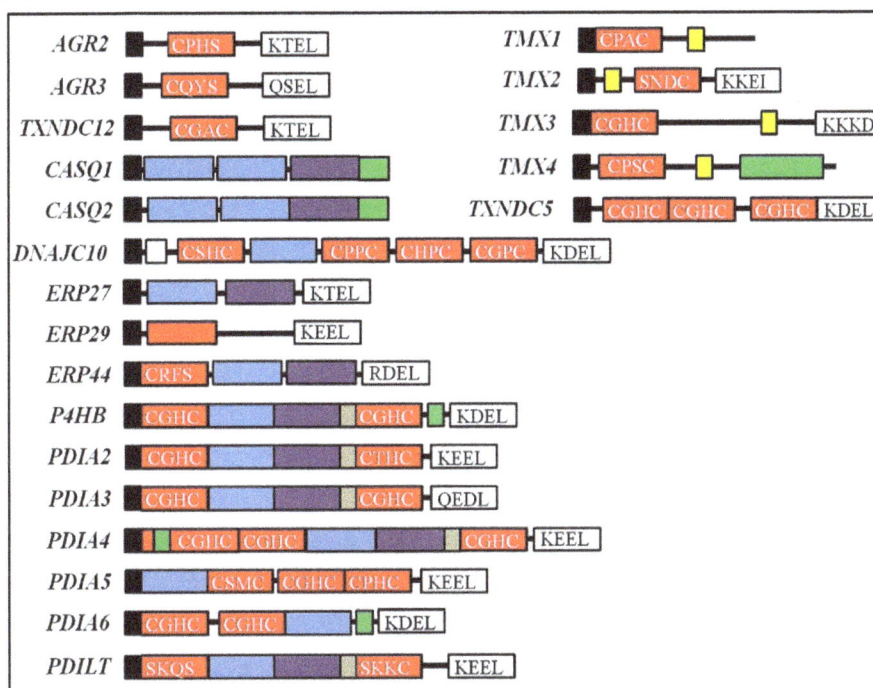

Figure 1 Schematic representation of the domain compositions of the 21 proteins in the *PDI* gene family. All proteins contain a short NH$_2$-terminal signal sequence designated in black. Catalytically active TRX-like domains (**a** or **a'**) are represented in *red* with active sites noted; inactive TRX domains in *blue* (**b**) and *purple* (**b'**); *green* represents the Asp/Glu rich Ca^{2+}-binding domains; linker regions (*gray*); transmembrane domains (*yellow*); COOH-terminal ER-retention sequences in *white* with a.a. composition denoted. Although ERP29 does contain an **a**-like domain, this is based on homology and not catalytic efficiency. Sequences and domain compositions were acquired and verified from the National Center for Biotechnology Information (NCBI) database. Figure was adapted and modified from [4].

theme between these proteins is the presence of at least one TRX-like domain, whether this be catalytically active (**a** or **a'**) or inactive (**b** or **b'**) [4]. These domains contain a TRX structural fold that has amino acids arranged in a conserved three-dimensional conformation [8]. While the enzymatic function of these domains is not conserved, the current theory proposes that all PDI family members evolved through domain duplications from an ancestral prokaryotic PDI which contained a single TRX domain [9]. Although all human PDIs possess a TRX-like domain, this remains one of the few similarities between these proteins, differing greatly in molecular mass and a.a. composition outside of the TRX fold. Phylogenetic analysis, presented as a dendrogram in Figure 2, outlines the evolutionary divergence of the human PDI proteins. Sequence analysis reveals a subset of genes within this family that are evolutionarily related, as shown by the calsequestrin (CASQ) and anterior gradient (AGR) branches (in red and blue, respectively). Supporting the hypothesis that these subsets of genes are phylogenetically related, a high degree of similarity was also observed with both domain architecture (Figure 1) and sequence identity (Table 2). Although similarities are evident, the overall homology between the proteins is quite minimal. This is supported by previous attempts

to cluster large sets of both eukaryotic and prokaryotic PDIs where marginal resolution of PDI domains was also observed [9]. Given the broad spectrum of both enzymatic functions and domain compositions, it is not surprising that the proteins share little sequence homology with one another (Table 2).

The human *PDI* gene family
DNAJC10
The *DNAJC10* gene is located at Chr 2q32.1 and encodes the 793-a.a. DNAJC10 protein (also commonly known as ERdj5 or MTHr) [10]. The *DNAJC10* gene consists of 3483 bp; transcription of two splice variants has been identified due to a skipped exon, resulting in a 138-bp (46 a.a.) absence, present between nucleotides 1243 and 1244 [10]. To date, a total of four missense single-nucleotide polymorphisms (SNPs) have been reported for ERdj5, located at a.a. 76 (Asp → Asn), 347 (Leu → Ile), 414 (Tyr → Cys) and 646 (His → Gln); ERdj5 also contains one potential Asn-linked glycosylation site, present at a.a. 530. Like other *PDI* family members, the *DNAJC10* promoter region contains a putative ER stress element (ERSE) box, yielding gene induction following ER stress [11]. Expression patterns of *DNAJC10* revealed ubiquitous expression with varying intensities and high

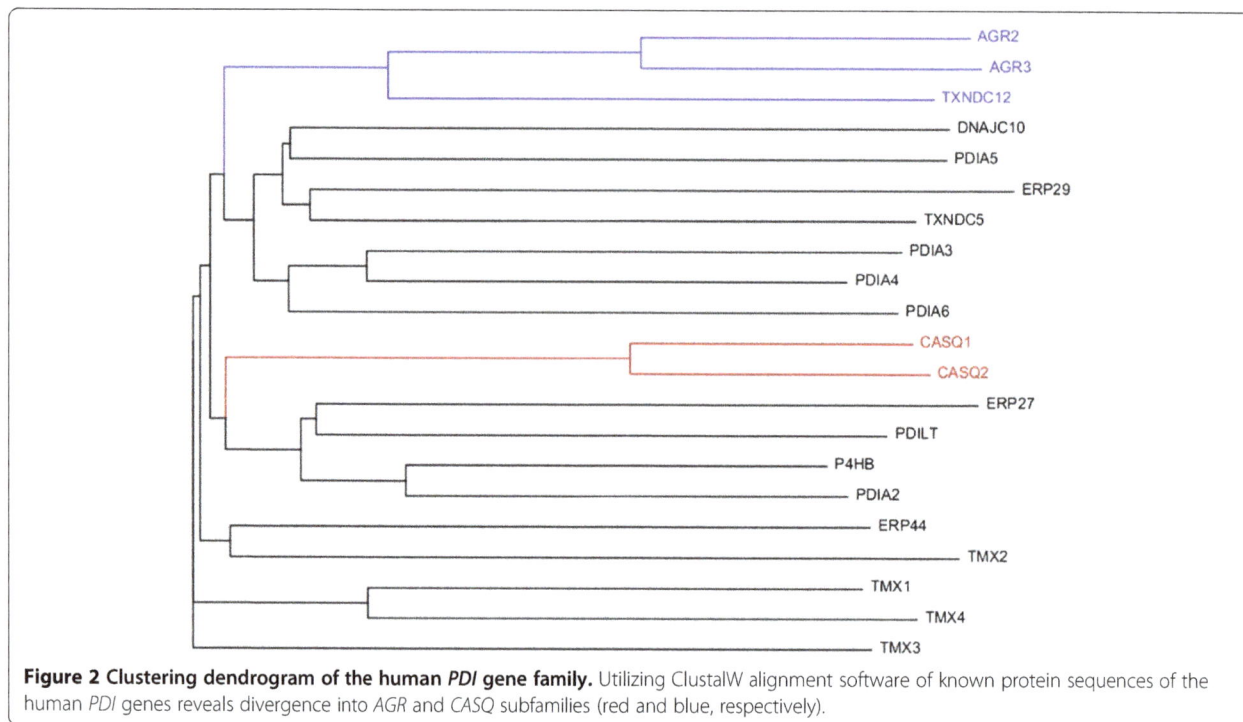

Figure 2 Clustering dendrogram of the human *PDI* gene family. Utilizing ClustalW alignment software of known protein sequences of the human *PDI* genes reveals divergence into *AGR* and *CASQ* subfamilies (red and blue, respectively).

levels of expression in secretory tissues such as the pancreas and testis [10,11]. In addition to the PDI family, DNAJC10 is also a member of the ERdj family, being comprised of an unverified NH_2-terminal signal peptide (32 a.a.), one DnaJ domain (which plays a major role in protein folding), five TRX domains (one **b** and four **a** type domains, active sites Cys-Ser-His-Cys, Cys-Pro-Pro-Cys, Cys-His-Pro-Cys, and Cys-Gly-Pro-Cys), and a COOH-terminal ER retention sequence (Lys-Asp-Glu-Leu) [10,12]. Despite the high number of TRX domains, DNAJC10 was found to possess roughly one-third the activity of P4HB and displayed no oxidase or isomerase activity [12]. These results were found to reflect an apparent redox equilibrium constant of 190 mM, roughly 100 times more reducing than that of the ER [12].

ERP27

Currently, little is known about the precise role and function of *ERP27*. *ERP27* was discovered in 2006 following a database search for novel human PDI family members [13]. The *ERP27* gene has been mapped to Chr 12p12.3 and encodes the 273-a.a protein. The ERP27 protein contains a cleavable NH_2-terminal signal sequence, leaving the mature protein to be 248 a.a, beginning at Glu-26 [13]. ERP27 does not contain a redox active Cys-X-X-Cys TRX motif, however, the protein does contain one **b** and one **b'** type domains and an ER retention sequence (Lys-Val-Glu-Leu) [13]. ERP27 was also found to directly interact with another PDI family member, PDIA3 [13]. Expression of ERP27 is fairly

ubiquitous, with highest expression found in pancreas [13]. To date, no knockout studies have been conducted and future studies are needed to fully understand the function and role of ERP27.

ERP29

The *ERP29* gene is located at Chr 12q24.13 and may represent a gene-duplication event with *ERP27*, given their close proximity. *ERP29* encodes a protein of 261 a. a. termed ERP29 (also known as ERP28). In a comprehensive study of the genomic organization of *ERP29*, Sargsyan et al. studied a 5'-flanking region consisting of ~2 kb, 3 exons, 2 introns and a 3'-flanking region of 0.31 kb [14]. ERP29 contains an NH_2-terminal signal peptide (32 a.a.), one **a** type TRX domain and a COOH-terminal ER-retention sequence (Lys-Glu-Glu-Leu) [15]. While the presence of one **a** type domain is present, ERP29 is unique in that it does not contain an active-site motif; assignment of the TRX domain is placed strictly on sequence homology with the **a**-type TRX domain, not on the activity of the domain. ERP29 contains two potential phosphorylation sites, both located on tyrosine residues, located at a.a. 64 and 66. ER localization of ERP29 was also confirmed using immunofluorescence and subcellular fractionation [15]. ERP29 is ubiquitously expressed with high levels being found in secretory tissues as well as the prostate, pancreas, and liver [14,16]. Although *ERP29* lacks any identified ERSE in its promoter region, ERP29 is described as an ER stress-inducible protein and has been shown to

Table 2 Amino acid similarity between human PDI proteins, as reported using EMBOSS Water Pairwise Alignment Algorithm (http://www.ebi.ac.uk/Tools/psa/emboss_water/)

	AGR2	AGR3	CASQ1	CASQ2	DNAJC10	ERP27	ERP29	ERP44	P4HB	PDIA2	PDIA3	PDIA4	PDIA5	PDIA6	PDILT	TMX1	TMX2	TMX3	TMX4	TXNDC5
AGR2																				
AGR3	65.0% (81.4%)																			
CASQ1	18.8% (34.8%)	17.4% (40.2%)																		
CASQ2	15.3% (37.4%)	17.40% (34.0%)	68.4% (86.1%)																	
DNAJC10	16.2% (37.2%)	19.0% (35.8%)	16.8% (33.3%)	18.2% (33.3%)																
ERP27	28.2% (41.0%)	20.5% (40.2%)	17.2% (34.4%)	19.3% (37.6%)	18.6% (35.8%)															
ERP29	20.6% (40.2%)	23.3% (39.5%)	18.2% (31.1%)	20.4% (33.0%)	18.1% (30.9%)	15.7% (38.9%)														
ERP44	22.2% (39.4%)	26.5% (47.0%)	19.2% (42.9%)	20.0% (40.8%)	20.4% (30.6%)	24.3% (40.5%)	25.3% (42.0%)													
P4HB	20.1% (34.3%)	21.7% (29.2%)	21.6% (39.3%)	22.9% (38.3%)	20.2% (36.0%)	32.5% (54.1%)	24.50% (39.9%)	23.5% (42.9%)												
PDIA2	23.5% (35.5%)	20.8% (38.7%)	21.2% (31.8%)	20.6% (38.7%)	20.8% (32.1%)	28.6% (49.0%)	23.3% (37.2%)	23.3% (37.7%)	47.3% (64.5%)											
PDIA3	21.7% (37.5%)	16.9% (37.9%)	21.5% (33.3%)	18.8% (31.3%)	21.4% (37.5%)	21.8% (38.9%)	22.3% (34.1%)	23.1% (40.6%)	33.6% (51.0%)	31.1% (49.5%)										
PDIA4	19.80% (31.6%)	19.7% (38.6%)	18.8% (38.2%)	21.4% (38.5%)	17.5% (30.8%)	18.3% (35.0%)	29.6% (48.1%)	24.0% (40.9%)	36.8% (55.1%)	29.1% (46.2%)	41.9% (60.2%)									
PDIA5	24.2% (45.1%)	22.1% (44.1%)	21.3% (35.1%)	21.7% (36.9%)	21.4% (40.0%)	19.6% (33.5%)	19.7% (33.9%)	22.6% (40.6%)	24.3% (38.3%)	23.1% (35.0%)	22.9% (37.1%)	21.2% (31.5%)								
PDIA6	27.6% (35.2%)	15.2% (57.6%)	24.0% (40.6%)	19.4% (33.1%)	32.7% (48.0%)	20.6% (34.8%)	17.0% (32.2%)	20.2% (38.4%)	23.5% (31.4%)	21.8% (28.6%)	24.2% (33.4%)	25.8% (40.0%)	28.1% (43.1%)							
PDILT	23.9% (45.7%)	16.2% (35.9%)	20.1% (43.4%)	21.1% (41.8%)	20.4% (36.1%)	29.7% (54.1%)	19.1% (38.5%)	21.9% (40.9%)	31.8% (52.4%)	32.1% (51.3%)	22.1% (37.1%)	26.0% (46.7%)	19.9% (35.3%)	23.5% (36.3%)						
TMX1	18.6% (36.1%)	24.8% (36.0%)	23.7% (37.8%)	17.2% (29.9%)	21.1% (38.2%)	23.9% (39.4%)	15.5% (32.3%)	25.8% (49.0%)	29.9% (46.9%)	25.7% (40.0%)	30.0% (50.0%)	25.1% (41.3%)	35.0% (56.4%)	36.2% (59.0%)	27.0% (39.3%)					
TMX2	23.0%	18.6%	20.2%	28.6%	23.3%	22.0%	25.6%	23.10%	25.0%	28.3%	25.8%	25.7%	16.4%	24.7%	24.7%	24.6%				

Table 2 Amino acid similarity between human PDI proteins, as reported using EMBOSS Water Pairwise Alignment Algorithm (http://www.ebi.ac.uk/Tools/psa/emboss_water/) (Continued)

	(36.1%)	(33.1%)	(39.9%)	(44.4%)	(44.8%)	(41.5%)	(43.6%)	(38.0%)	(40.9)	(44.6%)	(36.9%)	(43.7%)	(34.9%)	(38.9%)	(42.9%)	(43.1%)	
TMX3	20.0%	19.7%	20.3%	21.8%	22.4%	18.1%	20.5%	21.4%	27.8%	21.6%	23.1%	27.1%	23.5%	45.3%	19.2%	34.2%	18.7%
	(37.4%)	(36.5%)	(43.0%)	(43.6%)	(44.1%)	(33.3%)	(35.7%)	(34.7%)	(47.9%)	(40.3%)	(41.1%)	(42.7%)	(41.2%)	(64.2%)	(32.2%)	(50.5%)	(38.4%)
TMX4	23.5%	25.7%	27.0%	23.1%	23.7%	28.6%	19.5%	20.1%	21.8%	24.9%	24.3%	20.5%	27.2%	24.0%	25.0%	43.6%	24.4%	22.4%
	(39.5%)	(36.6%)	(62.2%)	(47.3%)	(39.4%)	(42.9%)	(36.0%)	(37.0%)	(34.4%)	(37.0%)	(45.7%)	(32.1%)	(41.6%)	(37.1%)	(42.5%)	(65.0%)	(42.2%)	(34.2%)
TXNDC5	21.6%	20.9%	22.9%	20.7%	25.4%	22.0%	20.8%	33.3%	26.2%	26.9%	24.4%	23.5%	27.3%	33.0%	23.3%	24.6%	29.5%	51.4%	21.6%
	(28.4%)	(39.5%)	(37.9%)	(34.0%)	(41.5%)	(37.9%)	(32.9%)	(50.0%)	(37.2%)	(37.5%)	(38.5%)	(34.1%)	(43.8%)	(53.0%)	(34.9%)	(33.8%)	(50.0%)	(71.6%)	(38.0%)
TXNDC12	34.2%	38.3%	20.0%	41.2%	18.5%	20.2%	27.5%	21.1%	26.2%	26.9%	24.1%	22.9%	22.7%	26.9%	23.1%	27.7%	21.6%	31.9%	18.7%	20.1%
	(50.6%)	(53.9%)	(30.8%)	(70.6%)	(36.3%)	(30.2%)	(40.0%)	(32.2%)	(37.2%)	(37.5%)	(39.7%)	(41.2%)	(43.8%)	(37.0%)	(38.2%)	(39.8%)	(37.4%)	(48.9%)	(32.1%)	(31.5%)

Sequence identity and similarity (in parentheses) are noted.

co-localize with other ER stress-associated chaperones, glucose-regulated protein 78 (GRP78 or BiP) [16]. ERP29 has been postulated to play a role in the progression of tumorigenesis in mice; following implantation of both knockdown and over-expressed null ERP29 MCF-7 cells, a significant decrease in tumor size and altered morphogenesis was observed in mice [17]. Currently, no knockout mouse is available for *Erp29*.

ERP44

The *ERP44* gene is located at Chr 9q22.33 and encodes the 406-a.a. ERP44 protein [18]. ERP44 was discovered following immunoprecipitation experiments with human endoplasmic reticulum oxidoreductase-1α (ERO1-Lα) and was originally identified as KIAA0573 [18]. The coding sequence of *ERP44* contains 12 exons and there is an ERSE in the promoter region of the gene [18]. The human ERP44 protein contains an unverified NH$_2$-terminal signal peptide (29 a.a.), three TRX-like domains (one being the catalytically active **a** domain with a Cys-Arg-Phe-Ser active site), and the ER-retention sequence (Arg-Asp-Glu-Leu) [18]. The precise physiological function of ERP44 has yet to be determined; however, oxidation of ERO1α has been observed, suggesting that ERP44 may control the function of the ERO1 proteins, thus controlling the redox state of the ER [18]. ERP44 is up-regulated during ER stress response and has also been proposed to play a role in adiponectin secretion which influences glucose regulation and fatty acid catabolism [18,19].

P4HB

P4HB is the first described member of the *PDI* gene family and was originally identified as the β-subunit of human prolyl-4-hydroxylase (P4H) [20,21]. The *P4HB* gene is located at Chr 17q25 and consists of 11 exons [20,22]. The promoter region of *P4HB* contains 11 protein binding sites, including an ERSE, underscoring the dynamic nature of *P4HB* transcriptional regulation [23]. Although the presence of an ERSE has been confirmed in the promoter region of the gene, P4HB is considered to be a weakly-induced ER stress protein, likely due to its high abundance. The promoter region also contains six CCAAT elements in the first 378 nt of the gene, and mutations introduced into any of these elements was found to reduce promoter activity by up to 50% [23].

The *P4HB* gene encodes a 508-a.a. protein containing a 17-a.a. signal peptide, four TRX domains with two **a** type (Cys-Gly-His-Cys, Cys-Gly-His-Cys), an Asp/Glu rich domain (a.a. 480 to 500), and a COOH-terminal ER retention sequence (Lys-Glu-Asp-Leu) [2]. P4HB is ubiquitously expressed in nearly all tissues and is very highly abundant; estimations predict P4HB to account for up to 0.8% of total cellular protein [24]. Currently, a

crystal structure has yet to be resolved for the full, intact protein, although multiple domains have been solved. Despite this, P4HB remains the most widely studied and understood protein in the family. P4HB is effective at oxidizing, reducing, and isomerizing disulfide bonds both *in vitro* and *in vivo* and exists as a homodimer [25,26]. Although its role in disulfide bond generation remains the most widely studied enzymatic action, P4HB has also been shown to exhibit chaperone-like activity, demonstrating an additional role in maturation of nascent proteins regardless of the presence of disulfide bonds [27,28]. P4HB has also been shown to be an essential subunit for microsomal triglyceride transfer protein and P4H [21,29]. To date, no viable knockout mouse strain for *P4hb* has been reported, likely due to its critical role in disulfide bond generation [3].

PDIA2

PDIA2 was identified in 1996 as a pancreas-specific member of the PDI family, resulting in its common name, PDIp [30]. Located on Chr 16p13.3, the initial characterization of the *PDIA2* gene revealed that it encodes a protein with an ORF of 511 a.a; these studies were unable to validate an in-frame stop codon located 5′ upstream of the ATG start site [30]. Due to this discrepancy, the PDIA2 sequence was verified in 2006 to have an NH$_2$-terminal signal sequence, generating a protein of 525 a.a. in length. The *PDIA2* gene also yields two splice variants encoding two isoforms of the mature PDIA2 protein, varying by three amino acids in length (isoform 2 does not contain a.a. 181 to 183). Multiple SNPs have been detected in the *PDIA2* gene, resulting in mutations at a.a. 39 (Pro → Ser), 119 (Thr → Arg), 185 (Glu → Lys), 286 (Thr → Met), 382 (Pro → Ala), 388 (Arg → Gln), and 502 (Pro → Ser).

The PDIA2 protein contains two **a** type (Cys-Gly-His-Cys and Cys-Thr-His-Cys active sites) and one **b** and one **b'** type domains; although redox active, PDIA2 was found to be less effective than P4HB in assays for both reduction and oxidation of disulfides [30]. In addition to its role as a folding catalyst, PDIA2 has been proposed to play a role in the production and secretion of digestive enzymes *in vivo* [31]. Evidence has also suggested a role for PDIA2 in the binding and regulation of intracellular 17β-estradiol levels, thus regulating estrogen synthesis [32]. PDIA2 contains three sites for Asn-linked glycosylation, located at residues 127, 284, and 516 [30]. To date, no knockout studies have been conducted with *Pdia2*.

PDIA3

Originally identified as phospholipase C alpha, the *PDIA3* gene, located at Chr 15q15, encodes the 505-a.a PDIA3 protein (also commonly known as ERp57, ERp60,

P58) [33]. Like other PDI family members, PDIA3 contains a signal peptide, corresponding to the first 24 a.a., yielding a mature protein of 481 a.a. in length [34]. PDIA3 contains two **a** type (Cys-Gly-His-Cys and Cys-Gly-His-Cys active sites), one b and one b' type domain and an ER retention sequence, Gln-Glu-Asp-Leu [35,36]. PDIA3 expression has been detected in liver, placenta, lung, pancreas, kidney, heart, skeletal muscle, brain, and spermatozoa [33,37]. Presently, only one missense SNP has been reported for *PDIA3*, resulting in a mutation of Lys → Arg at a.a. 415.

The precise physiological role of PDIA3 has come under much scrutiny. Following the initial characterization of the protein *in vitro*, PDIA3 (termed ERp60 or P58 at the time) was identified as a cysteine protease despite little sequence homology with other heavily studied cysteine proteases [35]. In 1995, however, PDIA3 was determined to be redox active, showing the ability to reduce insulin disulfides. Bourdi et al. were also able to definitively prove that PDIA3 possessed no protease activity [36]. PDIA3 has also been shown to play a role in the correct folding of glycoproteins when in a complex containing both calnexin and calreticulin [38]. The physiological role of PDIA3 has also been investigated in rodent models, utilizing *Pdia3(-/-)* knockout mice. Although ubiquitous deletion of *Pdia3* was found to be embryolethal, heterozygous knockouts were generated, revealing multiple bone abnormalities, most notably in the femur [39]. Ablation of *Pdia3* was found to abolish signaling induced by 1,25-dihydroxyvitamin D$_3$, a crucial regulator of bone and cartilage development, by eliminating signaling through protein kinase C [39]. Additional knockout studies in murine B cells revealed a critical role for PDIA3 in the presentation of antigens by major histocompatibility complex I molecules [40]. Although additional work is needed, these studies suggest a wide array of physiological roles for PDIA3.

PDIA4

PDIA4 is located at Chr 7q35 and encodes the 645-a.a. PDIA4 protein (commonly known as ERp72). Although not much is known about the physiological role of *PDIA4*, studies indicate that this gene is induced following ER stress; these results have been found to be the result of a putative ERSE in the promoter region of the gene [41,42]. Like other PDI family members, PDIA4 contains an NH$_2$-terminal signal peptide of 21 a.a., yielding a mature protein of 625 a.a. [43]. PDIA4 contains five TRX domains, three **a** type (with all three active sites being comprised of Cys-Gly-His-Cys), one b and one b' type; PDIA4 also contains the ER retention sequence (Lys-Glu-Glu-Leu) [43,44]. Mutagenesis studies to the active-site cysteines revealed varying degrees of decreased enzymatic activity, whereas mutagenesis to multiple domains revealed a more pronounced reduction

in enzymatic activity of the protein [44]. PDIA4 also contains a string of highly acidic residues near the NH$_2$-terminus of the protein; while the precise role of these residues remains unknown, they have been proposed to play a role in regulation of Ca^{2+}, yielding its rat homolog name calcium-binding protein-2 (CaBP2) [45]. PDIA4 is a fairly ubiquitously expressed protein, although less abundant than PDI, expression patterns are similar to those of PDIA3 [46]. One missense SNP has been reported in the *PDIA4* gene, resulting in a mutation located at residue 173 (Thr → Met). Studies analyzing *Pdia3* knockdown revealed partial functional restoration by the PDIA4 protein, although, to date, no *Pdia4* knockout mouse has been generated [47].

PDIA5

Although discovered in 1995, little is known about the precise role of the *PDIA5* gene. Located on Chr 3q21.1, *PDIA5* encodes the 519-a.a. PDIA5 (or PDI-related protein). PDIA5 contains four TRX domains (three **a** and one **b**-like domain), made up of active sites Cys-Ser-Met-Cys, Cys-Gly-His-Cys and Cys-Pro-His-Cys, a COOH-terminal ER retention sequence (Lys-Glu-Glu-Leu), and an unverified signal sequence comprising the first 21 a.a. [48]. Despite an additional Cys-X-X-Cys motif, Horibe et al. revealed that PDIA5 has significantly less enzymatic activity than that of P4HB [49]. The contributions of each Cys-X-X-Cys active site were also investigated, revealing varying degrees of altered activity following mutations to each, or multiple, active sites [49]. It was concluded that the second active site (Cys-Gly-His-Cys) was the most critical for isomerase activity and that all three motifs are not required for activity [49]. Much like P4HB, PDIA5 was also shown to exhibit chaperone-like activity by refolding denatured rhodanese, which does not contain any disulfide bonds [49]. PDIA5 mRNA has been detected in liver, kidney, lung, and brain––with the highest level of secretion being noted in the liver [48]. Although to date no ERSE has been identified, *PDIA5* has been shown to be moderately up-regulated following induction of the ER-stress response in cultured cells [48]. In a 2011 study by Carbone et al., a significant association was found between the SNP, rs11720822, and primary open-angle glaucoma in two separate populations [50]. No viable knockout mouse has been generated for *Pdia5*.

PDIA6

Much like *PDIA5*, little is known about the role of *PDIA6* both *in vitro* and *in vivo*. The *PDIA6* gene is located at Chr 2p25.1 and encodes the 440-a.a. PDIA6 protein (commonly reported as P5 or ERP5). PDIA6 contains an NH$_2$-terminal signal sequence of 19 a.a., three TRX domains (two **a** type and one **b**) consisting of

two Cys-Gly-His-Cys active sites, an Asp/Glu rich domain and a COOH-terminal Lys-Asp-Glu-Leu ER-retention sequence [51,52]. Recombinant PDIA6 demonstrates both isomerase and chaperone activities, although approximately 45% and 50% to 60% to that of P4HB, respectively [53]. Point mutations have also been conducted to the active-site cysteines, revealing that NH_2-terminal cysteines in each active site exhibit the majority of isomerase activity [53]. *PDIA6* contains an ERSE in its promoter region, which was recently validated *in vitro* utilizing over-expression of the ER-stress transcription factor, X-box protein-1 (XBP-1); *PDIA6* was found to be significantly increased, demonstrating inducibility by the unfolded protein response [54,55]. A complete expression profile for *PDIA6* has yet to be conducted; however, high levels of the protein were detected in platelets. Cell-surface expression of PDIA6 was found to be necessary for the proper development and function of platelets, whereas inhibition of the protein using anti- PDIA6 antibodies revealed inhibition of platelet aggregation [51]. PDIA6 has also been shown to directly interact with GRP78 (or BiP) suggesting a role for PDIA6 in the refolding of substrates that have been targeted to BiP [56]. Presently, one missense SNP has been identified (rs4807) resulting in a point mutation at a.a. 214 (Lys → Arg). At present, no viable *Pdia6* knockout mouse is available.

PDILT

Expression of the *PDILT* gene has been reported to be exclusively limited to the testis. *PDILT* is located on Chr 16p12.3, encoding the 584-a.a. PDILT protein [57]. Despite the presence of two **a** type TRX domains (with non-classical Ser-Lys-Gln-Ser and Ser-Lys-Lys-Cys motifs), PDILT does not exhibit the ability to oxidize or reduce disulfide bonds, although evidence has supported PDILT to engage in disulfide-bonded complexes *in vitro* [57,58]. PDILT contains a predicted NH_2-terminal signal peptide of 20 a.a. in length, a COOH-terminal ER-retention sequence (Lys-Glu-Glu-Leu) and is heavily glycosylated through nine potential Asn-linked glycosylation sites [57]. Much like P4HB, PDILT also interacts with the oxidoreductase ERo1α in cultured cells, suggesting a role in the shuffling of electrons in the ER lumen [57]. No knockout mouse is currently available and considerable research is needed to fully elucidate the precise role of *PDILT*.

TXNDC5

TXNDC5 is located at Chr 6p24.3 and encodes the 432-a.a. endothelial PDI, TXNDC5 (or EndoPDI) protein [59]. Despite its discovery in 2003, little research has been conducted on the role of *TXNCD5 in vivo*. TXNDC5 contains a signal peptide of 32 a.a., three **a**-type TRX domains all with Cys-Gly-His-Cys active sites, and a

COOH-terminal ER retention sequence, KDEL [59]. *TXNDC5* was originally identified in a screen for proteins highly expressed in endothelial cells, leading to its common name, EndoPDI. Subsequent studies revealed *TXNDC5* expression in a number of tissues with the highest expression being found in lymph nodes, stomach, pancreatic islets, and heart [59,60]. TXNDC5 is induced under conditions of hypoxia, and loss of TXNDC5 leads to an increase in apoptotic cell death in microvascular endothelial cells during hypoxia, but not normoxia [59]. Preliminary studies have also implicated a role for TXNDC5 in diabetes, noting a decrease in the expression of the protein in pancreatic islets in animals with consistently elevated glucose levels [60]. In a 2010 study, Jeong et al. investigated the role of *TXNDC5* in development of the skin disorder vitiligo [61]. A total of 230 Korean patients with non-segmental vitiligo were investigated for SNPs in the *TXNDC5* gene; in total, seven SNPs were identified in the *TXNDC5* gene, three of which (rs1043784, rs7764128, and rs8643) demonstrated an association with the vitiligo phenotype [61]. Although relevant *in vivo* studies have been conducted on the role of *TXNDC5* no biochemical parameters have been evaluated, with regard to its role in disulfide bond oxidation and reduction. A viable knockout mouse has not been generated for the study of *Txndc5*.

The anterior gradient homolog genes
AGR2

The *AGR2* gene is located at Chr 7p21.3 and encodes the 175-a.a. anterior gradient protein 2 homolog (AGR2) [62,63]. The AGR2 protein has an NH_2-terminal signal sequence of 20 a.a., one TRX domain (with active site Cys-Pro-His-Ser), and a COOH-terminal ER-retention sequence (Lys-Thr-Glu-Leu) [63]. Human AGR2 was originally identified in estrogen receptor-positive MCF7 cells using suppression subtractive hybridization [64]. Expression of *AGR2* transcripts has been detected in lung, pancreas, trachea, stomach, colon, prostate, and small intestine. *AGR2* has also been investigated as a potential biomarker for hormone-responsive breast cancer in estrogen receptor-α-positive breast cancer cell lines [64].

Utilizing knockout studies in mice, the *Agr2* gene was found to result in the inability to produce mucin, leading in an increased susceptibility to experimentally induced colitis and intestinal disease [65]. Due to its role in disulfide bond generation, it was hypothesized that AGR2 was responsible for the processing of MUC2, the major intestinal mucin. This protein contains >200 cysteine residues involved in various inter- and intra-protein disulfide bonds and has been found to directly associate with AGR2 [65]. Following this report, Zheng et al. investigated *AGR2* and *AGR3* as potential candidate genes for inflammatory bowel disease in humans [66]. A cohort of

2,540 patients having either ulcerative colitis or Chron's disease was investigated for SNPs in *AGR2* and *AGR3*; in total, 30 SNPs were identified, 25 were located in the *AGR2* gene, while 5 were located in the *AGR3* gene [66]. The promoter region of the *AGR2* gene was also found to contain binding sites for hepatic nuclear factor-1, hepatocyte nuclear factor 3-α (FOXA1) and hepatocyte nuclear factor 3-β (FOXA2)––transcription factors that have been reported to play a role in the morphogenesis of goblet cell differentiation during formation. In summary, two total SNPs in the 5′ promoter region of the *AGR2* gene were found to be associated with the risk haplotype of ulcerative colitis in two independent populations, providing further evidence for a role for *AGR2* in disease pathogenesis [66].

AGR3

The *AGR3* gene is located at Chr 7p21.1 and encodes the 166-a.a. anterior gradient protein-3 (AGR3) homolog [63]. AGR3 transcripts have been detected in lung and pancreas and, resembling AGR2, AGR3 has been reported to be co-expressed with estrogen receptor-α-containing breast cancer cell lines, suggesting it to be a marker for hormone-responsive breast cancer [63,64]. Unlike AGR2, however, little is known about the precise physiological role of AGR3. AGR3 protein was originally identified as breast cancer membrane protein 11 (BCMP11), following a proteomic screen of membrane proteins in breast cancer cell lines, and was later named AGR3 due to the high degree of sequence homology with AGR2 (see Table 2) [67]. Like AGR2, AGR3 contains one redox-active center, comprised of amino acids Cys-Gln-Tyr-Ser, an NH_2-terminal signal peptide composed of 21 a.a., and a COOH-terminal ER-retention sequence (Gln-Ser-Glu-Leu) [63]. Recently, an increase in AGR3 expression was observed in serous border-line ovarian tumors and low-grade serous ovarian carcinoma [68]. Utilizing Kaplan-Meier survival curves, King et al. also established that patients with AGR3-expressing tumors survived significantly longer than those patients lacking AGR3-expressing tumors [68].

TXNDC12

TXNDC12, also known as *AGR1*, *TLP19*, and *ERP18/19*, contains conserved intron positions with respect to amino acid sequence with the *AGR2* and *AGR3* genes [63]. Persson et al. also reported that several of the individual exon lengths are identical (or altered with one codon) to *AGR2* and *AGR3* [63]. *TXNDC12* has been mapped to Chr 1p32.3 and contains seven exons spanning more than 35 kb [69]. *TXNDC12* is ubiquitously expressed in all tissues, with the highest expression being found in the liver and placenta. *TXNDC12* in the placenta was found to express an additional transcript of 1.2 kb which is

associated with two poly(A) addition signals in its 3′-UTR [69]. The TXNDC12 protein contains 172 a.a. with 149 a.a. comprising the mature form of the protein (Ser^{24} – Leu^{172}); TXNDC12 has one active site comprised of Cys-Gly-Ala-Cys and an ER-retention sequence (Glu-Asp-Glu-Leu) [70]. Unlike the other members of the AGR subfamily, extensive work has been conducted on the biochemical and physiochemical actions of TXNDC12. The enzymatic activity of TXNDC12 has been found to be limited strictly to disulfide bond generation and not reduction; these studies were also confirmed with the use of point mutations to the active-site cysteines (Cys-Gly-Ala-Cys), after which no detectable activity was found [70]. Chemical denaturation curves were also found to favor greater protein stability in the reduced form over the oxidized form, a property consistent with other PDI family members [70].

The CASQ genes

The *CASQ* genes (1 and 2) are interesting members of the *PDI* family, possessing no cysteine containing redox-active sites and therefore playing no role in the formation or reduction of disulfide bonds. As indicated, many of the *PDI* family proteins bind Ca^{2+} with relatively high capacity and low affinity [71]. Following studies on CASQ, Shin et al. found that the COOH-terminal Asp-rich domain played a major role in storage of Ca^{2+} through interaction with ryanodine receptor (RYR), a protein involved in Ca^{2+} release from the sarcoplasmic reticulum (SR) [72]. These proteins, therefore, possess unique functions relating to the PDI family of proteins, despite limited sequence homology.

CASQ1

The *CASQ1* gene is located on Chr 1q21 and encodes the 396-a.a. CASQ1 protein [73]. Expression of the mature CASQ1 protein is primarily limited to the SR of fast-twitch skeletal muscles [74]. Studies in rabbit have revealed CASQ1 to be a high-capacity (40 to 50 Ca^{2+} per molecule of CASQ1), moderate-affinity (Kd = 1 mM) Ca^{2+}-binding protein that does not contain an ER-retention sequence [74-76]. Although experimentally unverified, the first 34 a.a. of CASQ1 encode the signal peptide, leaving the mature protein at 362 a.a. in length––which contains three TRX domains (two **b** and one **b'**) and a string of highly acidic residues from a.a. 353 – 396. Like other members of the PDI family, these Asp/Glu-rich stretches of a.a. are thought to be the primary binding regions for Ca^{2+}; CASQ1 plays a major role in Ca^{2+} flux through the regulation of Ca^{2+} channel activity and interaction with Ca^{2+} directly [72]. Systemic knockout studies in mice revealed hypersensitivity to heat and volatile anesthetics, along with a phenotypic resemblance to malignant hyperthermia [77]; these effects were found to be due to increased Ca^{2+} following an

increased exposure to heat [78]. CASQ1 also contains one potential Asn-linked glycosylation site, found at a.a. 350.

SNPs in the *CASQ1* gene have been reported in cases of diabetes in both Old-Order Amish and Northern European Caucasians [79,80]. Out of 26 identified SNPs, SNP CASQ-1404 (rs1186694) in the 5′ flanking region was found to have a statistically significant association with type-2 diabetes in Northern European Caucasians [79]. In a similar study analyzing type-2 diabetes susceptibility in Old-Order Amish populations, SNPs rs2275703 and rs617698 were defined as the 'at-risk alleles' [80]. Although mechanistically these correlations have not been confirmed, previous work has identified a putative role for Ca^{2+} release from the SR into the cytosol in regulating glucose transporter-4 (GLUT4), a key enzyme in regulation of glucose transport by insulin [80,81].

CASQ2

The *CASQ2* gene is located on Chr. 1p13.3-p11, and likely is the result of gene duplication with *CASQ1*--given their chromosomal location. *CASQ2* encodes the 399-a.a. CASQ2, which shares 91% identity with its homologue, *CASQ1* (see Table 2) [82,83] The CASQ2 protein is expressed exclusively in cardiac muscle and serves as the major Ca^{2+} reservoir in the SR of myocardium; CASQ2 also interacts with the RYR2 channel, regulating Ca^{2+} flux from the SR [76,84]. Much like CASQ1, CASQ2 contains an unverified NH_2-terminal signal peptide (19 a.a.), three **b**-like TRX domains (two **b** and one **b'**), no ER-retention sequence, and a string of highly acidic residues (a.a. 356 to 399); CASQ2 does, however, contain one potential Asn-linked glycosylation site at a.a. 355.

The *CASQ2* gene has become a heavily researched target for diseases associated with arrhythmic heartbeats. In 2001, Lahat et al. investigated missense mutations found in the coding region of the gene [83]. One SNP (G –> C) was found to result in an aspartic acid changed to a histidine at a.a. 307 of the mature protein, potentially altering the Ca^{2+}-chelating function of that region [83]. This SNP was found to be associated with Bedouin families from Israel susceptible to catecholine-induced polymorphic ventricular tachycardia [83]. These studies were later confirmed in *Casq2(-/-)* knockout mice, revealing susceptibility to polymorphic ventricular tachycardia following exposure to catecholamines [85].

Thioredoxin-related transmembrane proteins

The thioredoxin-related transmembrane (*TMX*) genes are newly discovered members of the *PDI* gene family. To date, little is known about the precise function of these genes; however, all four members in the PDI family consist of one TRX domain, one transmembrane domain, and non-conventional ER-retention sequences.

TMX1

Discovered in 2001, *TMX1 is* located at Chr 14q22.1 and encodes the 280-a.a. TMX1 protein [86]. TMX1 contains an NH_2-terminal signal sequence of 26 a.a., one **a**-type TRX domain with active site Cys-Pro-Ala-Cys, one transmembrane domain (a.a. 183 to 203), and lacks an ER-retention sequence [86]. Expression of TMX1 is fairly ubiquitous, with highest levels detected in liver, kidney, placenta, and lung [86]. Mature TMX1 possesses the ability to both oxidize and reduce disulfide bonds, although chaperone-like activity has yet to be investigated [86,87]. The *TMX1* gene does not contain a putative ERSE in the promoter region, supporting evidence that *TMX1* is not induced by numerous ER-stress-inducing agents; over-expression of the protein in cultured cells, however, has revealed amelioration of both Brefeldin A-induced apoptosis and tunicamycin-induced ER stress [86,88]. No knockout mouse has been generated for the study of *Tmx1*.

TMX2

Perhaps the least researched gene in the family, *TMX2* was discovered in 2003 [89]. Located on Chr 11cen-q22.3, *TMX2* encodes the 296-a.a. TMX2 protein [89]. Like TMX1, TMX2 contains a COOH-terminal signal peptide (48 a.a.), one **a**-type TRX domain (Ser-Asn-Asp-Cys active site), one transmembrane domain (a.a. 104 – 126), and an ER-retention sequence comprised of Lys-Lys-Glu-Ile [89]. Expression of TMX2 is fairly ubiquitous--with high levels detected in heart, brain, liver, kidney, and pancreas [89]. Although Meng et al. provided the initial characterization of the protein, sequence discrepancies have been found. The official NCBI sequence of the TMX2 protein reveals a protein of 296 a.a and a second isoform, lacking an in-frame exon in the central coding region, encoding a protein of 258 a.a (isoform 2 differs between a.a. 84 to 122). Future studies are required to fully elucidate the role of *TMX2 in vivo*. The availability of a *TMX2* knockout mouse has not been reported.

TMX3

The human *TMX3* gene is located at Chr 18q22 and encodes the 454-a.a. TMX3 protein [90]. Following cleavage of the 24-a.a. signal peptide, the 430-a.a. mature protein consists of one **a**-type TRX domain, with the active site being comprised of Cys-Gly-His-Cys, a transmembrane domain (located at a.a. 375 to 397), and the ER-retention sequence (Lys-Lys-Lys-Asp) [90]. Uncharacteristic to most PDI family members, TMX3 contains a luminal domain with weak sequence similar to that of the CASQ proteins [90]. Although no research has been conducted on the role of this domain in activity of the protein, it has been postulated to regulate Ca^{2+} in a manner similar to that of other CASQ proteins. NCBI

reports two isoforms for TMX3 (one encoding a 195-a.a. protein), though experimentally this has not been validated [90]. TMX3 has been detected in brain, testis, lung, skin, kidney, uterus, bone, stomach, liver, prostate, placenta, eye, and muscle, with highest levels detected in heart and skeletal muscle [90]. TMX3 contains two sites of Asn-linked glycosylation (a.a. 258 and 313), which have been validated *in vitro* and is not induced under conditions of ER stress [90]. Although far less efficient than P4HB, TMX3 does display the ability to oxidize disulfide bonds; this is likely due to the presence of only one Cys-Gly-His-Cys active site [90].

Although no knockout mouse has been generated for *TMX3*, studies have been conducted in mice, targeting *TMX3* transcripts using morpholinos [91]. Investigating the mechanisms behind microphthalmia in humans, a genetic disease associated with retarded growth of the eye, a 2.7-Mb deletion was found at Chr 18q22.1, leading to deletion of the *TMX3* gene [91]. Studies in mice using a targeted approach to delete the *TMX3* gene revealed a similar phenotype, which was rescued following injection of human TMX3 mRNA [91]. Sequencing of 162 patients with anopthalmia or microphthalmia revealed two missense mutations, leading to the missense SNPs, R39N, and D108N [91]. Future studies are required to fully elucidate the precise role of *TMX3*, although preliminary studies reveal exciting areas for research.

TMX4

The *TMX4* gene is located at Chr 20p12 and encodes the 349-a.a. TMX4 protein [92]. TMX4 consists of an NH_2-terminal signal sequence (23 a.a.), one a-type TRX domain (a. a. 39 – 136 with the active site comprised of Cys-Pro-Ser-Cys), a transmembrane domain (a.a. 188 to 210), a string of highly acidic a.a. (224 to 334), and (like the other TMX proteins) lacks an ER-retention sequence [93]. *TMX4* is ubiquitously expressed, with highest levels detected in heart [92]. Preliminary studies show that TMX4 is not induced following conditions of ER stress and does not contain a putative ERSE in the promoter region of the gene [92]. TMX4 contains one site of Asn-linked glycosylation and two sites of Ser phosphorylation, which have all been experimentally validated [93]. Enzymatic activity was confirmed by observing reduction of insulin disulfides; a dominant-negative mutant, with the active-site cysteines mutated to serine, displayed no enzymatic activity [92]. No knockout mouse has been generated and substantial work will be required to understand the role of *TMX4 in vivo*.

Conclusions

The PDI family of proteins consists of 21 members varying greatly in enzymatic activity, domain architecture, and tissue specificity. Although the predominant role of the PDI proteins is the regulation of protein folding

in vivo––through the oxidation, reduction and isomerization of disulfide bonds––these proteins have also been shown to regulate calcium homeostasis in the ER and induction of the unfolded protein response (UPR). Since its discovery over 40 years ago, PDI has become one of the most highly studied proteins and, despite these advances, extensive research is still needed to fully understand the role of PDI *in vivo*. The more recently characterized *TMX* genes have displayed promise in novel therapeutics, ranging from disorders of the eye to regulation of the ER-stress response. Many of the genes in the *PDI* family contain a putative ERSE sequence in the promoter region of the gene, suggesting a role in the UPR. Further research on the role of these proteins in the UPR is required before effective therapeutics can be generated for a plethora of disease states associated with the ER-stress response. The 21 members of the PDI family of proteins encompass many physiological responses, and these proteins will likely provide compelling avenues for future research.

Competing interests
The authors declare that they have no competing interests.

Authors' contributions
JJG carried out the sequence alignment and drafted the manuscript. DRP designed and funded the study. Both authors read and approved the final manuscript.

Acknowledgments
Studies were supported by the National Institutes of Health/National Institutes of Alcoholism and Alcohol Abuse under grant numbers R37AA09300 (DRP), R01DK074487-01 (DRP) and F31AA018606-01 (JJG).

Author details
[1]Department of Pharmacology, University of Colorado Anschutz Medical Campus, Aurora, CO 80045, USA. [2]Molecular Toxicology and Environmental Health Sciences Program, Department of Pharmacy and Pharmaceutical Sciences, University of Colorado Anschutz Medical Campus, Aurora, CO 80045, USA.

References
1. Malhotra JD, Kaufman RJ: The endoplasmic reticulum and the unfolded protein response. *Semin Cell Dev Biol* 2007, 18:716–731.
2. Appenzeller-Herzog C, Ellgaard L: The human PDI family: versatility packed into a single fold. *Biochim Biophys Acta* 2008, 1783:535–548.
3. Hatahet F, Ruddock LW: Protein disulfide isomerase: a critical evaluation of its function in disulfide bond formation. *Antioxid Redox Signal* 2009, 11:2807–2850.
4. Ellgaard L, Ruddock LW: The human protein disulphide isomerase family: substrate interactions and functional properties. *EMBO Rep* 2005, 6:28–32.
5. Jacquot JP, Gelhaye E, Rouhier N, Corbier C, Didierjean C, Aubry A: Thioredoxins and related proteins in photosynthetic organisms: molecular basis for thiol dependent regulation. *Biochem Pharmacol* 2002, 64:1065–1069.
6. Turano C, Coppari S, Altieri F, Ferraro A: Proteins of the PDI family: unpredicted non-ER locations and functions. *J Cell Physiol* 2002, 193:154–163.
7. Rapoport TA: Protein transport across the endoplasmic reticulum membrane: facts, models, mysteries. *FASEB J* 1991, 5:2792–2798.
8. Houston NL, Fan C, Xiang JQ, Schulze JM, Jung R, Boston RS: Phylogenetic analyses identify 10 classes of the protein disulfide isomerase family in

plants, including single-domain protein disulfide isomerase-related proteins. *Plant Physiol* 2005, **137**:762–778.

9. McArthur AG, Knodler LA, Silberman JD, Davids BJ, Gillin Fd, Sogin ML: **The evolutionary origins of eukaryotic protein disulfide isomerase domains: new evidence from the Amitochondriate protist *Giardia lamblia*.** *Mol Biol Evol* 2001, **18**:1455–1463.

10. Gu SH, Chen JZ, Ying K, Wang S, Jin W, Qian J, Zhao EP, Xie Y, Mao YM: **Cloning and identification of a novel cDNA which encodes a putative protein with a DnaJ domain and a thioredoxin active motif, human macrothioredoxin.** *Biochem Genet* 2003, **41**:245–253.

11. Cunnea PM, Miranda-Vizuete A, Bertoli G, Simmen T, Damdimopoulos AE, Hermann S, Leinonen S, Huikko MP, Gustafsson JA, Sitia R, Spyrou G: **ERdj5, an endoplasmic reticulum (ER)-resident protein containing DnaJ and thioredoxin domains, is expressed in secretory cells or following ER stress.** *J Biol Chem* 2003, **278**:1059–1066.

12. Ushioda R, Hoseki J, Araki K, Jansen G, Thomas DY, Nagata K: **ERdj5 is required as a disulfide reductase for degradation of misfolded proteins in the ER.** *Science* 2008, **321**:569–572.

13. Alanen HI, Williamson RA, Howard MJ, Hatahet FS, Salo KE, Kauppila A, Kellokumpu S, Ruddock LW: **ERp27, a new non-catalytic endoplasmic reticulum-located human protein disulfide isomerase family member, interacts with ERp57.** *J Biol Chem* 2006, **281**:33727–33738.

14. Sargsyan E, Baryshev M, Backlund M, Sharipo A, Mkrtchian S: **Genomic organization and promoter characterization of the gene encoding a putative endoplasmic reticulum chaperone, ERp29.** *Gene* 2002, **285**:127–139.

15. Ferrari DM, Van Nguyen P, Kratzin HD, Soling HD: **ERp28, a human endoplasmic-reticulum-lumenal protein, is a member of the protein disulfide isomerase family but lacks a CXXC thioredoxin-box motif.** *Eur J Biochem* 1998, **255**:570–579.

16. Mkrtchian S, Fang C, Hellman U, Ingelman-Sundberg M: **A stress-inducible rat liver endoplasmic reticulum protein, ERp29.** *Eur J Biochem* 1998, **251**:304–313.

17. Mkrtchian S, Baryshev M, Sargsyan E, Chatzistamou I, Volakaki AA, Chaviaras N, Pafiti A, Triantafyllou A, Kiaris H: **ERp29, an endoplasmic reticulum secretion factor is involved in the growth of breast tumor xenografts.** *Mol Carcinog* 2008, **47**:886–892.

18. Anelli T, Alessio M, Mezghrani A, Simmen T, Talamo F, Bachi A, Sitia R: **ERp44, a novel endoplasmic reticulum folding assistant of the thioredoxin family.** *EMBO J* 2002, **21**:835–844.

19. Wang ZV, Schraw TD, Kim JY, Khan T, Rajala MW, Follenzi A, Scherer PE: **Secretion of the adipocyte-specific secretory protein adiponectin critically depends on thiol-mediated protein retention.** *Mol Cell Biol* 2007, **27**:3716–3731.

20. Tasanen K, Parkkonen T, Chow LT, Kivirikko KI, Pihlajaniemi T: **Characterization of the human gene for a polypeptide that acts both as the beta subunit of prolyl 4-hydroxylase and as protein disulfide isomerase.** *J Biol Chem* 1988, **263**:16218–16224.

21. Pihlajaniemi T, Helaakoski T, Tasanen K, Myllyla R, Huhtala ML, Koivu J, Kivirikko KI: **Molecular cloning of the beta-subunit of human prolyl 4-hydroxylase. This subunit and protein disulphide isomerase are products of the same gene.** *EMBO J* 1987, **6**:643–649.

22. Pajunen L, Jones TA, Goddard A, Sheer D: **Regional assignment of the human gene coding for a multifunctional polypeptide (P4HB) acting as the beta-subunit of prolyl 4-hydroxylase and the enzyme protein disulfide isomerase to 17q25.** *Cytogenet Cell Genet* 1991, **56**:165–168.

23. Tasanen K, Oikarinen J, Kivirikko KI, Pihlajaniemi T: **Promoter of the gene for the multifunctional protein disulfide isomerase polypeptide. Functional significance of the six CCAAT boxes and other promoter elements.** *J Biol Chem* 1992, **267**:11513–11519.

24. Ferrari DM, Soling HD: **The protein disulphide-isomerase family: unravelling a string of folds.** *Biochem J* 1999, **339**(Pt 1):1–10.

25. Freedman RB, Brockway BE, Lambert N: **Protein disulphide-isomerase and the formation of native disulphide bonds.** *Biochem Soc Trans* 1984, **12**:929–932.

26. Freedman RB, Hirst TR, Tuite MF: **Protein disulphide isomerase: building bridges in protein folding.** *Trends Biochem Sci* 1994, **19**:331–336.

27. Song JL, Wang CC: **Chaperone-like activity of protein disulfide-isomerase in the refolding of rhodanese.** *Eur J Biochem* 1995, **231**:312–316.

28. Wang CC, Tsou CL: **Protein disulfide isomerase is both an enzyme and a chaperone.** *FASEB J* 1993, **7**:1515–1517.

29. Wetterau JR, Combs KA, Spinner SN, Joiner BJ: **Protein disulfide isomerase is a component of the microsomal triglyceride transfer protein complex.** *J Biol Chem* 1990, **265**:9800–9807.

30. Desilva MG, Lu J, Donadel G, Modi WS, Xie H, Notkins AL, Lan MS: **Characterization and chromosomal localization of a new protein disulfide isomerase, PDIp, highly expressed in human pancreas.** *DNA Cell Biol* 1996, **15**:9–16.

31. Klappa P, Stromer T, Zimmermann R, Ruddock LW, Freedman RB: **A pancreas-specific glycosylated protein disulphide-isomerase binds to misfolded proteins and peptides with an interaction inhibited by oestrogens.** *Eur J Biochem* 1998, **254**:63–69.

32. Fu XM, Zhu BT: **Human pancreas-specific protein disulfide isomerase homolog (PDIp) is an intracellular estrogen-binding protein that modulates estrogen levels and actions in target cells.** *J Steroid Biochem Mol Biol* 2009, **115**:20–29.

33. Koivunen P, Horelli-Kuitunen N, Helaakoski T, Karvonen P, Jaakkola M, Palotie A, Kivirikko KI: **Structures of the human gene for the protein disulfide isomerase-related polypeptide ERp60 and a processed gene and assignment of these genes to 15q15 and 1q21.** *Genomics* 1997, **42**:397–404.

34. Charnock-Jones DS, Day K, Smith SK: **Cloning, expression and genomic organization of human placental protein disulfide isomerase (previously identified as phospholipase C alpha).** *Int J Biochem Cell Biol* 1996, **28**:81–89.

35. Urade R, Oda T, Ito H, Moriyama T, Utsumi S, Kito M: **Functions of characteristic Cys-Gly-His-Cys (CGHC) and Gln-Glu-Asp-Leu (QEDL) motifs of microsomal ER-60 protease.** *J Biochem* 1997, **122**:834–842.

36. Bourdi M, Demady D, Martin JL, Jabbour SK, Martin BM, George JW, Pohl LR: **cDNA cloning and baculovirus expression of the human liver endoplasmic reticulum P58: characterization as a protein disulfide isomerase isoform, but not as a protease or a carnitine acyltransferase.** *Arch Biochem Biophys* 1995, **323**:397–403.

37. Kameshwari DB, Bhande S, Sundaram CS, Kota V, Siva AB, Shivaji S: **Glucose-regulated protein precursor (GRP78) and tumor rejection antigen (GP96) are unique to hamster caput epididymal spermatozoa.** *Asian J Androl* 2010, **12**:344–355.

38. Oliver JD, van der Wal FJ, Bulleid NJ, High S: **Interaction of the thiol-dependent reductase ERp57 with nascent glycoproteins.** *Science* 1997, **275**:86–88.

39. Wang Y, Chen J, Lee CS, Nizkorodov A, Riemenschneider K, Martin D, Hyzy S, Schwartz Z, Boyan BD: **Disruption of Pdia3 gene results in bone abnormality and affects 1alpha,25-dihydroxy-vitamin D3-induced rapid activation of PKC.** *J Steroid Biochem Mol Biol* 2010, **121**:257–260.

40. Garbi N, Tanaka S, Momburg F, Hammerling GJ: **Impaired assembly of the major histocompatibility complex class I peptide-loading complex in mice deficient in the oxidoreductase ERp57.** *Nat Immunol* 2006, **7**:93–102.

41. Paschen W, Gissel C, Linden T, Doutheil J: **Erp72 expression activated by transient cerebral ischemia or disturbance of neuronal endoplasmic reticulum calcium stores.** *Metab Brain Dis* 1998, **13**:55–68.

42. Roy B, Lee AS: **The mammalian endoplasmic reticulum stress response element consists of an evolutionarily conserved tripartite structure and interacts with a novel stress-inducible complex.** *Nucleic Acids Res* 1999, **27**:1437–1443.

43. Mazzarella RA, Srinivasan M, Haugejorden SM, Green M: **ERp72, an abundant luminal endoplasmic reticulum protein, contains three copies of the active site sequences of protein disulfide isomerase.** *J Biol Chem* 1990, **265**:1094–1101.

44. Satoh M, Shimada A, Keino H, Kashiwai A, Nagai N, Saga S, Hosokawa M: **Functional characterization of 3 thioredoxin homology domains of ERp72.** *Cell Stress Chaperones* 2005, **10**:278–284.

45. Nigam SK, Goldberg AL, Ho S, Rohde MF, Bush KT, Sherman MYu: **A set of endoplasmic reticulum proteins possessing properties of molecular chaperones includes Ca(2+)-binding proteins and members of the thioredoxin superfamily.** *J Biol Chem* 1994, **269**:1744–1749.

46. Marcus N, Shaffer D, Farrar P, Green M: **Tissue distribution of three members of the murine protein disulfide isomerase (PDI) family.** *Biochim Biophys Acta* 1996, **1309**:253–260.

47. Solda T, Garbi N, Hammerling GJ, Molinari M: **Consequences of ERp57 deletion on oxidative folding of obligate and facultative clients of the calnexin cycle.** *J Biol Chem* 2006, **281**:6219–6226.

48. Hayano T, Kikuchi M: **Molecular cloning of the cDNA encoding a novel protein disulfide isomerase-related protein (PDIR).** *FEBS Lett* 1995, **372**:210–214.

49. Horibe T, Gomi M, Iguchi D, Ito H, Kitamura Y, Masuoka T, Tsujimoto I, Kimura T, Kikuchi M: Different contributions of the three CXXC motifs of human protein-disulfide isomerase-related protein to isomerase activity and oxidative refolding. *J Biol Chem* 2004, **279**:4604–4611.

50. Carbone MA, Chen Y, Hughes GA, Weinreb RN, Zabriskie NA, Zhang K, Anholt RRH: Genes of the unfolded protein response pathway harbor risk alleles for primary open angle glaucoma. *PLoS One* 2011, **6**:e20649.

51. Jordan PA, Stevens JM, Hubbard GP, Barrett NE, Sage T, Authi KS, Gibbins JM: A role for the thiol isomerase protein ERP5 in platelet function. *Blood* 2005, **105**:1500–1507.

52. Hayano T, Kikuchi M: Cloning and sequencing of the cDNA encoding human P5. *Gene* 1995, **164**:377–378.

53. Kikuchi M, Doi E, Tsujimoto I, Horibe T, Tsujimoto Y: Functional analysis of human P5, a protein disulfide isomerase homologue. *J Biochem* 2002, **132**:451–455.

54. Belmont PJ, Chen WJ, San Pedro MN, Thuerauf DJ, Gellings Lowe N, Gude N, Hilton B, Wolkowicz R, Sussman MA, Glembotski CC: Roles for endoplasmic reticulum-associated degradation and the novel endoplasmic reticulum stress response gene Derlin-3 in the ischemic heart. *Circ Res* 2009, **106**:307–316.

55. Lee AH, Iwakoshi NN, Glimcher LH: XBP-1 regulates a subset of endoplasmic reticulum resident chaperone genes in the unfolded protein response. *Mol Cell Biol* 2003, **23**:7448–7459.

56. Jessop CE, Watkins RH, Simmons JJ, Tasab M, Bulleid NJ: Protein disulphide isomerase family members show distinct substrate specificity: P5 is targeted to BiP client proteins. *J Cell Sci* 2009, **122**:4287–4295.

57. van Lith M, Hartigan N, Hatch J, Benham AM: PDILT, a divergent testis-specific protein disulfide isomerase with a non-classical SXXC motif that engages in disulfide-dependent interactions in the endoplasmic reticulum. *J Biol Chem* 2005, **280**:1376–1383.

58. van Lith M, Karala AR, Bown D, Gatehouse JA, Ruddock LW, Saunders PT, Benham AM: A developmentally regulated chaperone complex for the endoplasmic reticulum of male haploid germ cells. *Mol Biol Cell* 2007, **18**:2795–2804.

59. Sullivan DC, Huminiecki L, Moore JW, Boyle JJ, Poulsom R, Creamer D, Barker J, Bicknell R: EndoPDI, a novel protein-disulfide isomerase-like protein that is preferentially expressed in endothelial cells acts as a stress survival factor. *J Biol Chem* 2003, **278**:47079–47088.

60. Alberti A, Karamessinis P, Peroulis M, Kypreou K, Kavvadas P, Pagakis S, Politis PK, Charonis A: ERp46 is reduced by high glucose and regulates insulin content in pancreatic beta-cells. *Am J Physiol Endocrinol Metab* 2009, **297**:E812–E821.

61. Jeong KH, Shin MK, Uhm YK, Kim HJ, Chung JH, Lee MH: Association of TXNDC5 gene polymorphisms and susceptibility to nonsegmental vitiligo in the Korean population. *Br J Dermatol* 2010, **162**:759–764.

62. Petek E, Windpassinger C, Egger H, Kroisel PM, Wagner K: Localization of the human anterior gradient-2 gene (AGR2) to chromosome band 7p21.3 by radiation hybrid mapping and fluorescencein situ hybridisation. *Cytogenet Cell Genet* 2000, **89**:141–142.

63. Persson S, Rosenquist M, Knoblach B, Khosravi-Far R, Sommarin M, Michalak M: Diversity of the protein disulfide isomerase family: identification of breast tumor induced Hag2 and Hag3 as novel members of the protein family. *Mol Phylogenet Evol* 2005, **36**:734–740.

64. Thompson DA, Weigel RJ: hAG-2, the human homologue of the *Xenopus laevis* cement gland gene XAG-2, is coexpressed with estrogen receptor in breast cancer cell lines. *Biochem Biophys Res Commun* 1998, **251**:111–116.

65. Park SW, Zhen G, Verhaeghe C, Nakagami Y, Nguyenvu LT, Barczak AJ, Killeen N, Erle D: The protein disulfide isomerase AGR2 is essential for production of intestinal mucus. *Proc Natl Acad Sci U S A* 2009, **106**:6950–6955.

66. Zheng W, Rosenstiel P, Huse K, Sina C, Valentonyte R, Mah N, Zeitlmann L, Grosse J, Ruf N, Nürnberg P, Costello CM, Onnie C, Mathew C, Platzer M, Schreiber S, Hampe J: Evaluation of AGR2 and AGR3 as candidate genes for inflammatory bowel disease. *Genes Immun* 2006, **7**:11–18.

67. Adam PJ, Boyd R, Tyson KL, Fletcher GC, Stamps A, Hudson L, Poyser HR, Redpath N, Griffiths M, Steers G¶, Harris AL, Patel S, Berry J, Loader JA, Townsend RR, Daviet L, Legrain P, Parekh R, Terrett JA: Comprehensive proteomic analysis of breast cancer cell membranes reveals unique proteins with potential roles in clinical cancer. *J Biol Chem* 2003, **278**:6482–6489.

68. King ER, Tung CS, Tsang YT, Zu Z, Lok GT, Deavers MT, Malpica A, Wolf JK, Lu KH, Birrer MJ, Mok SC, Gershenson DM, Wong KK: The anterior homolog 3 (AGR3) gene is associated with differentiation and survival in ovarian cancer. *Am J Surg Pathol* 2011, **35**:904–912.

69. Liu F, Rong YP, Zeng LC, Zhang X, Han ZG: Isolation and characterization of a novel human thioredoxin-like gene hTLP19 encoding a secretory protein. *Gene* 2003, **315**:71–78.

70. Alanen HI, Williamson RA, Howard MJ, Lappi AK, Jäntti HP, Rautio SM, Kellokumpu S, Ruddock LW: Functional characterization of ERp18, a new endoplasmic reticulum-located thioredoxin superfamily member. *J Biol Chem* 2003, **278**:28912–28920.

71. Coe H, Michalak M: Calcium binding chaperones of the endoplasmic reticulum. *Gen Physiol Biophys* 2009, **28**:F96–F103. Spec No Focus.

72. Shin DW, Pan Z, Kim EK, Lee JM, Bhat MB, Parness J, Kim DH, Ma J: A retrograde signal from calsequestrin for the regulation of store-operated Ca2+ entry in skeletal muscle. *J Biol Chem* 2003, **278**:3286–3292.

73. Fujii J, Willard HF, MacLennan DH: Characterization and localization to human chromosome 1 of human fast-twitch skeletal muscle calsequestrin gene. *Somat Cell Mol Genet* 1990, **16**:185–189.

74. MacLennan DH, Wong PT: Isolation of a calcium-sequestering protein from sarcoplasmic reticulum. *Proc Natl Acad Sci U S A* 1971, **68**:1231–1235.

75. Fliegel L, Ohnishi M, Carpenter MR, Khanna VK, Reithmeier RA, MacLennan DH: Amino acid sequence of rabbit fast-twitch skeletal muscle calsequestrin deduced from cDNA and peptide sequencing. *Proc Natl Acad Sci U S A* 1987, **84**:1167–1171.

76. Yano K, Zarain-Herzberg A: Sarcoplasmic reticulum calsequestrins: structural and functional properties. *Mol Cell Biochem* 1994, **135**:61–70.

77. Dainese M, Quarta M, Lyfenko AD, Paolini C, Canato M, Reggiani C, Dirksen RT, Protasi F: Anesthetic- and heat-induced sudden death in calsequestrin-1-knockout mice. *FASEB J* 2009, **23**:1710–1720.

78. Zhao X, Min CK, Ko JK, Parness J, do Kim H, Weisleder N, Ma J: Increased store-operated Ca2+ entry in skeletal muscle with reduced calsequestrin-1 expression. *Biophys J* 2010, **99**:1556–1564.

79. Das SK, Chu W, Zhang Z, Hasstedt SJ, Elbein SC: Calsquestrin 1 (CASQ1) gene polymorphisms under chromosome 1q21 linkage peak are associated with type 2 diabetes in Northern European Caucasians. *Diabetes* 2004, **53**:3300–3306.

80. Fu M, Damcott CM, Sabra M, Pollin TI, Ott SH, Wang J, Garant MJ, O'Connell JR, Mitchell BD, Shuldiner AR: Polymorphism in the calsequestrin 1 (CASQ1) gene on chromosome 1q21 is associated with type 2 diabetes in the old order Amish. *Diabetes* 2004, **53**:3292–3299.

81. Youn JH, Gulve EA, Holloszy JO: Calcium stimulates glucose transport in skeletal muscle by a pathway independent of contraction. *Am J Physiol* 1991, **260**:C555–C561.

82. Otsu K, Fujii J, Periasamy M, Difilippantonio M, Uppender M, Ward DC, MacLennan DH: Chromosome mapping of five human cardiac and skeletal muscle sarcoplasmic reticulum protein genes. *Genomics* 1993, **17**:507–509.

83. Lahat H, Pras E, Olender T, Avidan N, Ben-Asher E, Man O, Levy-Nissenbaum E, Khoury A, Lorber A, Goldman B, Lancet D, Eldar M: A missense mutation in a highly conserved region of CASQ2 is associated with autosomal recessive catecholamine-induced polymorphic ventricular tachycardia in Bedouin families from Israel. *Am J Hum Genet* 2001, **69**:1378–1384.

84. Gyorke I, Hester N, Jones LR, Gyorke S: The role of calsequestrin, triadin, and junctin in conferring cardiac ryanodine receptor responsiveness to luminal calcium. *Biophys J* 2004, **86**:2121–2128.

85. Knollmann BC, Chopra N, Hlaing T, Akin B, Yang T, Ettensohn K, Knollmann BE, Horton KD, Weissman NJ, Holinstat I, Zhang W, Roden DM, Jones LR, Franzini-Armstrong C, Pfeifer K: Casq2 deletion causes sarcoplasmic reticulum volume increase, premature Ca2+ release, and catecholaminergic polymorphic ventricular tachycardia. *J Clin Invest* 2006, **116**:2510–2520.

86. Matsuo Y, Akiyama N, Nakamura H, Yodoi J, Noda M, Kizaka-Kondoh S: Identification of a novel thioredoxin-related transmembrane protein. *J Biol Chem* 2001, **276**:10032–10038.

87. Matsuo Y, Nishinaka Y, Suzuki S, Kojima M, Kizaka-Kondoh S, Kondo N, Son A, Sakakura-Nishiyama J, Yamaguchi Y, Masutani H, Ishii Y, Yodoi J: TMX, a human transmembrane oxidoreductase of the thioredoxin family: the possible role in disulfide-linked protein folding in the endoplasmic reticulum. *Arch Biochem Biophys* 2004, **423**:81–87.

88. Matsuo Y, Masutani H, Son A, Kizaka-Kondoh S, Yodoi J: Physical and functional interaction of transmembrane thioredoxin-related protein with major histocompatibility complex class I heavy chain: redox-based protein quality control and its potential relevance to immune responses. *Mol Biol Cell* 2009, **20**:4552–4562.

89. Meng X, Zhang C, Chen J, Peng S, Cao Y, Ying K, Xie Y, Mao Y: Cloning and identification of a novel cDNA coding thioredoxin-related transmembrane protein 2. *Biochem Genet* 2003, **41**:99–106.

90. Haugstetter J, Blicher T, Ellgaard L: **Identification and characterization of a novel thioredoxin-related transmembrane protein of the endoplasmic reticulum.** *J Biol Chem* 2005, **280**:8371–8380.

91. Chao R, Nevin L, Agarwal P, Riemer J, Bai X, Delaney A, Akana M, JimenezLopez N, Bardakjian T, Schneider A, Chassaing N, Schorderet DF, FitzPatrick D, Kwok P, Ellgaard L, Gould DB, Zhang Y, Malicki J, Baier H, Slavotinek A: **A male with unilateral microphthalmia reveals a role for TMX3 in eye development.** *PLoS One* 2010, **5**:e10565.

92. Sugiura Y, Araki K, Iemura S, Natsume T, Jun Hoseki J, Nagata K: **Novel thioredoxin-related transmembrane protein TMX4 has reductase activity.** *J Biol Chem* 2010, **285**:7135–7142.

93. Roth D, Lynes E, Riemer J, Hansen HG, Althaus N, Simmen T, Ellgaard L: **A di-arginine motif contributes to the ER localization of the type I transmembrane ER oxidoreductase TMX4.** *Biochem J* 2010, **425**:195–205.

2

Rapid detection of genetic mutations in individual breast cancer patients by next-generation DNA sequencing

Suqin Liu[1], Hongjiang Wang[1], Lizhi Zhang[1], Chuanning Tang[2], Lindsey Jones[3], Hua Ye[2], Liying Ban[1], Aman Wang[1], Zhiyuan Liu[2], Feng Lou[2], Dandan Zhang[2], Hong Sun[2], Haichao Dong[2], Guangchun Zhang[2], Zhishou Dong[2], Baishuai Guo[2], He Yan[2], Chaowei Yan[2], Lu Wang[2], Ziyi Su[2], Yangyang Li[2], Xue F Huang[3], Si-Yi Chen[3*] and Tao Zhou[1*]

Abstract

Breast cancer is the most common malignancy in women and the leading cause of cancer deaths in women worldwide. Breast cancers are heterogenous and exist in many different subtypes (luminal A, luminal B, triple negative, and human epidermal growth factor receptor 2 (HER2) overexpressing), and each subtype displays distinct characteristics, responses to treatment, and patient outcomes. In addition to varying immunohistochemical properties, each subtype contains a distinct gene mutation profile which has yet to be fully defined. Patient treatment is currently guided by hormone receptor status and HER2 expression, but accumulating evidence suggests that genetic mutations also influence drug responses and patient survival. Thus, identifying the unique gene mutation pattern in each breast cancer subtype will further improve personalized treatment and outcomes for breast cancer patients. In this study, we used the Ion Personal Genome Machine (PGM) and Ion Torrent AmpliSeq Cancer Panel to sequence 737 mutational hotspot regions from 45 cancer-related genes to identify genetic mutations in 80 breast cancer samples of various subtypes from Chinese patients. Analysis revealed frequent missense and combination mutations in PIK3CA and TP53, infrequent mutations in PTEN, and uncommon combination mutations in luminal-type cancers in other genes including BRAF, GNAS, IDH1, and KRAS. This study demonstrates the feasibility of using Ion Torrent sequencing technology to reliably detect gene mutations in a clinical setting in order to guide personalized drug treatments or combination therapies to ultimately target individual, breast cancer-specific mutations.

Keywords: Breast cancer, Genetic mutations, Ion torrent sequencing, Targeted therapy, Personalized medicine

Introduction

Breast cancer is the second most common malignancy worldwide and the most frequent in women. Roughly 1.67 million new cases and 522,000 deaths were reported globally in 2012, making breast cancer the fifth leading cause of cancer death. Breast cancer incidence differs with population and geographic location, where China alone accounted for more than 187,000 cases and nearly 48,000 deaths in 2012, whereas over 230,000 cases and more than 43,000 deaths were reported in the US [1].

Patient screening is superior in the US than in China [2], which may account for a higher incidence despite a much smaller population. While risk factors for developing breast cancer include ethnicity, older age, and environmental factors, lifestyle and diet also play a significant role, where westernization in Asia is thought to have contributed to the rise in spontaneous breast cancer incidence in Chinese populations over the last 20 years [3-5].

Breast cancers are highly heterogenous and may display different characteristics of hormone receptors (HR) (estrogen receptor (ER) and progesterone receptor (PR)) and human epidermal growth factor receptor 2 (HER2) status, and together this information helps to distinguish different types of breast cancers: luminal A (HR+/HER2–, tumor grade 1 or 2), luminal B/HER2– (HR+/HER2–, tumor grade 3 or 4), luminal B/HER2+

* Correspondence: zhoutao1967@163.com; siyichen@usc.edu
[1]The First Affiliated Hospital of Dalian Medical University, Dalian, Liaoning, China
[3]Norris Comprehensive Cancer Center, Department of Molecular Microbiology and Immunology, Keck School of Medicine, University of Southern California, Los Angeles, CA, USA
Full list of author information is available at the end of the article

(HR+/HER2+), triple negative (HR−/HER2−), or HER2 overexpressing (HR−/HER2+) [6]. Together luminal A and B subtypes account for 65%–70% of all breast cancers, whereas 10%–15% are triple negative and 10%–20% are HER2 overexpressing [7].

These distinct types of breast cancers all have different characteristics, behaviors, and prognoses and also respond differently to drug treatments. Nearly three quarters of all breast cancers are ER+ and are therefore in some way dependent on estrogen for growth, providing a useful target for treating these cancers via ER modulators or downregulators or aromatase inhibitors [8]. But only 20%–40% of patients with advanced ER+ breast cancer have a response to endocrine therapy, which only averages 8 to 14 months [9]. Luminal A types tend to have the best outcome with a 95% 5-year survival rate, whereas luminal B tumors, which tend to have lower HR expression and subsequently less sensitivity to endocrine therapy but increased sensitive to chemotherapy, tend to have a worse outcome [10-12]. Typically found in younger patients, triple-negative breast cancers (TNBCs) are known to be particularly aggressive and are associated with germline *BRCA* mutations. TNBCs also have higher relapse rates and decreased overall patient survival than other breast cancer types [13]. HER2-overexpressing breast cancers also have poor prognoses and high metastases rates, and as they lack HR expression, they do not respond to endocrine therapies and are resistant to current chemotherapies [14]. These distinct breast cancer types can be further divided into a multitude of subtypes, and all of these types and subtypes exhibit distinct gene mutation patterns that have yet to be fully defined [10].

Currently, patient prognoses and treatment regimens for breast cancer are guided by the aforementioned characteristics of the tumor, but accumulating evidence suggests that this information is not enough; risk assessments, treatments, and patient outcomes are also influenced by both germline and somatic gene mutations. Known genetic factors like inherited *BRCA* mutations confer a lifetime risk of developing breast cancer of 60% to 85%; however, these mutations account for only 2%–3% of all breast cancer cases [15]. Spontaneous mutations in *PIK3CA* are a much more common event in breast cancers, with more than a quarter of breast cancer patients harboring a mutation in this gene [16,17]. While *PIK3CA* mutations have been shown to be associated with improved patient prognoses, these genetic aberrations have also been shown to impart resistance to trastuzumab, a common treatment option for HER2-overexpressing breast cancers [18]. Identifying gene mutations in patients with TNBC is especially important because these cancers currently do not have direct targets for treatments. Therefore, it is important to establish both the immunohistochemical properties and genetic profile of each breast cancer tumor to optimize treatment regimens and avoid unnecessary drug toxicities and ultimately to improve patient outcomes.

There are a number of different next-generation sequencing (NGS) platforms available, including Illumina, 454, and SOLiD, but these are typically expensive both in instrument and assay cost, and therefore, these tools are unrealistic for widespread clinical diagnostic use. But new technology like the Ion Torrent sequencing platform has been shown to be more cost and time effective with reliable results [19], which may help make cancer DNA sequencing and personalized treatments a reality for each cancer patient in the near future. In the present study, we have used Ion Torrent sequencing technology with the Ion Personal Genome Machine (PGM) and Ion Torrent AmpliSeq Cancer Panel as a rapid and affordable method to detect gene mutations in 80 clinical breast cancer samples of different types from Chinese patients.

Materials and methods

Ethics statement

The study has been approved by the Human Research Ethics Committee of the First Affiliated Hospital of Dalian Medical University, China. The institutional ethics committee waived the need for consent for formalin-fixed, paraffin-embedded (FFPE) tumor samples obtained from the tumor tissue bank at the hospital's Department of Pathology. All samples and medical data used in this study have been irreversibly anonymized.

Patient information

Tumor samples used in the study were collected from the First Affiliated Hospital of Dalian Medical University, China. A total of 80 FFPE tumor samples from Chinese breast cancer patients were analyzed (Table 1). Patients were an average of 55 years old, with a range of 30–75 years. American Joint Committee on Cancer (AJCC)/tumor, node, and metastasis (TNM) cancer staging was assessed, and tumor samples were also analyzed for immunohistochemical status of HR and HER2 (Table 1). Based on these, patients were categorized into five breast cancer subtypes: luminal A (HR+/HER2−, AJCC stage 1 or 2; 26.3%), luminal B/HER2− (HR+/HER2−, AJCC stage 3 or 4; 21.3%), luminal B/HER2+ (HR+/HER2+; 30.0%), triple negative (HR−/HER2−; 7.5%), and HER2 overexpressing (HR−/HER2+; 8.8%), and five samples (6.3%) were unclassifiable due to unknown HER2 status (Table 2).

DNA preparation

FFPE tissue samples were first deparaffinized in xylene, 3–5-μm-thick sections were extracted, and DNA was

Table 1 Clinical features of 80 breast cancer patients

Characteristic		n (%)
Age (years)	Median: 56	
	Range: 30–75	
HR status	HR+	65 (81.3%)
	HR−	15 (18.8%)
HER2 status	HER2+	12 (15.0%)
	HER2++	18 (22.5%)
	HER2+++	1 (1.3%)
	HER2−	44 (55.0%)
	Unknown	5 (6.3%)
AJCC/TNM stage	2a	19 (23.8%)
	2b	26 (32.5%)
	3a	23 (28.8%)
	3b	4 (8.8%)
	3c	7 (8.8%)
	4	1 (1.3%)
Pathological diagnosis of infiltrating ductal carcinoma	IDC2	71 (88.8%)
	IDC3	6 (7.5%)
	Others	3 (3.8%)

isolated using the QIAamp DNA Mini Kit (QIAGEN) as per the manufacturer's instructions.

Ion Torrent PGM library preparation and DNA sequencing

An Ion Torrent adapter-ligated library was constructed with the Ion AmpliSeq Library Kit 2.0 (Life Technologies, Part #4475345 Rev. A) following the manufacturer's protocol. Briefly, 50 ng of pooled amplicons were end-repaired, Ion Torrent adapters P1 and A were ligated, and the adapter-ligated products were then purified with AMPure beads (Beckman Coulter, Brea, CA, USA), nick-translated, and PCR-amplified for 5 cycles. The resulting library was purified with AMPure beads (Beckman Coulter), and the library concentration and size was determined with the Agilent 2100 Bioanalyzer and Agilent Bioanalyzer DNA High-Sensitivity LabChip (Agilent Technologies).

Sample emulsion PCR, emulsion breaking, and enrichment were performed with the Ion PGM 200 Xpress Template Kit (Life Technologies, Part #4474280 Rev. B), according to the manufacturer's instructions. Briefly, an input concentration of one DNA template copy/Ion Sphere Particles (ISPs) was added to emulsion PCR master mix, and the emulsion was generated with an IKADT-20 mixer (Life Technologies). Next, ISPs were recovered, and template-positive ISPs were enriched with Dynabeads MyOne Streptavidin C1 beads (Life Technologies). The Qubit 2.0 fluorometer (Life Technologies) was used to confirm ISP enrichment. Three-hundred sixteen chips were used to sequence barcoded samples on the Ion Torrent PGM for 65 cycles, and an Ion PGM 200 Sequencing Kit (Life Technologies, Part # 4474004 Rev. B) was used for sequencing reactions, as per the recommended protocol.

This Personalized Cancer Mutation Panel targets 737 mutational hotspot regions in the following 45 genes: *ABL1, AKT1, ALK, APC, ATM, BRAF, CDH1, CDKN2A, CSF1R, CTNNB1, EGFR, ERBB2, ERBB4, FBXW7, FGFR1, FGFR2, FGFR3, FLT3, GNAS, HNF1A, HRAS, IDH1, JAK3, KDR, KIT, KRAS, MET, MLH1, MPL, NOTCH1, NPM1, NRAS, PDGFRA, PIK3CA, PTEN,*

Table 2 Average patient age, average disease-free survival (DFS), and mutation frequency in breast cancer subtypes with or without mutations

Subtype	n (% total)	Average age (years)	Average DFS (months)	Samples with mutations (freq.)	Samples without mutations (freq.)	Average age with mutations (years)	Average age without mutations (months)	P value	Average DFS with mutations (months)	Average DFS without mutations (months)	P value
All	80 (100%)	54.5	38.3	32 (40.0%)	48 (60.0%)	54.9	54.2	0.757	38.8	38.0	0.741
Luminal A	21 (26.3%)	53.7	39.0	9 (42.9%)	12 (57.1%)	53.8	53.7	0.979	40.6	37.8	0.407
Luminal B/ HER2−	17 (21.3%)	55.8	39.4	6 (35.3%)	11 (64.7%)	57.3	54.9	0.626	40.5	38.7	0.799
Luminal B/ HER2+	24 (30.0%)	52.9	37.3	11 (45.8%)	13 (54.2%)	53.6	52.3	0.772	37.5	37.1	0.936
Triple negative	6 (7.5%)	53.8	37.5	4 (66.7%)	2 (33.3%)	58.8	44.0	0.165	39.8	33.0	0.508
HER2 overexpressing	7 (8.8%)	56.6	34.6	2 (28.6%)	5 (71.4%)	52.5	58.2	0.633	31.5	35.8	0.595
Unknown	5 (6.3%)	59.0	43.0	0 (0.0%)	5 (100%)	-	59.0	-	-	43.0	-

PTPN11, *RB1*, *RET*, *SMAD4*, *SMARCB1*, *SMO*, *SRC*, *STK11*, *TP53*, and *VHL*.

Variant calling

Initial data from the PGM runs were processed with the Ion Torrent platform-specific pipeline software Torrent Suite to generate sequence reads, then trim adapter sequences, filter, and remove poor signal-profile reads. Torrent Suite software v3.4 with a plug-in "variant caller v3.4" program was used to generate initial variant calling from the Ion AmpliSeq sequencing data. In order to eliminate erroneous base calling and generate final variant calling, several filtering steps were used: defining average total coverage depth, variant coverage, variant frequency of each sample, and P value <0.01; visually inspecting and removing DNA strand-specific errors; defining variants within hotspots; and eliminating variants in amplicon AMPL339432 (PIK3CA, exon13, chr3:178938822–178938906) which is not uniquely matched in the human genome.

Sequence coverage

From the 80 samples, the mean read length was 76 bp and the average reads were approximately 24 Mb of sequence per sample. With normalization to 300,000 reads per specimen, there was an average of 1,639 reads per amplicon (range: 22 to 6,020) (Figure 1A); 180/189 (95.2%) amplicons averaged at least 100 reads; and 170/189 (89.9%) amplicons averaged at least 300 reads (Figure 1B).

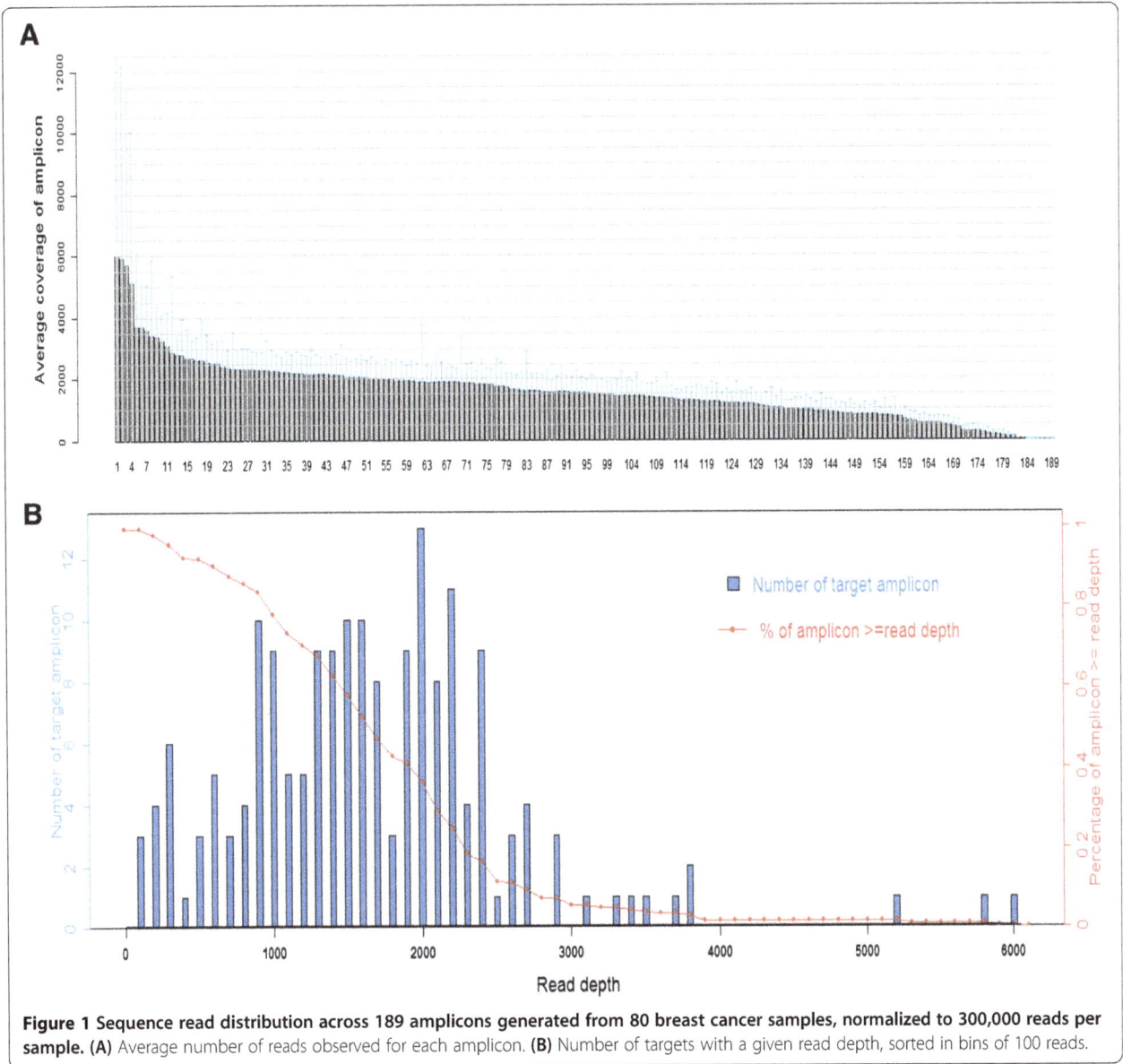

Figure 1 Sequence read distribution across 189 amplicons generated from 80 breast cancer samples, normalized to 300,000 reads per sample. (A) Average number of reads observed for each amplicon. **(B)** Number of targets with a given read depth, sorted in bins of 100 reads.

Somatic mutations

Detected mutations were compared to variants in the 1000 Genomes Project [20] and 6,500 exomes of the National Heart, Lung, and Blood Institute Exome Sequencing Project [21] to distinguish between somatic and germline mutations.

Bioinformatical and experimental validation

We used the COSMIC [22] (version 64), MyCancer-Genome database (http://www.mycancergenome.org/), and some publications to assess recurrent mutations in breast cancer (Additional file 1: Table S1). Additionally, detected missense mutations were confirmed by Sanger sequencing (data not shown). All mutations identified with Sanger sequencing were consistent with those identified with the Ion Torrent PGM.

Statistical analysis

The Fisher's exact test was used to define significant values in the detected mutated genes, and the total variants and odds ratios (OR) between samples with mutations and without mutations were determined using 2 × 2 contingency tables and the GraphPad QuickCalcs online calculator for Scientists (http://www.graphpad.com/quickcalcs/index.cfm). All P values are two-sided, and statistical significance was defined as $P < 0.05$.

Results and discussion

From the 45 genes screened in our study, 39 mutations were detected in 32 of 80 samples (40.0%) (Figure 2). Except for the five unclassified samples with no mutations, mutations were detected at different frequencies across all breast cancer subtypes (Table 2). Triple negative samples contained the highest mutation frequency (66.7%), whereas HER2-overexpressing samples contained the lowest mutation frequency (28.6%), and luminal A, luminal B/HER2−, and luminal B/HER2+ had similar mutation frequencies (42.9%, 35.3%, and 45.8%, respectively). Twenty-six samples (32.5%) contained one mutation, five samples (6.3%) contained two mutations, and one sample (1.3%) contained three mutations, and interestingly, combination mutations were only found in the luminal subtypes (Table 3). *PIK3CA* mutations and TP53 mutations were the most prevalent (32.5% and 10.0%, respectively), and mutations were also identified in *BRAF, GNAS, IDH1, KRAS,* and *PTEN* all at a frequency of 1.3% (Figure 3). Among each subtype, there was no statistically significant difference in age or disease-free survival (DFS) between patients with mutations and patients without mutations (Table 2).

PIK3CA mutations

Twenty-six samples (32.5%) harbored *PIK3CA* mutations, which accounted for 66.7% of all detected mutations in our study. Five different *PIK3CA* mutations were identified: p.N345K in the C2 domain encoded by exon 4, p.E542K and p.E545K in the helical domain encoded by exon 9, and p.H1047R and p.H1047L in the kinase domain encoded by exon 20. Mutations p.E542K, p.E545K, and p.H1047R have been found in previous studies to be the most prevalent in human breast cancers and are associated with an increase in kinase activity in the

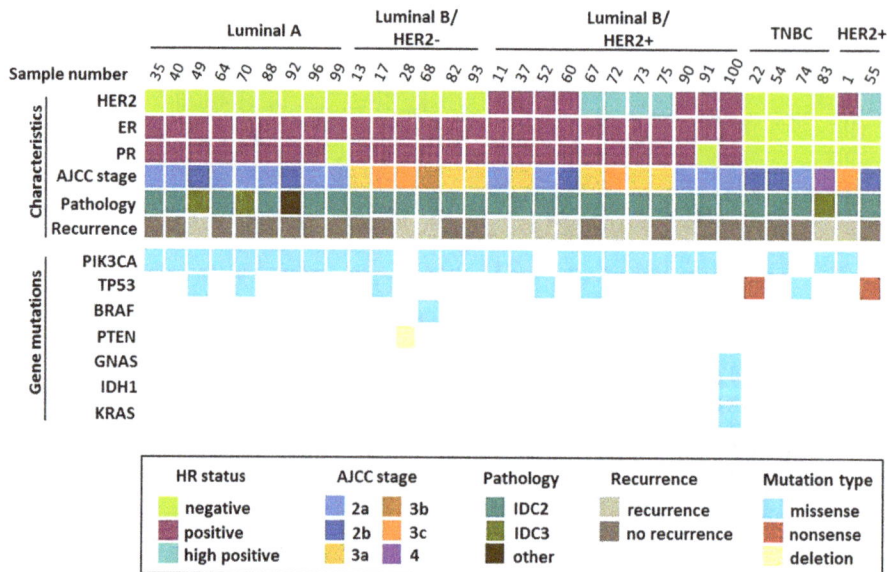

Figure 2 Summary of mutated genes detected in 80 breast cancer samples. Thirty-two samples harbor mutations in *PIK3CA, TP53, KRAS, BRAF, PTEN, GNAS,* and *IDH1.* Samples are classified by four methods: 1) Immunohistochemistry of ER, PR, and HER2; 2) pathologic type (IDC2, IDC3, other); 3) AJCC/TNM-staging (2a, 2b, 3a, 3b, 3c, 4); and 4) recurrence or no recurrence.

Table 3 Detected point mutations per breast cancer subtype

Breast cancer type	Gene(s)	Mutation(s)	Age (years)	DFS (months)	Recurrence	AJCC	ER	PR	HER2
Luminal A	PIK3CA	p.N345K	62	36	N	2a	+	+	−
	PIK3CA	p.E545K	46	36	N	2a	+	+	−
	PIK3CA	p.E545K	61	47	N	2a	+	+	−
	PIK3CA	p.H1047L	55	36	N	2b	+	+	−
	PIK3CA	p.H1047R	51	48	N	2a	+	+	−
	PIK3CA	p.H1047R	46	42	N	2a	+	+	−
	PIK3CA	p.H1047R	56	36	N	2a	+	−	−
	PIK3CA/TP53	p.H1047R/p.R248W	57	41	N	2a	+	+	−
	PIK3CA/TP53	p.H1047R/p.R175H	50	36	Y	2b	+	+	−
Luminal B/HER2−	PTEN	p.T321fs*23	55	28	Y	3c	+	+	−
	PIK3CA/TP53	p.E545K/ p.H193R	52	53	N	3c	+	+	−
	PIK3CA/BRAF	p.H1047R/ p.V600M	59	36	Y	3b	+	+	−
	PIK3CA	p.H1047R	52	36	N	3a	+	+	−
	PIK3CA	p.H1047R	58	36	N	3a	+	+	−
	PIK3CA	p.H1047R	68	54	N	3a	+	+	−
Luminal B/HER2+	GNAS/IDH1/KRAS	p.R201C/ p.R132C/ p.G12D	57	36	N	2a	+	+	+
	PIK3CA	p.E542K	50	47	N	3a	+	+	+
	PIK3CA	p.E545K	66	42	N	2b	+	+	+
	PIK3CA	p.H1047L	72	40	N	3a	+	+	++
	PIK3CA	p.H1047R	44	36	N	2a	+	+	+
	PIK3CA	p.H1047R	45	55	N	2a	+	+	+
	PIK3CA	p.H1047R	32	36	N	2a	+	−	+
	PIK3CA	p.H1047R	65	25	Y	3a	+	+	++
	PIK3CA	p.H1047R	63	12	Y	3c	+	+	++
	PIK3CA/TP53	p.H1047R/ p.P278L	49	41	N	3a	+	+	++
	TP53	p.Y220C	47	43	N	2a	+	+	+
Triple negative	PIK3CA	p.E545K	60	43	N	2b	−	−	−
	PIK3CA	p.E542K	61	26	Y	4	−	−	−
	TP53	p.Y163C	57	40	N	2a	−	−	−
	TP53	p.R196*	57	50	N	2b	−	−	−
HER2 overexpressing	PIK3CA	p.H1047R	65	20	Y	3c	−	−	+
	TP53	p.R213*	40	43	N	2b	−	−	++

DFS disease-free survival, del deletion, fs frameshift.
*Nonsense mutation resulting in a stop codon.

PI3K pathway [23,24]. These three mutations accounted for 88.5% of all PIK3CA mutations in our study. The remaining mutations, p.N345K and p.H1047L, are much less common and are found in less than 2% and 3.5% of breast cancers, respectively [22,25,26].

PIK3CA mutations are an early event in breast cancer development. Accordingly, we found mutations in this gene at all stages and all breast cancer subtypes in our study. While previous research has found PIK3CA mutations to be associated with older patient age [25], we did not find significant differences between age and PIK3CA mutations among all patients or breast cancer subtypes.

Consistent with a study by Kalinsky et al. [25], PIK3CA mutations were found in more HER2− tumors than HER2+ tumors (61.5% vs. 38.5%, respectively), although not significantly (OR: 1.20; $P = 0.81$). Also, patients with HR+ tumors had a higher likelihood of having a PIK3CA mutation than those with HR− tumors, but again, this was not statistically significant (OR: 2.19; $P = 0.36$). Others have found lower PIK3CA mutation frequencies in luminal B than luminal A breast cancers [10], but we found the opposite in our study with 53.8% of PIK3CA-mutated samples as luminal B types vs. 34.6% luminal A type. Drugs like everolimus, a rapamycin derivative that

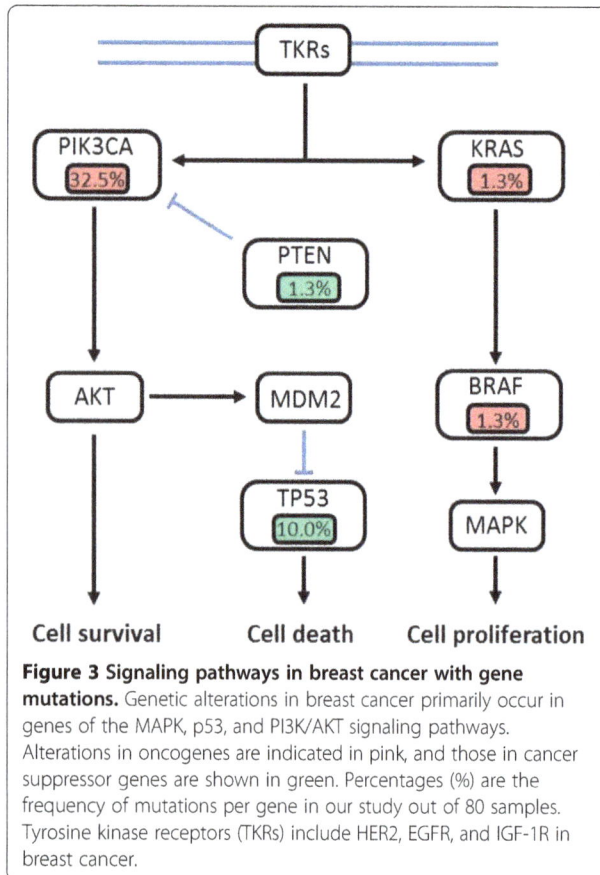

Figure 3 Signaling pathways in breast cancer with gene mutations. Genetic alterations in breast cancer primarily occur in genes of the MAPK, p53, and PI3K/AKT signaling pathways. Alterations in oncogenes are indicated in pink, and those in cancer suppressor genes are shown in green. Percentages (%) are the frequency of mutations per gene in our study out of 80 samples. Tyrosine kinase receptors (TKRs) include HER2, EGFR, and IGF-1R in breast cancer.

inhibits PI3K/ATK/mTOR signaling, have been approved to treat advanced HR+/HER2− breast cancer patients after other treatments have failed and have been shown to increase progression-free survival [27-29].

While the study of Kalinsky et al. found overall patient survival to be significantly improved in patients with *PIK3CA* mutations [25], we found that patients with *PIK3CA* mutations collectively had a roughly equal DFS time to patients with WT *PIK3CA* (38.6 vs. 38.2 months, respectively); however, patients with triple negative and HER2-overexpressing tumors and *PIK3CA* mutations had a shorter average DFS (34.5 and 20 months, respectively) than those with WT *PIK3CA* (39 and 37 moths, respectively), possibly suggesting that the other characteristics of the tumor may have greater prognostic value than *PIK3CA* mutations. Consistent with a clinical study on the impact of specific *PIK3CA* mutations on breast cancer patient prognosis [30], we found that patients with *PIK3CA* mutations in exon 20 had a slightly shorter DFS than those with *PIK3CA* mutations in exons 4 and 9 (37.4 vs. 41.3 months, respectively; $P = 0.37$). Because activation of the PI3K/ATK/mTOR pathway has been shown to confer resistance to trastuzumab treatment in HER2+ breast tumors [31,32], and because the specific mutation may offer prognostic value, it is important to identify *PIK3CA*

mutations in all breast cancer patients regardless of subtype.

TP53 mutations

Eight samples (10.0%) were found to harbor *TP53* mutations, and these mutations were identified in each subtype: 2 (9.5%) in luminal A, 1 (5.9%) in luminal B/HER2−, 2 (8.3%) in luminal B/HER2+, 2 (33.3%) in triple negative, and 1 (14.3%) in HER2-overexpressing tumors. There was no statistically significant difference between age and *TP53* mutations among all patients or breast cancer subtypes, nor was there a difference in DFS and *TP53* mutations and subtype. Among 80 samples, we found *TP53* mutations were more likely to occur in HR− tumors (OR: 3.0; $P = 0.17$) and tumors of a lower grade (75.0% at stage 2 vs. 25.0% at 3 or 4; OR: 2.54; $P = 0.46$). Five of the eight (62.5%) *TP53*-mutated tumors were HER2−, although study wide, there was no correlation between *TP53* mutations and HER2 status (OR: 1.25; $P = 1.0$). Others have reported that *TP53* is associated with worse overall patient survival [33-35], but we found this to be the opposite in our study; across all types, patients with *TP53* mutations had an average DFS of 44.3 vs. 37.7 months for those with WT *TP53* ($P = 0.09$).

The eight different *TP53* mutations detected in our study were found at known hotspot locations, all within the DNA binding domains: two in exon 5 (p.Y136C and p.R175H); four in exon 6 (p.H193R, p.R196*, p.R213*, and p.Y220C); one in exon 7 (p.R248W); and one in exon 8 (p.P278L). Mutations in exons 5 and 7 have previously been shown to correlate with poorer overall survival and disease-free progression in breast cancer patients [36,37]. Specifically, mutations in these exons which affect the L2 and L3 loop domains of the protein (codons 163–195 and 236–251, respectively) have been found to confer resistance to certain cytotoxic drugs, including 5-fluorouracil and mitomycin [38,39]. However, in our study, there were only four patients with *TP53* mutations in these exons and specific codons, and these patients had an equal DFS to patients with other *TP53* mutations (44.3 months), and a longer, albeit not significant, DFS than the study average of 38.3 months ($P = 0.26$).

Combination and less frequent mutations

One advanced luminal B/HER2− sample contained a deletion in codon 321, exon 8 of *PTEN*, resulting in a frameshift mutation (p.T321fs*23). This mutation has been found in other cancers of the endometrium and large intestine but has not yet been identified in breast cancers [22]. As an antagonist of PI3K signaling, improper *PTEN* function leads to uncontrolled activation of its downstream signals [16], and reduced *PTEN*

expression correlates with breast cancer progression and is associated with advanced disease [40,41]. Accordingly, the *PTEN*-mutated sample in our study was stage 3C, and the patient's DFS was 10 months shorter that the study average. Others have reported that *PTEN* mutations occur more often in HR– tumors [41,42], which was not the case for the *PTEN*-mutated sample in our study.

Six of the samples (7.5%) contained mutations in a combination of genes, and all co-mutations were found in the luminal subtypes (100% HR+, 66.7% HER2–) (Table 3). Five of the six combination mutations co-occurred with *PIK3CA*, and four of these were a combination of *PIK3CA* and *TP53*. Research suggests that the presence of *PIK3CA* and *TP53* co-mutations increase breast cancer sensitivity to some PI3K/AKT/mTOR inhibitors when compared to those with *PIK3CA* mutations alone [43,44]; regardless, promising drugs like BEZ235, a PI3K/mTOR kinase inhibitor, may only be effective in a specific subset of triple-negative breast cancers [45].

One luminal B/HER2– sample harbored a combination mutation in *PIK3CA* (p.H1047R) and *BRAF* (p.V600M). This combination is commonly found in melanoma but rarely in breast cancer [46]. The vast majority of *BRAF* mutations identified in all cancer types are activating mutations that occur at codon 600, most commonly p.V600E (84.6%), whereas p.V600M accounts for only 0.3% of *BRAF* mutations at this codon [47]. While this mutation is not common in breast cancers, several drugs have been developed to target *BRAF* mutations in other cancer types and are showing promising results in breast cancer models [48,49]. By combining treatments to target both *PIK3CA* and *BRAF* mutations, patients with combination mutations such as these may find benefit greater than a single treatment. Everolimus, however, would not be effective because research has shown this drug to be ineffective in the presence of a *KRAS* or *BRAF* mutation [27].

One stage 2A luminal B/HER2+ sample contained a unique combination of mutations in three different genes, *GNAS* (p.R201C), *IDH1* (p.R132C), and *KRAS* (p.G12D), a combination that, to our knowledge, has yet to be identified in breast cancer. *GNAS* mutations, which are commonly found in pituitary and pancreatic cancers, and *IDH1* mutations, common in gliomas, are uncommon events in breast cancers, both found in less than 1% [22,50]. *KRAS* mutations are only slightly more common, found in 2%–5% of breast cancers [22,51]. *KRAS* is involved in the *EGFR* signaling pathway, and anti-EGFR drugs like lapatinib used to treat advanced HER2+ breast cancers require WT *KRAS* to be effective [52,53]. For a patient with these tumor characteristics, knowledge of mutated *KRAS* status would prevent unnecessary drug toxicity.

Conclusion

Characterization of breast cancer tumors is critical in determining appropriate treatment options and predicting patient prognosis. The hormone receptor status and HER2 expression act as markers to direct drug treatments or to predict behavior of the disease. In addition to immunohistochemical properties, it is becoming increasingly evident that gene mutations play a role in breast cancer progression and response to treatment, making it critical to determine the genetic profile of each breast cancer tumor to personalize treatments and optimize patient outcomes. In our study, we used Ion Torrent sequencing technology to identify mutations in 80 clinical breast cancer tumors of various subtypes. Our study revealed not only uncommon and novel combination mutations but also mutations commonly found in breast cancers at frequencies similar to those previously reported, indicating the reliability of the Ion Torrent sequencing method to genotype cancer samples in a clinical setting. Ion Torrent sequencing technology has also been shown to be more cost and time effective than other traditional sequencing methods [19,54], and as such, it may be a feasible way to advance personalized patient treatments by providing clinicians a tool to characterize breast cancers beyond immunohistochemical markers and ultimately improve outcomes for breast cancer patients.

Competing interests
Authors Chuanning Tang, Hua Ye, Feng Lou, Dandan Zhang, Hong Sun, Haichao Dong, Guangchun Zhang, Zhiyuan Liu, Zhishou Dong, Baishuai Guo, He Yan, Chaowei Yan, Lu Wang, Ziyi Su, and Yangyang Li are employees of San Valley Biotechnology, Inc. in Beijing, China.

Authors' contributions
SL, HW, LZ, CT, HY, LB, AW, FL, DZ, HS, HD, GZ, ZL, ZD, BG, HY, CY, LW, ZS, YL, and TZ performed the breast cancer tumor sample sequencing, data analysis, and figure preparation. SL, HW, LZ, SYC, and TZ designed the study, performed the data analysis, and assisted with the manuscript preparation. LJ, XFH, and SYC performed the additional data analysis, wrote the main manuscript text, and prepared the manuscript tables. All authors read and approved the final manuscript.

Acknowledgements
We would like to thank Rong Shi at the Wu Jieping Foundation, Dr. Haibo Wang, Ying Li, and other members of San Valley Biotechnology Inc. Beijing for their assistance in sample and data collection. We would also like to thank the staffs at the Beijing Military Hospital for their generous support for DNA sequencing and data collection. This research was supported by the grants from the Wu Jieping Foundation and the National Institute of Health (R01 CA90427 & R01 AI084811 to SY Chen).

Author details
[1]The First Affiliated Hospital of Dalian Medical University, Dalian, Liaoning, China. [2]San Valley Biotechnology Incorporated, Beijing, China. [3]Norris

Comprehensive Cancer Center, Department of Molecular Microbiology and Immunology, Keck School of Medicine, University of Southern California, Los Angeles, CA, USA.

References

1. Ferlay J, Soerjomataram I, Ervik M, Dikshit R, Eser S, Mathers C, et al. GLOBOCAN 2012 v1.0, Cancer incidence and mortality worldwide: IARC CancerBase No. 11 [Internet]. Lyon, France: International Agency for Research on Cancer; 2013. Available from: http://globocan.iarc.fr, accessed on 9/10/2014.

2. Wang B, He M, Wang L, Engelgau MM, Zhao W, Wang L. Breast cancer screening among adult women in China, 2010. Prev Chronic Dis. 2013;10:E183.

3. He F-J, Chen J-Q. Consumption of soybean, soy foods, soy isoflavones and breast cancer incidence: differences between Chinese women and women in Western countries and possible mechanisms. Food Sci Hum Wellness. 2013;2:146–61.

4. Brody JG, Rudel RA, Michels KB, Moysich KB, Bernstein L, Attfield KR, et al. Environmental pollutants, diet, physical activity, body size, and breast cancer. Cancer. 2007;109:2627–34.

5. Fan L, Strasser-Weippl K, Li JJ, St Louis J, Finkelstein DM, Yu KD, et al. Breast cancer in China. Lancet Oncol. 2014;15:e279–289.

6. Parise CA, Caggiano V. Breast cancer survival defined by the ER/PR/HER2 subtypes and a surrogate classification according to tumor grade and immunohistochemical biomarkers. J Cancer Epidemiol. 2014;2014:469251.

7. Perou CM, Sorlie T, Eisen MB, van de Rijn M, Jeffrey SS, Rees CA, et al. Molecular portraits of human breast tumours. Nature. 2000;406:747–52.

8. Ciruelos Gil EM. Targeting the PI3K/AKT/mTOR pathway in estrogen receptor-positive breast cancer. Cancer Treat Rev. 2014;40:862–71.

9. Johnston SR. New strategies in estrogen receptor-positive breast cancer. Clin Cancer Res. 2010;16:1979–87.

10. Ades F, Zardavas D, Bozovic-Spasojevic I, Pugliano L, Fumagalli D, de Azambuja E, et al. Luminal B breast cancer: molecular characterization, clinical management, and future perspectives. J Clin Oncol. 2014;32:2794–803.

11. Zhang MH, Man HT, Zhao XD, Dong N, Ma SL. Estrogen receptor-positive breast cancer molecular signatures and therapeutic potentials (review). Biomed Rep. 2014;2:41–52.

12. Ignatiadis M, Sotiriou C. Luminal breast cancer: from biology to treatment. Nat Rev Clin Oncol. 2013;10:494–506.

13. Abramson VG, Lehmann BD, Ballinger TJ, Pietenpol JA. Subtyping of triple-negative breast cancer: implications for therapy. Cancer. 2015;121:8–16.

14. Slamon DJ, Clark GM, Wong SG, Levin WJ, Ullrich A, McGuire WL. Human breast cancer: correlation of relapse and survival with amplification of the HER-2/neu oncogene. Science. 1987;235:177–82.

15. Wooster R, Weber BL. Breast and ovarian cancer. N Engl J Med. 2003;348:2339–47.

16. Chalhoub N, Baker SJ. PTEN and the PI3-kinase pathway in cancer. Annu Rev Pathol. 2009;4:127–50.

17. Lee JW, Soung YH, Kim SY, Lee HW, Park WS, Nam SW, et al. PIK3CA gene is frequently mutated in breast carcinomas and hepatocellular carcinomas. Oncogene. 2004;24:1477–80.

18. Mukohara T. Mechanisms of resistance to anti-human epidermal growth factor receptor 2 agents in breast cancer. Cancer Sci. 2011;102:1–8.

19. Glenn TC. Field guide to next-generation DNA sequencers. Mol Ecol Resour. 2011;11:759–69.

20. Consortium GP. A map of human genome variation from population-scale sequencing. Nature. 2010;467:1061–73.

21. Server EV: NHLBI Go Exome Sequencing Project (ESP). Seattle, WA http://evs.gs.washington.edu/EVS/ 2013:[March 19 Accessed].

22. Bamford S, Dawson E, Forbes S, Clements J, Pettett R, Dogan A, et al. The COSMIC (Catalogue of Somatic Mutations in Cancer) database and website. Br J Cancer. 2004;91:355–8.

23. Mankoo PK, Sukumar S, Karchin R. PIK3CA somatic mutations in breast cancer: mechanistic insights from Langevin dynamics simulations. Proteins Struc Funct Bioinformatics. 2009;75:499–508.

24. Bachman KE, Argani P, Samuels Y, Silliman N, Ptak J, Szabo S, et al. The PIK3CA gene is mutated with high frequency in human breast cancers. Cancer Biol Ther. 2004;3:772–5.

25. Kalinsky K, Jacks LM, Heguy A, Patil S, Drobnjak M, Bhanot UK, et al. PIK3CA mutation associates with improved outcome in breast cancer. Clin Cancer Res. 2009;15:5049–59.

26. Gymnopoulos M, Elsliger M-A, Vogt PK. Rare cancer-specific mutations in PIK3CA show gain of function. Proc Natl Acad Sci. 2007;104:5569–74.

27. Di Nicolantonio F, Arena S, Tabernero J, Grosso S, Molinari F, Macarulla T, et al. Deregulation of the PI3K and KRAS signaling pathways in human cancer cells determines their response to everolimus. J Clin Invest. 2010;120:2858–66.

28. Atkins MB, Yasothan U, Kirkpatrick P. Everolimus. Nat Rev Drug Discov. 2009;8:535–6.

29. Baselga J, Campone M, Piccart M, Burris HA, Rugo HS, Sahmoud T, et al. Everolimus in postmenopausal hormone-receptor-positive advanced breast cancer. N Engl J Med. 2012;366:520–9.

30. Lai Y-L, Mau B-L, Cheng W-H, Chen H-M, Chiu H-H, Tzen C-Y. PIK3CA exon 20 mutation is independently associated with a poor prognosis in breast cancer patients. Ann Surg Oncol. 2008;15:1064–9.

31. Nahta R, Yu D, Hung M-C, Hortobagyi GN, Esteva FJ. Mechanisms of disease: understanding resistance to HER2-targeted therapy in human breast cancer. Nat Clin Prac Oncol. 2006;3:269–80.

32. Razis E, Bobos M, Kotoula V, Eleftheraki AG, Kalofonos HP, Pavlakis K, et al. Evaluation of the association of PIK3CA mutations and PTEN loss with efficacy of trastuzumab therapy in metastatic breast cancer. Breast Cancer Res Treat. 2011;128:447–56.

33. Powell B, Soong R, Iacopetta B, Seshadri R, Smith DR. Prognostic significance of mutations to different structural and functional regions of the p53 gene in breast cancer. Clin Cancer Res. 2000;6:443–51.

34. Andersen TI, Holm R, Nesland JM, Heimdal KR, Ottestad L, Borresen AL. Prognostic significance of TP53 alterations in breast carcinoma. Br J Cancer. 1993;68:540–8.

35. Soong R, Iacopetta BJ, Harvey JM, Sterrett GF, Dawkins HJS, Hahnel R, et al. Detection of p53 gene mutation by rapid PCR-SSCP and its association with poor survival in breast cancer. Int J Cancer. 1997;74:642–7.

36. Berns EM, van Staveren IL, Look MP, Smid M, Klijn JG, Foekens JA. Mutations in residues of TP53 that directly contact DNA predict poor outcome in human primary breast cancer. Br J Cancer. 1998;77:1130–6.

37. Cuny M, Kramar A, Courjal F, Johannsdottir V, Iacopetta B, Fontaine H, et al. Relating genotype and phenotype in breast cancer: an analysis of the prognostic significance of amplification at eight different genes or loci and of p53 mutations. Cancer Res. 2000;60:1077–83.

38. Geisler S, Lonning PE, Aas T, Johnsen H, Fluge O, Haugen DF, et al. Influence of TP53 gene alterations and c-erbB-2 expression on the response to treatment with doxorubicin in locally advanced breast cancer. Cancer Res. 2001;61:2505–12.

39. Geisler S, Borresen-Dale AL, Johnsen H, Aas T, Geisler J, Akslen LA, et al. TP53 gene mutations predict the response to neoadjuvant treatment with 5-fluorouracil and mitomycin in locally advanced breast cancer. Clin Cancer Res. 2003;9:5582–8.

40. Bose S, Crane A, Hibshoosh H, Mansukhani M, Sandweis L, Parsons R. Reduced expression of PTEN correlates with breast cancer progression. Hum Pathol. 2002;33:405–9.

41. Depowski PL, Rosenthal SI, Ross JS. Loss of expression of the PTEN gene protein product is associated with poor outcome in breast cancer. Mod Pathol. 2001;14:672–6.

42. Saal LH, Holm K, Maurer M, Memeo L, Su T, Wang X, et al. PIK3CA mutations correlate with hormone receptors, node metastasis, and ERBB2, and are mutually exclusive with PTEN loss in human breast carcinoma. Cancer Res. 2005;65:2554–9.

43. Kim N, He N, Kim C, Zhang F, Lu Y, Yu Q, et al. Systematic analysis of genotype-specific drug responses in cancer. Int J Cancer. 2012;131:2456–64.

44. Daemen A, Griffith OL, Heiser LM, Wang NJ, Enache OM, Sanborn Z, et al. Modeling precision treatment of breast cancer. Genome Biol. 2013;14:R110.

45. Moestue SA, Dam CG, Gorad SS, Kristian A, Bofin A, Maelandsmo GM, et al. Metabolic biomarkers for response to PI3K inhibition in basal-like breast cancer. Breast Cancer Res. 2013;15:R16.

46. Davies H, Bignell GR, Cox C, Stephens P, Edkins S, Clegg S, et al. Mutations of the BRAF gene in human cancer. Nature. 2002;417:949–54.

47. Ihle MA, Fassunke J, Konig K, Grunewald I, Schlaak M, Kreuzberg N, et al. Comparison of high resolution melting analysis, pyrosequencing, next generation sequencing and immunohistochemistry to conventional Sanger sequencing for the detection of p.V600E and non-p.V600E BRAF mutations. BMC Cancer. 2014;14:13.

48. Santarpia L, Qi Y, Stemke-Hale K, Wang B, Young EJ, Booser DJ, et al. Mutation

profiling identifies numerous rare drug targets and distinct mutation patterns in different clinical subtypes of breast cancers. Breast Cancer Res Treat. 2012;134:333–43.

49. Nagaria TS, Williams JL, Leduc C, Squire JA, Greer PA, Sangrar W. Flavopiridol synergizes with sorafenib to induce cytotoxicity and potentiate antitumorigenic activity in EGFR/HER-2 and mutant RAS/RAF breast cancer model systems. Neoplasia. 2013;15:939–51.

50. Bleeker FE, Lamba S, Leenstra S, Troost D, Hulsebos T, Vandertop WP, et al. IDH1 mutations at residue p.R132 (IDH1R132) occur frequently in high-grade gliomas but not in other solid tumors. Hum Mutat. 2009;30:7–11.

51. Karnoub AE, Weinberg RA. Ras oncogenes: split personalities. Nat Rev Mol Cell Biol. 2008;9:517–31.

52. Lv N, Lin S, Xie Z, Tang J, Ge Q, Wu M, et al. Absence of evidence for epidermal growth factor receptor and human homolog of the Kirsten rat sarcoma-2 virus oncogene mutations in breast cancer. Cancer Epidemiol. 2012;36:341–6.

53. Geyer CE, Forster J, Lindquist D, Chan S, Romieu CG, Pienkowski T, et al. Lapatinib plus capecitabine for HER2-positive advanced breast cancer. N Engl J Med. 2006;355:2733–43.

54. Loman NJ, Misra RV, Dallman TJ, Constantinidou C, Gharbia SE, Wain J, et al. Performance comparison of benchtop high-throughput sequencing platforms. Nat Biotech. 2012;30:434–9.

3

Comparative sequence- and structure-inspired drug design for PilF protein of *Neisseria meningitidis*

Abijeet Singh Mehta[1†], Kirti Snigdha[1*†], M Sharada Potukuchi[2] and Panagiotis A Tsonis[1*]

Abstract

Serogroup A of *Neisseria meningitidis* is the organism responsible for causing epidemic diseases in developing countries by a pilus-mediated adhesion to human brain endothelial cells. Type IV pilus assembly protein (PilF) associated with bacterial adhesion, aggregation, invasion, host cell signaling, surface motility, and natural transformation can be considered as a candidate for effective anti-meningococcal drug development. Since the crystal structure of PilF was not available, in the present study, it was modeled after the Z2491 strain (CAM09255.1) using crystal structure of chain A of *Vibrio cholerae* putative Ntpase EpsE (Protein Data Bank (PDB) ID: 1P9R) and then we based this analysis on sequence comparisons and structural similarity using *in silico* methods and docking processes, to design a suitable inhibitor molecule. The ligand 3-{(4S)-5-{[(1R)-1-cyclohexylethyl]amino}-4-[(5S)-5-(prop-2-en-1-yl) cyclopent-1-en-1-yl]-1,4-dihydro-7H-pyrrolo[2,3-d] pyrimidin-7-yl}-1,2-dideoxy-b-L-erythro-hex-1-en-3-ulofuranosyl binds to the protein with a binding energy of −8.10 kcal and showed a drug likeness of 0.952 with no predicted health hazard. It can be utilized as a potent inhibitor of *N. meningitidis* pilus-mediated adhesion to human brain endothelial cells preventing meningeal colonization.

Keywords: Type IV pilus assembly protein (PilF), Homology modeling, *In silico* drug designing, Molecular docking, ADME analysis

Introduction

Neisseria meningitidis or *meningococcus* known for causing meningitis and other forms of meningococcal diseases is one of the prime causes of child mortality in industrialized countries and epidemics in Asian and African countries. *N. meningitidis* is an aerobic, gram-negative, non-endospore forming, non-motile (although has pilus), coccal parasitic bacterium [1]. Most strangely, this bacterium thrives in the human throat commensally and colonizes in a distinct number of populations as a healthy carrier [2]. Of the five common serogroups (A, B, C, Y, W135) responsible for about 90% of infections, serogroups A, B, and C account for most cases throughout the world, with the serogroup A strain responsible for epidemic diseases in developing countries [2,3]. The strain Z2491 [GenBank: CAM09255.1] is included under serogroup A.

N. meningitidis has shown widespread resistance to the currently used drugs and thus demands an advanced drug strategy to combat it. New drug targets will be helpful in the discovery of more diverse and potent drugs. Mainly, enzymes involved in macromolecular and metabolite synthesis appear to be promising potential targets, few of which have already been validated in many microorganisms. The two component signal transduction systems of bacteria which enable it to sense, respond, and adapt to their environmental or intracellular state variations are quite essential for the growth and survival in adverse environmental conditions. This system is ubiquitous in bacteria and has been reported to be involved in virulence. The translocation of proteins that is associated with the virulence factors across the outer membrane is supported by the type II secretion pathway. For the pathogen-specific type II secretion pathway, five virulent factors i.e., type IV pilus assembly protein (PilF), pilus assembly

* Correspondence: kirtisnigdha@gmail.com; ptsonis1@udayton.edu
†Equal contributors
[1]Department of Biology, University of Dayton, Dayton, OH, USA

protein (PilG), twitching motility protein1 (PilT-1), twitching motility protein2 (PilT-2), and competence protein (comA) are found to be involved. However, the major components of the secretion pathway are type IV pilus assembly proteins. Type IV pili are responsible for bacterial adhesion, aggregation, invasion, host cell signaling, surface motility, and natural transformation [4]. As reported by Coureuil et al. [5] the initiation of signaling cascades by *meningococcus* follows the pilus-mediated adhesion to human brain endothelial cells that eventually leads to opening of intercellular junctions and allows meningeal colonization. Hence, for the development of an effective anti-meningococcal drug, type IV pilus assembly protein (PilF) can serve as a potential target [3].

Since the crystal structure of PilF was not available, the present study was conducted to model after the Z2491 strain using the crystal structure of chain A of *Vibrio cholerae* Putative Ntpase EpsE [PDB: 1P9R] as a template and then using *in silico* methods and docking processes. Thus, a suitable inhibitor molecule has been

designed. Further, this molecule has been checked for its possible health effects and drug likeness using bioinformatic tools.

Methods

Sequence alignment and structure prediction

The FASTA sequence of the target protein type IV pilus assembly protein (PilF) of the *Neisseria meningitidis* Z2491 strain [GenBank: CAM09255.1] was retrieved from the NCBI Entrez sequence search [http://www.ncbi.nlm.nih.gov]. Following the BLASTp run [http://blast.ncbi.nlm.nih.gov/Blast.cgi] against the Protein Data Bank database [http://www.rcsb.org/pdb/], the crystal structure of chain A of *Vibrio cholerae* Putative Ntpase EpsE [PDB: 1P9R] was selected as a template sequence. The 3D structure of the query protein was predicted by the automated homology modeling program, Modeller 9.10 [6]. For modeling, the template and target sequences were carefully aligned to remove potential alignment errors. Structural models were visualized by Swiss Pdb Viewer (SPDBV)

```
Score =   340 bits (871),   Expect = 4e-112, Method: Compositional matrix adjust.
 Identities = 174/388 (45%), Positives = 246/388 (63%), Gaps = 7/388 (2%)

Query  165  EEAEDGPVPRFIHKTLSDALRSGASDIHFEFYEHNARIRFRVDGQLREVVQPPIAVRGQL  224
            E +D P+ + I+ L +A++ GASDIH E +E    IRFRVDG LREV+ P   +   L
Sbjct   26  ESEDDAPIIKLINAXLGEAIKEGASDIHIETFEKTLSIRFRVDGVLREVLAPSRKLSSLL   85

Query  225  ASRIKVMSRLDISEKRIPQDGRMQLTFQKGGKPVDFRVSTLPTLFGEKVVMRILNSDAAS  284
             SR+KV ++LDI+EKR+PQDGR+ L   GG+ VD RVST P+  GE+VV R+L+ +A
Sbjct   86  VSRVKVXAKLDIAEKRVPQDGRISLRI--GGRAVDVRVSTXPSSHGERVVXRLLDKNATR  143

Query  285  LNIDQLGFEPFQKKLLLEAIHRPYGMVLVTGPTGSGKTVSLYTCLNILNTESVNIATAED  344
            L++  LG      I RP+G++LVTGPTGSGK+ +LY L LN+  NI T ED
Sbjct  144  LDLHSLGXTAHRHDNFRRLIKRPHGIILVTGPTGSGKSTTLYAGLQELNSSERNILTVED  203

Query  345  PAEINLPGINQVNVNDKQGLTFAAALKSFLRQDPDIIMVGEIRDLETADIAIKAAQTGHM  404
            P E ++ GI Q VN +    TFA  L++ LRQDPD++ VGEIRDLETA IA++A+ TGH+
Sbjct  204  PIEFDIDGIGQTQVNPRVDXTFARGLRAILRQDPDVVXVGEIRDLETAQIAVQASLTGHL  263

Query  405  VFSTLHTNNAPATLSRMLNMGVAPFNIASSVSLIMAQRLLRRLCSSCKQEVERPSASALK  464
            V STLHTN A   ++R+ + G+ PF I+SS+  ++AQRL +R LC  CK+    P + +
Sbjct  264  VXSTLHTNTAVGAVTRLRDXGIEPFLISSSLLGVLAQRLVRTLCPDCKE----PYEADKE  319

Query  465  EVGFTDEDLAKDWKLYRAVGCDRCRGQGYKGRAGVYEVMPISEEMQRVIMNNGTEVDILD  524
            +    D    + LYRA GC +C   +GY+GR G++E++  + + +Q +I +   E    +
Sbjct  320  QRKLFDSKKKEPLILYRATGCPKCNHKGYRGRTGIHELLLVDDALQELIHSEAGE-QAXE  378

Query  525  VAYKEGMVDLRRAGILKVMQGITSLEEV  552
            +      +R  G+ KV QGITSLEEV
Sbjct  379  KHIRATTPSIRDDGLDKVRQGITSLEEV  406
```

Figure 1 Structural evaluation of PilF protein. Secondary sequence alignment between CAM09255.1 and 1P9R by the BLASTp run [http://blast.ncbi.nlm.nih.gov/Blast.cgi] against the Protein Data Bank database [http://www.rcsb.org/pdb/]. It provided with maximum identity score and hence the crystal structure of chain A of *Vibrio cholerae* Putative Ntpase EpsE [PDB: 1P9R] was selected as a template sequence.

version v 4.0.4 [http://www.expasy/spdbv.org]. Validation of the model was performed by the Ramachandran plot analysis [7], using SPDBV and the online tool, PROCHECK [8] [http://nihserver.mbi.ucla.edu/SAVES/]. Further evaluation was carried out using ProSA-web [https://prosa.services.came.sbg.ac.at/prosa.php] and Verify_3D [http://nihserver.mbi.ucla.edu/SAVES/].

Active site, lead prediction, and ligand building

The active domains of the developed model were determined using the software LIGSITE^{csc} [9] [http://projects.biotec.tu-dresden.de/pocket/]. LIGSITE^{csc} is a web server for the automatic identification of pockets on a protein surface using the Connolly surface and the degree of conservation. From the study of interaction between different chemical components of ligands with type II secretion system in bacteria, a potential lead molecule was determined and modified using ChemSketch version 12.01 [http://www.acdlabs.com], which was then integrated with a modeled protein structure using the Hex 6.3 [http://hex.loria.fr/]. *Hex* is an interactive protein docking and molecular superposition program. The protein-lead complex was then used to develop the ligand using the software LigBuilder version 1.3 [http://www.chem.ac.ru/Chemistry/Soft/LIGBUILD.en.html], which is a general-purpose program package written for a structure-based drug-design procedure. Based on the 3D structure of the target protein, it can automatically build ligand molecules within the binding pocket.

Autodocking

The ligands developed were chosen for docking studies. The docking study was carried out using the AutoDock software version 1.5.4, which uses genetic algorithm (GA). PilF [GenBank: CAM09255.1] was loaded into AutoDock Tools (ADT) [http://autodock.scripps.edu/resources] as a receptor and made ready for docking by the addition of charges and hydrogen which any PDB file of the molecule usually does not contain, using the edit option in ADT. All the ligands were separately and individually docked with the PilF model; a grid for dock search was built for the whole molecule to find the most probable binding site in CAM09255.1 and to measure its interaction parameters with the ligands. The docking process was carried out in the default parameters of ADT.

ADME and toxicity analysis

The absorption, digestion, metabolism, and excretion (ADME) properties are directly related to the biological effect of drugs and their fate in an organism and therefore need to be evaluated in medicinal chemistry. The ADME properties and its possible effects on health are determined using different bioinformatic tools like Molinspiration [http://www.molinspiration.com/], PASS (Prediction of Activity Spectra for Substances) [http://www.genexplain.com/pass], ACD/Labs I-Lab 2.0 [http://www.acdlabs.com/resources/ilab/index.php], and Toxtree [10].

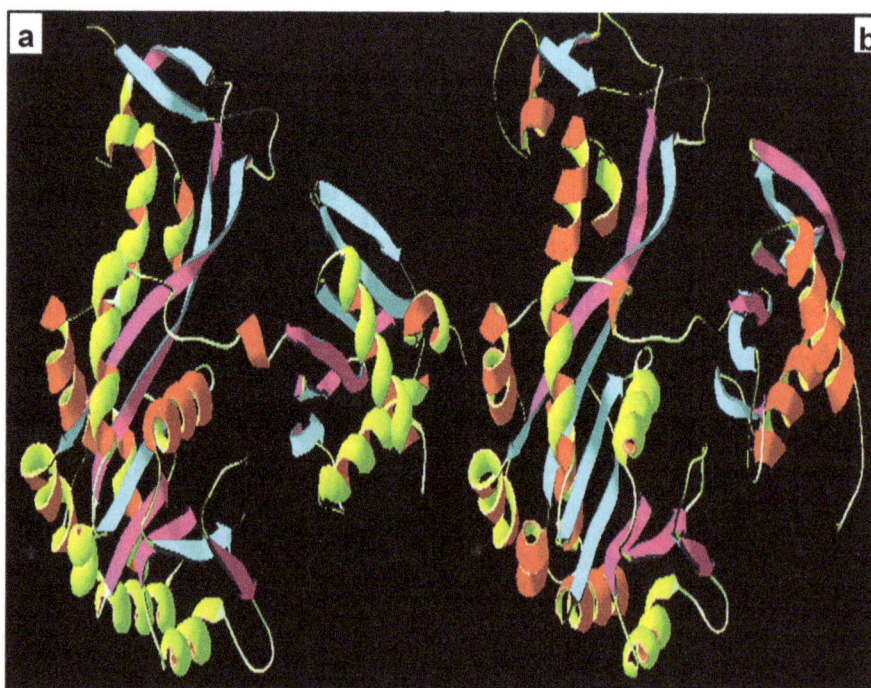

Figure 2 Homology modeling. **(a)** 3D structure of 1P9R and **(b)** theoretical model of PilF (CAM 09255.1) obtained after homology modeling.

Molinspiration offers free online services for calculation of important molecular properties (log$P$, polar surface area, number of hydrogen bond donors and acceptors, and others) as well as prediction of bioactivity score for the most important drug targets (GPCR ligands, kinase inhibitors, ion channel modulators, nuclear receptors).

Results and discussion
Structural evaluation of PilF protein

Structural information about the type IV pilus assembly protein (PilF) of the *Neisseria meningitidis* Z2491strain is currently unavailable. The present study reports the homology modeling and structural interaction between type IV pilus assembly protein of the *Neisseria meningitidis* Z2491 strain with an inhibitor. Chain A of *Vibrio cholerae* putative Ntpase Epse [PDB: 1P9R] having a high degree of homology with CAM09255.1 was used as a template with an identity of 45% (and conserved amino acid change of 63%). Its x-ray crystal structure has an

atomic resolution of 2.50 Å and an R value of 0.217. The secondary structure alignment obtained between the query and template sequence is shown in Figure 1. Based on the Discrete Optimized Protein Energy (DOPE) score, the selected model with minimum score is shown in Figure 2b. Root mean square deviation (RMSD) value of a C alpha carbon atom between the template and predicted model was found to be 2.49A˙ determined by using the SPDBV software. The predicted model was subjected to PROCHECK analysis to determine psi and phi torsion angles. Good overall stereochemistry is obtained for the model with 89.1% of the residue psi/phi angles falling in the most favored regions and 10.3% in the allowed region. The Ramachandran plot is shown in Figure 3. The G-factors indicating the quality of covalent and bond angle distance were –0.25 for dihedrals, 0.19 for covalent, and overall –0.07. The interaction energy per residue is also calculated by program PROSA and Verify_3D. Figure 4 displays the PROSA profile calculated for the PilF model. The comparable Ramachandran plot characteristics and

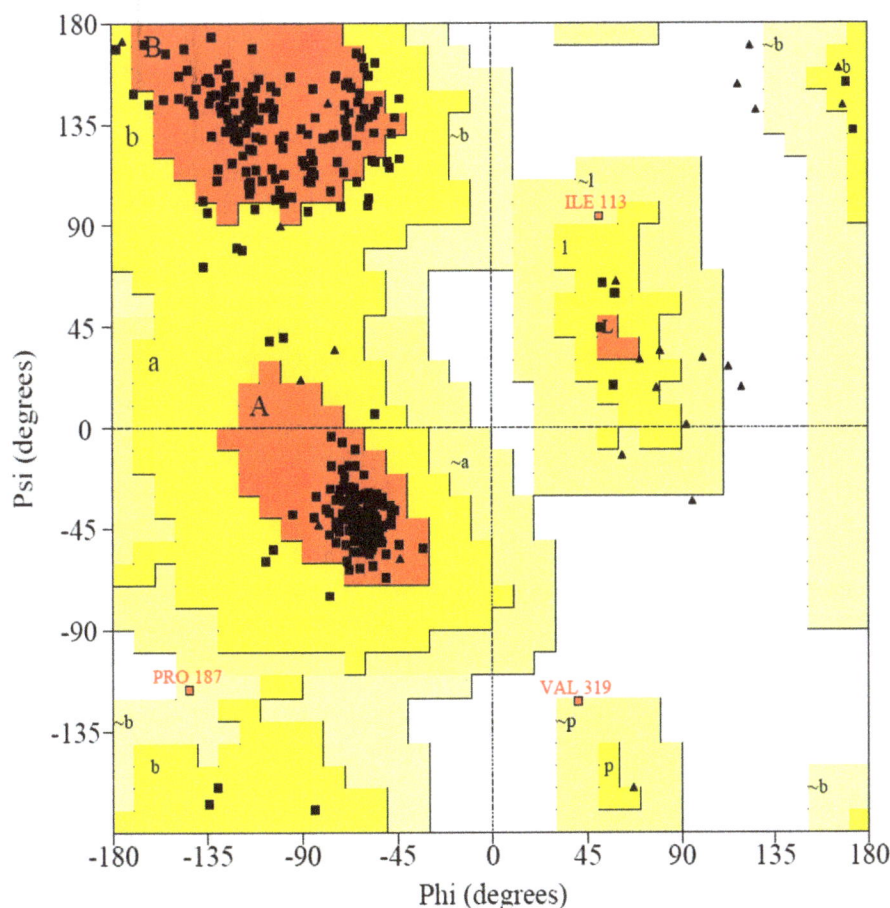

Figure 3 Ramachandran plot of the model. Validation of the model was done by Ramachandran plot analysis using Swiss PDB Viewer (SPDBV) version v 4.0.4 [http://www.expasy/spdbv.org]. Good overall stereochemistry is obtained for the model with 89.1% of the residue psi/phi angles falling in the most favored regions and 10.3% in the allowed region.

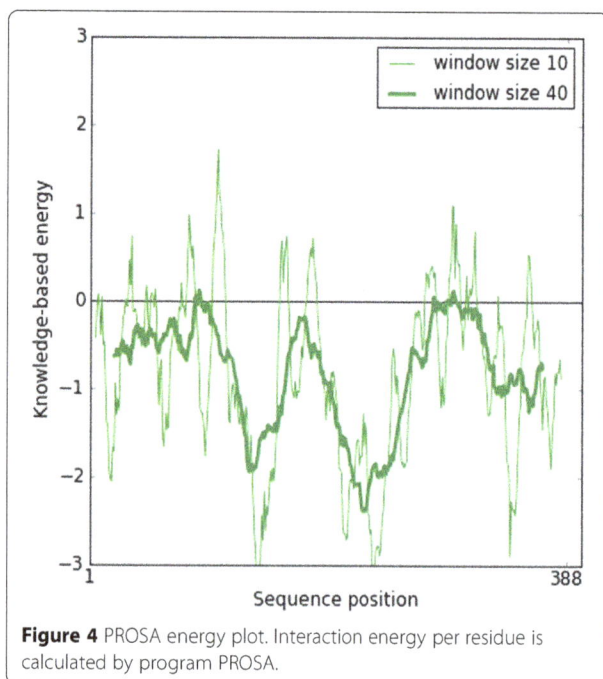

Figure 4 PROSA energy plot. Interaction energy per residue is calculated by program PROSA.

G-factors confirm the quality of the predicted model (Table 1).

Functional site prediction, lead identification, and ligand generation

The software LIGSITE^{csc} was utilized for functional site prediction. On the basis of interaction with the surrounding atoms, out of the three pockets predicted as shown in Figure 5, the pocket PKT 1032 with the best interaction result was identified as the possible functional site of the protein molecule. The molecule 7-[5-O-(hydroxy{[hydroxyl(phosphonooxy)phosphoryl]oxy}phosphoryl)pentofuranosyl]-7H-pyrrolo[2,3-d]pyrimidin-5-amine (Figure 6) was identified as the lead based on the interaction study

Table 1 Validation statistics for theoretical model of PilF by PROCHECK analysis

	Plot statistics	
Residues in the most favored regions [A,B,L]	304	89.1%
Residues in additional allowed regions [a,b,l,p]	35	10.3%
Residues in generously allowed regions [~a,~b,~l,~p]	2	0.6%
Residues in disallowed regions	0	0.0%
Number of non-glycine and non-proline residues	341	100.0%
Number of end-residues (exd. Gly and Pro)	2	
Number of glycine residues	28	
Number of proline residues	17	
Total number of residues	388	

between different chemical components of ligands with the type II secretion system in bacteria. Using the Hex 6.3 [http://hex.loria.fr/], it was introduced into the protein binding site. The ten best ligands were selected for docking out of the many ligands developed by the software LigBuilder based on the Lipinski's "Rule of 5" properties [11].

Docking study

Docking was performed using AutoDock Tools-1.5.4 for the receptor of 1P9R with all the ten ligands loaded individually into ADT to analyze ten best conformations. The study showed that the conformation of ligand, 3-{(4S)-5-{[(1R)-1-cyclohexylethyl]amino}-4-[(5S)-5-(prop-2-en-1-yl)cyclopent-1-en-1-yl]-1,4-dihydro-7H-pyrrolo[2,3-d]pyrimidin-7-yl}-1,2-dideoxy-b-L-*erythro*-hex-1-en-3-ulofuranosyl ($C_{28}H_{44}N_4O_3$) with PilF protein has a binding energy of −8.10 kcal M^{-1} and inhibitory constant of 1.16 μM. It was concluded to be the best docked ligand molecule in comparison to other nine molecules. The ligand molecule after docking is shown in Figure 7b. The PilF protein and docked ligand in the pocket is shown in Figure 7c.

ADME and toxicity analysis of the ligand

The pharmacokinetic features of the ligand taken into account in this work were preliminarily investigated. It is extremely advantageous if information about the ADME properties of the studied molecules could be produced in the early stages of the drug discovery process. The use of *in silico* methods to predict ADME properties is intended as a first step in this direction, and the results of such an analysis are herein reported and discussed. The molecular properties of the ligand 3-{(4S)-5-{[(1R)-1-cyclohexylethyl]amino}-4-[(5S)-5-(prop-2-en-1-yl)cyclopent-1-en-1-yl]-1,4-dihydro-7H-pyrrolo[2,3-d]pyrimidin-7-yl}-1,2-dideoxy-b-L-*erythro*-hex-1-en-3-ulofuranosyl as predicted by the Molinspiration property engine v2011.04 and ACD/Labs I-Lab 2.0 were found to be in accordance with the Lipinski's "Rule of 5" properties [11], which states, that most "drug-like" molecules have $\log P$ <= 5, molecular weight < = 500, number of hydrogen bond acceptors <= 10, and number of hydrogen bond donors <= 5. The bioactivity of the ligand was predicted using Molinspiration Bioactivity engine v2011.06 (Tables 2 and 3). Oral bioavailability as determined by ACD/Labs I-Lab was between 30%–70%. Drug likeness of the ligand was found to be very high i.e., 0.952 calculated by PASS (http://www.ibmc.msk.ru/PASS/). Moreover, no toxic or harmful effect of the ligand over human health was predicted by PASS and Toxtree (http://toxtree.sourceforge.net/) software. Possible health effects of the ligand molecule over blood, cardiovascular system, gastrointestinal system, kidney, liver, and lungs are shown in Figure 8 and

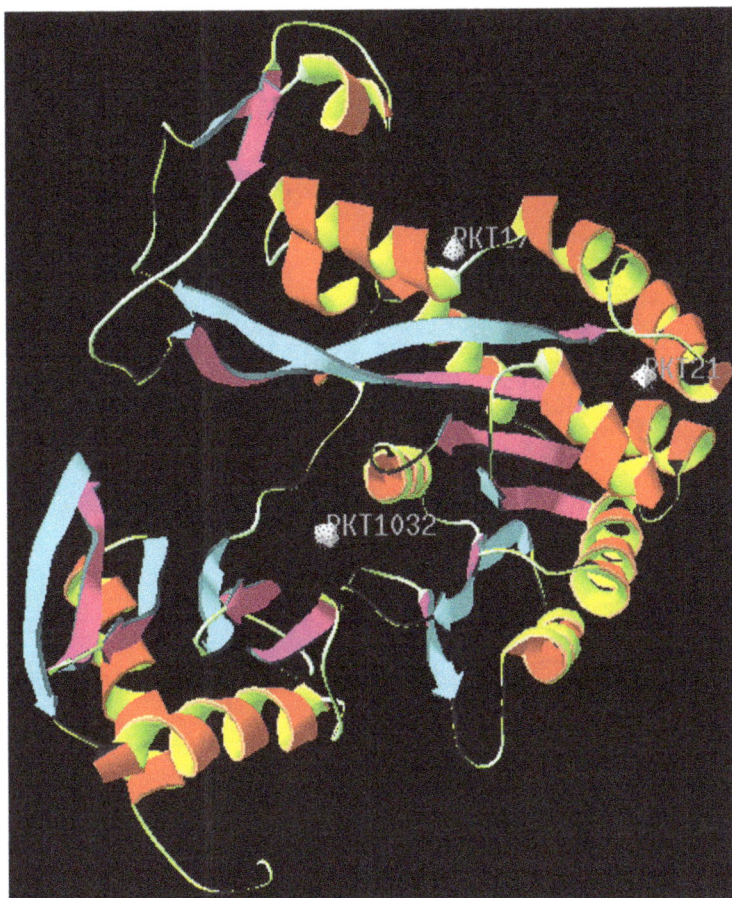

Figure 5 Functional site prediction using LIGSITE*csc*. On the basis of interaction with the surrounding atoms, out of the three pockets predicted, the pocket PKT 1032 with the best interaction result was identified as the possible functional site of the protein molecule.

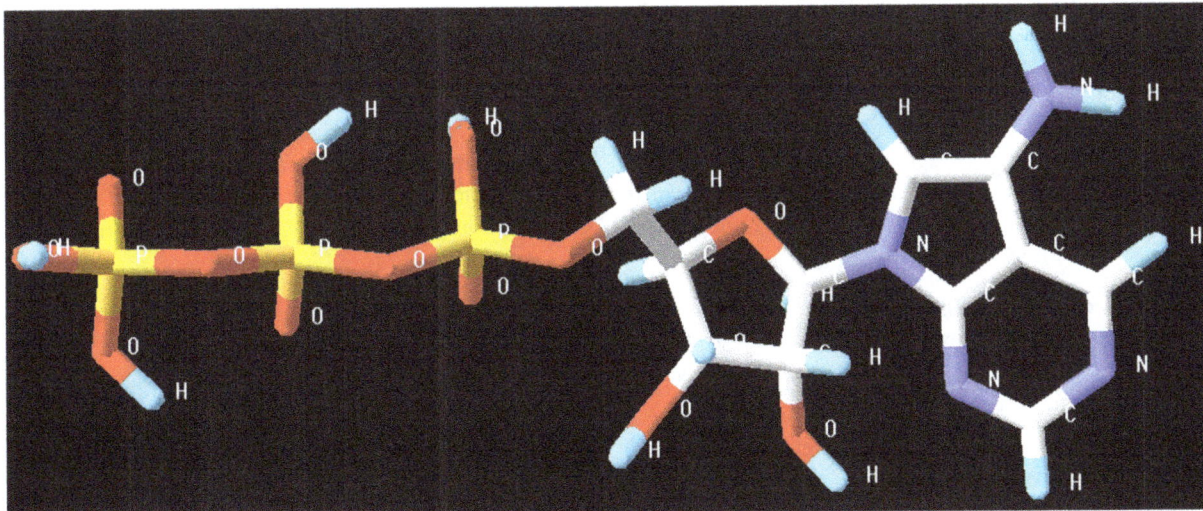

Figure 6 Lead introduction in protein binding site. The molecule, 7-[5-*O*(hydroxy{[hydroxyl(phosphonooxy)phosphoryl]oxy}phosphoryl)pentofuranosyl]-7*H*-pyrrolo[2,3-*d*]pyrimidin-5-amine was identified as the lead and using the Hex 6.3 [http://hex.loria.fr/] it was introduced into the protein binding site.

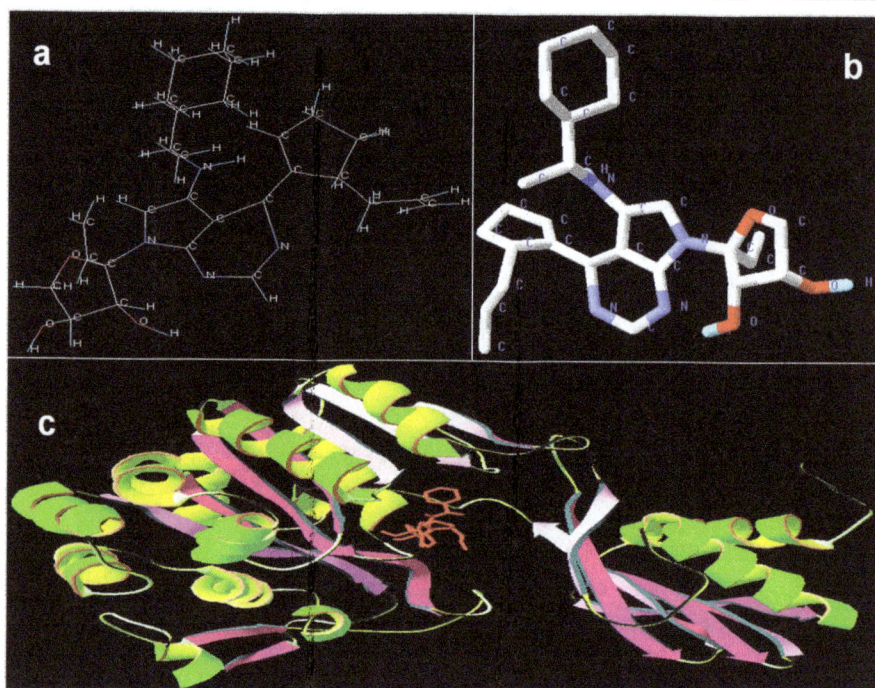

Figure 7 Ligand molecule and docking. Ligand molecule as build by the software LigBuilder version 1.3. **(a)** After docking using AutoDock Tools-1.5.4 for the receptor of 1P9R. **(b)** Protein-ligand complex, ligand is shown in red color. **(c)** The conformation of ligand, 3-{(4S)-5-{[(1R)-1-cyclohexylethyl] amino}-4-[(5S)-5-(prop-2-en-1-yl)cyclopent-1-en-1-yl]-1,4-dihydro-7H-pyrrolo[2,3-d]pyrimidin-7-yl}-1,2-dideoxy-b-L-erythro-hex-1-en-3-ulofuranosyl ($C_{28}H_{44}N_4O_3$) with PilF protein has binding energy of −8.10 kcal M^{-1} and inhibitory constant of 1.16 μM, the best docked ligand molecule.

were determined by ACD/Labs I-Lab (Algorithm Version: v5.0.0.184). The software facilitates in determining specific structural toxicophores that are responsible for the organ-specific toxic effect on the following: blood, cardiovascular system, gastrointestinal system, kidney, liver, and lungs. These fragmental contribution maps illustrate the role of individual atoms and fragments of the ligand in a color-coded manner; red color indicates a positive contribution to the final predicted value or toxic action whereas green color means the atom/fragment has a negative coefficient in the regression equation i.e., unrelated to the health effects under investigation. These estimates are based on data from over 100,000 compounds from chronic, sub-chronic, and acute toxicity and carcinogenicity

studies with specific to particular organ systems. Evaluation of the safety of new chemicals and pharmaceuticals requires the combination of information from various sources (e.g., in silico, in vitro, and in vivo) to provide an assessment of risk to human health and the environment [12], but in silico techniques can enable compounds to be deselected earlier in development, thereby limiting the need for animal testing. Previously, it has been proved that in silico drug design is a potent tool to screen a drug and virtually suggests its efficiency and any type of associated health hazards before evaluating the same in vitro [13] and/or in vivo [14]. Therefore, it is more likely that in the future, performing the wet lab experiments involving chemical synthesis of designed drug and testing the same

Table 2 Molecular properties of the ligand determined by Molinspiration property engine v2011.04

	Values
LogP	4.07
Molecular weight	480.64
No. of hydrogen bond donors	4
No. of hydrogen bond acceptors	7
Topological polar surface area	91.042
No. of rotatable bonds	8

Table 3 Molecular Bioactivities of the ligand determined by Molinspiration property engine v2011.04

	Values
GPCR ligand	0.16
Ion channel modulator	−0.13
Kinase inhibitor	0.03
Nuclear receptor ligand	0.09
Protease inhibitor	0.10
Enzyme inhibitor	0.14

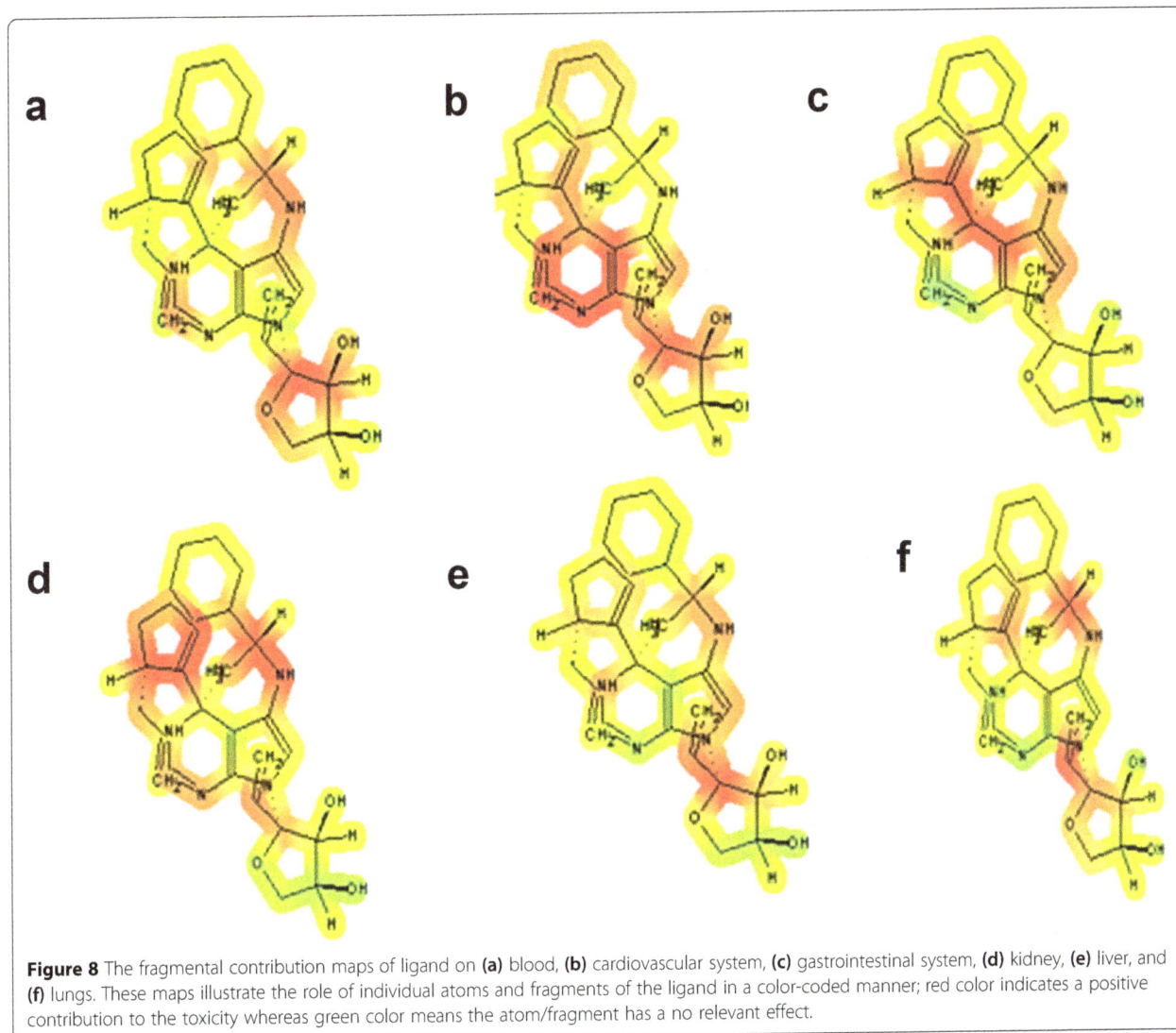

Figure 8 The fragmental contribution maps of ligand on (a) blood, (b) cardiovascular system, (c) gastrointestinal system, (d) kidney, (e) liver, and (f) lungs. These maps illustrate the role of individual atoms and fragments of the ligand in a color-coded manner; red color indicates a positive contribution to the toxicity whereas green color means the atom/fragment has a no relevant effect.

in vivo using specific cell lines would give promising results.

Conclusions

The homology modeling of PilF protein of the *N. meningitidis* Z2491strain based on the crystal structure of chain A of *Vibrio cholerae* Putative Ntpase Epse [PDB: 1P9R] was performed. Using a structure-based drug designing method, a potent inhibitor molecule was designed. Docking studies of 3-{(4S)-5-{[(1R)-1-cyclohexylethyl]amino}-4-[(5S)-5-(prop-2-en-1-yl)cyclopent-1-en-1-yl]-1,4-dihydro-7H-pyrrolo[2,3-d]pyrimidin-7-yl}-1,2-dideoxy-b-L-*erythro*-hex-1-en-3-ulofuranosyl into the active site of PilF resulted in lowest binding energy signifying highest binding affinity. The *in silico* ADME and toxicity analysis confirmed its drug likeness and no predicted health hazard. Also, it can be expected that this study will be helpful in forming a base regarding the modeling of PilF protein and the development of anti-meningococcal drugs.

Competing interests
The authors declare that they have no competing interests.

Authors' contributions
The work was conceived, planned, and executed by ASM and KS. The manuscript was written by ASM and KS. MSP helped in editing the draft. PAT participated in designing the work, helped in analysis of results and writing the paper. All authors read and approved the final manuscript.

Acknowledgements
We thank Mr. Rajnikant Namdeo of IBI Biosolutions Pvt. Ltd., who guided us through the modeling and designing process.

Author details
[1]Department of Biology, University of Dayton, Dayton, OH, USA. [2]School of Biotechnology, Shri Mata Vaishno Devi University, Katra, J&K, India.

References

1. Genco C, Wetzler BL. Neisseria: molecular mechanisms of pathogenesis. Caister Academic Press; 2010. ISBN 978-1-904455-51-6.

2. Shovana K, Thiagarajan B. *In silico* identification of putative drug targets in *Neisseria meningitidis* through metabolic pathway analysis. Scholar's Res J. 2010;1(1):13–24.

3. Manachanda V, Gupta S, Bhalla P. Meningococcal disease: history, epidemiology, pathogenesis, clinical manifestations, diagnosis, antimicrobial susceptibility and prevention. Indian J Med Microbiol. 2006;24(1):7–19.

4. Rothbard JB, Fernandez R, Wang L, Teng NNH, Schoolnik GK. Antibodies to peptides corresponding to a conserved sequence of gonococcal pilins block bacterial adhesion. Proc Natl Acad Sci U S A. 1985;82:915–9.

5. Coureuil M, Herve L, Scott MGH, Boularan CD, Enslen H, Soyer M, et al. Meningococcus hijacks a β2-adrenoceptor/β-arrestin pathway to cross brain microvasculature endothelium. Cell. 2010;143(7):1149–60.

6. Eswar N, Marti-Renom MA, Webb B, Madhusudhan MS, Eramian D, Shen M, et al. Comparative protein structure modeling with MODELLER. Curr Protoc Bioinformatics, John Wiley & Sons, Inc. 2006;15(Suppl):5.6.1-5.6.30.

7. Ramachandran GN. Stereochemistry of polypeptide chain configurations. J Mol Biol. 1963;7:9.

8. Laskowski RA, MacArthur MW, Moss DS, Thornton JM. PROCHECK - a program to check the stereochemical quality of protein structures. J Appl Cryst. 1993;26:283–91.

9. Huang B, Schroeder M. LIGSITE^csc: predicting protein binding sites using the Connolly surface and degree of conservation. BMC Struct Biol. 2006;6:19.

10. Patlewicz G, Jeliazkova N, Safford RJ, Worth AP, Aleksiev B. An evaluation of the implementation of the Cramer classification scheme in the Toxtree software. SAR QSAR Environ Res. 2008;19(5–6):495–524.

11. Lipinski CA, Lombardo F, Dominy BW, Feeney PJ. Experimental and computational approaches to estimate solubility and permeability in drug discovery and development settings. Adv Drug Deliv Rev. 1997;23:4–25.

12. Chapman KL, Holzgrefe H, Black EL, Brown M, Chellman G, Copeman C, et al. Pharmaceutical toxicology: designing studies to reduce animal use, while maximizing human translation. Regul Toxicol Pharmacol. 2013;66(1):88–103.

13. Berzal CF, Rojas FA, Escario JA, Kouznetsov VV, Gómez-Barrio A. In vitro phenotypic screening of 7-chloro-4-amino(oxy)quinoline derivatives as putative anti-Trypanosoma cruzi agents. Bioorg Med Chem Lett. 2014;24(4):1209–13.

14. de Candia M, Fiorella F, Lopopolo G, Carotti A, Romano MR, Lograno MD, et al. Synthesis and biological evaluation of direct thrombin inhibitors bearing 4-(piperidin-1-yl) pyridine at the P1 position with potent anticoagulant activity. Med Chem. 2013;256:8696–711.

4

Whole-genome sequencing targets drug-resistant bacterial infections

N. V. Punina[1][*] , N. M. Makridakis[2], M. A. Remnev[3] and A. F. Topunov[1]

Abstract

During the past two decades, the technological progress of whole-genome sequencing (WGS) had changed the fields of Environmental Microbiology and Biotechnology, and, currently, is changing the underlying principles, approaches, and fundamentals of Public Health, Epidemiology, Health Economics, and national productivity. Today's WGS technologies are able to compete with conventional techniques in cost, speed, accuracy, and resolution for day-to-day control of infectious diseases and outbreaks in clinical laboratories and in long-term epidemiological investigations. WGS gives rise to an exciting future direction for personalized Genomic Epidemiology. One of the most vital and growing public health problems is the emerging and re-emerging of multidrug-resistant (MDR) bacterial infections in the communities and healthcare settings, reinforced by a decline in antimicrobial drug discovery. In recent years, retrospective analysis provided by WGS has had a great impact on the identification and tracking of MDR microorganisms in hospitals and communities. The obtained genomic data are also important for developing novel easy-to-use diagnostic assays for clinics, as well as for antibiotic and therapeutic development at both the personal and population levels. At present, this technology has been successfully applied as an addendum to the real-time diagnostic methods currently used in clinical laboratories. However, the significance of WGS for public health may increase if: (a) unified and user-friendly bioinformatics toolsets for easy data interpretation and management are established, and (b) standards for data validation and verification are developed. Herein, we review the current and future impact of this technology on diagnosis, prevention, treatment, and control of MDR infectious bacteria in clinics and on the global scale.

Introduction

Human genomics is inseparably linked to the genomics of bacteria. Bacteria share a long history with humans and play a major role in our life [152, 200]. Beneficial utilization of bacterial products can provide key solutions to many pressing problems on the planet, from environmental pollution to human diseases. Investigation of bacterial pathogens remains agenda priority mainly due to two additional reasons: (i) over 13 % of the world's deaths are related to bacterial infectious disease (including respiratory diseases and tuberculosis (TB)) every year [79, 250], and (ii) the growth of ancient pathogen re-emergence is driven by steadily increasing resistance to multiple widely used antimicrobial agents [59, 60, 249]. Despite the importance and utility of bacteria, until quite recently, little was known about their genomic structure.

During the last two decades, bacteria genomics is rapidly changing, mostly through the evolution of whole-genome sequencing (WGS) technologies. Recent technical advantages significantly reduced the cost of WGS and improved its power and resolution. Since WGS tools (both chemistry and bioinformatics-wise) are changing rapidly, we will not dwell in the details of individual technologies and equipment. The variety and applicability of the major high-throughput sequencing platforms are well presented in several reviews (e.g., [149, 199, 263]).

The advent and ever-growing use of the novel WGS technologies resulted in a rapid intensification in the scope and speed of the completion of bacterial genome sequencing projects. This explosion in bacterial genomics has greatly expanded our view of the genetic and physiological diversity of bacteria. To date, over 39,000 genome projects have been started, approximately 3,000 microbes' whole-genome sequences were completed and published [134, 181, 229], and more than 500 new species are being described every year [68, 112]. However,

* Correspondence: hin-enkelte@yandex.ru
[1]Bach Institute of Biochemistry, Russian Academy of Science, Moscow 119071, Russia
Full list of author information is available at the end of the article

most of these projects were driven by the potential practical applications of the investigated microorganisms and thus missed most of the microbial diversity on the planet [133, 134, 180].

Although researchers have only scratched the surface of microbial biodiversity, the information gained has already resulted in the discovery of large numbers of pathogenic bacteria in humans. WGS technologies granted access to potential virulence determinants, disruptive targets, candidate drug compounds [85], mechanisms of pathogenicity, drug resistance and spread [62], and their evolution in pathogens. In addition, WGS analysis provided information about uncultured or difficult-to-grow bacterial strains isolated from clinical specimens [15]. Knowledge of the enormous range of microbial capacities and functional activity can address many epidemiological questions and will have broad and far-reaching implications for personalized and public healthcare in the future. In this field, potential applications of WGS can be essential for:

i. Detection, identification, and characterization of infectious microorganisms
ii. Design of novel diagnostic assays for laboratory use
iii. Assessment of multidrug resistance (MDR) or virulence repertoires in pathogens, as well candidate antimicrobial compounds in beneficial microorganisms
iv. Monitoring the emergence and spread of bacterial infectious agents in different healthcare settings [46, 69, 126]

The WGS technology is very likely to become an alternative to the traditional methods of fighting DR bacteria. Even today, this technology is already used globally as an addendum to complement conventional laboratory approaches (microscopy, pathogenic tests, mass spectrometry, conventional molecular diagnostics, techniques for vaccine and antibiotic design) in routine clinical workflow and scientific investigations [93, 96, 149]. In the future, WGS may simplify the diagnostic laboratory workflow and sample trace, as well as reduce the number and type of collected biological specimens [11, 46, 126, 138, 201]. Deploying WGS into individual genome sequencing (IGS) technology has great potential to become a part of routine personalized clinical practice (e.g., TruGenome Clinical Sequencing tests™ by Illumina Clinical Services Laboratory; Complete Genomics Platform™ by Complete Genomics BGI, Helicos Helicope™ by SeqLL; Personal Genome Project) [92]. It is further expected that WGS will permit a deep understanding of infection mechanisms, allow for more rational preventive measures [24], and reduce the risk of unnecessary infection-control interventions [228].

The growing incidence of bacterial resistance to a broad range of antibacterial drugs in hospitals and communities is a major public health threat today and a compelling reason for WGS application. MDR pathogens complicate efforts of infection control and result in considerable morbidity and mortality around the world [111, 131, 217]. Today, MDR infections are recognized as multidimensional global challenge by many health organizations [26, 232, 251]. This complex problem requires comprehensive measures to be solved [42]. It was postulated that effective problem-solving strategies should include: (i) revealing and monitoring infectious agents, (ii) tracking antibiotic resistance, (iii) developing new antimicrobial drugs, (iv) providing rational antimicrobial stewardship program in healthcare institutions in order to avoid inappropriate or excessive antibiotic use, and (v) developing unified toolsets and standards for effective worldwide data management [42, 221, 224].

Taking into account the growing concern about emerging infections, in this review, we detail the main uses and hurdles of WGS technologies in clinical practice and public health with regard to MDR bacterial infections.

Main directions of WGS applications in MDR bacterial infections (review scope)

There are numerous possible applications of WGS in dealing with infectious disease of MDR bacteria. WGS can be used as a primary tool for:

i. Detection of multidrug susceptibility
ii. Monitoring MDR evolution and transmission dynamics of MDR pathogen
iii. Diagnosis and control of MDR infections locally and regionally
iv. Development of new tests and assays for accurate and rapid MDR bacterial diagnostics in clinics and points-of-care
v. Discovery of novel antibacterial drugs and therapeutics and assessment of their preventability

Each of these tasks is important for clinical and public health and requires methods with different levels of typing resolution. Theoretically, this problem can be addressed by reliable, quick, and low-cost WGS technology in the near future.

Detection of MD susceptibility

Recently introduced into routine clinical microbiological analysis, WGS has had a great impact on the study of the spectrum of genetic factors involved in MDR to microorganisms and, consequently, on the cost-effectiveness of subsequent disease treatment [214]. Rapid and accurate identification and characterization of known and new

antibiotic resistance determinants and their arrangements play a key role in preventing the emergence and spread of MDR pathogenic microorganisms in any healthcare setting [214]. Current knowledge of the type of pathogen and its antibiotic resistance profile is essential for selection of therapy and development of new antibacterial drugs [106, 123, 214] and for reducing the high mortality rate in infected patients. This knowledge also has particular significance for the pathogens causing most frequent and severe types of healthcare-associated and community-acquired infections such as bloodstream (BSI), urinary tract (UTI), and wound stream infections (WSI) [170]. The MDR bacterial pathogens of international concern [36, 161, 252] are presented in Table 1.

Many chromosome- and plasmid-mediated resistance determinants were successfully identified for most severe pathogenic bacteria using WGS technologies (Table 1). Together with data obtained by classic antimicrobial susceptibility tests [118] and genotyping methods [66], these determinants were deposited into the Antibiotic Resistance Genes Database (ARDB) [146]. Currently, there is an open catalog of more 13,000 antibiotic resistance genes, composing the resistome [253], with rich information, including resistance profile, mechanisms, requirements, epidemiology, coding sequences, and their mutations for more than 250 bacterial genera.

Revelation of the links between genetic and phenotypic traits of bacteria still remains one of the most critical issues that thwart implementation of WGS in clinical and public health practice. Determination of the genetic components of antibiotic resistance (resistant genotypes) and their correlation to resistant bacterial phenotypes can potentially promote its practical application. The possibility to ascertain the phenotypic antimicrobial resistance on the basis of genomic data has been extensively studied [196, 261]. The resistance phenotypes determined based on WGS data were compared to the results of phenotypic tests for methicillin-resistant *Staphylococcus aureus* (MRSA) [82, 103], *Clostridium difficile* [53], *Escherichia coli*, *Klebsiella pneumonia* [100, 218], and *Pseudomonas aeruginosa* [41, 124]. The analyses showed that data obtained for these bacteria through WGS can reliably predict antibiotic susceptibility phenotype, with overall sensitivity and specificity more than 95 % [53, 82, 218]. Hence, WGS may be applied as first-line antibiotic resistance screening method in clinical practice of these pathogens. However, it is important to remember that in some cases, bacterial MDR depends on the mode and level of the resistance gene expression [118]. Thus, presence of the genetic resistance determinants does not solely determine MDR phenotype and success/failure of the antibiotic therapy.

Owing to this and other facts (discussed herein), current WGS technology can be clinically applicable only as an integral part of a comprehensive state/government-approved workflow for the clinically relevant cases, e.g., typing of linezolid-resistant *Enterococcus faecium* or screening of carbapenem-resistant Enterobacteriaceae [101, 194]. Future investigations of pathogen resistance mechanisms together with establishment of robust links between genetic components and phenotypic traits in MDR bacteria will help the development of successful WGS-based antibiotic resistance tests. Development of standardized procedures for validation and verification of WGS data, as well user-friendly bioinformatics tools for quick handling and analysis of the genomic information will speed up the implementation of WGS technologies into laboratory practice. For example, one of these tools is provided by the Center for Genomic Epidemiology [136].

Investigation of MDR evolution and emergence dynamics
WGS has been used for the study of the evolution of resistance (or proto-resistance) to multiple drugs and its emergence in different healthcare settings [182]. Large-scale worldwide studies showed that this method could be applied to elucidate historical antibiotic resistance patterns in pathogen populations and study infection transmission mechanisms and emergence dynamics. Specifically, WGS technologies allowed uncovering the genetic basis behind the emergence/re-emergence of successful clones in outbreaks and measuring the rates at which resistance emerges. In addition, WGS also elucidated some of the etiologic factors that allow pathogenesis and spreading MDR bacteria [93, 143, 190].

WGS revealed that the speed of bacterial MDR evolution depends on the genome plasticity and epidemiology of the pathogen, as well as type and duration of applied antibacterial treatment in healthcare settings. For example, the number of SNPs and structural variations (SVs) was higher in MRSA clones in under-resourced healthcare settings where barriers to transmission were lower [227]. Furthermore, the number of SNP differences between isolates belonging to the same outbreaks positively correlated to the time of their isolation in case of MRSA and *Mycobacterium tuberculosis*, pathogens which are transmitted strictly from human to human within a hospital community [52, 95, 127, 227, 258]. In contrast, studies of *Salmonella enterica* subsp. *enterica* and subsp. *typhimurium*, pathogens which can be transmitted from human to human indirectly through various sources, did not show any impact on the accumulated SNP numbers [141, 178]. Genomic analysis also extended our knowledge about the origin of MDR evolution in bacterial populations demonstrating that evolution is acquired through at least three ways:

Table 1 Common MDR bacterial agents of epidemiological importance causing severe infections in hospitals (H) and communities (C)

Bacterial agent	Diseases	Resistance	Example of main resistance determinants revealed in whole sequenced genomes
Escherichia coli (H, C)	UTI, BSI	β-Lactams (cephalosporins)	*ampC*, 2 copies of *bla_T* [74]
		Quinolones (fluoroquinolones)	*gyrA* (Ser83Leu,Asp87Asn), *parC* (Ser80Ile,Glu84Gly) [74, 188]
Klebsiella pneumonia (H, C)	UTI, BSI, pneumonia	β-Lactams (cephalosporins, carbapenems)	*bla_SHV-75*, *bla_SHV-60*, *bla_KPC-2*, *bla_TEM-1*, *bla_TEM-12*, *bla_P1*, *bla_CTX-M* [132, 145]
		Quinolones (fluoroquinolones)	*qnrA1*, *qnrB4*, *oqxAB*, *gyrA* (Ser 83Phe), *parC* (Ala339Gly, Asp641Tyr) [209]
		Amynoglycosides	*armA*, *aph* [209]
		Colistin	IS1 insertion in the *mgrB* [95, 209]
Staphylococcus aureus (H, C)	WSI, BSI	β-Lactams (methicillin)	*mecC* [157]
		Aminoglycosides	*aadD* [258]
		Mupirocines	*ileS-2* [258]
		Mercury resistance	*mer* operon [258]
		Antiseptic resistance	*qacA* [258]
Streptococcus pneumonia (C)	Pneumonia, meningitis, otitis	β-Lactams	*pbp2a*, *pbp2b*, *pbp2x*, *spr1238* [56]
		Tetracycline	*rpsJ*, *patA*, *patB* [153]
Salmonella enterica subsp. *enterica*, *Typhimurium*, *Choleraesuis* (C)	Salmonellosis, foodborne diarrhea, BS	β-Lactams	*bla_OXA-30*, *ampC*, *bla_TEM-1*, *bla_TEM-67* [31, 91]
		Quinolones (fluoroquinolones)	*gyrA* (Ser83Leu), *parC* (Ser80Leu), *acrAB-tolC* [31]
		Aminoglycosides	2 copies *aadA1*, 3 copies *aadA3*, *aac3*, *aph*, *strA*, *strB*, *sat-1* [91, 109]
Shigella spp. (C)	"Bacillary dysenteria"	β-Lactams	*bla_TEM-1*, *bla_OXA-1* [257]
		Fluoroquinolones	mutated *parC* and *gyrA* [257]
		Aminoglycosides	*aadA1*, *aadA2*, *sat-1* [257]
Neisseria gonorrhoeae (C)	Gonorrhea	β-Lactams (3rd gen. cephalosporins)	*mtrR* (G45D, A39T), *mtrCDE* (del1327932), *penB* (G101K, A102D), *penA* (mosaic) [54, 83, 240]
		Tetracycline	*rpsJ* (V57M), *tetM* including its promoter, *penB* [54]
Coagulase-negative *Staphylococci* spp. (CoNS) (H, C)	SSI, endocarditis, and BSI	β-Lactams	*blaZ*, *mecC* [25, 173, 184]
Enterobacter aerogenes (H)	SSI and BSI	β-Lactams	*bla_TEM-24*, 2 copies *ampC*, 3 copies *M-bla*, 4 copies *bla* [48]
		Quinolones	*gyrA* (Thr83Ile), *parC* (Ser80Ile) [47, 48]
		Aminoglycosides	*aadA1*, *aac(6')* [47, 48]
		Rifampicin	*rpoB* (Asp252Tyr) [47, 48]
Acinetobacter baumannii (H, C)	BSIs, VAP, HAP, SSI, CA-UTI, ventilator-associated pneumonia	β-Lactams (3rd gen. cephalosporins)	*bla* class A, *ISAba1*, *bla_OXA-23*, *bla_OXA-10*, *bla_OXA-69*, *bla_ampC*, *bla_OXA-23*, *bla_OXA-66*, *bla_ADC* (Nigro et al. 2013 [173]); [2, 76, 83, 173, 195, 240]), *ampC*, *bla_OXA-51*-like [74], *bla_TEM-1* [2], 5 copies *M-bla*, 2 copies *ampC*, *bla_OXA-82* [204]
		Amynoglycosides	Modified *armA*, *aac(3')*, *aac(3)-Ia*, *aac(6')*, *aac(2')-Ib*, *aadA1*, *aadAB*, *aphA1*, *aph(3')*, *aph6*, *strAB* [2, 71, 204, 254, 265] *adeT*, *aadA2* [204]
		Quinolones	*gyrA* (Ser83Leu), *parC* (Ser80Leu) [2, 71, 204]
	BSIs, VAP, HAP, SSI, CA-UTI, ventilator-associated pneumonia	Colistin	*pmrB* [90]
		Tetracyclines	*tetAR*, *adeB* [2, 265], *bcr* [204]
		Chloramphenicol	*cmlA*, *cmlA5*, *cat* [71], *catB6* [265], *catB8* [254], *cmr* [204]
Pseudomonas aeruginosa (H, C)	BSIs, VAP, HAP, SSI, CA-UTI, cystic fibrosis (CF)	β-Lactams (3rd gen. cephalosporins)	*blaI_MP-1*, *oprD* [163] *ampCDR* [124]
		Quinolones	*gyrA* (Thr83Ile), *parC* (Ser87Leu) [247]

Table 1 Common MDR bacterial agents of epidemiological importance causing severe infections in hospitals (H) and communities (C) (*Continued*)

		Aminoglycosides	*aac(6')* [163], *aph, ant(4')-IIb, strAB* [247], *aacA29a/aacA29b* [124]
		Colistin	*pmrAB, phoPQ* [247]
		Wide range of antibacterial agents	*mexAB-oprM, mexXY, mexCD-oprJ, mexEF-oprN, mexHI-opmD, mexR, nfxB, mexT, mexG* [124]
Mycobacterium tuberculosis (H, C)	Tuberculosis	Rifampicin	*rpoB* (S450L) [52]
		Isoniazid	*katG* (P7 frameshift), *ptrBa, fadD15, ppsB, atsH* [88]
		Fluoroquinolone ethambutol amikacin para-aminosalicylic acid	*gyrB* (T500N), *embB* (D1024N), *rrs*(A514C, A1401G), *thyA* (P17L) [52]

BSI bloodstream infection, *SSI* surgical-site infection, *CA-UTI* catheter-associated urinary tract infection, *VAP* ventilator-associated pneumonia, *HAP* hospital-acquired pneumonia, *WSI* wound stream infection, *UTI* urinary tract infection

i. Transmission of plasmids bearing diverse antimicrobial resistance genes between pathogens/or horizontal gene transfer with the help of mobile genetic elements (MGEs) [12, 37, 179]
ii. Mutations in bacterial drug-related genes and intergenic regions [2, 47, 48, 52, 71, 74, 99, 247]
iii. Differential expression of genes which mediate drug effects [262]

Acquisition of new resistance genes and virulent determinants by horizontal transfer via conjugation, transduction, or transformation usually is associated with pathogen adaptation to new niches or lifestyles and affects the evolution of its genomic content, leading to clinically significant effects. This evolution mostly underlies the success of the MDR emerging strains and may be a major reason of the outbreaks all over the world. Transmissible plasmids and phages frequently bear resistance genes/cassettes conferring bacterial resistance to one or several different antibiotics and facilitate their transfer through different genera. For example, it was revealed that IncA/C plasmids carry extended-spectrum β-lactamases, AmpC β-lactamases, and carbapenemases among broad host range pathogenic Enterobacteriaceae [63, 73, 100, 158, 210, 212]. They are considered the most common reason of hospital MDR of these bacteria for many old and new generations of the β-lactams, including cephalosporins, penicillins, cephamycins, and monobactams [110, 162] (Table 1). Other clinically relevant plasmids include pTW20_1, harboring *qacA* (encoding antiseptic resistance) and *mer* operon (mercury resistance), and pPR9-like carrying *aadD* (aminoglycoside resistance) and *ileS-2* (resistance to mupirocin) genes, are conjugated between MRSA ST 239 isolates [227] and, possibly, can be transmitted between other staphylococcal strains and species [9, 17].

The horizontal gene transfer of chromosomal genes with the help of MGEs is also important in conferring resistance to a wide variety of antibiotics, particularly towards new ones. For instance, recent retrospective studies of *S. aureus* showed that all emergent MRSA populations differed from methicillin-sensitive *S. aureus* (MSSA) not only in plasmid replacement and content but also in such genetic features as small deletion/insertion polymorphisms (DIPs) and presence of MGEs and resistance genes on the chromosome [230, 231, 241]. Further, it was shown that MDR genes are often associated with the MGEs and, with their help, can be transferred to other bacteria between the same or different species [225, 254]. For example, it was shown that the evolution of methicillin resistance in nosocomial and community-acquired MRSA was mostly arisen by acquisition of the staphylococcal cassette chromosome (SCC*mec* type IV cassette) integrated into the chromosome and carrying the *mecA* or *mecC* genes encoding penicillin-binding proteins, which reduced affinity for β-lactam antibiotics [76, 205].

Other recent large-scale studies extended our knowledge about resistance evolution of *S. aureus* CC398 lineage, the most prevalent emerging pathogen with broad host tropism in many European countries [157, 245]. These works shed light on the nature of MDR in CC398 and questioned its origin and the major reasons of its emergence in clinics. All human-specific MSSA and MRSA isolates carried two unique genetic markers: *φ7* bacteriophage and *φ3* bacteriophage with human-specific immune evasion genes *chp, scn*, and *sak* (only in MRSA) [157]. Based on these studies, it was hypothesized that livestock-associated MRSA has diverged from the human-associated MSSA and that it acquired tetracycline and methicillin resistance genes and lost phage-carried human virulence genes [157, 192, 213]. However, further discrete-trait analyses provided for this lineage did not support the hypothesis about its human origin and left the question about evolutionary routes open [245]. This discrepancy may be explained by the lack of unified and standardized computational methods and interpretative algorithms applied for the WGS data analysis.

The WGS data, accumulating for various bacterial species, also showed that specific acquired determinants (mostly, virulence-related genes or islands) can also be the key reasons of the emergence of MDR pathogens causing outbreaks. For instance, it was shown that Panton-Valentine toxin and *sasX* gene, encoding a surface protein, contributed to the outbreaks caused recently by MRSA in the UK and China, respectively [93, 143]. Further, the *mgrB* gene, encoding a transmembrane protein produced upon activation of the PhoPQ signaling system, was found to be associated with colistin resistance in re-emergent *K. pneumonia* causing nosocomial outbreaks worldwide [190].

Antibiotic resistance can also be caused by spontaneous and induced missense mutations within the antibiotic targets or their binding sites, e.g., gyrase subunits A and B, *gyrA* and *gyrB* (targets of quinolones), RNA polymerase subunit B, *rpoB* (target of rifampicin), dihydrofolate reductase, *alr* (rimethoprim), protein biotin ligase, *birA* (Bio-AMS), or membrane proteins (e.g., multidrug efflux protein *norM*) (Table 1) [99]. For example, WGS revealed the mutations in *blaI*, *blaR1*, as well as in the *mecA* regulone (*mecI-mecR1-mecA*) in MRSA [16]. Similarly, it was demonstrated that the major mechanism of MDR in re-emergent *M. tuberculosis* is primarily arisen by point mutations in *rpoB* (S450L), *katG* (P7 frameshift), *gyrB* (T500N), *embB* (D1024N), *rrs* (A514C and A1401G), and *thyA* (P17L) genes [22, 52, 88, 186, 242].

The genomic information together with powerful bioinformatics tools made it possible to distinguish the molecular pathways responsible for MDR-caused diversity. For example, Darch and colleagues have demonstrated that distinct recombination events were the dominant driver of phenotypic diversity in extant population of *P. aeruginosa* obtained from a single cystic fibrosis (CF) patient (with a weight of recombination relative to mutation, r/m, rate approaching 10) [41]. Other retrospective studies identified the exact unique genetic SNVs in main virulence-related genetic factors of *P. aeruginosa* associated with epidemic CF infection [81]. The increased resistance of emerging MDR *P. aeruginosa* to antibiotics was explained by SNPs enrichment of the efflux pumps which actively transport the toxic compound out of the bacterial cell to avoid contact with the target site [45, 113]. Similarly, the revealed genome-wide recombination events in chromosomal β-lactamase genes bla_{ADC} and $bla_{OXA-51-like}$, plasmid-borne resistance genes, as well transposon- and integron-derived modules were also proposed as major drug resistance diversification drivers for epidemic strains of *Acinetobacter baumannii* [215, 254].

Revealed SNPs and SNVs can be potentially used as a molecular clock to prognose new or potentially emerging/re-emerging outbreak strains, precise tracking, early warning, and targeted infection control of pathogenic bacteria. For instance, the time frame for the emergence of a bacterial pathogen clone and its evolution during epidemic spread had been estimated for MRSA [95]. On the basis of the WGS data, the level of nucleotide substitutions was estimated at 1.68×10^{-6} substitutions per site per year in the BEAST analysis, or 2.72 mutations per megabase per year in the parsimony [245, 258]. This translates to approximately one mutation per genome every 6–10 weeks [95]. Taking into account that 1–3 point mutations or large genetic rearrangements (recombination more than 100 bp) in targets related with drug resistance are enough to make differences in antibiotic susceptibility, the provisional prediction of an emergence of novel MRSA clones in clinical settings can be afforded [53, 95]. However, another work demonstrated that using a simple threshold of a maximal number of mutations to rule out direct transmission and emergence of MDR *M. tuberculosis* led to inaccurate interpretation of the data [52]. These authors showed that about 38 % of all individual SNPs were involved in resistance of MDR *M. tuberculosis* and made an important contribution to evolution and emergence of MDR in the bacteria within a single patient [52].

In summary, together with developed tools for WGS data analysis (e.g., Rainbow [264]) and unifying genome-wide database (e.g., *M. tuberculosis* Variation Database (GMTV) [29], The Bacterial Isolate Genome Sequence Database (BIGSdb) [116]) containing the broad spectrum information about individual mutations of pathogens, WGS can be a powerful tool for the preliminary prediction of drug resistance, geographical origin, as well clinical strategies and outcomes.

Diagnostics and control of MDR bacterial infections

Successful containment and prevention of MDR infections involves (i) timely identification and characterization of the MDR infectious/outbreak cause, and (ii) discovery of its source and transmission pathways [86, 222, 249]. A significant transformation in MDR infectious disease diagnostics has occurred during the past few decades, including key changes in basic concepts, data analysis approaches, and, especially, methods of exposure measurement and pathogen surveillance [10]. Today, diagnosis of DR pathogenic bacteria are mainly done by means of expensive and time-consuming experimental approaches, including complex phenotypic and genotypic standardized methods [68, 169, 205, 206, 222, 235] (Fig. 1). The techniques applied for this task are mostly based on the detection of phenotypic and genetic traits related to drug resistance, pathogenicity or survival mechanisms of pathogens. Standardized culture-based methods [235], traditional typing (such as biotyping, antibiograms, resistograms), and molecular typing techniques [68, 206, 222] are widely used

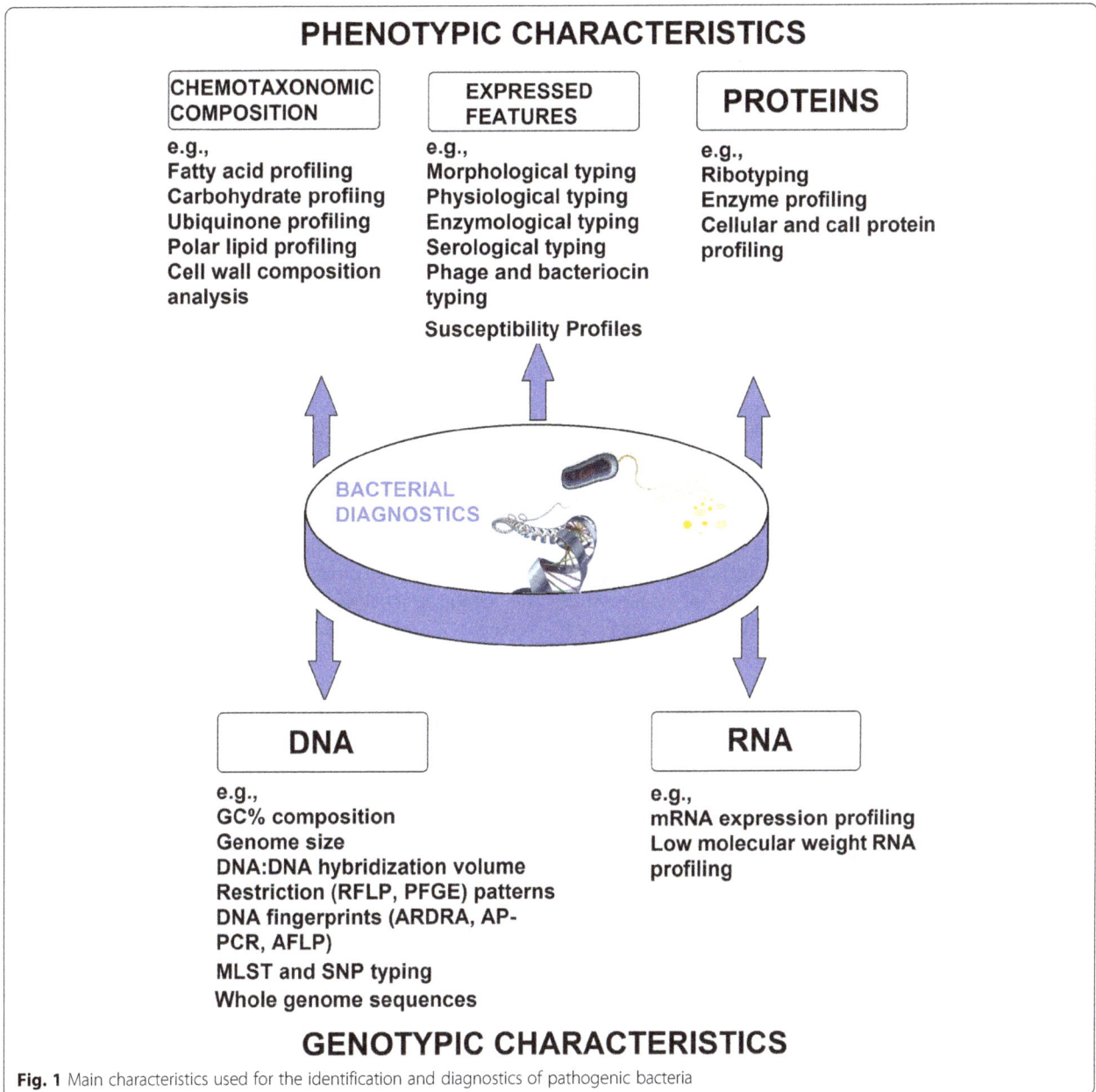

PHENOTYPIC CHARACTERISTICS

CHEMOTAXONOMIC COMPOSITION

e.g.,
Fatty acid profiling
Carbohydrate profiling
Ubiquinone profiling
Polar lipid profiling
Cell wall composition analysis

EXPRESSED FEATURES

e.g.,
Morphological typing
Physiological typing
Enzymological typing
Serological typing
Phage and bacteriocin typing

Susceptibility Profiles

PROTEINS

e.g.,
Ribotyping
Enzyme profiling
Cellular and call protein profiling

BACTERIAL DIAGNOSTICS

DNA

e.g.,
GC% composition
Genome size
DNA:DNA hybridization volume
Restriction (RFLP, PFGE) patterns
DNA fingerprints (ARDRA, AP-PCR, AFLP)
MLST and SNP typing
Whole genome sequences

RNA

e.g.,
mRNA expression profiling
Low molecular weight RNA profiling

GENOTYPIC CHARACTERISTICS

Fig. 1 Main characteristics used for the identification and diagnostics of pathogenic bacteria

to detect and identify the cause and course of the outbreaks in the clinical laboratories. Over the last few years, these methods have improved dramatically: they have incorporated automation to increase speed, discrimination power, and throughput, and reduce cost. However, none of these methods is considered optimal for all forms of research and infections. Choice of the method significantly depends on the epidemiological problem to solve, time constrains, its reliability, accuracy, and geographical scale of its use [206]. Furthermore, almost all available approaches have limitations detecting pathogenic microorganisms with rapid transmission dynamics and mutational rates [169], or mixed MDR infections involving multiple unrelated strains or outbreaks caused by closely related isolates [201]. As a result, existing integrated approaches are laborious, time-consuming, expensive, and can lead to misdiagnosis.

Although most of the WGS investigations were retrospective, they demonstrated that WGS technology may make real-time genomic diagnostics a reality [117]. In contrast to multifaceted algorithms used in standard testing, genomic data can provide rapid and accurate detection and control of emerging MDR pathogenic strains in a single process, reducing unnecessary infection-control measures [228]. The genomic information affords unprecedented and detailed insight into microevolution of pathogenicity factors, antibiotic resistance, and transmission mechanisms of pathogens, and, thus, allows robust detection and control of the spread of closely related

pathogenic isolates in the clinics [14, 130, 142, 239], communities [30, 72, 77, 84, 159, 203], and globally [15, 94, 95, 168, 227].

The first application of WGS technology was for MRSA, the leading cause of healthcare-associated infections worldwide [45, 55, 171, 172, 258]. WGS techniques detected closely related MRSA clones associated with putative outbreaks, which could not be confirmed with conventional methods, and allowed the reconstruction of local and intercontinental spread of MRSA lineages [53, 93, 95, 127, 130, 258]. For instance, Harris and colleagues studied a putative MRSA outbreak on a special care baby unit at a National Health Service Foundation Trust in Cambridge, UK. During these studies, the cause of a persistent outbreak, a new type ST2371 with Panton-Valentine leucocidin encoding gene, was revealed. WGS technique provided the best discrimination between closely related bacterial clones of the same MRSA lineage, compared to multilocus sequencing typing (MLST) analysis [93]. Importantly, this study resulted in a fundamental shift in the understanding of transmission dynamics and sources of successful epidemic MRSA clones between healthcare facilities and communities. WGS provided strong evidence that community-associated MRSA can be carried for a long period by healthy people [75, 93] and become the cause of healthcare-acquired MRSA infections replacing dominant healthcare-associated lineages [80]. These data facilitated improved infection-control measures for the infectious sources (e.g., workers, visitors, equipment). Later, this study was complemented with more detailed investigations of cause and sources of hospital- and community-associated MRSA lineages in settings with extensive and poor infection-control practices all over the world [157, 227, 248]. It was shown that low resource countries can be the main source of the global emerging MRSA [227]. Thus, the population of MRSA ST239 lineage, aka the Brazilian clone most prevalent across the globe, was significantly more variable (evolved faster) in countries with low-cost prevention planning and implementation than in those with well-resourced healthcare settings [200, 227]. Another work provided evidence for frequent transfer of most prevalent human- and animal-associated MDR MRSA CC398 lineage and indicated that livestock and animals could be the main source of infection in humans [245]. The fact that S. aureus could be transferred between humans, animals, and livestock (probably in all directions) raised the main concern for clinicians. Together with evidence for higher levels of MDR in the livestock-associated clades, this raised the need to change the existing biosecurity control in agricultural settings.

Pallen and colleagues were the first who applied WGS technology to study the prolonged hospital outbreak of MDR A. baumannii in Birmingham, England, between July 2011 and February 2013. With the help of WGS, a novel isolate, the causative outbreak agent was revealed [142, 177]. This clone could not be detected by conventional methods. As in the case of MRSA, it was revealed that early transmission events can occur through the ward-based contact and environmental contamination of the hospital environment [177]. This knowledge led to tighter ward decontamination procedures and infection-control interventions with the purpose of reducing the risk of further transmission.

WGS has shown potential for elucidation of the transmission dynamics of the MDR Salmonella species [6, 177] and for the detection of various epidemic S. enterica subspecies [141, 174, 175]. MDR and highly clonal lineages of K. pneumonia, an important opportunistic pathogen associated with nosocomial and community-acquired infections [189], can be also successfully detected through WGS [151]. In addition to results for MRSA and A. baumannii which showed strong evidence of transmission via alternative routes (e.g., silent transmission vectors), the retrospective genomic analysis of the nosocomial carbapenem-resistant K. pneumonia isolates together with epidemiological data revealed unexpected transmission, perhaps through asymptomatic carriers or inanimate objects (ventilators, equipment). In addition, it was concluded that combination of the genomic and patient trace data together with algorithms which accounted for K. pneumoniae's capacity for silent colonization can be used for more effective control of the outbreaks and reconstruction of the most likely pathogen transmission routes [216].

WGS analysis allowed identification and tracing of MDR M. tuberculosis more precisely than the currently used conventional typing methods [67, 77, 121, 152, 202, 209, 242]. Using WGS technology, Walker and colleagues first analyzed TB cases of the community outbreaks in the UK Midlands. Only genomic data allowed elucidation of the genetic diversity and detection of closely related mycobacterial genotypes causing these outbreaks [242].

Due to the complexity of antibiotic susceptibility regulation mechanisms in P. aeruginosa and the high level of its diversity, the most indisputable WGS implication was usually related to diagnostic and control of CF infections [41, 165]. A number of recent studies of MDR P. aeruginosa from a single patient have shown that this technology has a great potential for routine diagnostics and antibiotic susceptibility detection in a clinically relevant time frame [41, 124, 247]. It was proposed that further investigation of the enabling gene pool and resistance mechanisms of MDR P. aeruginosa populations could improve clinical outcomes of antibiotic sensitivity and detection testing in the future [41].

Besides the retrospective studies, the real-time WGS analysis was successfully applied for rapid detection of

infections and outbreaks caused by neonatal MRSA [53, 130], verocytotoxin-producing *E. coli* (VTEC) [114, 120], *Legionella* sp. [198], carbapenem-resistant *K. pneumoniae* [216], *C. difficile* [53], and *A. baumannii* [204]. For instance, in 2011, real-time WGS clarified the cause of a very mysterious outbreak in a farm in Germany. The outbreak was caused by enteroaggregative *E.coli* O104:H4 clone, epidemiologically linked to human cases and transmitted via contaminated seeds [84, 195, 203, 233]. Another modification of real-time WGS analysis, direct real-time WGS (sequencing clinical specimens without the need for culture), was successfully applied for identification and characterization of slowly growing and difficult-to-culture pathogens in clinical samples [7, 98, 150, 211]. Whereas direct WGS is considered as not cost-effective and less sensitive for some clinical workflows (e.g., in the case of fecal samples or mixed infections) [126], single-colony sequencing is considered a very promising epidemiological tool which can address multiple clinically relevant questions more accurately and faster in the future [129]. A simple WGS protocol has been developed and tested for the detection of a broad range of pathogenic bacteria (17 most clinically important pathogens) from a single bacterial colony [3, 129]. Once the procedure is validated, this method has a lot of advantages for clinical practice [3]. However, the single-colony WGS method may be difficult to optimize in the case of difficult-to-grow pathogens [41].

Although it is presumed that WGS may become the primary tool to provide pathogen diagnostics and control in clinical and healthcare settings in the nearest future, many obstacles remain [126]. Today, real-time genomic diagnosis is mostly based on the detection of SNP, SNV, and SV of relevant multiple genetic loci selected for typing. The housekeeping, structural, and functional genes and intergenic regions [11, 30, 53, 77, 95, 126, 136, 140, 142, 156, 168, 195, 203, 260], as well as the virulent and resistance factors are considered as clinically important markers and are applicable for benchtop typing [206]. Growing WGS data and advances in sequencing technologies constantly lead to the discovery of new genetic or genomic variations important for bacterial growth, pathogenesis, antibiotic resistance, and survival. However, before being applied for diagnostics, this plethora of biomarkers requires intensive study of their functions and associations with particular phenotypic changes. Subsequently, the simple and unified analytical tools/platforms to readily extract relevant information from the genome and interpret it without complex and computer-intensive analysis should be developed, and the clinical health personnel should have a quick access to them [135, 136, 140, 256]. One example of this strategy is the study of *Neisseria meningitidis* outbreak [57, 78, 115] which occurred at the University of Southampton, UK, in 1997. Jolley and colleagues developed an integrated analysis platform and applied it for the robust interpretation

and analysis of WGS data obtained for *N. meningitidis*. As a result, this analysis took only a few minutes and permitted complete resolution of the meningococcal outbreak. While these tools are being developed for self-contained laboratory workflow, the integration of the WGS technology with phenotypic, molecular typing methods [39, 40], new strategies of sample and culture selection [68], and epidemiologic data analysis is already enhancing our ability to control and prevent nosocomial or healthcare-associated infections.

Development of new diagnostics markers and assays

While WGS sequencing is highly informative, it is not cheap, fast or readily available for screening DR bacterial isolates in various healthcare settings today. For example, current WGS technologies may be too slow for point-of-care diagnostics. As a result, target-specific PCR, real-time PCR, and related technologies [160, 223] still remain the most common methods used in clinical practice. However, it still remains critical to select specific sequences (signatures/targets) for designing molecular assays for the pathogen of interest [5]. In this case, WGS can act as a precursor to generate specific diagnostic tests for timely case definition [102, 193, 219]. The genomic data should be analyzed using computational methods (e.g., KPATH, TOFI, Insignia, TOPSI, ssGeneFinder, or alignment-free methods) in order to identify pathogen signatures, estimate their evolutionary rates across the group, and design highly specific diagnostic assays for target groups of pathogens [104, 193]. Due to the obtained WGS data, numerous novel diagnostic genetic targets have been suggested for routine diagnostics of several pathogenic bacteria over the last few years. An extensive list of putative markers is presented in Table 2. WGS technology can also provide robust information about the reliability of the existing and implemented diagnostic markers and thus can help in avoiding false-negative and false-positive results. For example, the obtained WGS data improved the current diagnostic, cultural, and molecular tests for several pathogens: *S. aureus* [184], TB [125], *E. coli* [51], and *K. pneumoniae* [48].

Developing new antibacterial drugs

Today, a lot of strategies are applied to optimize the identification of new targets and their inhibitors (antibacterial compounds, hits) for the discovery of new antibacterial drugs [50, 214] and predict the mechanisms of their action and their effects in patients. However, clinical management of drug-resistant strains still remains cumbersome. At the same time, the number of newly approved drugs per year has been decreasing, and only five new antibiotics were approved since 2003 [18, 49]. WGS can assist this effort by accelerating the discovery of novel antibacterial inhibitors and targets overlooked by conventional discovery platforms, e.g., sputum smear, culture, and drug susceptibility

Table 2 List of the putative genetic markers obtained by WGS for diagnostics of the bacterial agents of epidemiological importance

Potential target	Location	Target identity	Pathogen	Ref.
mecA/mecC	Chr	Adapter protein/Penicillin-binding protein 2a	MRSA/S. aereus	[184]
tetM	Chr	Tetracycline resistance protein	Livestock-associated S. aereus CC398	[157]
φ3/φ7	Chr	Bacteriophages	Human-specific S. aereus CC398	[157]
Chp		Chemotaxis inhibitory protein		
Scn		Staphylococcal complement inhibitor		
Sak		Staphylokinase		
gp20	Chr	Putative prophage DNA transfer protein	Verocytotoxin-producing E. coli O104:H4	[102, 193]
impB	PI	DNA polymerase type Y		
usid000007 (contig 69, 14714:14853)	Chr	Sequence positions 47–69 similar to Ricinus communis putative receptor serine-threonine kinase mRNA (XM_002525007.1)		
usid000002 (contig 43, 1486:1633)	Chr	Positions 4–34 similar to Pseudomonas putida BIRD-1 major facilitator transporter protein coding sequence (ADR60257.1)		
ISAba1	Chr	Transposase of ISAba1, IS4 family	MDR A. baumannii	[166, 254]
csuE	Chr	Chaperone-usher pili assembly system	MDR A. baumannii,	[254]
bla$_{OXA-51}$		Bla$_{OXA-51-like}$ beta-lactamase	GC2 (SG1)	
Coding SNP:	Chr		Colistin-resistant K. pneumonia KPNIH1	[216]
ind(GA) 321 in KPNIH1_08595 CTG→ATG		Microcin B17 transporter		
(L→M) at 130 in KPNIH1_18808 ACC→ATC		Putative membrane protein		
(T→I) at 1106 in KPNIH1_07189 GGC→TGC		L-Ala-D/L-Glu epimerase and methyl viologen resistance protein SmvA		
(G → C) at 811 in KPNIH1_05438		Putative transport protein		
ampC	PI/	β-Lactamase	β-Lactam resistant Enterobacteriaceae and P. aeruginosa	[110, 162, 247]
ampD	Chr	N-acetylmuramyl-L-alanine amidase, negative regulator of AmpC,AmpD		
ampR		HTH-type transcriptional activator AmpR		
ampG		Putative transporter		
gyrA (Thr83Ile)	Chr	DNA gyrase subunit A	P. aeruginosa resistant to fluoroquinolones	[247]
parC (Ser87Leu)		DNA topoisomerase 4 subunit A		
bla$_{VIM-2}$	PI	Beta-lactamase class B VIM-2	P. aeruginosa resistant to β-lactams except monobactams	[247]
aacA29a/aacA29b putative aph ant(4')-IIb	Chr[a]	6'-N-aminoglycoside acetyltransferase type I, phosphotransferases	Aminoglycoside resistant P. aeruginosa	[247]
mexAB-oprM	Chr	Efflux pumps and multidrug resistance operon repressors	MDR P. aeruginosa	[247]
mexCD-oprJ				
mexEF-oprN				
mexHI-opmD				
mexR		Operons respective regulator Genes		
nfxB				
mexT				
mexG				
mexMN				
mexVW				

Table 2 List of the putative genetic markers obtained by WGS for diagnostics of the bacterial agents of epidemiological importance (Continued)

mexXY				
muxABC-opmB				
emrAB-opmG				
SMR		Small multidrug resistance family of proteins		
triABC	Chr	Presence triclosan efflux pump operon	Triclosan resistant P. aeruginosa	[247]
mexJKL		Resistance nodulation cell division efflux pump		
czcCBA	Chr	Cobalt-zinc-cadmium efflux resistance operon	Heavy metal resistant P. aeruginosa	[247]
pmrAB	Chr	Membrane bound sensor	Colistin and polymyxin resistant P. aeruginosa, S. enterica subsp. Typhimurium, and A. baumannii	[139, 247]
phoPQ	Chr	Kinases and cytosolic response regulator		
60 SNPs		Intergenic regions, enzymes, regulatory and membrane proteins	S. enterica subsp. Enteritidis	[87, 176]
iniBAC	Drug efflux operon		M.tuberculosis resistant to rifampicin, isoniazid, fluoroquinolone ethambutol, amikacin, para-aminosalicylic acid	[52]
rpoB (S450L)				
katG (P7 frameshift)				
gyrB (T500N)				
embB (D1024N)				
rrs (A514C, A1401G)				
thyA (P17L)				

Ch chromosome, Pl plasmid
[a]Except plasmid location in rifampin-resistant P. aeruginosa PU21

testing. The innovative WGS technologies can be successfully applied for clinical trials to evaluate the potential antibacterial targets, inhibitors, efficacy of the drugs, and therapeutic alteration of the microbiome in a range of conditions for rational structure-based drug design in a single step (Fig. 2). An important point is that the WGS strategies of screening for novel "drugable" classes of molecules and targets are easily compatible with natural product discovery programs and existing phenotypic high-throughput screening and thus can significantly improve and speed up current practical outcomes [13, 35, 108, 148].

Inhibitor-first approach (reverse pharmacology) The inhibitor-first strategies are more effective than target-driven ones [220] and remain the main approaches of choice for delivering antibacterial drugs to the clinics [20]. WGS screening can be applied to identify molecules that inhibit bacterial growth by diverse mechanisms, including those that engage multiple targets. An extensive list of the antimicrobial drugs discovered recently via WGS data is presented by Deane and Mitchell [44]. As a whole, most of these natural products are essential components of the metabolic pathways for the vitamin biosynthetic (B1, B3, B9), fatty acid synthesis (FASII), and isoprenoid biosynthesis (fosmidomycin, 6-fluoromevalonate). Genomic analysis can also help to reveal genes or gene clusters that

are important for biosynthesis of natural antibacterial inhibitors but remain silent under laboratory growth conditions or in the environment. For example, induced expression analysis of environmental DNA gene clusters revealed that tetarimycin A, a new class of tetracyclic MRSA-active antibiotic isolated from the culture broth extract of *Streptomyces albus*, was encoded by the *tam* gene cluster [119]. Screening of libraries of complete genomes of the soil microbial community extended the potential value of this compound having revealed numerous silent *tam*-like gene clusters that possibly encode other members of tetarimycin family in the environment [119]. *Streptomyces coelicolor* is another example. Before completion of its genome sequence, only three gene clusters coding natural products had been identified for actinorhodin [154], prodiginine [58], and lipopeptide calcium-dependent antibiotic [32]. WGS revealed that *S. coelicolor* carries clusters of new "cryptic" genes which have a potential for biosynthesis of 29 structurally complex unknown natural products that can be potentially applied as antimicrobials [38].

Target-driven approaches Knowledge of the three-dimensional structure of the drugable targets can also be used for generating or discovering novel-specific inhibitors. Traditionally, a target-driven approach starts from high-

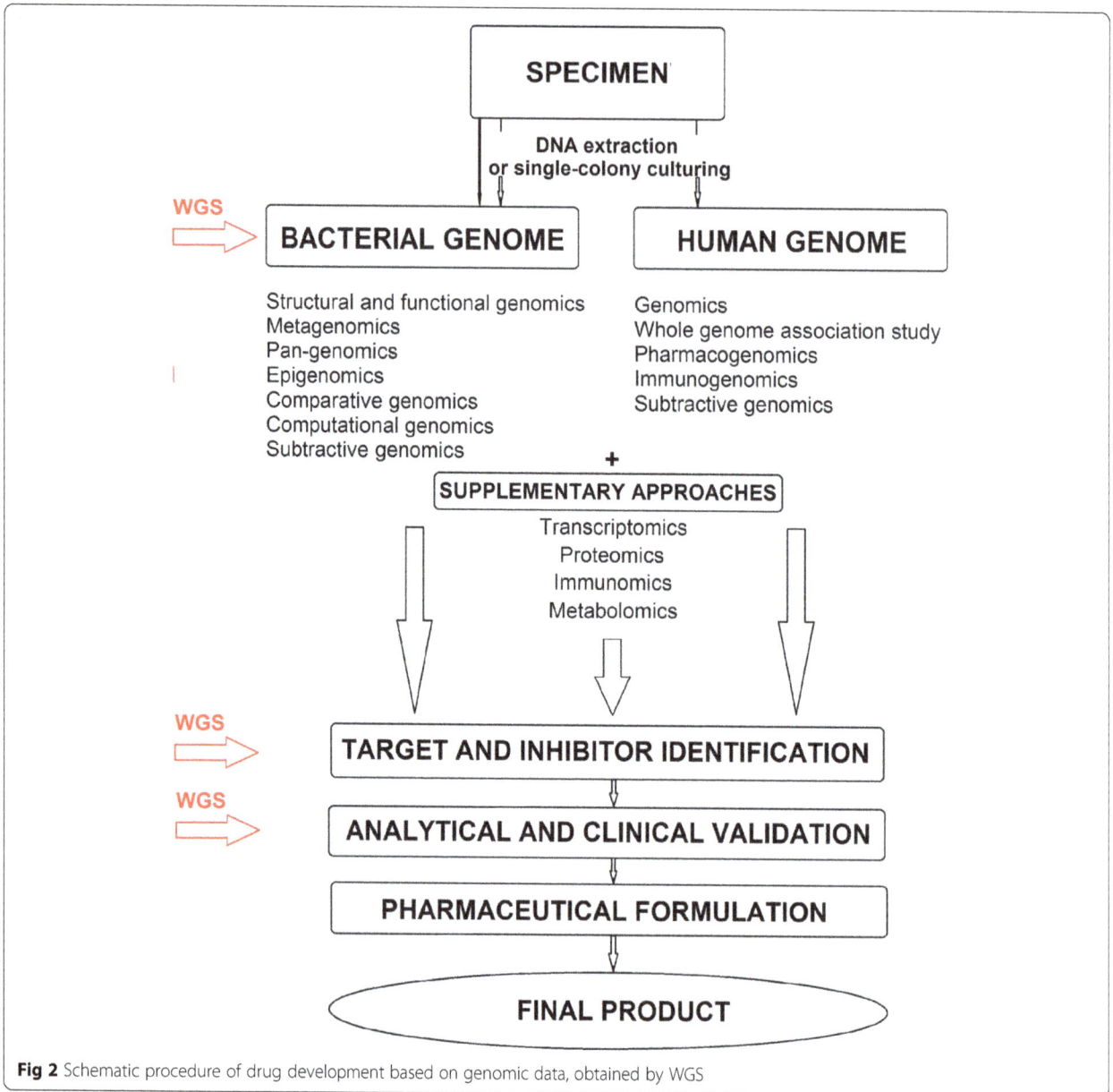

Fig 2 Schematic procedure of drug development based on genomic data, obtained by WGS

throughput screening for inhibitors of a purified target protein. Unfortunately, most inhibitors identified in high-throughput screenings are not active against live bacteria or are not safe for use in humans [185]. WGS can contribute to the de bene esse discovery of the candidate genetic targets for both inhibitors of known or entirely novel mechanism of action (MOAs) before conventional screening for DR bacteria. Determination of resistance mutations in the targets by WGS can also be used for evaluation and estimation of the resistance of the bacterial population to the drug. The target-driven WGS approach was first applied for target FabF, an enzyme required for fatty acid biosynthesis (FAS) [122]. Four novel type II FAS (FASII) inhibitors with broad spectrum activity against Gram-positive bacteria, including MRSA, Platensimcyin, Plantencin, BABX,

and Phomallenic acid C, were developed using this method [19, 122, 207, 244, 259]. Recently, several novel antibiotics, fasamycin A and B, with specific activity against FabF of MRSA and vancomycin-resistant *Enterococcus faecalis* were also revealed [61].

Studies performed on a collection of several human pathogens suggested that on the average, about 15–25 % of all genes in a genome are potential drugable targets [33, 164, 238]. These studies concluded that the potential targets are regions whose products/structures are important for bacterial growth and survival under a variety of conditions (e.g., the synthetic machinery of the bacterial membranes, peptidoglycans, lipopolysaccharides, the DNA replication machinery, the nucleic acid synthesis pathway, and ribosomal structures) but do not prevent growth in

animals or humans [243]. Thus, WGS screening identified mutations correlating with mycobacterial MDR in genes involved in respiration, fatty acid biosynthesis *kasA* [137], *qcrB* [1, 187], protein synthesis *aspS* [89, 107], protein secretion *eccB3* [107], polyketide biosynthesis *pks13* [107, 246], mycolic acid transport *mmpL3* [197], and arabinogalactan synthesis *dprE1* [34]. Another study of pathogenic bacteria revealed other candidate structures e.g., amino-acyl-tRNA binding site (A-site) and components of the 2-C-methyl-D-erythritol 4-phosphate (MEP) pathway which are also potential targets for the development of new antibiotics for various emerging pathogens [105, 186]. Screening of bacterial genomes for the presence of this ligand can be used for the development of drugs which are active against a wide range of pathogens [64, 105, 236].

However, the target-driven method has some limitations. For example, it can only be applied if resistant strains were obtained. Furthermore, it is important to remember that the target-mechanism identified (such as efflux pump expression, chemical inactivation, or malfunction of transforming an inactive prodrug into the active derivative) can be just one of the existing mechanisms by which mutations can impart resistance. Presence of several candidate targets, which belong to the same protein family with conserved inhibitor binding, can also complicate their further interpretation and evaluation by overexpression analyses [21, 234]. In addition, mutations in nonessential genes can also significantly modulate the main target's structure or functionality resulting in partial activity of antibiotics [147].

Clinical trials WGS can be applied to design clinical trials more efficiently. First, it can be used at the early phases of drug development to screen a phylogenetically diverse collection of the pathogens for the presence and variability of the candidate drug's target. Such analysis will prove that this target and its variations are valid and important for all species and lineages of the pathogenic genus and, thus, reduce the chance to miss any resistant strains [128].

Second, WGS can be used to determine drug's MOA directly. Although it is not mandatory to define an antibacterial compound's MOA for use in humans, this knowledge can help developing novel drugs for a broad range of bacteria and evaluate their toxicity and specificity a priori. Knowledge of MOAs will also reduce time for clinical trials of chemically redundant putative compounds that fail for the same mechanistic reasons. Further, identification of the MOA and candidate targets can give another chance to existing antibiotics. For example, bottromycins, antibacterial peptides with activity against several Gram-positive bacteria and mycoplasma, were discovered more than 50 years ago. Later on, it was revealed that these peptides' binding

A-site on the 50S ribosome lead to the inhibition of protein synthesis and thus can become a novel promising class of antibiotics applied against vancomycin-resistant Enterococci (VRE) and MRSA [105].

Third, knowledge about resistance mechanisms at the genetic level is very important for determining and avoiding cross-resistance of the pathogen, when multiple antibiotics should be applied for treatment [167]. Fourth, sequencing of pathogens during clinical trials has the potential to distinguish exogenous re-infection from the primary infection. This is crucial in order to assess the efficacy of study drugs and estimate the therapeutic effect in a range of conditions [22, 23, 127, 237].

However, as the field of the genomic drug and target discovery moves forward, the problem stemming from the elucidation of novel unknown classes of gene products remains significant. It is important to remember that no single method is sufficient to define the MOAs of most antibacterial drugs, but a complex approach is required [27]. The detailed genomic analysis of the human pathogens (microbiota), as well as gene expression and drug susceptibility analyses of pathogens, together with powerful bioinformatics tools, can provide new applications to "old" drugs and invigorate the discovery process for novel antibiotics [43, 191]. In this regard, the discovery of the novel anti-TB inhibitors (e.g., bedaquiline, pyridomycin, SQ109, BM212, adamantyl ureas, benzimidazole, BTZ, TCA, and imidazol[1,2-a]pyridine related derivatives) succeeded by a combination of high-throughput screening and WGS analysis of spontaneous resistant mutants for target identification, combined with modern bioinformatics tools [8, 97, 183]. Zomer and colleagues also demonstrated that the combination of high-density transposon mutagenesis, WGS, and integrative genomics has a great potential for reliable identification of potential drug targets in *Streptococcus pneumoniae*, *Haemophilus influenzae*, and *Moraxella catarrhalis* [164]. This complex analysis predicted 249 potential drug targets, 67 of which were targets for 75 FDA-approved antimicrobials and 35 other researched small molecule inhibitors.

Conclusions

What does the future hold for WGS? Herein, we showed that WGS may be well poised to make a decisive impact on the study and control of MDR in pathogenic bacteria (Table 1) [126]. However, although not reviewed here, studies have shown that WGS can also contribute to the investigation of various pathogenic and beneficial resistant microorganisms: bacteria [70, 155], fungi (*Candida* spp., *Cryptococcus neoformans*, *Pneumocystis* spp., and *Aspergillus* spp.) [208], and viruses (HIV virus, hepatitis B, hepatitis C, influenza, herpes viruses) [144, 255]. Of course, we should not neglect the potential importance of the human genome sequencing and investigation of

host–pathogen interaction for patient management and drug development. The combination of the MDR bacterial and human WGS data together with genome-wide association studies and expanding computational capacity offers new power to elucidate host immune traits and genetic factors/variants contributing/altering to susceptibility to MDR bacterial diseases in humans [28]. Such studies have been extensively published [4, 28, 65, 226].

Technical development promises portable, single-molecule, long-read, and user-friendly sequencing platforms, with high functionality and cost-effectiveness. These novel technologies will provide unprecedented opportunities for clinics and public health and may soon change our lifestyle. However, there are still many difficulties to overcome. There is a call for conceptual change of rational sampling strategies, experiment design, and data analysis management. The proper collection, processing, and storage of biological specimens are also critical. The pathway from sequencing the DNA of a specimen to a clinical treatment plan of the patient depends on the integration of each sample's genomic information with databases that contain known genotype–phenotype correlations and clinical associations obtained from large sample sets. Well curated and regularly updated databases of resistance genotype–phenotype links of MDR pathogens and computational tools to interrogate the ever-increasing information in a robust way are urgently required for MDR pathogen identification and control as well as for novel drug development. These improvements will help to solve many of the critical issues of WGS applicability for both public health and scientific purposes.

Abbreviations

AGST: antigen gene sequence typing; A-site: amino-acyl-tRNA binding site; BSI: bloodstream infection; CA-UTI: catheter-associated urinary tract infection; CF: cystic fibrosis; CI: confidence interval; DIPs: deletion/insertion polymorphisms; DR: drug resistance (or resistant); FAS: fatty acid synthesis; HAP: hospital-acquired pneumonia; IGS: individual genome sequencing; MALDI-TOF MS: matrix-assisted laser desorption ionization–time of flight mass spectrometry; MDR: multidrug resistance (or resistant); MGEs: mobile genetic elements; MIRU-VNTR: mycobacterial interspersed repetitive unit variable number tandem repeat genotyping; MLST: multilocus sequencing typing; MRSA: meticillin-resistant *Staphylococcus aureus*; MSSA: methicillin-sensitive *S. aureus*; NGS: next-generation sequencing; PBP: penicillin-binding protein; SCC: staphylococcal cassette chromosome; SNP: single-nucleotide polymorphism (a single nucleotide aberration which can be found in more than at least 1 % members of bacterial population); SNV: single-nucleotide variation (a single nucleotide aberration without any limitations of frequency e.g., was not validated for population and can be found in one individual); SSI: surgical-site infection; SV: structural variations (large genomic variations, including insertions, deletions, inversions, translocations, and duplications); TB: tuberculosis; UTI: Urinary tract infection; VAP: ventilator-associated pneumonia; WGS: whole-genome sequencing; WSI: wound stream infection.

Competing interests

The authors declare that they have no competing interests.

Authors' contributions

NVP and MAR conceived and drafted the study. NMM and AFT contributed knowledge and expertise and helped in the writings. NVP and NMM wrote

the manuscript. All authors read and approved the final version of the manuscript.

Acknowledgements

We thank V Punina and V Kadnikov for critically reading this manuscript. We are grateful for past and present support from A.N. Bach Institute of Biochemistry of Russia Academy of Science, Moscow, Russia. This work was supported by Russian Foundation for Humanities (RFH), № 15-36-01024, DOD grant PC094628, and NIH grant 8 P20 GM103518 (to NMM).

Author details

[1]Bach Institute of Biochemistry, Russian Academy of Science, Moscow 119071, Russia. [2]Tulane University School of Public Health and Tropical Medicine, New Orleans, LA 70112, USA. [3]The Federal State Unitary Enterprise All-Russia Research Institute of Automatics, Moscow 127055, Russia.

References

1. Abrahams KA, Cox JA, Spivey VL, Loman NJ, Pallen MJ, Constantinidou C, et al. Identification of novel imidazo[1,2-a]pyridine inhibitors targeting *M. tuberculosis* QcrB. PLoS One. 2012. doi:10.1371/journal.pone.0052951.
2. Adams MD, Goglin K, Molyneaux N, Hujer KM, Lavender H, Jamison JJ, et al. Comparative genome sequence analysis of multidrug-resistant *Acinetobacter baumannii*. J Bacteriol. 2008. doi:10.1128/JB.00834-08.
3. Adey A, Morrison HG, Xun X, Kitzman JO, Turner EH, Stackhouse B, et al. Rapid, low-input, low-bias construction of shotgun fragment libraries by high-density in vitro transposition. Genome Biol. 2010. doi:10.1186/gb-2010-11-12-r119.
4. Ahn S-H, Deshmukh H, Johnson N, Cowell LG, Rude TH, Scott WK, et al. Two genes on A/J chromosome 18 are associated with susceptibility to Staphylococcus aureus infection by combined microarray and QTL analyses. PLoS Pathog. 2010;6(9), e1001068. doi:10.1371/journal.ppat.1001088.
5. Albuquerque P, Mendes MV, Santos CL, Moradas-Ferreira P, Tavares FF. DNA signature-based approaches for bacterial detection and identification. Sci Total Environ. 2008;26:3641–51.
6. Allard MW, Luo Y, Strain E, Li C, Keys CE, Son I, et al. High resolution clustering of *Salmonella enterica* serovar Montevideo strains using a next-generation sequencing approach. BMC Genomics. 2012;13:32.
7. Andersson P, Klein M, Lilliebridge RA, Giffard PM. Sequences of multiple bacterial genomes and a Chlamydia trachomatis genotype from direct sequencing of DNA derived from a vaginal swab diagnostic specimen. Clin Microbiol Infect. 2013. doi:10.1111/1469-0691.12237.
8. Andries K, Verhasselt P, Guillemont J, Göhlmann HW, Neefs JM, Winkler H, et al. A diarylquinoline drug active on the ATP synthase of *Mycobacterium tuberculosis*. Science. 2005;307(5707):223–7.
9. Anthonisen IL, Sunde M, Steinum TM, Sidhu MS, Sorum H. Organization of the antiseptic resistance gene qacA and Tn552-related beta-lactamase genes in multidrug-resistant *Staphylococcus haemolyticus* strains of animal and human origins. Antimicrob Agents Chemother. 2002;46(11):3606–12.
10. Aragon TJ, Reingold A. Epidemiologic concepts for the prevention and control of infectious diseases. In Baker S, Hanage WP, Holt KE, editors. Navigating the future of bacterial molecular epidemiology. Curr Opin Microbiol. 2010;13:640–5.
11. Baker S, Holt KE, Clements AC, Karkey A, Arjyal A, Boni MF, et al. Combined high-resolution genotyping and geospatial analysis reveals modes of endemic urban typhoid fever transmission. Open Biol. 2011. doi:10.1098/rsob.110008.
12. Batra R, Cooper BS, Whiteley C, Patel AK, Wyncoll D, Edgeworth JD. Efficacy and limitation of a chlorhexidine-based decolonization strategy in preventing transmission of methicillin-resistant *Staphylococcus aureus* in an intensive care unit. Clin Infect Dis. 2010;50:210–7.
13. Bax BD, Chan PF, Eggleston DS, Fosberry A, Gentry DR, Gorrec F, et al. Type IIA topoisomerase inhibition by a new class of antibacterial agents. Nature. 2010;466:935–40.
14. Ben Zakour NL, Venturini C, Beatson SA, Walker MJ. Analysis of a *Streptococcus pyogenes* puerperal sepsis cluster by use of whole-genome sequencing. J Clin Microbiol. 2012;50(7):2224–8.
15. Beres SB, Carroll RK, Shea PR, Sitkiewicz I, Martinez-Gutierrez JC, Low DE, et al. Molecular complexity of successive bacterial epidemics deconvoluted by comparative pathogenomics. Proc Natl Acad Sci U S A. 2010;107:4371–6.

16. Berger-Bachi B, Dyke K, Gregory P. Resistance to beta-lactam antibiotics. Resistance not mediated by beta-lactamase (methicillin resistance). In: Crossley KB, Archer GL, editors. The staphylococci in human disease. New York, NY: Churchill Livingstone; 1997. p. 139–57.

17. Berger-Bachi B, Rohrer S. Factors influencing methicillin resistance in staphylococci. Arch Microbiol. 2002;178:165–71.

18. Bologa CG, Ursu O, Oprea TI, Melançon CE, Tegos GP. Emerging trends in the discovery of natural product antibacterials. Curr Opin Pharmacol. 2013;13(5):678–87.

19. Bourgogne A, Garsin DA, Qin X, Singh KV, Sillanpaa J, Yerrapragada S, et al. Large scale variation in *Enterococcus faecalis* illustrated by the genome analysis of strain OG1RF. Genome Bio. 2008. doi:10.1186/gb-2008-9-7-r110.

20. Brotz-Oesterhelt H, Sass P. Postgenomic strategies in antibacterial drug discovery. Future Microbiol. 2010;5:1553–79.

21. Brown AK, Taylor RC, Bhatt A, Futterer K, Besra GS. Platensimycin activity against mycobacterial b-ketoacyl-ACP synthases. PLoS One. 2010. doi:10.1371/journal.pone.0006306.

22. Bryant JM, Harris SR, Parkhill J, Dawson R, Diacon AH, van Helden P, et al. Whole-genome sequencing to establish relapse or re-infection with *Mycobacterium tuberculosis*: a retrospective observational study. The Lancet Respiratory Medicine. 2013;1(10):786–92.

23. Bryant JM, Schurch AC, van Deutekom H, Harris SR, de Beer JL, de Jager V, et al. Inferring patient to patient transmission of *Mycobacterium tuberculosis* from whole genome sequencing data. BMC Infect Dis. 2013;13:110.

24. Caskey CT. Using genetic diagnosis to determine individual therapeutic utility. Annu Rev Med. 2010;61:1–15.

25. Cavanagh JP, Hjerde E, Holden MT, Kahlke T, Klingenberg C, Flægstad T, et al. Whole-genome sequencing reveals clonal expansion of multiresistant *Staphylococcus haemolyticus* in European hospitals. J Antimicrob Chemother. 2014;69(11):2920–7.

26. Centers for Disease Control and Prevention, US Department of Health and Human Services. Antibiotic resistance threats in the United States, 2013. Report. 2013. http://www.cdc.gov/drugresistance/threat-report-2013/. Accessed 23 April 2013.

27. Chakraborty S, Gruber T, Barry CE, Boshoff HI, Rhee KY. Para-aminosalicylic acid acts as an alternative substrate of folate metabolism in *Mycobacterium tuberculosis*. Science. 2013;339:88–91.

28. Chapman SJ, Hill AV. Human genetic susceptibility to infectious disease. Nature Rev Genet. 2012;13:175–88.

29. Chernyaeva EN, Shulgina MV, Rotkevich MS, Dobrynin PV, Simonov SA, Shitikov EA, et al. Genome-wide *Mycobacterium tuberculosis* variation (GMTV) database: a new tool for integrating sequence variations and epidemiology. BMC Genomics. 2014;15:308.

30. Chin CS, Sorenson J, Harris JB, Robins WP, Charles RC, Jean-Charles RR, et al. The origin of the Haitian cholera outbreak strain. N Engl J Med. 2011;364(1):33–42.

31. Chiu CH, Tang P, Chu C, Hu S, Bao Q, Yu J, et al. The genome sequence of *Salmonella enterica* serovar Choleraesuis, a highly invasive and resistant zoonotic pathogen. Nucleic Acids Res. 2005;33(5):1690–8.

32. Chong PP, Podmore SM, Kieser HM, Redenbach M, Turgay K, Marahiel M, et al. Physical identification of a chromosomal locus encoding biosynthetic genes for the lipopeptide calcium-dependent antibiotic (CDA) of *Streptomyces coelicolor* A3(2). Microbiology. 1998;144(1):193–9.

33. Christen B, Abeliuk E, Collier JM, Kalogeraki VS, Passarelli B, Coller JA, et al. The essential genome of a bacterium. Mol Syst Biol. 2011;7:528.

34. Christophe T, Jackson M, Jeon HK, Fenistein D, Contreras-Dominguez M, Kim J, et al. High content screening identifies decaprenyl-phosphoribose 2′ epimerase as a target for intracellular antimycobacterial inhibitors. PLoS Pathog. 2009. doi:10.1371/journal.ppat.1000645.

35. Clark RB, He M, Fyfe C, Lofland D, O'Brien WJ, Plamondon L, et al. 8-Azatetracyclines: synthesis and evaluation of a novel class of tetracycline antibacterial agents. J Med Chem. 2011;54:1511–28.

36. Cleven BEE, Palka-Santini M, Gielen J, Meembor S, Krönke M, Krut O. Identification and characterization of bacterial pathogens causing bloodstream infections by DNA microarray. J Clin Microbiol. 2006;44(7):2389–97.

37. Cooper BS, Kypraios T, Batra R, Wyncoll D, Tosas O, Edgeworth JD. Quantifying type-specific reproduction numbers for nosocomial pathogens: evidence for heightened transmission of an Asian sequence type 239 MRSA clone. PLoS Comput Biol. 2012. doi:10.1371/journal.pcbi.1002454.

38. Craney A, Ahmed S, Nodwell J. Towards a new science of secondary metabolism. J Antibiot. 2013;66:387–400.

39. Croxatto A, Prodhom G, Greub G. Applications of MALDI-TOF mass spectrometry in clinical diagnostic microbiology. FEMS Microbiol Rev. 2012;36:380–407.

40. Cunningham SA, Sloan LM, Nyre LM, Vetter EA, Mandrekar J, Patel R. Three-hour molecular detection of *Campylobacter*, *Salmonella*, *Yersinia*, and *Shigella* species in feces with accuracy as high as that of culture. J Clin Microbiol. 2010;48:2929–33.

41. Darch SE, McNally A, Harrison F, Corander J, Barr HL, Paszkiewicz K, et al. Recombination is a key driver of genomic and phenotypic diversity in a *Pseudomonas aeruginosa* population during cystic fibrosis infection. Scientific Reports. 2015;5:7649.

42. Davies SC, Fowler T, Watson J, Livermore DM, Walker D. Annual report of the chief medical. The Lancet. 2013;381(9878):1606–9.

43. de Jong A, van Hijum SA, Bijlsma JJ, Kok J, Kuipers OP. BAGEL: a web-based bacteriocin genome mining tool. Nucleic Acids Res. 2006;34:273–9.

44. Deane CD, Mitchell DA. Lessons learned from the transformation of natural product discovery to a genome-driven endeavor. J Ind Microbiol Biotechnol. 2014;41(2):315–31.

45. Dettman JR, Rodrigue N, Aaron SD, Kassen R. Evolutionary genomics of epidemic and nonepidemic strains of *Pseudomonas aeruginosa*. Proc Natl Acad Sci U S A. 2013;110(52):21065–70.

46. Didelot X, Bowden R, Wilson DJ, Peto TE, Crook DW. Transforming clinical microbiology with bacterial genome sequencing. Nature Rev Genet. 2012;13:601–12.

47. Diene SM, Merhej V, Henry M, El Filali A, Roux V, Robert C, et al. The rhizome of the multidrug-resistant *Enterobacter aerogenes* genome reveals how new "killer bugs" are created because of a sympatric lifestyle. Mol Biol Evol. 2013;30(2):369–83.

48. Diene SM, Rolain JM. Investigation of antibiotic resistance in the genomic era of multidrug-resistant Gram-negative bacilli, especially *Enterobacteriaceae*, *Pseudomonas* and *Acinetobacter*. Expert Rev Anti Infect Ther. 2013;11:277–96.

49. Donadio S, Maffioli S, Monciardini P, Sosio M, Jabes D. Antibiotic discovery in the twenty-first century: current trends and future perspectives. J Antibiot (Tokyo). 2010;63(8):423–30.

50. Dougherty TJ, Barrett JF, Pucci MJ. Microbial genomics and novel antibiotic discovery: new technology to search for new drugs. Curr Pharm Des. 2002;8:1119–35.

51. Edwards JS, Ibarra RU, Palsson BO. *In silico* predictions of *Escherichia coli* metabolic capabilities are consistent with experimental data. Nat Biotechnol. 2001;19(2):125–30.

52. Eldholm V, Norheim G, von der Lippe B, Kinander W, Dahle UR, Caugant DA, et al. Evolution of extensively drug-resistant *Mycobacterium tuberculosis* from a susceptible ancestor in a single patient. Genome Biol. 2014;15(11):490.

53. Eyre DW, Golubchik T, Gordon NC, Bowden R, Piazza P, Batty EM, et al. A pilot study of rapid benchtop sequencing of *Staphylococcus aureus* and *Clostridium difficile* for outbreak detection and surveillance. BMJ Open. 2012. doi:10.1136/bmjopen-2012-001124.

54. Ezewudo MN, Joseph SJ, Castillo-Ramirez S, Dean D, Del Rio C, Didelot X, et al. Population structure of *Neisseria gonorrhoeae* based on whole genome data and its relationship with antibiotic resistance. PeerJ. 2015. doi:10.7717/peerj.806.

55. Falagas ME, Karageorgopoulos DE, Leptidis J, Korbila IP. MRSA in Africa: filling the global map of antimicrobial resistance. PLoS One. 2013. doi:10.1371/journal.pone.0068024.

56. Fani F, Leprohon P, Zhanel GG, Bergeron MG, Ouellette M. Genomic analyses of DNA transformation and penicillin resistance in *Streptococcus pneumoniae* clinical isolates. Antimicrob Agents Chemother. 2014;58(3):1397–403.

57. Feavers IM, Gray SJ, Urwin R, Russell JE, Bygraves JA, Kaczmarski EB, et al. Multilocus sequence typing and antigen gene sequencing in the investigation of a meningococcal disease outbreak. J Clin Microbiol. 1999;37(12):3883–7.

58. Feitelson JS, Malpartida F, Hopwood DA. Genetic and biochemical characterization of the red gene cluster of *Streptomyces coelicolor* A3(2). J Gen Microbiol. 1985;131(9):2431–41.

59. Felmingham D. The need for antimicrobial resistance surveillance. J Antimicrob Chemother. 2002;50 Suppl S1:1–7.

60. Felmingham D. Evolving resistance patterns in community-acquired respiratory tract pathogens: first results from the PROTEKT global surveillance

study. Prospective resistant organism tracking and epidemiology for the ketolide telithromycin. J Infect. 2002;44 Suppl A:3–10.

61. Feng Z, Chakraborty D, Dewell SB, Reddy BV, Brady SF. Environmental DNA-encoded antibiotics fasamycin A and B inhibit FabF in type II fatty-acid biosynthesis. J Am Chem Soc. 2012;134:29817.

62. Feng J, Lupien A, Gingras H, Wasserscheid J, Dewar K, Légaré D, et al. Genome sequencing of linezolid-resistant Streptococcus pneumoniae mutants reveals novel mechanisms of resistance. Genome Res. 2009;19:1214–23.

63. Fernandez-Alarcon C, Singer RS, Johnson TJ. Comparative genomics of multidrug resistance-encoding IncA/C plasmids from commensal and pathogenic Escherichia coli from multiple animal sources. PLoS ONE. 2011. doi:10.1371/journal.pone.0023415.

64. Fischbach MA, Walsh CT. Antibiotics for emerging pathogens. Science. 2009;325:1089–93.

65. Flores J, Okhuysen PC. Genetics of susceptibility to infection with enteric pathogens. Curr Opin Infect Dis. 2009;22(5):471–6.

66. Fluit ADC, Visser MR, Schmitz FJ. Molecular detection of antimicrobial resistance. Clin Microbiol Rev. 2011;14(4):836–71.

67. Ford CB, Lin PL, Chase MR, Shah RR, Iartchouk O, Galagan J, et al. Use of whole genome sequencing to estimate the mutation rate of Mycobacterium tuberculosis during latent infection. Nat Genet. 2011;43(5):482–6.

68. Fournier PE, Drancourt M, Colson P, Rolain JM, La Scola B, Raoult D. Modern clinical microbiology: new challenges and solutions. Nat Rev Microbiol. 2013;11(8):574–85.

69. Fournier PE, Drancourt M, Raoult D. Bacterial genome sequencing and its use in infectious diseases. Lancet Infect Dis. 2007;7:711–23.

70. Fournier PE, Dubourg G, Raoult D. Clinical detection and characterization of bacterial pathogens in the genomics era. Genome Medicine. 2014;6(11):114.

71. Fournier PE, Vallenet D, Barbe V, Audic S, Ogata H, Poirel L, et al. Comparative genomics of multidrug resistance in Acinetobacter baumannii. PLoS Genet. 2006. doi:10.1371/journal.pgen.0020007.

72. Frank C, Werber D, Cramer JP, Askar M, Faber M, der Heiden M, et al. Epidemic profile of Shiga-toxin-producing Escherichia coli O104:H4 outbreak in Germany. N Engl J Med. 2011;365(19):1771–80.

73. Fricke WF, Welch TJ, McDermott PF, Mammel MK, LeClerc JE, White DG, et al. Comparative genomics of the IncA/C multidrug resistance plasmid family. J Bacteriol. 2009;191:4750–7.

74. Fricke WF, Wright MS, Lindell AH, Harkins DM, Baker-Austin C, Ravel J, et al. Insights into the environmental resistance gene pool from the genome sequence of the multidrug-resistant environmental isolate Escherichia coli SMS-3-5. J Bacteriol. 2008;190(20):6779–94.

75. Fridkin SK, Hageman JC, Morrison M. Meticillin-resistant Staphylococcus aureus disease in three communities. N Engl J Med. 2005;352:1436–44.

76. Garcia-Alvarez L. Meticillin-resistant Staphylococcus aureus with a novel mecA homologue in human and bovine populations in the UK and Denmark: a descriptive study. Lancet Infect Dis. 2011;11:595–603.

77. Gardy JL, Johnston JC, Ho Sui SJ, Cook VJ, Shah L, Brodkin E, et al. Whole-genome sequencing and social-network analysis of a tuberculosis outbreak. N Engl J Med. 2011;364:730–9.

78. Gilmore A, Jones G, Barker M, Soltanpoor N, Stuart JM. Meningococcal disease at the University of Southampton: outbreak investigation. Epidemiol Infect. 1999;123(2):185–92.

79. Global Burden Diseases (GBD) 2013 Mortality and Causes of Death Collaborators. Global, regional, and national age-sex specific all-cause and cause-specific mortality for 240 causes of death, 1990-2013: a systematic analysis for the Global Burden of Disease Study 2013. Lancet. 2015;385(9963):117–71.

80. Gonzalez BE, Rueda AM, Shelburne SA, Musher DM, Hamill RJ, Hulten KG. Community-associated strains of meticillin-resistant Staphylococcus aureus as the cause of healthcare-associated infection. Infect Control Hosp Epidemiol. 2006;27:1051–6.

81. Gooderham WJ, Hancock REW. Regulation of virulence and antibiotic resistance by two-component regulatory systems in Pseudomonas aeruginosa. FEMS Microbiol Rev. 2009;33:279–94.

82. Gordon NC, Price JR, Cole K, Everitt R, Morgan M, Finney J, et al. Prediction of Staphylococcus aureus antimicrobial resistance by whole-genome sequencing. J Clin Microbiol. 2014;52(4):1182–91.

83. Grad YH, Kirkcaldy RD, Trees D, Dordel J, Harris SR, Goldstein E, et al. Genomic epidemiology of Neisseria gonorrhoeae with reduced susceptibility to cefixime in the USA: a retrospective observational study. Lancet Infect Dis. 2014;14(3):220–6.

84. Grad YH, Lipsitch M, Feldgarden M, Arachchi HM, Cerqueira GC, Fitzgerald M, et al. Genomic epidemiology of the Escherichia coli O104:H4 outbreaks in Europe, 2011. Proc Natl Acad Sci U S A. 2012;109(8):3065–70.

85. Green S, Studholme DJ, Laue BE, Dorati F, Lovell H, Arnold D, et al. Comparative genome analysis provides insights into the evolution and adaptation of Pseudomonas syringae pv. aesculi on Aesculus hippocastanum. PLoS One. 2010. doi:10.1371/journal.pone.0010224.

86. Grundmann H. Towards a global antibiotic resistance surveillance system: a primer for a roadmap. Ups J Med Sci. 2014;119:87–95.

87. Guard J, Morales CA, Fedorka-Cray P, Gast RK. Single nucleotide polymorphisms that differentiate two populations of Salmonella enteritidis within phage type. BMC Res Notes. 2011;4:369.

88. Guerra-Assunção JA, Houben RM, Crampin AC, Mzembe T, Mallard K, Coll F, et al. Recurrence due to relapse or reinfection with Mycobacterium tuberculosis: a whole-genome sequencing approach in a large, population-based cohort with a high HIV infection prevalence and active follow-up. J Infect Dis. 2015;211(7):1154–63.

89. Gurcha SS, Usha V, Cox JAG, Fütterer K, Abrahams KA, Bhatt A, et al. Biochemical and structural characterization of mycobacterial aspartyl-tRNA synthetase AspS, a promising TB drug target. PLoS One. 2014. doi:10.1371/journal.pone.0113568.

90. Halachev MR, Chan JZ, Constantinidou CI, Cumley N, Bradley C, Smith-Banks M, et al. Genomic epidemiology of a protracted hospital outbreak caused by multidrug-resistant Acinetobacter baumannii in Birmingham. England Genome Med. 2014;6(11):70.

91. Hall RM. Salmonella genomic islands and antibiotic resistance in Salmonella enterica. Future Microbiol. 2010;5(10):1525–38.

92. Hamburg MA, Collins FS. The path to personalized medicine. N Engl J Med. 2010;363:301–4.

93. Harris SR, Cartwright EJ, Torok ME, Holden MT, Brown NM, Ogilvy-Stuart AL, et al. Whole-genome sequencing for analysis of an outbreak of meticillin-resistant Staphylococcus aureus: a descriptive study. Lancet Infect Dis. 2013;13:130–6.

94. Harris SR, Feil EJ, Holden MT, Quail MA, Nickerson EK, Chantratita N, et al. Evolution of MRSA during hospital transmission and intercontinental spread. Science. 2011;327:469–74.

95. Harris SR, Feil EJ, Holden MT. Evolution of MRSA during hospital transmission and intercontinental spread. Science. 2010;327:469–74.

96. Harris SR, Török ME, Cartwright EJ, Quail MA, Peacock SJ, Parkhill J. Read and assembly metrics inconsequential for clinical utility of whole-genome sequencing in mapping outbreaks. Nat Biotechnol. 2013;31:592–4.

97. Hartkoorn RC, Sala C, Neres J, Pojer F, Magnet S, Mukherjee R, et al. Towards a new tuberculosis drug: pyridomycin—nature's isoniazid. EMBO Mol Med. 2012;4:1032–42.

98. Hasman H, Saputra D, Sicheritz-Ponten T, Lund O, Svendsen CA, Frimodt-Møller N, et al. Rapid whole-genome sequencing for detection and characterization of microorganisms directly from clinical samples. J Clin Microbiol. 2014;52(1):139–46.

99. Hassan M, Kjos M, Nes IF, Diep DB, Lotfipour F. Natural antimicrobial peptides from bacteria: characteristics and potential applications to fight against antibiotic resistance. J Appl Microbiol. 2012;113(4):723–36.

100. Hazen TH, Zhao L, Boutin MA, Stancil A, Robinson G, Harris AD, et al. Comparative genomics of an IncA/C multidrug resistance plasmid from Escherichia coli and Klebsiella isolates from intensive care unit patients and the utility of whole-genome sequencing in health care settings. Antimicrob Agents Chemother. 2014;58(8):4814–25.

101. Health Protection Agency. Weekly report. Potentially transferable linezolid resistance in Enterococcus faecium in Scotland. Eurosurveillance. 2012;46(33):276–83.

102. Ho CC, Yuen KY, Lau SKP, Woo PCY. Rapid identification and validation of specific molecular targets for detection of Escherichia coli O104:H4 outbreak strain by use of high-throughput sequencing data from nine genomes. J Clin Microbiol. 2011;49(10):3714–6.

103. Holden MT, Hsu LY, Kurt K, Weinert LA, Mather AE, Harris SR, et al. A genomic portrait of the emergence, evolution, and global spread of a methicillin-resistant Staphylococcus aureus pandemic. Genome Res. 2013;23:653–64.

104. Hung GC, Nagamine K, Li B, Lo SC. Identification of DNA signatures suitable for use in development of real-time PCR assays by whole-genome sequence approaches: use of Streptococcus pyogenes in a pilot study. J Clin Microbiol. 2012;50(8):2770–3.

105. Huo L, Rachid S, Stadler M, Wenzel SC, Müller R. Synthetic biotechnology to study and engineer ribosomal bottromycin biosynthesis. Chem Biol. 2012;19(10):1278–87.

106. Ibrahim EH, Sherman G, Ward S, Fraser VJ, Kollef MH. The influence of inadequate antimicrobial treatment of bloodstream infections on patient outcomes in the ICU setting. Chest. 2000;118(1):146–55.

107. Ioerger TR, O'Malley T, Liao R, Guinn KM, Hickey MJ, Mohaideen N, et al. Identification of new drug targets and resistance mechanisms in *Mycobacterium tuberculosis*. PLoS One. 2013. doi:10.1371/journal.pone.0075245.

108. Ishii Y, Eto M, Mano Y, Tateda K, Yamaguchi K. In vitro potentiation of carbapenems with ME1071, a novel metallo-β-lactamase inhibitor, against metallo-β-lactamase-producing *Pseudomonas aeruginosa* clinical isolates. Antimicrob Agents Chemother. 2010;54:3625–9.

109. Izumiya H, Sekizuka T, Nakaya H, Taguchi M, Oguchi A, Ichikawa N, et al. Whole-genome analysis of *Salmonella enterica* serovar Typhimurium T000240 reveals the acquisition of a genomic island involved in multidrug resistance via IS1 derivatives on the chromosome. Antimicrob Agents Chemother. 2011;55(2):623–30.

110. Jacoby GA. AmpC beta-lactamases. Clin Microbiol Rev. 2009;22:161–82.

111. Jain A, Awasthi A, Kumar M. Etiological and antimicrobial susceptibility profile of nosocomial blood stream infections in neonatal intensive care unit. Indian J Med Microbiol. 2007;25(3):299–300.

112. Janda JM, Abbott SL. 16S rRNA gene sequencing for bacterial identification in the diagnostic laboratory: pluses, perils, and pitfalls. J Clin Microbiol. 2007;45(9):2761–4.

113. Jeukens J, Boyle B, Kukavica-Ibrulj I, Ouellet MM, Aaron SD, Charette SJ, et al. Comparative genomics of isolates of a *Pseudomonas aeruginosa* epidemic strain associated with chronic lung infections of cystic fibrosis patients. PLoS One. 2014. doi:10.1371/journal.pone.0087611.

114. Joensen KG, Scheutz F, Lund O, Hasman H, Kaas RS, Nielsen EM, et al. Real-time whole-genome sequencing for routine typing, surveillance, and outbreak detection of verotoxigenic *Escherichia coli*. J Clin Microbiol. 2014;52(5):1501–10.

115. Jolley KA, Hill DM, Bratcher HB, Harrison OB, Feavers IM, Parkhill J, et al. Resolution of a meningococcal disease outbreak from whole-genome sequence data with rapid web-based analysis methods. J Clin Microbiol. 2012;50(9):3046–53.

116. Jolley KA, Maiden MC. BIGSdb: scalable analysis of bacterial genome variation at the population level. BMC Bioinformatics. 2010;11:595.

117. Jones K, Patel NG, Levy MA, Storeygard A, Balk D, Gittleman JL, et al. Global trends in emerging infectious diseases. Nature. 2008;451:990–3.

118. Jorgensen JH, Ferraro MJ. Antimicrobial susceptibility testing: a review of general principles and contemporary practices. Clin Infect Dis. 2009;49(11):1749–55.

119. Kallifidas D, Kang HS, Brady SF. Tetarimycin A, an MRSA-active antibiotic identified through induced expression of environmental DNA gene clusters. J Am Chem Soc. 2012;134(48):19552–5.

120. Karch H, Tarr PI, Bielaszewska M. Enterohaemorrhagic *Escherichia coli* in human medicine. Int J Med Microbiol. 2005;295:405–18.

121. Kato-Maeda M, Ho C, Passarelli B, Banaei N, Grinsdale J, Flores L, et al. Use of whole genome sequencing to determine the microevolution of *Mycobacterium tuberculosis* during an outbreak. PLoS One. 2013. doi:10.1371/journal.pone.0058235.

122. Kodali S, Galgoci A, Young K, Painter R, Silver LL, Herath KB, et al. Determination of selectivity and efficacy of fatty acid synthesis inhibitors. J Biol Chem. 2005;280:1669–77.

123. Kollef MH, Sherman G, Ward S, Fraser VJ. Inadequate antimicrobial treatment of infections: a risk factor for hospital mortality among critically Ill patients. Chest. 1999;115(2):462–74.

124. Kos VN, Déraspe M, McLaughlin RE, Whiteaker JD, Roy PH, Alm RA, et al. The resistome of *Pseudomonas aeruginosa* in relationship to phenotypic susceptibility. Antimicrob Agents Chemother. 2015;59(1):427–36.

125. Köser CU, Bryant JM, Comas I, Feuerriegel S, Niemann S, Gagneux S, et al. Comment on: characterization of the *embB* gene in *Mycobacterium tuberculosis* isolates from Barcelona and rapid detection of main mutations related to ethambutol resistance using a low-density DNA array. J Antimicrob Chemother. 2014;69(8):2298–9.

126. Köser CU, Ellington MJ, Cartwright EJ, Gillespie SH, Brown NM, Farrington M, et al. Routine use of microbial whole genome sequencing in diagnostic and public health microbiology. PLoS Pathog. 2012. doi:10.1371/journal.ppat.1002824.

127. Köser CU, Ellington MJ, Peacock SJ. Whole-genome sequencing to control antimicrobial resistance. Trends Genet. 2014;30(9):401–7.

128. Köser CU, Feuerriegel S, Summers DK, Archer JA, Niemann S. Importance of the genetic diversity within the *Mycobacterium tuberculosis* complex for the development of novel antibiotics and diagnostic tests of drug resistance. Antimicrob Agents Chemother. 2012;56(12):6080–7.

129. Köser CU, Fraser LJ, Ioannou A, Becq J, Ellington MJ, Holden MT, et al. Rapid single-colony whole-genome sequencing of bacterial pathogens. J Antimicrob Chemother. 2014;69(5):1275–81.

130. Köser CU, Holden MTG, Ellington MJ, Cartwright EJP, Brown NM, Ogilvy-Stuart AL, et al. Rapid whole-genome sequencing for investigation of a neonatal MRSA outbreak. N Engl J Med. 2012;366:2267–75.

131. Kumar S, Rizvi M, Vidhani S, Sharma VK. Changing face of septicaemia and increasing drug resistance in blood isolates. Indian J Pathol Microbiol. 2004;47(3):441–6.

132. Kumar V, Sun P, Vamathevan J, Li Y, Ingraham K, Palmer L, et al. Comparative genomics of *Klebsiella pneumoniae* strains with different antibiotic resistance profiles. Antimicrob Agents Chemother. 2011;55(9):4267–76.

133. Kyrpides NC. Fifteen years of microbial genomics: meeting the challenges and fulfilling the dream. Nature Biotech. 2009;27:627–32.

134. Kyrpides NC, Hugenholtz P, Eisen JA, Woyke T, Goker M, Parker CT, et al. Genomic encyclopedia of bacteria and Archaea: sequencing a myriad of type strains. PLoS Biol. 2014. doi:10.1371/journal.pbio.1001920.

135. Larsen MV. Internet-based solutions for analysis of next-generation sequence data. J Clin Microbiol. 2013;51(9):3162.

136. Larsen MV, Cosentino S, Rasmussen S, Friis C, Hasman H, Marvig RL, et al. Multilocus sequence typing of total-genome-sequenced bacteria. J Clin Microbiol. 2012;50:1355–61.

137. Larsen MH, Vilcheze C, Kremer L, Besra GS, Parsons L, Salfinger M, et al. Overexpression of inhA, but not kasA, confers resistance to isoniazid and ethionamide in *Mycobacterium smegmatis*, *M. bovis* BCG and *M. tuberculosis*. Mol Microbiol. 2002;46:453–66.

138. Le VT, Diep BA. Selected insights from application of whole-genome sequencing for outbreak investigations. Curr Opin Crit Care. 2013;19:432–9.

139. Lee JY, Ko KS. Mutations and expression of PmrAB and PhoPQ related with colistin resistance in *Pseudomonas aeruginosa* clinical isolates. Diagn Microbiol Infect Dis. 2014;78(3):271–6.

140. Leekitcharoenphon P, Kaas RS, Thomsen MCF, Friis C, Rasmussen S, Aarestrup FM. SnpTree–a web-server to identify and construct SNP trees from whole genome sequence data. BMC Genomics. 2012; Suppl 7. doi:10.1186/1471-2164-13-S7-S6.

141. Leekitcharoenphon P, Nielsen EM, Kaas RS, Lund O, Aarestrup FM. Evaluation of whole genome sequencing for outbreak detection of *Salmonella enterica*. PLoS One. 2014. doi:10.1371/journal.pone.0087991.

142. Lewis T, Loman NJ, Bingle L, Jumaa P, Weinstock GM, Mortiboy D, et al. High-throughput whole-genome sequencing to dissect the epidemiology of *Acinetobacter baumannii* isolates from a hospital outbreak. J Hosp Infect. 2010;75(1):37–41.

143. Li M, Du X, Villaruz AE, Diep BA, Wang D, Song Y, et al. MRSA epidemic linked to a quickly spreading colonization and virulence determinant. Nat Med. 2012;18:816–9.

144. Lipkin WI, Firth C. Viral surveillance and discovery. Curr Opin Virol. 2013;3:199–204.

145. Liu P, Li P, Jiang X, Bi D, Xie Y, Tai C, et al. Complete genome sequence of *Klebsiella pneumoniae* subsp. *pneumoniae* HS11286, a multidrug-resistant strain isolated from human sputum. J Bacteriol. 2012;194(7):1841–2.

146. Liu B, Pop M. ARDB-antibiotic resistance genes database. Nucleic Acids Res. 2009. doi:10.1093/nar/gkn656.

147. Liu A, Tran L, Becket E, Lee K, Chinn L, Park E, et al. Antibiotic sensitivity profiles determined with an *Escherichia coli* gene knockout collection: generating an antibiotic bar code. Antimicrob Agents Chemother. 2010;54:1393–403.

148. Livermore DM. Discovery research: the scientific challenge of finding new antibiotics. J Antimicrob Chemother. 2011;66(9):1941–4.

149. Loman NJ, Constantinidou C, Chan JZ, Halachev M, Sergeant M, Penn CW, et al. High-throughput bacterial genome sequencing: an embarrassment of choice, a world of opportunity. Nat Rev Microbiol. 2012;10:599–606.

150. Loman NJ, Constantinidou C, Christner M, Rohde H, Chan JZ, Quick J, et al. A culture-independent sequence-based metagenomics approach to the investigation of an outbreak of Shiga-toxigenic *Escherichia coli* O104:H4. JAMA. 2013;309:1502–10.

151. López-Camacho E, Gómez-Gil R, Tobes R, Manrique M, Lorenzo M, Galván B, et al. Genomic analysis of the emergence and evolution of multidrug resistance during a *Klebsiella pneumonia* outbreak including carbapenem and colistin resistance. J Antimicrob Chemother. 2014;69(3):632–6.

152. Luckey TD. Introduction to intestinal microecology. Am J Clin Nutr. 1972;25:1292–4.

153. Lupien A, Gingras H, Bergeron MG, Leprohon P, Ouellette M. Multiple mutations and increased RNA expression in tetracycline-resistant *Streptococcus pneumonia* as determined by genome-wide DNA and mRNA sequencing. J Antimicrob Chemother. 2015. doi:10.1093/jac/dkv060.

154. Malpartida F, Hopwood DA. Physical and genetic characterisation of the gene cluster for the antibiotic actinorhodin in *Streptomyces coelicolor* A3(2). Mol Gen Genet. 1986;205(1):66–73.

155. Mardanov MM, Babykin AV, Beletsky AV, Grigoriev AI, Zinchenko VV, Kadnikov VV, et al. Metagenomic analysis of the dynamic changes in the gut microbiome of the participants of the MARS-500 experiment, simulating long term space flight. Acta Nat. 2013;5(3):116–25.

156. McAdam PR, Templeton KE, Edwards GF, Holden MT, Feil EJ, Aanensen DM, et al. Molecular tracing of the emergence, adaptation, and transmission of hospital-associated methicillin-resistant *Staphylococcus aureus*. Proc Natl Acad Sci U S A. 2012;109(23):9107–12.

157. McCarthy AJ, van Wamel W, Vandendriessche S, Larsen J, Denis O, Garcia-Graells C, et al. *Staphylococcus aureus* CC398 clade associated with human-to-human transmission. Appl Environ Microbiol. 2012;78(24):8845–8.

158. McGann P, Hang J, Clifford RJ, Yang Y, Kwak YI, Kuschner RA, et al. Complete sequence of a novel 178-kilobase plasmid carrying *bla*(NDM-1) in a *Providencia stuartii* strain isolated in Afghanistan. Antimicrob Agents Chemother. 2012;56:1673–9.

159. Mellmann A, Harmsen D, Cummings CA, Zentz EB, Leopold SR, Rico A, et al. Prospective genomic characterization of the German enterohemorrhagic *Escherichia coli* O104:H4 outbreak by rapid next-generation sequencing technology. PLoS One. 2011. doi:10.1371/journal.pone.0022751.

160. Milillo M, Kwak YI, Snesrud E, Waterman PE, Lesho E, McGann P. Rapid and simultaneous detection of blaKPC and blaNDM by use of multiplex real-time PCR. J Clin Microbiol. 2013;51(4):1247–9.

161. Minnesota Department of Health. Antimicrobial susceptibilities of selected pathogens (MDH antibiogram). 2013. Accessed 4 December 2014.

162. Miró E, Segura C, Navarro F, Sorlí L, Coll P, Horcajada JP, et al. Spread of plasmids containing the *bla*(VIM-1) and *bla*(CTX-M) genes and the qnr determinant in *Enterobacter cloacae*, *Klebsiella pneumoniae* and *Klebsiella oxytoca* isolates. J Antimicrob Chemother. 2010;65(4):661–5.

163. Miyoshi-Akiyama T, Kuwahara T, Tada T, Kitao T, Kirikae T. Complete genome sequence of highly multidrug-resistant *Pseudomonas aeruginosa* NCGM2.S1, a representative strain of a cluster endemic to Japan. J Bacteriol. 2011;193(24):7010.

164. Mobegi FM, van Hijum SA, Burghout P, Bootsma HJ, de Vries SP, van der Gaast-de Jongh CE, et al. From microbial gene essentiality to novel antimicrobial drug targets. BMC Genomics. 2014;15(1):958.

165. Mowat E, Paterson S, Fothergill JL, Wright EA, Ledson MJ, Walshaw MJ, et al. *Pseudomonas aeruginosa* population diversity and turnover in cystic fibrosis chronic infections. Am J Res Crit Care Med. 2011;183:1674–9.

166. Mugnier PD, Poirel L, Nordmann P. Functional analysis of insertion sequence IS*Aba1*, responsible for genomic plasticity of *Acinetobacter baumannii*. J Bacteriol. 2009;191:2414–8.

167. Muller B, Borrell S, Rose G, Gagneux S. The heterogeneous evolution of multidrug resistant *Mycobacterium tuberculosis*. Trends Genet. 2013;29:160–9.

168. Mutreja A, Kim DW, Thomson NR, Connor TR, Lee JH, Kariuki S, et al. Evidence for several waves of global transmission in the seventh cholera pandemic. Nature. 2011;477(7365):462–5.

169. Nair D. Whole-genome sequencing and infectious disease: a novel application of sequencing technology. Genetic testing and molecular biomarkers. 2013;17(10):719–20.

170. Custodio HT. Hospital Acquired Infections. 2014. http://emedicine.medscape.com/article/967022-overview. Accessed 14 July 2015.

171. Nickerson EK, Hongsuwan M, Limmathurotsakul D, Wuthiekanun V, Shah KR, Srisomang P, et al. *Staphylococcus aureus* bacteraemia in a tropical setting: patient outcome and impact of antibiotic resistance. PLoS One. 2009. doi:10.1371/journal.pone.0004308.

172. Nickerson EK, West TE, Day NP, Peacock SJ. *Staphylococcus aureus* disease and drug resistance in resource-limited countries in South and East Asia. Lancet Infect Dis. 2009;9:130–5.

173. Nigro SJ, Farrugia DN, Paulsen IT, Hall RM. A novel family of genomic resistance islands, AbGRI2, contributing to aminoglycoside resistance in *Acinetobacter baumannii* isolates belonging to global clone 2. J Antimicrob Chemother. 2013;68(3):554–7.

174. Octavia S, Wang Q, Tanaka MM, Kaur S, Sintchenko V, Lan R. Delineating community outbreaks of *Salmonella enterica* serovar Typhimurium using whole genome sequencing: insights into genomic variability within an outbreak. J Clin Microbiol. 2015;53(4):1063–71.

175. Ogunremi D, Devenish J, Amoako K, Kelly H, Dupras AA, Belanger S, et al. High resolution assembly and characterization of genomes of Canadian isolates of *Salmonella Enteritidis*. BMC Genomics. 2014;15(1):713.

176. Ogunremi D, Kelly H, Dupras AA, Belanger S, Devenish J. Development of a new molecular subtyping tool for *Salmonella enterica* serovar Enteritidis based on single nucleotide polymorphism genotyping using PCR. J Clin Microbiol. 2014;52(12):4275–85.

177. Okoro CK, Kingsley RA, Connor TR, Harris SR, Parry CM, Al-Mashhadani MN, et al. Intracontinental spread of human invasive *Salmonella Typhimurium* pathovariants in sub-Saharan Africa. Nat Genet. 2012;44:1215–21.

178. Okoro CK, Kingsley RA, Quail MA, Kankwatira AM, Feasey NA, Belanger S, et al. High-resolution single nucleotide polymorphism analysis distinguishes recrudescence and reinfection in recurrent invasive nontyphoidal *Salmonella* Typhimurium disease. Clin Infect Dis. 2012;54:955–63.

179. Otter JA, Patel A, Cliff PR, Halligan EP, Tosas O, Edgeworth JD. Selection for qacA carriage in CC22, but not CC30, methicillin-resistant *Staphylococcus aureus* bloodstream infection isolates during a successful institutional infection control programme. J Antimicrob Chemother. 2013;68:992–9.

180. Pace NR. Mapping the tree of life: progress and prospects. Microbiol Mol Biol Rev. 2009;73:565–76.

181. Pagani I, Liolios K, Jansson J, Chen IM, Smirnova T, Nosrat B, et al. The Genomes OnLine Database (GOLD) v.4: status of genomic and metagenomic projects and their associated meta-data. Nucleic Acids Res. 2012. doi:10.1093/nar/gkr1100.

182. Palmer AC, Kishony R. Understanding, predicting and manipulating the genotypic evolution of antibiotic resistance. Nat Rev Genetics. 2013;14:243–8.

183. Palomino JC, Martin A. TMC207 becomes bedaquiline, a new anti-TB drug. Future Microbiol. 2013;8(9):1071–80.

184. Paterson GK, Harrison EM, Holmes MA. The emergence of *mecC* methicillin-resistant *Staphylococcus aureus*. Trends Microbiol. 2014;22(1):42–7.

185. Payne DJ, Gwynn MN, Holmes DJ, Pompliano DL. Drugs for bad bugs: confronting the challenges of antibacterial discovery. Nat Rev Drug Discov. 2007;6:29–40.

186. Perez-Lago L, Comas I, Navarro Y, González-Candelas F, Herranz M, Bouza E, et al. Whole genome sequencing analysis of intrapatient microevolution in *Mycobacterium tuberculosis*: potential impact on the inference of tuberculosis transmission. J Infect Dis. 2014;209:98–108.

187. Pethe K, Bifani P, Jang J, Kang S, Park S, Ahn S, et al. Discovery of Q203, a potent clinical candidate for the treatment of tuberculosis. Nat Med. 2013;19:1157–60.

188. Petty NK, Ben Zakour NL, Stanton-Cook M, Skippington E, Totsika M, Forde BM, et al. Global dissemination of a multidrug resistant *Escherichia coli* clone. Proc Natl Acad Sci U S A. 2014;111(15):5694–9.

189. Podschun R, Ullmann U. *Klebsiella* spp. as nosocomial pathogens: epidemiology, taxonomy, typing methods, and pathogenicity factors. Clin Microbiol Rev. 1998;11:589–603.

190. Poirel L, Jayol A, Bontron S, Villegas MV, Ozdamar M, Türkoglu S, et al. The *mgrB* gene as a key target for acquired resistance to colistin in *Klebsiella pneumoniae*. J Antimicrob Chemother. 2015;70(1):75–80.

191. Pontali E, Matteelli A, Migliori GB. Drug-resistant tuberculosis. Curr Opin Pulm Med. 2013;19(3):266–72.

192. Price LB, Stegger M, Hasman H, Aziz M, Larsen J, Andersen PS, et al. *Staphylococcus aureus* CC398: host adaptation and emergence of methicillin resistance in livestock. MBio. 2012. doi:10.1128/mBio.00305-11.

193. Pritchard L, Holden NJ, Bielaszewska M, Karch H, Toth IK. Alignment-free design of highly discriminatory diagnostic primer sets for *Escherichia coli* O104:H4 outbreak strains. PLoS One. 2012. doi:10.1371/journal.pone.0034498.

194. Public Health England. Carbapenemase-producing Enterobacteriaceae: early detection, management and control toolkit for acute trusts. 2014. https://www.gov.uk/government/publications/carbapenemase-producing-entero-bacteriaceae-early-detection-management-and-control-toolkit-for-acute-trusts. Accessed 15 December 2014.

195. Rasko DA, Webster DR, Sahl JW, Bashir A, Boisen N, Scheutz F, et al. Origins of the E. coli strain causing an outbreak of hemolytic–uremic syndrome in Germany. N Engl J Med. 2011;365(8):709–17.

196. Read TD, Massey RC. Characterizing the genetic basis of bacterial phenotypes using genome-wide association studies: a new direction for bacteriology. Genome Med. 2014;6(11):109.

197. Remuinan MJ, Perez-Herran E, Rullas J, Alemparte C, Martinez-Hoyos M, Dow DJ, et al. Tetrahydropyrazolo[1,5-a]pyrimidine-3-carboxamide and N-benzyl-60,70-dihydrospiro[piperidine-4,4'-thieno[3,2-c]pyran] analogues

with bactericidal efficacy against *Mycobacterium tuberculosis* targeting MmpL3. PLoS One. 2013. doi:10.1371/journal.pone.0060933.

198. Reuter S, Harrison TG, Köser CU, Ellington MJ, Smith GP, Parkhill J, et al. A pilot study of rapid whole-genome sequencing for the investigation of a *Legionella* outbreak. BMJ Open. 2013;3:1–6.

199. Rieber N, Zapatka M, Lasitschka B, Jones D, Northcott P, Hutter B, et al. Coverage bias and sensitivity of variant calling for four whole-genome sequencing technologies. PLoS One. 2013. doi:10.1371/journal.pone.0066621.

200. Riley DR, Sieber KB, Robinson KM, White JR, Ganesan A, Nourbakhsh S, et al. Bacteria-human somatic cell lateral gene transfer is enriched in cancer samples. PLoS Comput Biol. 2013. doi:10.1371/journal.pcbi.1003107.

201. Robinson ER, Walker TM, Pallen MJ. Genomics and outbreak investigation: from sequence to consequence. Genome Medicine. 2013;5(4):36.

202. Roetzer A, Diel R, Kohl TA, Rückert C, Nübel U, Blom J, et al. Whole genome sequencing versus traditional genotyping for investigation of a Mycobacterium tuberculosis outbreak: a longitudinal molecular epidemiological study. PLoS Med. 2013. doi:10.1371/journal.pmed.1001387.

203. Rohde H, Qin J, Cui Y, Li D, Loman NJ, Hentschke M, et al. Open-source genomic analysis of Shiga-toxin-producing *E. coli* O104:H4. N Engl J Med. 2011;365(8):718–24.

204. Rolain JM, Diene SM, Kempf M, Gimenez G, Robert C, Raoult D. Real-time sequencing to decipher the molecular mechanism of resistance of a clinical pan-drug-resistant *Acinetobacter baumannii* isolate from Marseille, France. Antimicrob Agents Chemother. 2013;57(1):592–6.

205. Sabat AJ. Detection of new methicillin-resistant *Staphylococcus aureus* strains that carry a novel genetic homologue and important virulence determinants. J Clin Microbiol. 2012;50:3374–7.

206. Sabat AJ, Budimir A, Nashev D, Sá-Leão R, van Dijl JM, Laurent F, et al. On behalf of the ESCMID Study Group of Epidemiological Markers (ESGEM). Overview of molecular typing methods for outbreak detection and epidemiological surveillance. Euro Surveill. 2013;18(4):20380.

207. Satoshi O, Yoko T, Yong-Pil K, Hideaki H, Hiroshi K, Makoto S, et al. New KB-3346-5 substance and method for producing the same. Japanese patent JP2009046404. 2009.

208. Scazzocchio C. Fungal biology in the post-genomic era. Fungal Biology and Biotechnology. 2014;1(1):7.

209. Schürch AC, Kremer K, Daviena O, Kiers A, Boeree MJ, Siezen RJ, et al. High-resolution typing by integration of genome sequencing data in a large tuberculosis cluster. J Clin Microbiol. 2010;48(9):3403–6.

210. Sekizuka T, Matsui M, Yamane K, Takeuchi F, Ohnishi M, Hishinuma A, et al. Complete sequencing of the bla(NDM-1)-positive IncA/C plasmid from *Escherichia coli* ST38 isolate suggests a possible origin from plant pathogens. PLoS One. 2011. doi:10.1371/journal.pone.0025334.

211. Seth-Smith HM, Harris SR, Skilton RJ, Radebe FM, Golparian D, Shipitsyna E, et al. Whole-genome sequences of *Chlamydia trachomatis* directly from clinical samples without culture. Genome Res. 2013;23(5):855–66.

212. Shahada F, Sekizuka T, Kuroda M, Kusumoto M, Ohishi D, Matsumoto A, et al. Characterization of *Salmonella enterica* serovar typhimurium isolates harboring a chromosomally encoded CMY-2 β-lactamase gene located on a multidrug resistance genomic island. Antimicrob Agents Chemother. 2011;55(9):4114–21.

213. Shepheard MA, Fleming VM, Connor TR, Corander J, Feil EJ, Fraser C, et al. Historical zoonoses and other changes in host tropism of *Staphylococcus aureus*, identified by phylogenetic analysis of a population dataset. PLoS One. 2013. doi:10.1371/journal.pone.0062369.

214. Silver LL. Challenges of antibacterial discovery. Clin Microbiol Rev. 2011;24(1):71–109.

215. Snitkin ES, Zelazny AM, Montero CI, Stock F, Mijares L. Genome-wide recombination drives diversification of epidemic strains of Acinetobacter baumannii. Proc Natl Acad Sci U S A. 2011;108(33):13758–63.

216. Snitkin ES, Zelazny AM, Thomas PJ, Stock F. Tracking a hospital outbreak of carbapenem-resistant *Klebsiella pneumoniae* with whole-genome sequencing. Sci Transl Med. 2012. doi:10.1126/scitranslmed.3004129.

217. Solomon SL, Oliver KB. Antibiotic resistance threats in the United States: stepping back from the brink. Am Fam Physician. 2014;89(12):938–41.

218. Stoesser N, Batty EM, Eyre DW, Morgan M, Wyllie DH, Del Ojo EC, et al. Predicting antimicrobial susceptibilities for *Escherichia coli* and *Klebsiella pneumoniae* isolates using whole genomic sequence data. J Antimicrob Chemother. 2013;68(10):2234–44.

219. Sundsfjord A, Simonsen GS, Haldorsen BC, Haaheim H, Hjelmevoll SO, Littauer P, et al. Genetic methods for detection of antimicrobial resistance. APMIS. 2004;112:815–37.

220. Swinney DC, Anthony J. How were new medicines discovered? Nat Rev Drug Discov. 2011;10:507–19.

221. Sydnor ERM, Perl TM. Hospital epidemiology and infection control in acute-care settings. Clin Microbiol Rev. 2011;24(1):141–73.

222. Tang YW, Procop GW, Persing DH. Molecular diagnostics of infectious diseases. Clin Chem. 1997;43(11):2021–38.

223. Tenover FC, Canton R, Kop J, Chan R, Ryan J, Weir F, et al. Detection of colonization by carbapenemase-producing Gram-negative Bacilli in patients by use of the Xpert MDRO assay. J Clin Microbiol. 2013;51(11):3780–7.

224. The Regional Committee for the Eastern Mediterranean Region. Annual report of the Regional Director for 2012 and progress reports. Resolution EM/RC60/R.1. 2013. http://www.emro.who.int/about-who/rc60/index.html. Accessed 20 December 2014.

225. Thomas CM, Nielsen KM. Mechanisms of, and barriers to, horizontal gene transfer between bacteria. Nat Rev Microbiol. 2005;3(9):711–21.

226. Thye T, Vannberg FO, Wong SH, Owusu-Dabo E, Osei I, Gyapong J, et al. Genome-wide association analyses identifies a susceptibility locus for tuberculosis on chromosome 18q11.2. Nature Genet. 2010;42:739–41.

227. Tong SY, Holden MT, Nickerson EK, Cooper BS, Köser CU, Cori A, et al. Genome sequencing defines phylogeny and spread of methicillin-resistant *Staphylococcus aureus* in a high transmission setting. Genome Res. 2015;25(1):111–8.

228. Török ME, Harris SR, Cartwright EJ, Raven KE, Brown NM, Allison ME, et al. Zero tolerance for healthcare-associated MRSA bacteraemia: is it realistic? J Antimicrob Chemother. 2014;69:2238–45.

229. Uchiyama I, Mihara M, Nishide H, Chiba H. MBGD update 2015: microbial genome database for flexible ortholog analysis utilizing a diverse set of genomic data. Nucl Acids Res. 2014. doi:10.1093/nar/gku1152.

230. Uhlemann AC, Dumortier C, Hafer C, Taylor BS, Sánchez J, Rodriguez-Taveras C, et al. Molecular characterization of *Staphylococcus aureus* from outpatients in the Caribbean reveals the presence of pandemic clones. Eur J Clin Microbiol Infect Dis. 2011;31:505–11.

231. Uhlemann AC, Porcella SF, Trivedi S, Sullivan SB, Hafer C, Kennedy AD, et al. Identification of a highly transmissible animal-independent *Staphylococcus aureus* ST398 clone with distinct genomic and cell adhesion properties. MBio. 2012. doi:10.1128/mBio.00027-12.

232. UK Department of Health. UK five year antimicrobial resistance strategy 2013 to 2018. 2013. https://www.gov.uk/government/publications/uk-5-year-antimicrobial-resistance-strategy-2013-to-2018. Accessed 15 December 2014.

233. Underwood AP, Dallman T, Thomson NR, Williams M, Harker K, Perry N, et al. Public health value of next-generation DNA sequencing of enterohemorrhagic *Escherichia coli* isolates from an outbreak. J Clin Microbiol. 2013;51(1):232–7.

234. Usha V, Gurcha SS, Lovering AL, Lloyd AJ, Papaemmanouil A, Reynolds RC, et al. Identification of novel diphenyl urea inhibitors of Mt-GuaB2 active against *Mycobacterium tuberculosis*. Microbiology. 2011;157:290–9.

235. van Belkum A. Rapid clinical bacteriology and its future impact. Ann Lab Med. 2013;33:14–27.

236. van der Westhuyzen R, Hammons JC, Meier JL, Dahesh S, Moolman WJ, Pelly SC, et al. The antibiotic CJ-15,801 is an antimetabolite that hijacks and then inhibits CoA biosynthesis. Chem Biol. 2012;19(5):559–71.

237. van Nood E, Vrieze A, Nieuwdorp M, Fuentes S, Zoetendal EG, de Vos WM, et al. Duodenal infusion of donor feces for recurrent *Clostridium difficile*. N Engl J Med. 2013;368:407–15.

238. van Opijnen T, Bodi KL, Camilli A. Tn-seq: high-throughput parallel sequencing for fitness and genetic interaction studies in microorganisms. Nat Methods. 2009;6(10):767–72.

239. Vernet G, Saha S, Satzke C, Burgess DH, Alderson M, Maisonneuve JF, et al. Laboratory-based diagnosis of pneumococcal pneumonia: state of the art and unmet needs. Clin Microbiol Infect. 2011;17 Suppl 3:1–13.

240. Vidovic S, Caron C, Taheri A, Thakur SD, Read TD, Kusalik A, et al. Using crude whole-genome assemblies of *Neisseria gonorrhoeae* as a platform for strain analysis: clonal spread of gonorrhea infection in Saskatchewan, Canada. J Clin Microbiol. 2014;52(10):3772–6.

241. Vogel V, Falquet L, Calderon-Copete SP, Basset P, Blanc DS. Short term evolution of a highly transmissible methicillin-resistant *Staphylococcus*

aureus clone (ST228) in a tertiary care hospital. PLoS One. 2012. doi:10.1371/journal.pone.0038969.

242. Walker TM, Ip CL, Harrell RH, Evans JT, Kapatai G, Dedicoat MJ, et al. Whole-genome sequencing to delineate *Mycobacterium tuberculosis* outbreaks: a retrospective observational study. Lancet Infect Dis. 2013;13:137–46.

243. Walsh C. Where will new antibiotics come from? Nat Rev Microbiol. 2003;1(1):65–70.

244. Wang J, Kodali S, Lee SH, Galgoci A, Painter R, Dorso K, et al. Discovery of platencin, a dual FabF and FabH inhibitor with in vivo antibiotic properties. Proc Natl Acad Sci. 2007;104:7612–6.

245. Ward MJ, Gibbons CL, McAdam PR, van Bunnik BA, Girvan EK, Edwards GF, et al. Time-scaled evolutionary analysis of the transmission and antibiotic resistance dynamics of *Staphylococcus aureus* CC398. Appl Environ Microbiol. 2014. doi:10.1128/AEM.01777-14.

246. Wilson R, Kumar P, Parashar V, Vilcheze C, Veyron-Churlet R, Freundlich JS, et al. Antituberculosis thiophenes define a requirement for Pks13 in mycolic acid biosynthesis. Nat Chem Biol. 2013;9:499–506.

247. Witney AA, Gould KA, Pope CF, Bolt F, Stoker NG, Cubbon MD, et al. Genome sequencing and characterization of an extensively drug-resistant sequence type 111 serotype O12 hospital outbreak strain of *Pseudomonas aeruginosa*. Clin Microbiol Infect. 2014;20(10):609–18.

248. Worby CJ, Lipsitch M, Hanage WP. Within-host bacterial diversity hinders accurate reconstruction of transmission networks from genomic distance data. PLoS Comput Biol. 2014. doi:10.1371/journal.pcbi.1003549.

249. World Health Organization (WHO). The evolving threat of antimicrobial resistance: options for action. 2012. http://www.who.int/patientsafety/implementation/amr/publication/en/. Accessed 1 April 2015.

250. World Health Organization (WHO). The global burden of disease (GBD). 2014. http://www.who.int/healthinfo/global_burden_disease/GBD_report_2004update_full.pdf Accessed 4 December 2014.

251. World Health Organization (WHO). Antimicrobial resistance: global report on surveillance. 2014. http://www.who.int/drugresistance/documents/surveillancereport/en/. Accessed 5 December 2014.

252. World Health Organization (WHO). Global Health Observatory (GHO) data. World Health Statistics (full report). 2014. http://www.who.int/gho/publications/world_health_statistics/2014/en/. Accessed 15 April 2015.

253. Wright GD. The antibiotic resistome: the nexus of chemical and genetic diversity. Nat Rev Microbiol. 2007;5(3):175–86.

254. Wright MS, Haft DH, Harkins DM, Perez F, Hujer KM, Bajaksouzian S, et al. New insights into dissemination and variation of the health care-associated pathogen *Acinetobacter baumannii* from genomic analysis. MBio. 2014. doi:10.1128/mBio.00963-13.

255. Wylie KM, Weinstock GM, Storch GA. Virome genomics: a tool for defining the human virome. Curr Opin Microbiol. 2013;16:479–84.

256. Wyres KL, Conway TC, Garg S, Queiroz C, Reumann M, Holt K, et al. WGS analysis and interpretation in clinical and public health microbiology laboratories: what are the requirements and how do existing tools compare? Pathogens. 2014;3:437–58.

257. Ye C, Lan R, Xia S, Zhang J, Sun Q, Zhang S, et al. Emergence of a new multidrug-resistant serotype X variant in an epidemic clone of *Shigella flexneri*. J Clin Microbiol. 2010;48(2):419–26.

258. Young BC, Golubchik T, Batty EM, Fung R, Larner-Svensson H, Votintseva AA, et al. Evolutionary dynamics of *Staphylococcus aureus* during progression from carriage to disease. Proc Natl Acad Sci U S A. 2012;109:4550–5.

259. Young K, Jayasuriya H, Ondeyka JG, Herath K, Zhang CW, Kodali S, et al. Discovery of FabH/FabF inhibitors from natural products. Antimicrob Agents Chemother. 2006;50:519–26.

260. Zankari E, Hasman H, Cosentino S, Vestergaard M, Rasmussen S, Lund O, et al. Identification of acquired antimicrobial resistance genes. J Antimicrob Chemother. 2012;67:2640–4.

261. Zankari E, Hasman H, Kaas RS, Seyfarth AM, Agersø Y, Lund O, et al. Genotyping using whole-genome sequencing is a realistic alternative to surveillance based on phenotypic antimicrobial susceptibility testing. J Antimicrob Chemother. 2013;68(4):771–7.

262. Zavascki AP, Carvalhaes CG, Picão RC, Gales AC. Multidrug-resistant *Pseudomonas aeruginosa* and *Acinetobacter baumannii*: resistance mechanisms and implications for therapy. Expert Rev Anti Infect Ther. 2010;8(1):71–93.

263. Zhao J, Grant SF. Advances in whole genome sequencing technology. Curr Pharm Biotechnol. 2011;12(2):293–305.

264. Zhao S, Prenger K, Smith L, Messina T, Fan H, Jaeger E, et al. Rainbow: a tool for large-scale whole-genome sequencing data analysis using cloud computing. BMC Genomics. 2013;14:425.

265. Zhou H, Zhang T, Yu D, Pi B, Yang Q, Zhou J, et al. Genomic analysis of the multidrug-resistant *Acinetobacter baumannii* strain MDR-ZJ06 widely spread in China. Antimicrob Agents Chemother. 2011;55(10):4506–12.

Performance evaluation of indel calling tools using real short-read data

Mohammad Shabbir Hasan[1] (ID), Xiaowei Wu[2] and Liqing Zhang[1*]

Abstract

Background: Insertion and deletion (indel), a common form of genetic variation, has been shown to cause or contribute to human genetic diseases and cancer. With the advance of next-generation sequencing technology, many indel calling tools have been developed; however, evaluation and comparison of these tools using large-scale real data are still scant. Here we evaluated seven popular and publicly available indel calling tools, GATK Unified Genotyper, VarScan, Pindel, SAMtools, Dindel, GTAK HaplotypeCaller, and Platypus, using 78 human genome low-coverage data from the 1000 Genomes project.

Results: Comparing indels called by these tools with a known set of indels, we found that Platypus outperforms other tools. In addition, a high percentage of known indels still remain undetected and the number of common indels called by all seven tools is very low.

Conclusion: All these findings indicate the necessity of improving the existing tools or developing new algorithms to achieve reliable and consistent indel calling results.

Keywords: Indel calling, Variant calling, HaplotypeCaller, Next-generation sequencing, Deep sequencing, Software evaluation

Introduction

Insertion and deletion (indel), is a common form of polymorphism corresponding to the addition or removal of base pairs in the DNA sequence of an organism. Indels have been recognized as the second most abundant source of genetic variation in human populations [1–3]. Studies have shown that in the human body, 16 to 25 % of all sequence polymorphisms are indels [4]. Furthermore, indels have been identified to play a key role in causing diseases. For example, cystic fibrosis, a common genetic disease, is frequently caused by deletion of three nucleotides in the coding region of the CFTR gene [5]. Diseases such as fragile X syndrome [6], trinucleotide repeat disorders [7], Mendelian disorders [8], Bloom syndrome [9], acute myeloid leukemia [10–12], and lung cancer [13] are often caused by short repeats/insertions in the DNA sequence. Moreover, insertion of transposable elements such as Alu, L1, and SVA can interrupt gene function and cause diseases like hemophilia, neurofibromatosis, muscular dystrophy, and cancer [14]. In addition, indels can also change gene expression by altering phasing and spacing of DNA sequences in the promoter regions [15]. For example, a small insertion of 5 bps can rotate the binding site to the opposite face of the DNA helix, whereas a long insertion of 100 bps can increase the spacing between two binding sites [3]. Therefore, indels in the promoter regions might explain certain difference in gene expression observed in humans [15] and can be used as genetic markers in natural populations [16]. Since indels influence human traits and diseases, detection of indels in a reliable manner is a prerequisite to develop effective treatment and medicine [17, 18].

In recent time, next-generation sequencing (NGS) has become more convenient because of its high efficiency, improved sensitivity of different sequencing platforms, and reduced cost as compared to Sanger sequencing [19, 20]. By applying NGS in a large scale, whole genome sequencing (WGS) is now possible at an individual level [21–23] and it has revealed a significant number of structural variants that were not reported previously. Since indels can alter human traits and cause diseases, the result of indel calling from individual WGS can be used to predict the future health of

* Correspondence: lqzhang@vt.edu
[1]Department of Computer Science, Virginia Tech, Blacksburg, VA 24061, USA
Full list of author information is available at the end of the article

sampled individuals and to develop customized medical treatments.

A good number of indel calling tools have been developed so far that can be divided into four major categories: alignment-based methods, split read mapping methods, paired-end read mapping methods, and haplotype-based methods. Alignment-based methods firstly map the reads to the reference sequence using read mapping software such as BWA [24] and Novoalign [25], and then call indels using the alignment data by applying some filtering steps to separate true indels from common sequence alignment errors (Fig. 1 (A)). In Fig. 1 (A), "True Call" refers to the indels that passed after the filters are applied to separate indels from sequence alignment errors. Therefore, "False Calls" are those variants which are probably not indels but caused due to the alignment errors. Many indel calling tools belong to this category including Dindel [26], Stampy [27], SAMtools [28], Genome Analysis Tool Kit (GATK Unified Genotyper) [29], and VarScan [30, 31]. The main difference among these tools is in the model they use to distinguish true indel calls from alignment errors. Some use the Bayesian probabilistic model (GATK Unified Genotyper, SAMtools, and Dindel), whereas others (VarScan) use the heuristic approach. Split read mapping methods, on the other hand, firstly identify discordant paired-end reads for which one end maps completely to the reference sequence and the other end does not. The unmapped ends of these reads are then clustered or aligned by de novo assembly to determine indels (Fig. 1 (B)). Tools in this category include Pindel [32] that uses a pattern growth approach to detect breakpoints of indels, and SV-M [33] that performs a discriminative classification based on features of split read alignment profiles and then filters the result against empirically derived training set data to reduce the false-positive rate. Paired-end read mapping methods compare the expected distance to the actual mapped distance to determine whether there is any indel in the sequence (Fig. 1 (C)). Tools belonging to this category include PEMer [34], Hydra [35], and BreakDancer [36]. Haplotype-based methods first identify the regions of interest where the reads show substantial evidence of having indels relative to the reference sequence. These regions are also known as active regions. For each active region, the callers build a De Bruijn graph to reassemble the active regions and yield the possible haplotypes present in the reads. After that, each read is realigned to the possible haplotypes and the likelihood of the haplotypes are calculated given the read data. Later, Baye's rule or EM algorithms are applied to calculate the posterior probabilities, and indels are called where the posterior probability exceeds a certain threshold value. In addition to that, some other filters are also applied to produce a fine-grained result. GATK HaplotypeCaller [37] and Platypus [38] belong to this category. Figure 1 (D) shows the general overview of the haplotype-based indel callers.

Despite many indel calling tools, evaluation of the tools objectively, particularly using large-scale real data, is sparse. There is an evaluation of four indel (Dindel, VarScan, GATK Unified Genotyper, and SAMtools) tools done by Neuman et al.; however, it was based on simulated data [39]. Instead of repeating the same experiment, here we performed the evaluation of the tools as well as three additional tools and we use real data to get the actual insight. In this study, we investigated seven indel calling tools, GATK Unified Genotyper [29], VarScan [30], Pindel [32], SAMtools [28], Dindel [26], GATK HaplotypeCaller [37], and Platypus [38], using 78 human genome data from different populations in the 1000 Genomes project. All these tools are publicly available and are commonly used for benchmarking. Another reason for choosing these tools is that GATK Unified Genotyper, VarScan, SAMtools, Dindel, GATK HaplotypeCaller, and Platypus can deal with short indels (<50 bps), whereas Pindel can call medium to large indels ranging from 50 to 10,000 bps. Therefore, altogether they cover indels of various lengths. Among these seven tools, four of them (GATK Unified Genotyper, VarScan, SAMtools, and Dindel) fall into the alignment-based method category, one (Pindel) implements the split read mapping method, and two (GATK HaplotypeCaller and Platypus) are haplotype-based methods. We did not consider tools that are based on paired-end read mapping because in most cases, they are insensitive to small indels, making it difficult to separate small perturbations in read pair distance from the normal background variability [40]. Moreover, the exact inserted or deleted sequence cannot be known from the results of tools that belong to this category [40]. We also note that only one of the two commonly used tools (Pindel and SV-M) from the split read mapping method category was included in this study. We did not consider SV-M mainly because this tool does not use BAM file as input. As described in the README file of SV-M, the input file requires the start and end position of each chromosome along with several features corresponding to that chromosome such as the number of uniquely mapped reads (UMRs) overlapping the deletion candidate, single position variation (SPV) from split read alignment, and number of split reads supporting the same indel location. For reason of consistency and to eliminate possible factors that could bias the comparison, we decide to exclude SV-M from this study.

Methods
Tools investigated
We investigated seven indel calling tools, GATK Unified Genotyper, VarScan, Pindel, SAMtools, Dindel, GATK HaplotypeCaller, and Platypus. A brief introduction of each tool and the commands for execution are provided below.

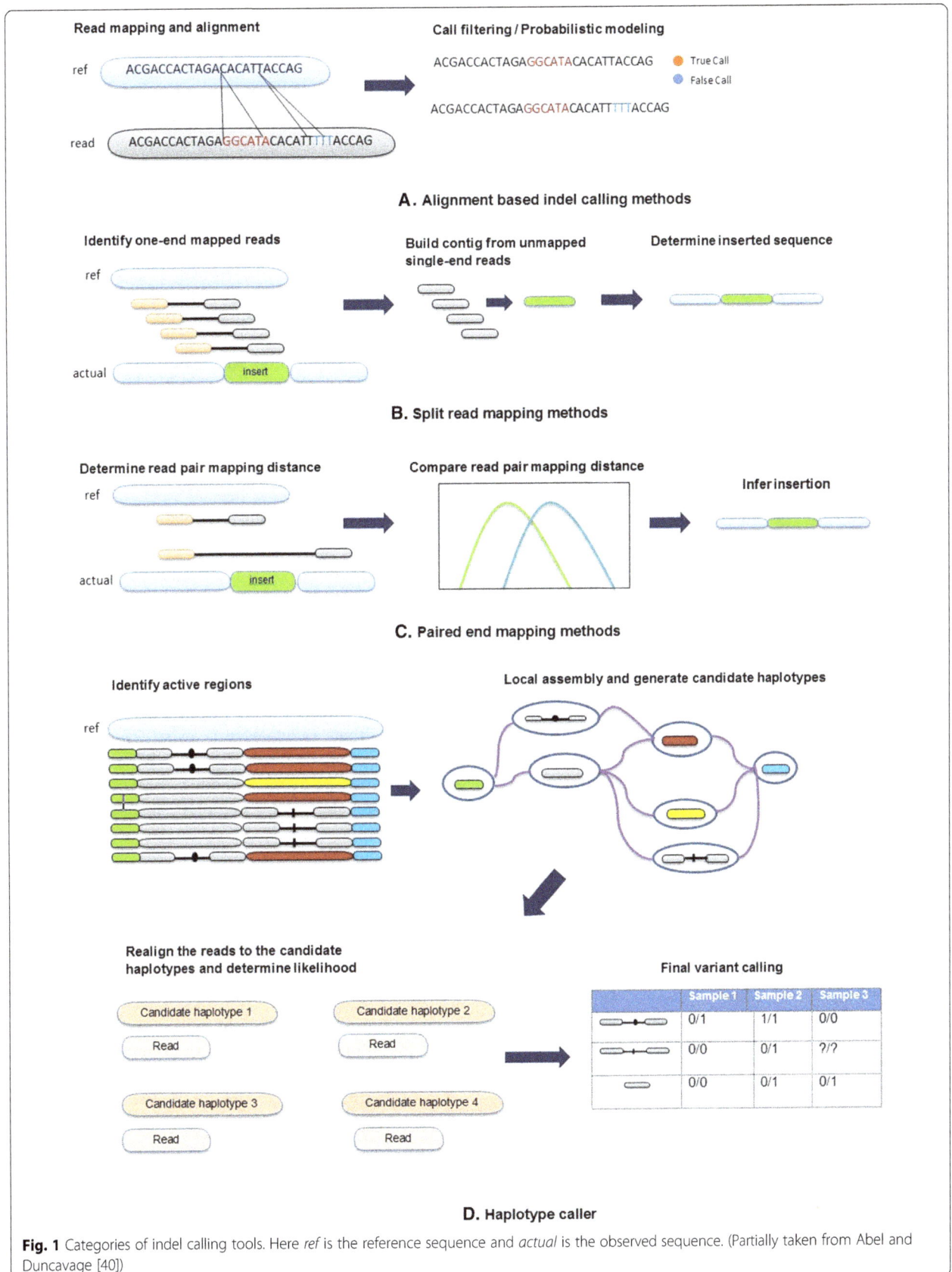

Fig. 1 Categories of indel calling tools. Here *ref* is the reference sequence and *actual* is the observed sequence. (Partially taken from Abel and Duncavage [40])

Human Molecular Genetics

GATK Unified Genotyper (GATK_UG) [29] (version 2.7) is a tool developed by the Broad Institute of MIT and Harvard. For indel calling, it incorporates realistic read mapping error and base miscall models. Using a Bayesian genotype likelihood model, GATK_UG estimates the most likely genotypes and allele frequency in the sample while emitting an accurate posterior probability of having a segregating variant allele at each locus. We called indels by GATK_UG for each sample using the following command with default settings:

```
java -jar GenomeAnalysisTK.jar -R <reference.fasta> -T
UnifiedGenotyper -I <input.bam> -glm INDEL -o <output.vcf>
```

VarScan [30] (version 2.2.2) is a platform-independent software tool developed by the Genome Institute of Washington University. It uses the *mpileup* file generated by SAMtools [28] for scoring and sorting sequence alignments. The reads mapped uniquely to one location in the reference sequence are kept, whereas the unmapped and ambiguous mapped reads are discarded. The uniquely mapped reads are further filtrated on read depth, base quality, and variant allele frequency in downstream analysis and then used to call indels by a heuristic approach. Indels were called by VarScan with its default settings using the following commands:
Generating the mpileup file using SAMtools:

```
./samtools mpileup -f <reference.fasta> <input.bam> >
out.mpileup
```

Calling indel from the mpileup file:

```
java -jar varscan.jar mpileup2indel out.mpileup --output-vcf 1
> <output.vcf>
```

Pindel [32] (version 0.2.4) is a pattern growth approach-based tool that detects breakpoints of large deletions, medium-sized insertions, and other structural variants from NGS data at single-based resolution. In Pindel, all reads are initially mapped to the reference genome. The mapping results are then inspected to select paired reads that are mapped with indels or have only one end mapped. Based on the mapped reads, Pindel determines the anchor point on the reference genome as well as the direction of unmapped reads or the reads mapped with indels. Using this information and user-defined maximum deletion size, a sub-region in the reference genome is located where the unmapped reads are broken into fragments and then the fragments are mapped separately. Pindel was executed with its default settings using the following commands:
The configuration file:

```
<input.bam>    250   <sample_name>
```

Here 250 is the insert size, i.e., the length of the region between the paired-end adapters in paired-end sequence.
Generating the output file:

```
./pindel -f <reference.fasta> -i <pindel_config.txt> -o
<pindel_output_file_name>
```

Creating the VCF file from the output file:

```
./pindel2vcf -r <reference.fasta> -R HUMAN_G1K_v37 -d <date> -
p <pindel_output_file_name_D> -e 5
```

SAMtools [28] (version 0.1.19) is a software package used for parsing and manipulating alignments in SAM/BAM format. For indel calling, it uses a Bayesian model for local realignment and base quality assessment. We called indels by SAMtools with its default settings using the following command:
Generating the mpile file:

```
java -jar GenomeAnalysisTK.jar -R <reference.fasta> -T
HaplotypeCaller -I <input.bam> -o <output.vcf>
```

Calling indel from the mpileup file:

```
./bcftools view <output.bcf> | ./vcfutils.pl varFilter -D100 >
<output.vcf>
```

Dindel [26] (version 1.01), developed by the Wellcome Trust Sanger Institute in UK, is a software tool that uses Bayesian network for calling indels from NGS data. First, a number of candidate haplotypes, each containing at least 120 bps, are generated according to the hypothesis that indel events exist in pre-specified genomic segments. After realigning all reads to the candidate haplotypes using a hidden Markov model, the posterior probability of a haplotype is calculated using the Bayesian approach and used to determine the presence of indels in the sample. Dindel assumes that all differences between the read and the candidate haplotype are caused by sequencing errors. By realigning reads to the candidate haplotype, it separates the indels from sequencing errors. Dindel uses mapping quality as the prior probability that a read should align to any of the candidate haplotypes, and thus, it effectively reduces the weight of reads that cannot be confidently mapped to that location in the genome. We used Dindel with default settings to call indels by the following commands:
Step 1: Extract candidate indels from the alignment file.

```
./dindel --analysis getCIGARindels --bamFile <input.bam> --
outputFile <dindel_output> --ref <reference.fasta>
```

Step 2: Create realignment windows.

```
./makeWindows.py --inputVarFile <dindel_output.variants.txt> -
-windowFilePrefix <dindel_output.realign_windows> --
numWindowsPerFile 1000
```

Step 3: For every window, generate candidate haplotypes from the candidate indels and realign the reads to these candidate haplotypes.

For each file created in step 2

```
./dindel --analysis indels --doDiploid --bamFile <input.bam> -
-ref <reference.fa> --varFile
<dindel_output.realign_windows.X.txt> --libFile
<dindel_output.libraries.txt> --outputFile
<dindel_output_windows.X> [Here X = window number currently
being analyzed]
```

Step 4: Create the final output.
Merging results from all realignment windows:

```
ls | grep ".glf.txt" > <list.txt>
mergeOutputDiploid.py --inputFiles <list.txt> --outputFile
<output.vcf> --ref <reference.fasta>
```

GATK HaplotypeCaller (GATK_HC) [37] (version 3.30) is a tool developed by the Broad Institute of MIT and Harvard. For indel calling, at first it determines the regions of the genome where there are significant evidences of variation. Therefore, regions that do not show any variation beyond the expected levels of background noise are skipped. After this step, the resulting regions having significant evidence of variations are passed to the next step. These regions are known as "Active Regions." For each active region, in the second step, GATK_HC builds a De Bruijn graph to reassemble the active regions and identifies the candidate haplotypes present in the reads of the given sample. Additionally, each haplotype is locally realigned to the reference haplotype to identify the potentially variant sites. In the next step, for each active region, each read is then pairwise aligned to each of the candidate haplotype using the PairHMM algorithm. This produces a matrix of likelihoods of haplotypes for the reads in the given sample. These likelihoods are then marginalized to obtain the likelihoods of the alleles per read for each potentially variant site. For each potentially variant site, in the next step, Baye's rule is applied to determine the posterior likelihoods of each genotype per sample using the likelihoods of alleles obtained in the previous step. The most likely genotype is then assigned to the given sample. We called indels using GATK_HC with default settings using the following command:

```
java -jar GenomeAnalysisTK.jar -R <reference.fasta> -T
HaplotypeCaller -I <input.bam> -o <output.vcf>
```

Platypus [38] (version 0.7.9.1) is a haplotype-based variant calling tool developed by the Wellcome Trust Sanger Institute in UK. In this tool, at the beginning, candidate variants are obtained from read alignments, local assembly, and external sources, and then candidate haplotypes are formed. After haplotypes are generated from candidate variants, their frequencies are estimated on the basis of their likelihood. These likelihoods are calculated by aligning a read to the haplotype sequence with an underlying hidden Markov model (HMM). The forward algorithm is used to calculate the likelihood of a read given haplotype. After the likelihood is calculated for all combinations of reads and haplotypes, an EM algorithm is used to estimate the frequency of each haplotype under a diploid genotype model. In the next step, the posterior support for any variant is computed by comparing the likelihood of the data given all haplotypes and the likelihood given only those haplotypes that do not include a particular variant. Later, indels are called when their posterior support exceeds a threshold using these frequencies as a prior. The variants are also filtered based on allele bias, strand bias, mapping quality, quality over depth, posterior quality, and sequence context. We called indel with the default settings of Platypus using the following command:

```
python Platypus.py callVariants --bamFiles=<input.bam> --
refFile=<reference.fasta> --output=<output.vcf>
```

Dataset

The dataset consists of low-coverage (~3X to ~12X) alignment profiles from 78 humans that belong to 26 populations and were collected for the 1000 Genomes project [41]. We used the alignment files of chromosome 11 as input for the tools we investigated. These short reads were sequenced on Illumina Genome Analyzer platform [42] and mapped using BWA [24]. We used hs37d5 as the human reference genome, which is an extended version of the Build37 dataset of the 1000 Genomes project with additional sequences. Note that this reference genome was used by the 1000 Genomes project in the final phase. Additional file 1: Table S1 lists the samples we used with their corresponding ethnic background and coverage.

Ideally, a benchmarking dataset for evaluating indel calling tools would consist of a list of known indels for the samples. However, such kind of benchmarking dataset is not available in large quantity [43]. Hence, for evaluation purpose, we used the indels identified in Mills et al. [43] as the gold standard. To call indels, Mills et al. [43] examined 98 million Applied Biosystems (Sanger) DNA re-sequencing traces from the trace archive of NCBI which has been proved to be sufficient for accurate indel calling [4]. After some pre-processing of the traces based on the quality scores, they were compared to the human reference genome to call indels. Details about the indel calling procedure and some post processing to

generate the gold standard dataset can be found in [4]. The called indels were validated using PCR-based methodologies, and the validation rate was 97.2 %. This dataset reports almost two million small and large indels found in all 24 chromosomes of 79 diverse humans with length ranging from 1 to 10,000 bps. Moreover, it has been confirmed that the sequence traces used in Mills et al. provide excellent coverage of the human genome [43]. Note that the samples we used here are sequenced on Illumina Genome Analyzer platform and the indels listed in the "gold standard" dataset are called using the Applied Biosystem (Sanger) DNA re-sequencing traces. In spite of these differences, the indels identified in the gold standard dataset are considered to be most likely reliable, and they have been used as the gold standard in other studies [44, 45]. In Mills et al. [43], 58,811 indels were identified for chromosome 11, and in the current study, we used this set as the gold standard. Note that we did not use simulated data for benchmarking because though simulated data are valuable, they do not always represent the actual phenomena. We could also use the sample benchmark dataset available in "Genome in a Bottle Consortium" [46], but that one relies on a single dataset from one human only (NA12878).

Evaluation criteria

We evaluated the tools using the criteria including running time, number of indels called, comparison with the set of gold standard indels, similarity among the tools, hierarchical clustering, and ranking of the tools.

For each sample, we executed the tools and recorded the number of indels called by each tool as well as the running time. To see the relation between running time and coverage of the read, besides the low-coverage samples, we also included the sample NA12878 with ~64X coverage. All analyses were done on a Linux machine with Intel Core i7-2600 CPU @

3.40 GHz * 8 processors, 16 GB RAM and Ubuntu 12.04 LTS operating system.

Indels called by the seven tools were compared with those identified in Mills et al. [43]. From this comparison, we calculated the corresponding recall and precision for each of the tools using formulas (1) and (2).

$$\text{Recall} = \frac{\text{TP}}{\text{TP} + \text{FN}} \qquad (1)$$

$$\text{Precision} = \frac{\text{TP}}{\text{TP} + \text{FP}} \qquad (2)$$

For comparing the accuracy of the tools, we used F-measure, the harmonic mean of the precision and recall, where an F-measure reaches its best value at 1 and worst score at 0. The F-measure was calculated using formula (3).

$$\text{F-measure} = \frac{2 \times \text{Recall} \times \text{Precision}}{\text{Recall} + \text{Precision}} \qquad (3)$$

Note that the position of an indel with respect to the reference sequence sometimes cannot be defined unambiguously by a single coordinate [20, 47]. As shown in Fig. 2, the insertion of a guanine into the local sequence of $T_i G_{i+1} G_{i+2} C_{i+3}$ after position i produces the same mutated sequence as inserting guanine after position $i + 1$ or $i + 2$. Hence, these insertions have identical biological meaning, and therefore, an unambiguous annotation for this insertion should list all equivalent indel positions, i.e., $+G \{i, i+1, i+2\}$ [20]. For this reason, while comparing an indel called by each tool with the indel in position i in the gold standard data, we treated the indel called by the tool as true positive if it is within the range of $i \pm 5$ positions.

Based on the indel calling results, these tools were ranked in the receiver operating characteristic (ROC) space [48], where the X and Y axes are denoted by false-positive rate (FPR) and true-positive rate (TPR),

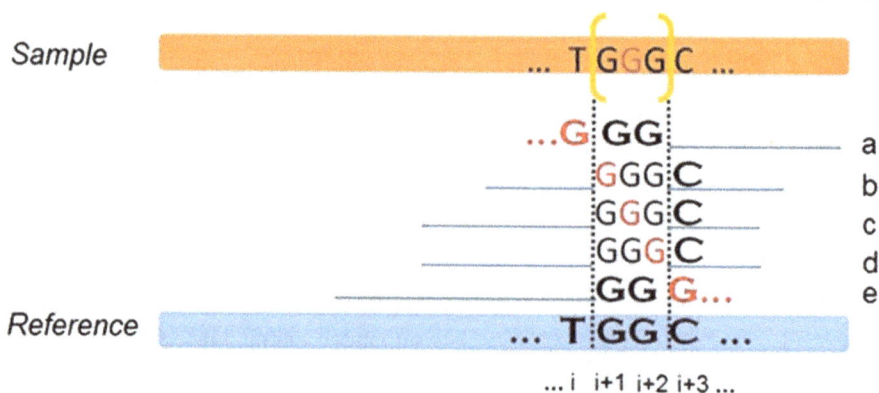

Fig. 2 Example of identical indels taking place in relative positions. (Adapted from Krawitz et al. [20])

respectively. Here TPR is equivalent to recall and FPR is simply (1 – precision) as calculated using formulas (1) and (2). In the ROC space, each point represents the prediction result or instance of a confusion matrix. The diagonal ($Y = X$) that divides the ROC space represents the decision from a "Random Guess." Points above the diagonal represent good classification results, whereas points below the line represent poor results. For each sample, we first calculated the TPR and FPR for each tool and plotted as a point in the ROC space, then ranked the tools based on the perpendicular distance of each point from the diagonal.

We also examined the similarity among the results produced by different tools. Jaccard index, also known as Jaccard similarity coefficient, is used to compare the similarity between indel predictions. For two finite sets A and B, the Jaccard index can be calculated using

$$J(A, \ B) = \frac{|A \cap B|}{|A \cup B|}, 0 \leq J(A, \ B) \leq 1. \tag{4}$$

The maximum value of the Jaccard index is 1 when two indel sets are the same, whereas the minimum is 0 when two indel sets are completely different.

Another interesting question to ask is "how are the seven indel calling tools related to one another on the whole"? To answer this question, we clustered the tools using the following three steps: (1) Divide the reference sequence into windows of equal size. We tested with different window sizes (1000, 10,000, 100,000, and 1,000,000 bps) and found that the window size does not affect the clustering result. For computational convenience, we set the window size to 1,000,000 bps. (2) For each window, calculate the number of indels called by each tool. (3) Construct a vector of indel counts of all windows for each tool and apply the UPGMA hierarchical clustering algorithm to the seven vectors.

Results
Running time
We compared the tools on the average running time taken to call indels for a sample. Table 1 shows the

average running time for samples with low coverage (average coverage ~6X) and high coverage (~64X). For both high- and low-coverage data, Platypus is the fastest and Dindel is the slowest of all the tools investigated. Clearly, indel calling is more time consuming for high-coverage data than for low-coverage data, which is especially evident for Dindel due to its complicated model for realignment. Since Dindel tests all indels identified by the read mapper, many of which might be sequencing errors, with the increase of number of reads and increase in sequencing errors, the computation time increases quadratically [26].

Number of indels called
Figure 3 shows the number of indels called by each tool for each sample. The seven tools under consideration call different numbers of indels. The numbers of indels called across the 78 samples range from 1431 to 15,585 for GATK_UG, from 114 to 10,619 for VarScan, from 1845 to 11,455 for Pindel, from 9351 to 20,245 for SAMtools, from 9864 to 19,876 for Dindel, from 10,915 to 24,786 for GATK_HC, and from 15,062 to 34,600 for Platypus. On average, Platypus calls the maximum number of indels (average number = 23,321), whereas VarScan calls the minimum (2,775). The average number of indels called by SAMtools (14,719) follows closely to that by Dindel across the samples. Similarly, the average numbers of indels called by GATK_UG (6733) and Pindel (6382) are very similar to each other across the samples. As we can see from these results, VarScan is evidently the most conservative one in calling indels. It calls much fewer indels than others. This might be due to its rather stringent filtering step during which all the unmapped and ambiguous reads are discarded. Although this step is helpful in keeping the false positives down, it also reduces the power of detecting true indels.

The lengths of indels called
We examined the distributions of lengths of indels called by the seven tools and compared them to that of the gold standard dataset. All the indel distributions based

Table 1 Average running time spent in calling indels for samples with low/high coverage

Tool	Time	
	Low Coverage (~6X)	High Coverage (~64X)
GATK_UG	16 minutes 43 seconds	24 minutes 19 seconds
VarScan	16 minutes 1 second	84 minutes 02 seconds
Pindel	25 minutes 36 seconds	139 minutes 09 seconds
SAMtools	11 minutes 26 seconds	64 minutes 14 seconds
Dindel	165 minutes 22 seconds	1549 minutes 18 seconds
GATK_HC	58 minutes 27 seconds	91 minutes 13 seconds
Platypus	3 minutes 36 seconds	5 minutes 59 seconds

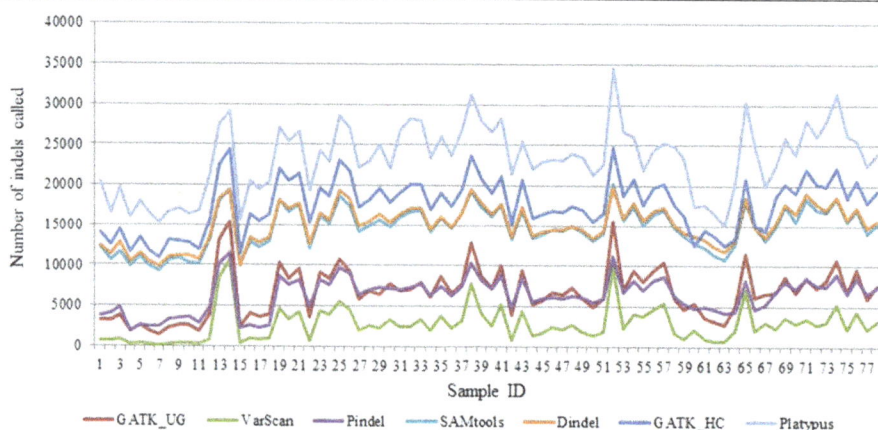

Fig. 3 Number of indels called by the seven tools for the 78 humans

on lengths are shown in Additional file 1: Figure S1 which shows that 96.2 % of the indels in the benchmark dataset are 1–10 bps, 98.6 % in GATK_UG, 99.0 % in VarScan, 93.4 % in Pindel, 92.2 % in SAMtools, 95.1 % in Dindel, 94.01 % in GATK_HC, and 97.19 % in Platypus. Therefore, most of the indels in the benchmark and the ones called by the tools are ≤10 bps. Chi-square statistical tests show that the distributions of indel sizes are not significantly different between the calling results of the tools and the gold standard (p values for comparing the gold standard with GATK_UG, VarScan, Pindel, SAMtools, Dindel, GATK_HC, and Platypus are 0.89, 0.81, 0.96, 0.28, 0.94, 0.95, and 0.99, respectively). Note that Pindel is known for calling medium to large indels, but here most of the indels called by Pindel are small indels.

Regardless of the gold standard indels, we are interested to see the similarity/dissimilarity of the distribution of indel sizes among the tools themselves. From chi-square statistical test between intra-tools, we see that the

distributions of indel sizes are not significantly different among the tools. The p values for intra-tool comparisons are showed in Additional file 1: Table S2.

Effect of the depth of coverage on the number of indels called

To see how the number of indels called by these tools is affected by the depth of coverage, we estimated the depth of coverage for each human sample (shown in Additional file 1: Table S1). Figure 4 shows the relationship between the number of indels called by the seven tools and the coverage depth. Overall, the higher the coverage is, the more indels are called. Pearson correlation coefficients between the coverage and the number of indels called by GATK_UG, VarScan, Pindel, SAMtools, Dindel, GATK_HC, and Platypus are 0.97 (p value $= 8.02 \times 10^{-48}$), 0.97 (p value $= 5.25 \times 10^{-48}$), 0.91 (p value $= 3.10 \times 10^{-30}$), 0.89 (p value $= 4.64 \times 10^{-27}$), 0.88 (p value $= 2.76 \times 10^{-26}$), 0.86 (p value $= 2.24 \times 10^{-23}$), and 0.82 (p value $= 6.84 \times 10^{-20}$), respectively. Thus, consistent with previous

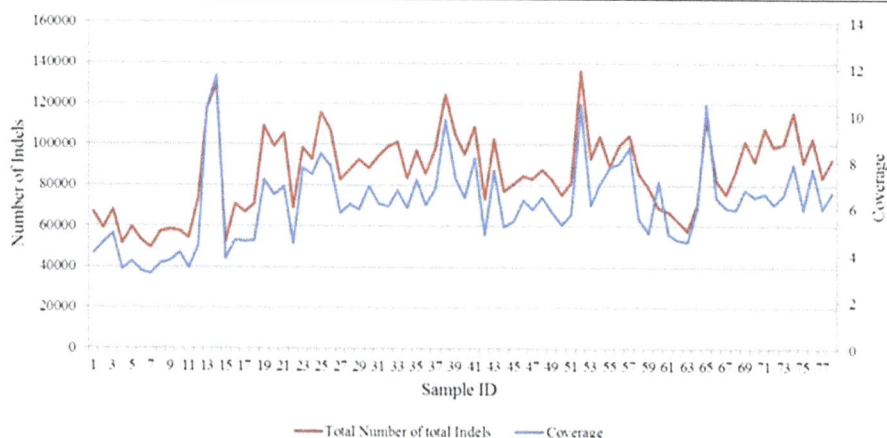

Fig. 4 Relationship between coverage and the pooled number of indels by the seven tools

findings [49], the number of indels called, regardless of the tools, is significantly positively correlated with the coverage depth.

Comparison with the set of "gold standard" indels and ranking of the tools

Figure 5 shows the percentage of gold standard indels called by the tools across the 78 samples. For chromosome 11, on average, only about 1.51 % of the gold standard indels are called by all seven tools, whereas about 76.91 % are undetected by any of the tools. The remaining ~21.58 % are called by at least one tool. We also compared the tools for the percentage of their own indels called by others regardless of the gold standard indels. For this purpose, we picked up Dindel, SAMtools, GATK_HC, and Platypus as they call more indels than the other three tools. The Venn diagram in Additional file 1: Figure S2 shows that only 15.64 % of the indels were called by all of these four tools revealing that regardless of the gold standard indels, a major percentage of indels remain undetected.

We also examined the overall performance of each tool on the 78 samples. The average F-measure values for GATK_UG, VarScan, Pindel, SAMtools, Dindel, GATK_HC, and Platypus are 0.14, 0.06, 0.12, 0.26, 0.27, 0.28, and 0.31, respectively.

In the ROC analysis, we ranked the seven tools based on their distance from the "Random Guess" line in the ROC space. Table 2 shows the frequency of the ranks of the tools based on the 78 samples. Platypus ranked the best for all 78 samples, and GATK_HC ranked the second best. VarScan performed poorly, ranking the worst for 76 samples. Pindel performed also poorly, ranking the worst for 2 samples and second worst for 60

samples. In addition to the ranking, we also computed the average recall, precision, and F-measure for the tools in Table 2. For the average recall of the 78 samples, Platypus ranks the highest (0.22), followed closely by GATK_HC (0.18), and VarScan the lowest (0.03). For the average precision, GATK_UG ranks the highest (0.72), followed closely by VarScan (0.71). GATK_HC (0.61) and Platypus (0.56) have slightly lower average precision. For the average F-measure, Platypus (0.31) ranks the highest and VarScan (0.06) the lowest. To get a clear idea about how the performance of the tools depends on the indel types, i.e., insertion and deletion, we split the benchmark dataset based on the indel types and results are shown in Additional file 1: Figure S3. Results show that except Pindel, performance of the other tools remains consistent regardless of the indel type. Pindel shows better performance in calling deletion than insertion.

Performance of the tools on indels of different lengths

A natural question to ask is whether the seven tools' performance changes with different indel sizes. We computed the average F-measure (Fig. 6), false-negative rate (Fig. 7), recall (Additional file 1: Figure S4), and precision (Additional file 1: Figure S5) of the seven tools for indels of lengths 1–10 bps. Results show that for all the tools, the performance of calling indels correctly shows a slight decrease with the increase of indel lengths. Platypus, GATK_HC, Dindel, and SAMtools show highly similar patterns for four metrics (i.e., F-measure, false-negative rate, recall, and precision) with respect to indel lengths. Altogether, this comparison based on indels of different lengths shows that these tools achieve similar performance for different subcategories of indels with

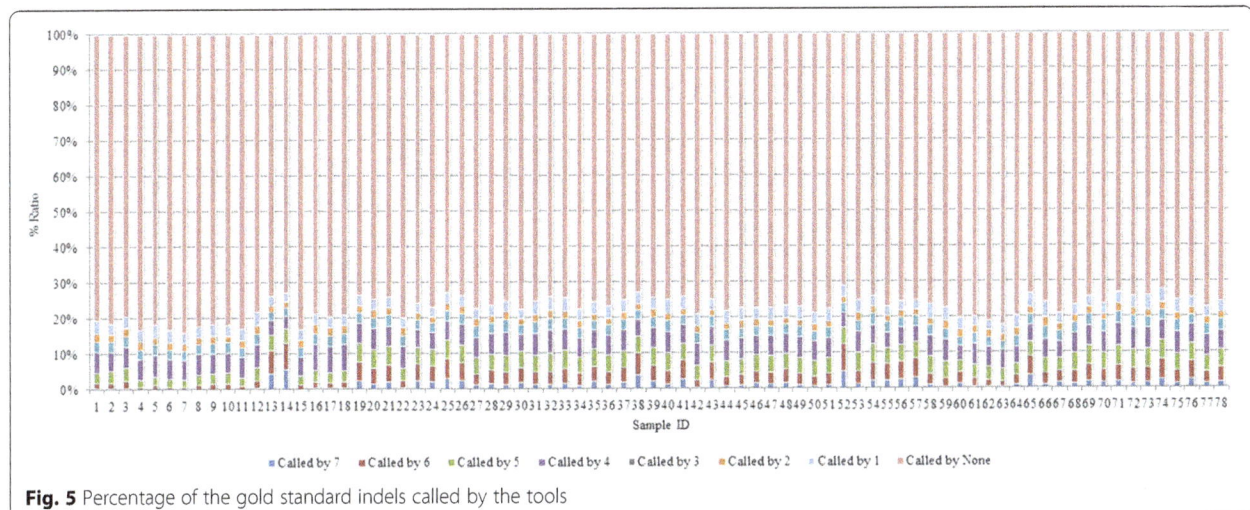

Fig. 5 Percentage of the gold standard indels called by the tools

Table 2 Frequency of the ranks of the tools based on the ROC curve for the 78 samples. Average recall, precision, and F-measure across the samples are also provided

Rank	1	2	3	4	5	6	7	Average Recall	Average Precision	Average F-measure
Name										
GATK_UG	0	0	0	0	62	16	0	0.081884	0.72141	0.14435
VarScan	0	0	0	0	0	2	76	0.033987	0.717315	0.063333
Pindel	0	0	0	0	16	60	2	0.068635	0.636704	0.122591
SAMtools	0	0	1	77	0	0	0	0.160989	0.645108	0.256343
Dindel	0	5	73	0	0	0	0	0.170076	0.662287	0.269404
GATK_HC	0	73	4	1	0	0	0	0.181928	0.608907	0.278323
Platypus	78	0	0	0	0	0	0	0.220391	0.559842	0.314071

certain length. In other words, indel length is not a confounding factor that affects the performance of these calling tools.

Similarity among the tools

We also compared the tools for their similarity regardless of the gold standard. For each sample, the Jaccard index of each pair of the tools is shown in Fig. 8, and the average Jaccard index across all samples is listed in Table 3. From the Jaccard index, we found high similarity between SAMtools and Dindel. A possible reason is that both tools use the Bayesian approach for calling indels. SAMtools calculates the Bayesian prior probability and uses it to calculate the actual genotype for the variants detected. Dindel, on the other hand, calls indels by realigning the reads against candidate haplotypes for which prior probabilities calculated using the Bayesian approach are already known. Both SAMtools and Dindel perform local realignment and base quality assessment for calling indels, and that is also another possible reason for their similarity. Similarly, Platypus and GATK_HC also have high Jaccard index value that represents their strong similarity. Being a haplotype caller, they have underlying similarity such as generating candidate haplotypes and

then realigning reads to each of these candidate haplotypes for variant calling which explains the reason of their similarity.

Figure 9 shows the dendrogram on hierarchical clustering of the tools. Again we see that Dindel and SAMtools group together and Platypus and GATK_HC group together which is supporting our previous observation of similarity between these tools.

Discussion

In this paper, we investigated seven tools that are publicly available and well known for calling indels from short reads. Using 78 whole genome short-read data from the 1000 Genomes project, we evaluated these tools based on several criteria, including running time, number of indels called, recall, precision, F-measure based on the "gold standard" data, and ranking and clustering of the tools. Results show that Platypus outperforms other tools in most of the aspects.

The low percentage of the called indels over the "gold standard" indels indicates that all these tools exhibit limited power in detecting indels. Several factors could contribute to the low true-positive rate. Firstly, since existing read mappers map each read to the reference

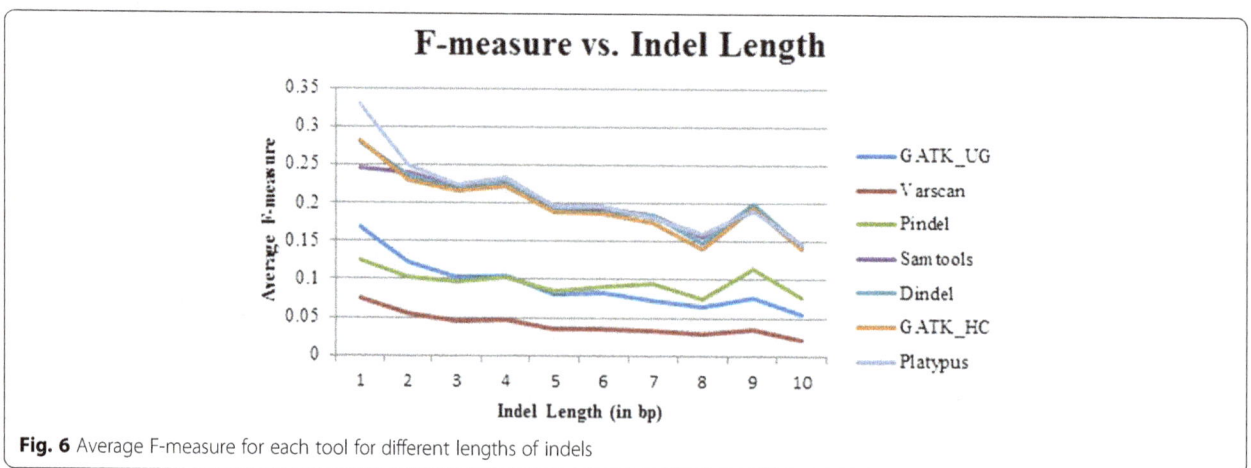

Fig. 6 Average F-measure for each tool for different lengths of indels

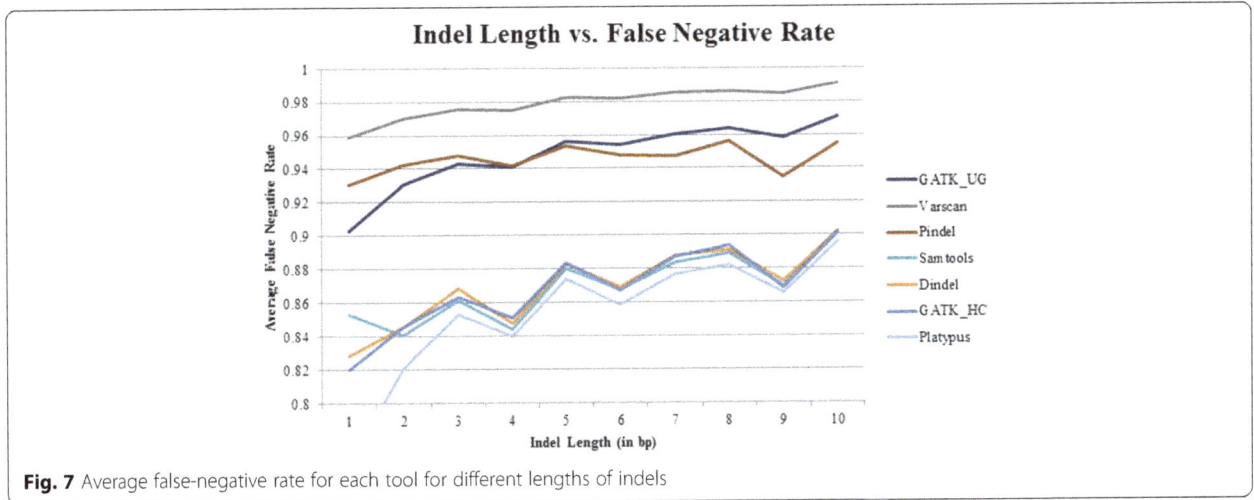

Fig. 7 Average false-negative rate for each tool for different lengths of indels

sequence independently of other reads, due to the alignment artifacts, insertions and deletions can be improperly placed relative to their true positions and it affects the indel calling results greatly. Secondly, most of the indel calling tools do not have sophisticated methods for checking sequencing errors before calling indels. Though Platypus, GATK_HC, and Dindel realign the candidate indels to the known haplotypes, especially for Dindel and GATK_HC, due to their high computational time, it is not an efficient way when the depth of coverage of the reads is high. Therefore, indel calling results can be improved if these factors are considered. Thirdly, the indels we used as gold standard were identified from the DNA traces obtained from the trace archive at NCBI [50], and though it is more reliable than

using short reads, indels identified in this way nevertheless can still be false positives, which could lead to an artificial decrease of true-positive rate. Fourthly, the set of gold standard indels is the pooled result of indels from 79 individuals, which naturally has more indels than individual humans. However, this might not be the dominating factor causing the low false-positive rate as the number of pooled indels for 78 humans is still very low compared to the "gold standard". Finally, the low true-positive rate might also be due to the chromosome-specific behavior of the calling tools. Although we have no particular reason to suspect that the indel calling results for chromosome 11 should be different from those for other autosomes, we examined the performance of the seven tools on chromosome 20 to see

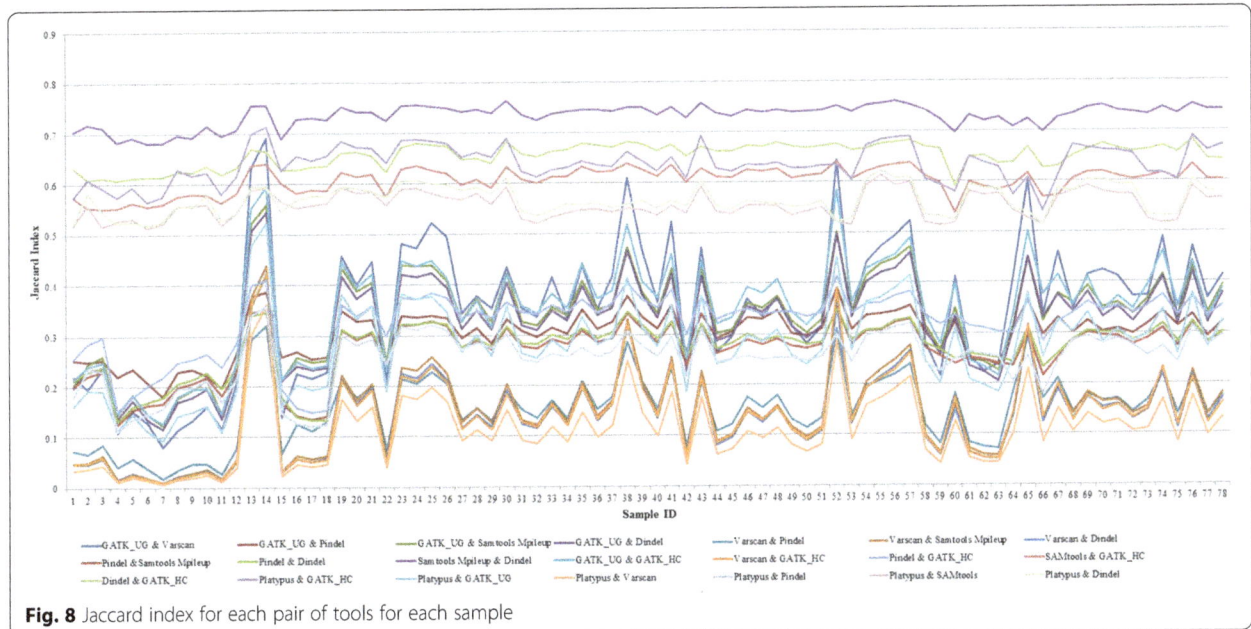

Fig. 8 Jaccard index for each pair of tools for each sample

Table 3 Average Jaccard index for each pair of the tools (Jaccard index is computed for each pair of the tools for each human sample and then averaged across all the 78 samples)

	GATK_UG	VarScan	Pindel	SAMtools	Dindel	GATK_HC	Platypus
GATK_UG	1	0.35	0.30	0.34	0.32	0.35	0.28
VarScan		1	0.15	0.15	0.14	0.14	0.11
Pindel			1	0.27	0.27	0.32	0.25
SAMtools				1	0.73	0.60	0.56
Dindel					1	0.65	0.57
GATK_HC						1	0.64
Platypus							1

whether the result is chromosome specific. Results show that all the metrics (i.e., recall, precision, and F-measures) follow closely those of chromosome 11 (Additional file 1: Figure S6), and therefore, the poor performance of the tools evaluated by the gold standard indels is not chromosome specific.

Clearly, an important issue in evaluating various indel calling tools is the lack of a gold standard dataset or benchmark dataset. In the current study, the performance comparison is done based on the "gold standard" dataset that is the best possible resource available. Although it lists two million short and long indels extracted from the genomes of 79 diverse human, it does not list all the indels that take place in the genomes of the human samples we considered here. Though we can say that Platypus performs better than other tools based on the "gold standard" dataset, however, in general, we cannot make a decision about which tool is the best

unless we have the list of true indels for each sample. So developing a list of indels for individual humans will be a good direction for future research, and that list will be a useful resource for validating the existing as well as newly developed indel calling tools. Moreover, people from the same ethnic group tend to have common indels [51, 52]. Therefore, creating a list of known indels for the same ethnic group and comparing the tools based on the indels called for the samples from that ethnic group would be a better way to evaluate the performance of the tools.

Besides improving the indel calling tools, another strategy to improve the indel calling result is increasing the depth of coverage of the reads. For each of the tools, the performance shows positive correlation with the coverage of the reads. Pearson correlation coefficients between coverage and F-measure for GATK_UG, VarScan, Pindel, SAMtools, Dindel, GATK_HC, and Platypus are 0.96 (p value = 6.75×10^{-44}), 0.98 (p value = 1.38×10^{-54}), 0.89 (p value = 1.56×10^{-27}), 0.87 (p value = 1.22×10^{-24}), 0.85 (p value = 3.65×10^{-23}), 0.85 (p value = 9.47×10^{-23}), and 0.81 (p value = 1.15×10^{-19}), respectively. Moreover, we also performed down-sampling of the individual that has 64X coverage to create a 5X coverage sample and conducted indel calling using the seven tools. Results further confirm that higher coverage yields better results, reflected by higher F-measures for all seven tools in the 64X coverage. However, for all seven tools, precision is higher in the 5X coverage sample than in the 64X coverage sample. Detailed results are shown in Additional file 1: Figure S7. Hence, the performance of the tools can be significantly improved by increasing the depth of coverage of the reads. Consistent with our finding, a previous evaluation of indel calling tools based on simulation data has shown that the sensitivity of indel calling tools increases with coverage depth [39]. Joint sample calling is another strategy to call indels from low-coverage data, and greater sensitivity can be achieved through this. However, it has a few limitations as follows: (i) Since it calls variants simultaneously across all samples, computational expense increases exponentially with the increase of the number of samples, and (ii) every time a new sample

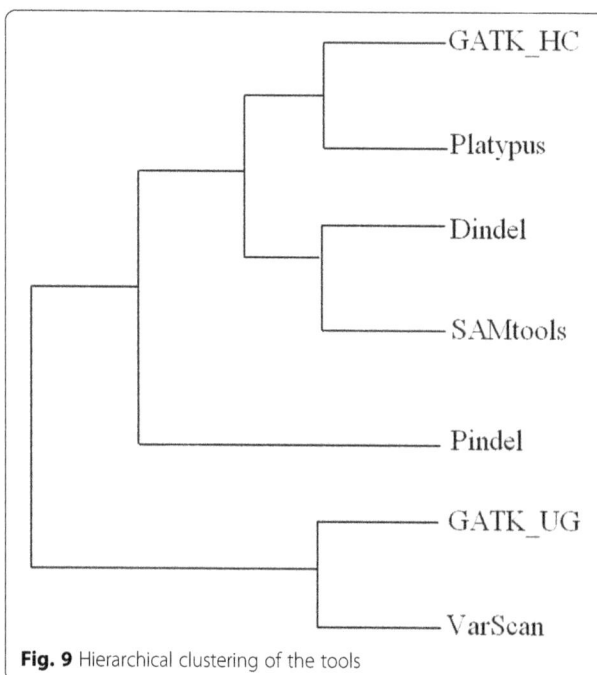

Fig. 9 Hierarchical clustering of the tools

is added to the cohort, the process of variant calling needs to start again from the scratch; this is known as the $(N+1)$ problem [53]. HaplotypeCaller like GATK_HC and Platypus are free from these limitations; however, the other tools are yet to overcome these limitations.

Finally, although the indel calling results produced by the tools show great discrepancy, these tools can show strengths in different aspects such as running time, the number of indels identified, and indels of different lengths. Thus, integrating the strength of existing tools to call indels and then passing the results to an aggregating machine learning model to increase true positives and reduce false positives might be a good solution. Similar ideas were discussed in [54] for creating highly confident SNP, indel, and homozygous reference genotype calls.

Conclusion

Indel is one of the main types of disease-causing variation in humans. Detecting indels in an efficient manner is necessary for discovering proper medication. The advent of NGS technology has made it possible to sequence human genomes at an individual level. We have investigated seven well-known tools, GATK Unified Genotyper (GATK_UG), VarScan, Pindel, SAMtools, Dindel, GATK HaplotypeCaller (GATK_HC), and Platypus that call indels using NGS data. Based on the benchmark dataset we used, Platypus outperformed other tools. However, all of these tools have limitations as a large number of indels listed in the benchmark dataset remain undetected. A sophisticated method to check sequencing errors before calling indels and an integrative approach to combine the strengths of existing indel calling tools might be a good solution to overcome this problem. Using reads with high coverage is another strategy to obtain better results. Although the benchmark dataset we used for comparing the tools contain a large number of short and long indels that take place in diverse human genomes, it may not contain all the indels occurring in the genome of the samples we considered here. Hence, developing a list of known indels at an individual level will be helpful for validating the existing and newly developed tools.

Additional file

Additional file 1: Supplementary materials for Performance evaluation of indel calling tools using real short-read data. This additional file includes List of the input samples used in this study with corresponding population and read coverage, P-value for Chi-square statistical test of indel size distribution between the tools, Distribution of indels based on lengths (1 to 10 bp) for the tools, Intra-tool comparison among GATK_HC, Dindel, SAMtools, and Platypus for percentage of their own indels called by others, Average Recall, Precision, and F-Measure of each tool for insertion and deletion, average recall and precision for each tool for different lengths of indels, Comparison between Chromosome 11 and Chromosome 20 for HG00157, and Comparison between High coverage samples for NA12878.

Abbreviations
HC: HaplotypeCaller; Indel: insertion and deletion; NGS: next-generation sequencing; ROC: receiver operating characteristic; SNP: single nucleotide polymorphism; WGS: whole genome sequencing.

Competing interests
The authors declare that they have no competing interests.

Authors' contributions
MSH performed the data processing, coding, and computational experiments. XW performed the statistical analysis. LZ participated in the design of the study and supervised the project. MSH, XW, and LZ wrote the paper. All authors read and approved the final manuscript.

Acknowledgements
We thank anonymous reviewers whose feedbacks are really helpful to improve the quality of the paper. The work is partially supported by an NSF grant OCI-1124123 to L. Zhang.

Author details
[1]Department of Computer Science, Virginia Tech, Blacksburg, VA 24061, USA. [2]Department of Statistics, Virginia Tech, Blacksburg, VA 24061, USA.

References
1. Bhangale TR, Rieder MJ, Livingston RJ, Nickerson DA. Comprehensive identification and characterization of diallelic insertion–deletion polymorphisms in 330 human candidate genes. Hum Mol Genet. 2005;14(1):59–69.
2. Dawson E, Chen Y, Hunt S, Smink LJ, Hunt A, Rice K, et al. A SNP resource for human chromosome 22: extracting dense clusters of SNPs from the genomic sequence. Genome Res. 2001;11(1):170–8.
3. Mullaney JM, Mills RE, Pittard WS, Devine SE. Small insertions and deletions (INDELs) in human genomes. Hum Mol Genet. 2010;19(R2):R131–R6.
4. Mills RE, Luttig CT, Larkins CE, Beauchamp A, Tsui C, Pittard WS, et al. An initial map of insertion and deletion (INDEL) variation in the human genome. Genome Res. 2006;16(9):1182–90.
5. Collins FS, Drumm ML, Cole JL, Lockwood WK, Woude GV, Iannuzzi MC. Construction of a general human chromosome jumping library, with application to cystic fibrosis. Science. 1987;235(4792):1046–9.
6. Warren ST, Zhang F, Licameli GR, Peters JF. The fragile X site in somatic cell hybrids: an approach for molecular cloning of fragile sites. Science. 1987;237(4813):420–3.
7. Usdin K. The biological effects of simple tandem repeats: lessons from the repeat expansion diseases. Genome Res. 2008;18(7):1011–9.
8. MacArthur DG, Tyler-Smith C. Loss-of-function variants in the genomes of healthy humans. Hum Mol Genet. 2010;19(R2):R125–R30.
9. Kaneo T, Tahara S, Matsuo M. Non-linear accumulation of 8-hydroxy-2′-deoxyguanosine, a marker of oxidized DNA damage, during aging. Mutat Res. 1996;316(5):277–85.
10. Paschka P, Marcucci G, Ruppert AS, Mrózek K, Chen H, Kittles RA, et al. Adverse prognostic significance of KIT mutations in adult acute myeloid leukemia with inv(16) and t(8; 21): a Cancer and Leukemia Group B Study. J Clin Oncol. 2006;24(24):3904–11.
11. Falini B, Mecucci C, Tiacci E, Alcalay M, Rosati R, Pasqualucci L, et al. Cytoplasmic nucleophosmin in acute myelogenous leukemia with a normal karyotype. N Engl J Med. 2005;352(3):254–66.
12. Nakao M, Yokota S, Iwai T, Kaneko H, Horiike S, Kashima K, et al. Internal tandem duplication of the flt3 gene found in acute myeloid leukemia. Leukemia. 1996;10(12):1911–8.
13. Sequist LV, Martins RG, Spigel D, Grunberg SM, Spira A, Jänne PA, et al. First-line gefitinib in patients with advanced non–small-cell lung cancer harboring somatic EGFR mutations. J Clin Oncol. 2008;26(15):2442–9.
14. Ostertag EM, Kazazian Jr HH. Biology of mammalian L1 retrotransposons. Annu Rev Genet. 2001;35(1):501–38.
15. Cheung VG, Spielman RS. Genetics of human gene expression: mapping DNA variants that influence gene expression. Nat Rev Genet. 2009;10(9):595–604.

16. Lee S, Mun HS, Kim H, Lee HK, Kim BJ, Hwang ES, et al. Naturally occurring hepatitis B virus X deletions and insertions among Korean chronic patients. J Med Virol. 2011;83(1):65–70.

17. Hasan MS, Zhang L. P-Dindel: A multi-thread based tool for calling indels from short reads. In Short abstract of the 11th International Symposium on Bioinformatics Research and Applications; June 7-10, 2015; Norfolk, Virginia. P. 71-74. Available from http://www.cs.gsu.edu/isbra15/sites/default/files/ISBRA12ShortAbstractsFinal.pdf.

18. Hasan MS, Zhang L. SPAI: Single Platform for Analyzing Indels. In Short abstract of the 11th International Symposium on Bioinformatics Research and Applications; June 7-10, 2015; Norfolk, Virginia. P. 75-78. Available from http://www.cs.gsu.edu/isbra15/sites/default/files/ISBRA12ShortAbstractsFinal.pdf.

19. Ding L, Ellis MJ, Li S, Larson DE, Chen K, Wallis JW, et al. Genome remodelling in a basal-like breast cancer metastasis and xenograft. Nature. 2010;464(7291):999–1005.

20. Krawitz P, Rödelsperger C, Jäger M, Jostins L, Bauer S, Robinson PN. Microindel detection in short-read sequence data. Bioinformatics. 2010;26(6):722–9.

21. Ct G. Primer: sequencing—the next generation. Nat Methods. 2008;5(1):15.

22. Metzker ML. Sequencing technologies—the next generation. Nat Rev Genet. 2010;11(1):31–46.

23. Mardis ER. Next-generation DNA, sequencing methods. Annu Rev Genomics Hum Genet. 2008;9:387–402.

24. Li H, Durbin R. Fast and accurate long-read alignment with Burrows–Wheeler transform. Bioinformatics. 2010;26(5):589–95.

25. Matsumura H, Yoshida K, Luo S, Kimura E, Fujibe T, Albertyn Z, et al. High-throughput SuperSAGE for digital gene expression analysis of multiple samples using next generation sequencing. PLoS One. 2010;5(8), e12010.

26. Albers CA, Lunter G, MacArthur DG, McVean G, Ouwehand WH, Durbin R. Dindel: accurate indel calls from short-read data. Genome Res. 2011;21(6):961–73.

27. Lunter G, Goodson M. Stampy: a statistical algorithm for sensitive and fast mapping of Illumina sequence reads. Genome Res. 2011;21(6):936–9.

28. Li H, Handsaker B, Wysoker A, Fennell T, Ruan J, Homer N, et al. The sequence alignment/map format and SAMtools. Bioinformatics. 2009;25(16):2078–9.

29. DePristo MA, Banks E, Poplin R, Garimella KV, Maguire JR, Hartl C, et al. A framework for variation discovery and genotyping using next-generation DNA sequencing data. Nat Genet. 2011;43(5):491–8.

30. Koboldt DC, Chen K, Wylie T, Larson DE, McLellan MD, Mardis ER, et al. VarScan: variant detection in massively parallel sequencing of individual and pooled samples. Bioinformatics. 2009;25(17):2283–5.

31. Koboldt DC, Zhang Q, Larson DE, Shen D, McLellan MD, Lin L, et al. VarScan 2: somatic mutation and copy number alteration discovery in cancer by exome sequencing. Genome Res. 2012;22(3):568–76.

32. Ye K, Schulz MH, Long Q, Apweiler R, Ning Z. Pindel: a pattern growth approach to detect break points of large deletions and medium sized insertions from paired-end short reads. Bioinformatics. 2009;25(21):2865–71.

33. Grimm D, Hagmann J, Koenig D, Weigel D, Borgwardt K. Accurate indel prediction using paired-end short reads. BMC Genomics. 2013;14(1):132.

34. Korbel JO, Abyzov A, Mu XJ, Carriero N, Cayting P, Zhang Z, et al. PEMer: a computational framework with simulation-based error models for inferring genomic structural variants from massive paired-end sequencing data. Genome Biol. 2009;10(2):R23.

35. Quinlan AR, Clark RA, Sokolova S, Leibowitz ML, Zhang Y, Hurles ME, et al. Genome-wide mapping and assembly of structural variant breakpoints in the mouse genome. Genome Res. 2010;20(5):623–35.

36. Chen K, Wallis JW, McLellan MD, Larson DE, Kalicki JM, Pohl CS, et al. BreakDancer: an algorithm for high-resolution mapping of genomic structural variation. Nat Methods. 2009;6(9):677–81.

37. GATK HaplotypeCaller. https://www.broadinstitute.org/gatk/guide/article?id=4148. Accessed 30 April 2015.

38. Rimmer A, Phan H, Mathieson I, Iqbal Z, Twigg SR, Wilkie AO, et al. Integrating mapping-, assembly- and haplotype-based approaches for calling variants in clinical sequencing applications. Nat Genet. 2014;46(8):912–8.

39. Neuman JA, Isakov O, Shomron N. Analysis of insertion–deletion from deep-sequencing data: software evaluation for optimal detection. Brief Bioinform. 2013;14(1):46–55.

40. Abel HJ, Duncavage EJ. Detection of structural DNA variation from next generation sequencing data: a review of informatic approaches. Cancer Genet. 2013;206(12):432–40.

41. Via García M, Consortium GP. An integrated map of genetic variation from 1,092 human genomes. Nature. 2012;491:56–65. 2012.

42. Bentley DR, Balasubramanian S, Swerdlow HP, Smith GP, Milton J, Brown CG, et al. Accurate whole human genome sequencing using reversible terminator chemistry. Nature. 2008;456(7218):53–9.

43. Mills RE, Pittard WS, Mullaney JM, Farooq U, Creasy TH, Mahurkar AA, et al. Natural genetic variation caused by small insertions and deletions in the human genome. Genome Res. 2011;21(6):830–9.

44. Whelan C. Detecting and Analyzing Genomic Structural Variation Using Distributed Computing. In Scholar Archive of OHSU Digital Commons, Paper 3482; February 2014; Available from http://digitalcommons.ohsu.edu/cgi/viewcontent.cgi?article=7928&context=etd.

45. Whelan CW, Tyner J, L'Abbate A, Storlazzi CT, Carbone L, Sönmez K. Cloudbreak: accurate and scalable genomic structural variation detection in the cloud with MapReduce. arXiv preprint arXiv:13072331; 2013. Available from: http://arxiv.org/abs/1307.2331.

46. Zook JM, Salit M. Genomes in a bottle: creating standard reference materials for genomic variation - why, what and how? Genome Biol. 2011;12:1–27.

47. Li Z, Wu X, He B, Zhang L. Vindel: a simple pipeline for checking indel redundancy. BMC Bioinformatics. 2014;15(1):359. doi:10.1186/s12859-014-0359-1.

48. Receiver operating characteristic. 2014. http://en.wikipedia.org/wiki/Receiver_operating_characteristic. Accessed 20 April 2014.

49. Fang H, Narzisi G, Rawe JA, Wu Y, Rosenbaum J, Ronemus M, et al. Reducing INDEL errors in whole-genome and exome sequencing. Genome Med. 2014;6(10):89.

50. DNA trace archive. http://www.ncbi.nlm.nih.gov/Traces/trace.cgi. Accessed 28 November 2014.

51. Meng H-T, Zhang Y-D, Shen C-M, Yuan G-L, Yang C-H, Jin R, et al. Genetic polymorphism analyses of 30 InDels in Chinese Xibe ethnic group and its population genetic differentiations with other groups. Sci Rep. 2015;5.

52. Ahn S-M, Kim T-H, Lee S, Kim D, Ghang H, Kim D-S, et al. The first Korean genome sequence and analysis: full genome sequencing for a socio-ethnic group. Genome Res. 2009;19(9):1622–9.

53. Should I analyze my samples alone or together? 2014. https://www.broadinstitute.org/gatk/guide/article?id=4150. Accessed 29 November 2014.

54. Zook JM, Chapman B, Wang J, Mittelman D, Hofmann O, Hide W, et al. Integrating human sequence data sets provides a resource of benchmark SNP and indel genotype calls. Nat Biotechnol. 2014;32(3):246–51.

6

Integrative DNA methylation and gene expression analysis to assess the universality of the CpG island methylator phenotype

Matahi Moarii[1,2,3], Fabien Reyal[4,5,6] and Jean-Philippe Vert[1,2,3]*

Abstract

Background: The CpG island methylator phenotype (CIMP) was first characterized in colorectal cancer but since has been extensively studied in several other tumor types such as breast, bladder, lung, and gastric. CIMP is of clinical importance as it has been reported to be associated with prognosis or response to treatment. However, the identification of a universal molecular basis to define CIMP across tumors has remained elusive.

Results: We perform a genome-wide methylation analysis of over 2000 tumor samples from 5 cancer sites to assess the existence of a CIMP with common molecular basis across cancers. We then show that the CIMP phenotype is associated with specific gene expression variations. However, we do not find a common genetic signature in all tissues associated with CIMP.

Conclusion: Our results suggest the existence of a universal epigenetic and transcriptomic signature that defines the CIMP across several tumor types but does not indicate the existence of a common genetic signature of CIMP.

Background

Epigenetic modifications have been recognized as important players in cancer etiology and development and constitute promising therapeutic targets for diagnosis or treatment due to their possible reversibility [1–3]. In particular, aberrant methylation of CpG islands (CGIs) located in promoter regions of tumor suppressor and DNA repair genes, leading to their silencing, is now considered a hallmark of cancer playing an important role in neoplasia [1–6].

The CpG island methylator phenotype (CIMP) was first defined and observed by [7] in a subset of colorectal cancers as the joint methylation of several promoter regions, leading to the inactivation of the corresponding genes. The stratification of patients based on CIMP was shown to be clinically relevant, as CIMP-positive patients had better prognosis than CIMP-negative ones, and could lead to

personalized treatments. Since the identification of CIMP in colorectal cancers, many studies have tried to replicate the analysis to find CIMP in different types of cancers including but not limited to colon [8–12], breast [13, 14], lung [15], stomach [16], and glioblastoma [17–19]. While most of these works concluded in the existence of a CIMP in different cancers, other studies did not yield the same conclusions [20, 21], and the genes whose promoter CGI methylation are considered to define the CIMP differ between studies. This raises the question of whether the CIMP is tissue specific or is a universal phenomenon with common biological causes affecting common genes across cancers. A recent review of CIMP-related studies across different cancers pointed out the diversity of methods and measurement technologies used to define CIMP, which hinders the establishment of a molecular basis for CIMP in spite of growing evidence linking mutations in specific genes and CIMP in several cancers [22].

In the present study, we investigate the existence and universality of CIMP by performing a systematic genome-wide methylation analysis on several large datasets of different cancer types simultaneously. We propose a simple methodology to assess the existence of a CIMP phenotype

*Correspondence: jean-philippe.vert@mines-paristech.fr
[1]CBIO-Centre for Computational Biology, Mines Paristech, PSL-Research University, 35 Rue Saint-Honore, F-77300 Fontainebleau, France
[2]Department of Bioinformatics, Biostatistics and System Biology, Institut Curie, 11-13 Rue Pierre et Marie Curie, F-75248 Paris, France
Full list of author information is available at the end of the article

in each cancer and to identify a set of genes whose promoter methylation is a marker for the CIMP. This allows us to compare the different cancer types in search for a cross-cancer CIMP signature and to analyze the link between CIMP and gene expression in different cancers. Finally, we assess the clinical relevance of CIMP on the overall survival.

Results

A cross-cancer CIMP signature

We first assess with a common methodology whether a CIMP can be detected on different cancers and whether CIMP in different cancers share a common signature in terms of which gene promoters are hypermethylated in CIMP-positive patients. For that purpose, we collected high-density methylation datasets from the cancer genome atlas (TCGA) data portal providing more than 485,000 CpG methylation levels for more than 2000 samples from five tissues of origin: bladder, breast, colon, lung, and stomach (Table 1). For each sample, we aggregate the methylation levels of CpG probes by CGI, including the CGI itself and its shores and shelves, resulting in a single methylation level for each of 21,176 CGIs in each sample.

A CIMP corresponds to the joint hypermethylation of a subset of CGIs in a subset of samples [7]. To characterize from whole-genome methylation data whether a CIMP exists for a cancer and which CGIs characterize it, we follow a standard methodology: (i) select the 5 % most variant CGIs in the set of samples, which we call the *CIMP signature* and (ii) check by unsupervised classification whether the samples cluster into two main clusters (CIMP-positive and -negative clusters) when we restrict them to the methylation values they take on the CGIs in the CIMP signature.

We apply this methodology to each of the five families of tumors, cutting the tree obtained by hierarchical clustering to two clusters in order to enforce a classification of all samples into two subgroups based on the methylation

of CGIs in the CIMP signature. Interestingly, in all five cases, one of the two clusters is clearly characterized by an overall hypermethylation of most CGIs in the signature compared to the second cluster, allowing us to characterize it as the CIMP-positive cluster, the second one being the CIMP-negative cluster (Additional file 1). The proportion of CIMP-positive samples according to this definition varies from about 20 % for breast and colon cancers to 30 % for bladder and about 60 and 70 % for stomach and lung cancers (Table 2). Proportion of the CIMP-positive group in each tissue is similar to previously reported studies [22]. Varying the size of the CIMP signature from 1 to 10 % of all CGIs had a small impact on the clustering stability (Additional file 2).

Comparing the epigenetic signatures that define CIMP for each tissue, we find a common set of 89 CGIs associated with 51 genes (Fig. 1a). If the signatures were random subsets of 5 % of all CGIs independent from each other, the overlap would contain on average $(5\%)^5 \simeq 3.10^{-5}$ % of all CGIs, namely 0.006 CGI. This provides a strong evidence that a common set of genes is involved in CIMP in different cancers. We call these 89 CGIs the *cross-cancer CIMP signature* (Table 3). A hierarchical clustering on all samples restricted to this cross-cancer CIMP signature is able to cluster CIMP-positive and CIMP-negative patients independently of the tissue of origin (Fig. 1b), suggesting that CIMP observed in each individual cancer share in common a significant proportion of genes whose promoter CGIs are hypermethylated in all CIMP-positive cancers. A functional enrichment analysis of the cross-cancer CIMP signature reveals that it is significantly enriched in genes involved in cell differentiation and neuronal developmental and immune response processes (Fig. 1c).

Are there 2 or 3 CIMP classes?

Several studies suggest the existence of a third class in CIMP phenotype that corresponds to an intermediate level of methylation [12, 23, 24]. While we enforced an

Table 1 Patients' dataset. Number of samples available for the different cancer types (first column) for methylation (Meth) and gene expression (GE). The "Meth/GE" column summarizes the number of samples with both methylation and gene expression, while the "Meth/Mutations" column shows the number of samples with both methylation and DNA mutation data

Tissue	Meth	GE	Meth/GE	Meth/Mutations
Bladder	373	56	43	28
Breast	626	778	478	468
Colon	291	193	34	219
Lung	452	125	82	411
Stomach	338	373	309	199
Overall	2090	1525	941	1325

Table 2 CIMP proportion. For each cancer type, this table shows the number of samples clustered in the CIMP-negative and CIMP-positive clusters and the percentage of CIMP-positive samples

Tissue	Negative	Positive	Ratio (%)
Bladder	262	111	30
Breast	509	117	19
Colon	232	59	20
Lung	136	316	70
Stomach	144	194	57
Overall	1283	797	38

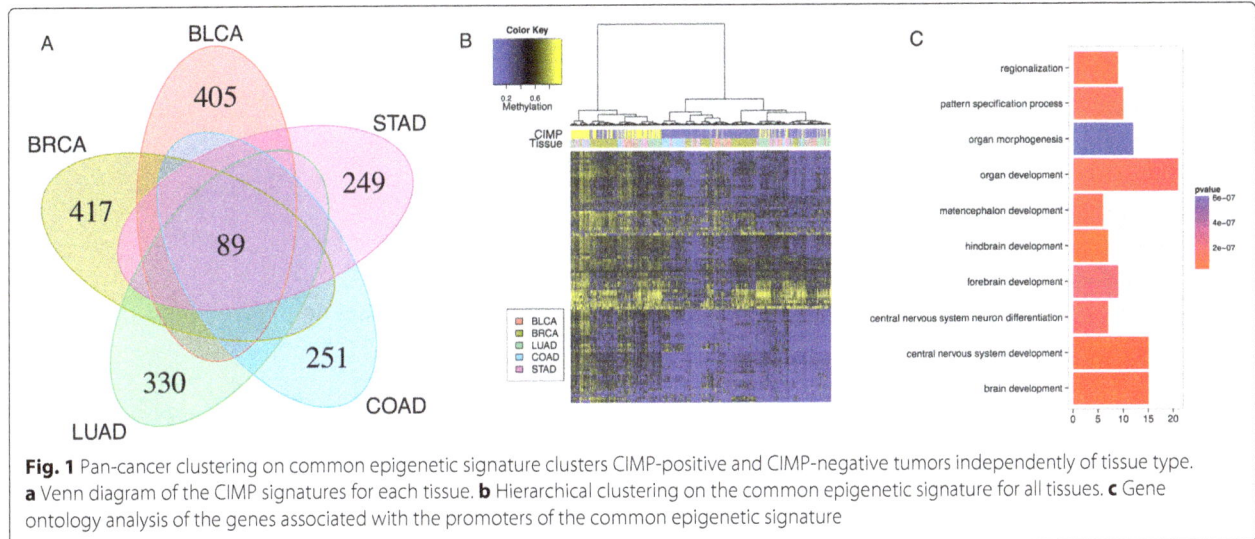

Fig. 1 Pan-cancer clustering on common epigenetic signature clusters CIMP-positive and CIMP-negative tumors independently of tissue type. **a** Venn diagram of the CIMP signatures for each tissue. **b** Hierarchical clustering on the common epigenetic signature for all tissues. **c** Gene ontology analysis of the genes associated with the promoters of the common epigenetic signature

analysis with 2 classes to define the CIMP of each sample as positive or negative in the previous section, we now examine whether the data call for a third class. Following [25], we assess the existence of an intermediate CIMP phenotype for each tissue by comparing the increase in empirical cumulative distributive distribution $\Delta(K)$ for different values of $K = 2, \ldots, 5$ where K is the number of clusters considered for CIMP.

Figure 2 shows how $\Delta(K)$ varies as a function of K for each cancer, suggesting how many clusters exist in each case. We observe that the existence of a third class is not clear-cut. While colon and breast tissues show a significant increase in $\Delta(K)$ for $K = 3$ suggesting a possible third cluster in CIMP, the bladder is flat between 2 and 3 clusters, while lung and gastric cancers do not support

the presence of 3 classes. In addition, we assess the stability of 3 clusters by varying the number of CGIs that define CIMP and observed that while CIMP clusters are highly robust for $K = 2$, there is some high variability in the cluster definitions for $K = 3$ (Additional file 2). In summary, the presence of 2 clusters is well supported by the data in all cancers, while the third cluster is much more debatable.

Similar gene expression variations are predictive of CIMP

To shed light on the relationship between methylation and transcription, we now assess to what extent a transcriptomic signature can classify the samples as CIMP positive or negative. For that purpose, we collected for each family of cancer samples with both methylation and gene expression data available, leading to a subset of samples with an overall proportion of CIMP-positive samples comparable to that of the original dataset (Table 4). We measure by cross-validation how well expression data alone can recover the two CIMP classes.

We first perform a multivariate regression analysis using the lasso technique to assess whether gene expression of a few genes can be predictive of the CIMP status for each tissue separately. The cross-validation accuracies for each family of cancer are shown in Table 5. We observe that while a classifier based on gene expression performs significantly better than random to recover CIMP-positive samples in breast, lung, and stomach cancers, the performance on the bladder and colon is not different from a random classifier. Moreover, we compare the lists of genes selected in the transcriptomic signature after bootstrap resampling of the samples in order to assess their robustness and potential biological significance (Fig. 3a). We observe that very few genes are robustly selected in the signatures, and in particular that no gene is associated with BLCA-CIMP and COAD-CIMP prediction in

Table 3 List of genes associated with the common set of CGIs that define CIMP in each tissue

Epigenetic Signature	LOC339524, GSTM1, CD1D, LMX1A
	CACNA1E ,NR5A2, WNT3A, GNG4
	EMX1, CTNNA2 ,LRRTM1, DLX1
	EVX2, HOXD13, GBX2, SYN2
	HAND2, NBLA00301, EBF1, HIST1H2BB
	HIST1H3C, HLA-DRB1, C6orf186, IKZF1
	CDKN2A, HMX3, KNDC1, KLHL35
	HOTAIR, SLC6A15, ALX1, RFX4
	CLDN10, ADCY4, RIPK3, NID2
	OTX2, OTX2OS1, GSC, KIF26A
	GREM1, SEC14L5, HS3ST3B1, IGF2BP1
	HOOK2, NFIX, ZNF577, ZNF649
	CPXM1, CDH22, CHRNA4

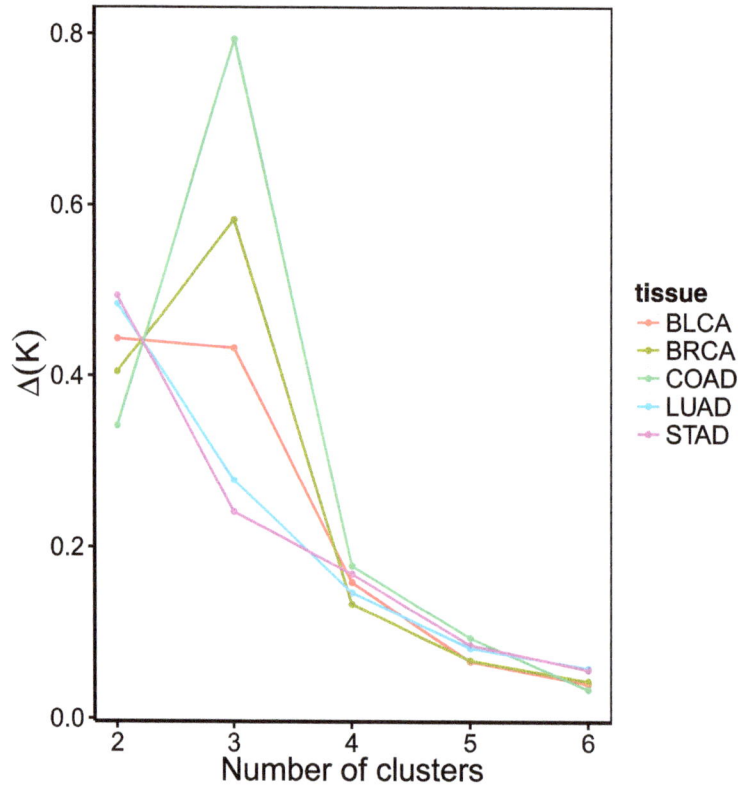

Fig. 2 Variation in the empirical cumulative distributive function $\Delta(K)$ for each tissue. The empirical cumulative distributive function $\Delta(K)$ is a data-driven criterion which can indicate the number of clusters K in the data when it reaches its maximum [25]. This plots shows how $\Delta(K)$ varies as a function of K, for the different tissues

more than 15 % of the bootstrap resampling . In addition, the transcriptomic signatures of different cancers are very diverse, and no gene is present in all of them (Fig. 3b). Overall, these results suggest that there is information in the transcriptome related to the CIMP status, but that a robust signature across cancers is difficult to obtain.

However, the poor accuracy as well as the non-robustness of genetic signatures to predict CIMP may be

Table 4 CIMP Proportion in samples with both methylation and gene expression data. This table shows the number of CIMP-positive and CIMP-negative samples characterized by both methylation and gene expression data, for each cancer type, as well as the proportion of CIMP-positive samples

Tissue	Negative	Positive	Ratio (%)
Bladder	27	16	37
Breast	385	93	20
Colon	27	7	20
Lung	22	60	75
Stomach	131	178	58
Overall	592	354	37

due to the small size of some datasets ($n_{BLCA} = 43$, $n_{COAD} = 34$). To overcome the lack of statistical power due to small sample size, we combine in a second analysis the different datasets into a single multivariate regression analysis, based on the assumption that the CIMP signatures of different cancers may share the same genes. We train classifiers to predict CIMP status from gene expression data jointly across cancers using two methods, based on two different assumptions: (i) assuming that all tissues share the same gene signature and coefficients for the prediction task, we run a single lasso classification on the combined datasets ("Combined-Lasso" prediction) or (ii) assuming that all tissues share the same gene signature but with different coefficients, we jointly train several models with a group lasso approach to constrain the selected genes to be the same across cancers without imposing their coefficients to coincide ("Group-Lasso" prediction) (see supplementary methods in Additional file 3). The rationale for the group lasso approach is that while CIMP may be caused by a common subset of genes, the specific contribution of each gene may vary between tissues. Our results show that both methods significantly outperform the tissue-specific predictions ($P \leq 2.10^{-16}$,

Table 5 Accuracy of CIMP prediction using gene expression profiles

	Accuracy			
	Random	Lasso	Combined lasso	Group lasso
Bladder	62.8	62.9 $(p = 1)$	74.2 $(p \leq 2.10^{-16})$	72.1 $(p \leq 2.10^{-16})$
Breast	80.5	83.9 $(p \leq 2.10^{-16})$	84.7 $(p \leq 2.10^{-16})$	85.5 $(p \leq 2.10^{-16})$
Colon	79.4	79.5 $(p = 1)$	95.0 $(p \leq 2.10^{-16})$	94.2 $(p \leq 2.10^{-16})$
Lung	73.2	84.2 $(p \leq 2.10^{-16})$	76.2 $(p \leq 2.10^{-16})$	86.6 $(p \leq 2.10^{-16})$
Stomach	57.6	81.2 $(p \leq 2.10^{-16})$	83.0 $(p \leq 2.10^{-16})$	84.8 $(p \leq 2.10^{-16})$
Overall	71.9	82.4	82.6	85.0

This table shows the accuracy, assessed by threefold cross-validation repeated 100 times over each tissue (first column), of sample classification in CIMP-positive and CIMP-negative classes from gene expression data using random classification (second column), lasso logistic regression (third column), combined lasso (fourth column), or group lasso logistic regression (fifth column)

Fig. 3c, Table 5) in particular for the bladder and colon where the size of the initial datasets could not give sufficient statistical power to predict CIMP accurately. There is overall little difference between both methods, with the notable exception of lung cancer where the combined lasso approach is significantly worse than the group lasso (and even the single lasso) model, suggesting that in that case, the weights of the genes in the CIMP signature may differ from other cancers. More importantly, each method allows to identify a common genetic signature (51 genes for the "Combined" prediction and 58 genes for the "Group-Lasso" prediction) that distinguishes CIMP-positive and CIMP-negative class for each tumors which is more robust than all the tissue-specific signatures (Fig. 3d). In addition, these signatures share a large common set of genes (25 common genes, Table 6). We represented the gene expression distribution for this common set of genes on the different datasets and observe a clear separation between CIMP-positive and CIMP-negative classes for all tissues (Additional file 4). Gene ontology analysis on the intersection of the two predictive gene signatures showed specific enrichment only for genetic regulatory processes.

A genetic signature is associated to CIMP only for colon and gastric cancers

Several somatic mutations have been found to be tightly associated with epigenetic aberrations in CIMP. Recent studies have pointed out the causal role of IDH1 mutations in Glioblastoma-CIMP [17, 19] and tight associations between IDH2 and TET2 mutations with other CIMPs (leukemia [26], enchondroma, and spindle cell hemangioma [27, 28]). In the colon, BRAF and KRAS mutations are associated with microsatellite instability and COAD-CIMP [9].

We re-assess the association between mutations in these genes and CIMP in the different types of cancers (Fig. 4a). We recover a strong association between BRAF mutation and CIMP-positive colon tumors but no specific

association with other tumor types. We also find no coordinated association between IDH1, IDH2, KRAS, BRAF, or TET2 mutations and CIMP phenotypes for all tissues. In addition, we perform genome-wide mutation analysis to assess whether specific gene mutations are associated with CIMP. We find no significant gene mutation association for bladder, breast nor lung CIMPs. For colon and gastric cancer, we find respectively 459 and 1070 gene mutations associated with CIMP with a common intersection of 195 genes (Additional file 5 panel A). Gene ontology analysis of this set of genes shows significant enrichment for extracellular matrix organization and cell adhesion but also neuronal developmental processes (Additional file 5 panel B).

Finally, we also look at the rate of mutations in each tissue given the CIMP phenotype. We observe a significant association between the number of mutations and the CIMP status for colon and gastric cancer (Fig. 4b), in accordance with the tight association between CIMP and microsatellite instability for these two tissues [9, 29–31]. However, the same observation could not be made for the bladder, breast, and lung.

Clinical impact of CIMP
Survival analysis in several CIMP studies has often shown distinct outcome between CIMP-positive and CIMP-negative tumors. However, there is no consensus in the general survival associated with CIMP: while CIMP has been associated with improved survival and lower risk of metastasis in breast [14], colorectal [9], leukemia [32–35], or gliomas [17], it has also been reportedly associated with poor survival for bladder [36], lung [15, 37], or prostate cancers [38], and prognosis even remains unclear for gastric cancers [39–43].

We perform a systematic survival analysis on the different tissues to assess the clinical impact of CIMP. However, we observe no significant association between CIMP and survival, in any of the tissues (Table 7 and Additional file 6).

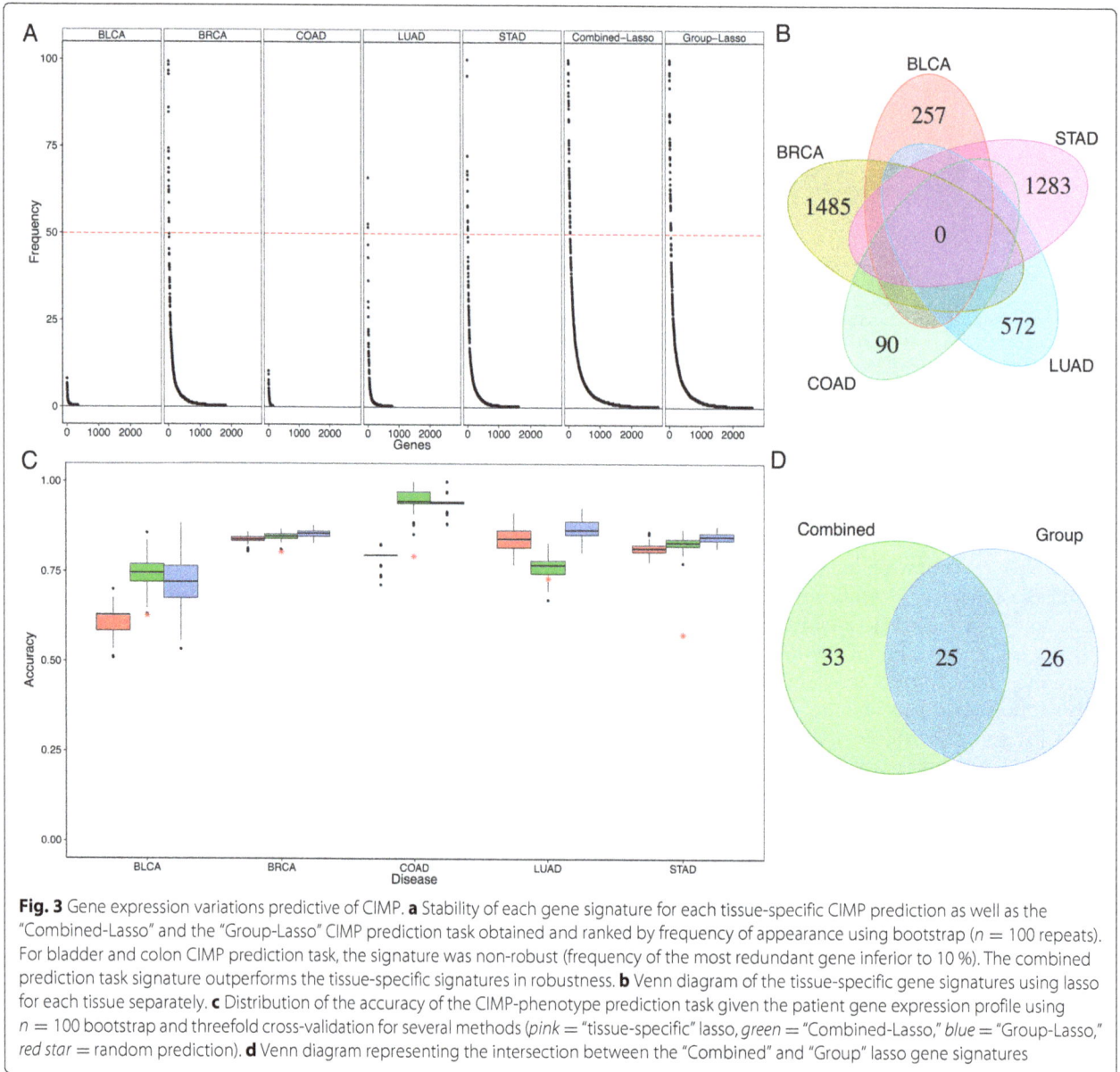

Fig. 3 Gene expression variations predictive of CIMP. **a** Stability of each gene signature for each tissue-specific CIMP prediction as well as the "Combined-Lasso" and the "Group-Lasso" CIMP prediction task obtained and ranked by frequency of appearance using bootstrap ($n = 100$ repeats). For bladder and colon CIMP prediction task, the signature was non-robust (frequency of the most redundant gene inferior to 10 %). The combined prediction task signature outperforms the tissue-specific signatures in robustness. **b** Venn diagram of the tissue-specific gene signatures using lasso for each tissue separately. **c** Distribution of the accuracy of the CIMP-phenotype prediction task given the patient gene expression profile using $n = 100$ bootstrap and threefold cross-validation for several methods (*pink* = "tissue-specific" lasso, *green* = "Combined-Lasso," *blue* = "Group-Lasso," *red star* = random prediction). **d** Venn diagram representing the intersection between the "Combined" and "Group" lasso gene signatures

Other clinical parameters have been associated with CIMP such as microsatellite instability (MSI) in the colon [9] and hormone receptor statuses in the breast [14]. We therefore assess the association between the CIMP status and eight clinical annotations provided in the TCGA, namely, age, MSI, ER status, PR status, HER2 status, tumor size, lymph node invasion, and presence of metastasis. We first observe that CIMP is significantly associated with a higher age in the breast, colon, and stomach ($P_{breast} = 2.10^{-4}$, $P_{colon} = 2.10^{-3}$, $P_{stomach} = 0.036$, student test, Additional file 7 panel A) but not in the bladder and lung. In the colon, we recover a significant association between CIMP and MSI ($P = 5.10^{-6}$, chi-squared test, Additional file 7 panel B). We also recover a significant association between CIMP and ER, PR, and HER2 statuses in breast ($P_{ER} = 2.10^{-5}$, $P_{PR} = 0.03$, $P_{HER2} = 5.10^{-8}$, chi-squared test, Additional file 7 panel C). However, we observed no significant association between CIMP and either tumor size, lymph node invasion, or metastasis in any tissue.

Discussion

CIMP has been thoroughly studied over the past few years in several tissue types but the heterogeneity of the methods and measurement technologies has hindered the assessment of a common epigenetic and genetic signature predictive of CIMP across all cancer sites [22]. In the present study, we analyze a large dataset of over

A

B

Fig. 4 Mutation analysis. **a** Association between specific mutations (*IDH1, IDH2, BRAF,* and *KRAS*) with the CIMP phenotype for all tissues (*yellow* = CIMP positive, *blue* = CIMP negative). **b** Significantly higher mutation rate for CIMP-positive (*yellow*) compared to CIMP-negative (*blue*) tumors is observed for colon and gastric cancers only and is concordant with CIMP association with microsatellite instability for these tissues

Table 6 Intersection of the genetic signatures for "Combined-Lasso" and "Group-Lasso" predictive of CIMP ranked by decreasing level of robustness

Over-expressed	ZIC2, AMH, LHX1, ZIC3, XKR9,TNNT1, CAMK2N2,PCDHB9, RAET1K, HIST1H2AB, C2CD4C, FBXL20, TFCP2L1
Under-expressed	MAGEC2, ZNF300,SLC15A1,TSPYL5, MLF1, GATA2, MAGEA12, LOC441666, MAGEA2, LOC389493, H2AFY2, LDHC

Other genes present in the cross-cancer CIMP signature such as *HOTAIR*, which is known to reprogram the chromatin state and is associated with breast cancer metastasis [45], might on the contrary be repressed in CIMP tumors and be linked with a better prognosis for breast cancer patients. *GREM1* is another gene present in the CIMP signature and is associated with tumor cell proliferation [46]. Less documented genes present in the CIMP signature could potentially be investigated for a biological validation of their role in tumor development.

Recent studies have pointed out that epigenetic aberrations could be derived from genetic aberrations [47]. By combining the different datasets into a single prediction task, we are able to identify a common set of genes whose expression levels can predict the CIMP status for each tissue. This gene list is enriched mostly in genetic regulatory pathways, suggesting that the epigenetic reprogramming and thus CIMP might be an intermediate step in the regulatory mechanism. Among the genes contained in the signature, *ZIC2*, which is robustly selected in each bootstrap of the CIMP prediction task and is significantly more expressed in CIMP-positive tumors for each tissue, has been known to act as a Wnt/β-catenin signalling inhibitor [48] which is

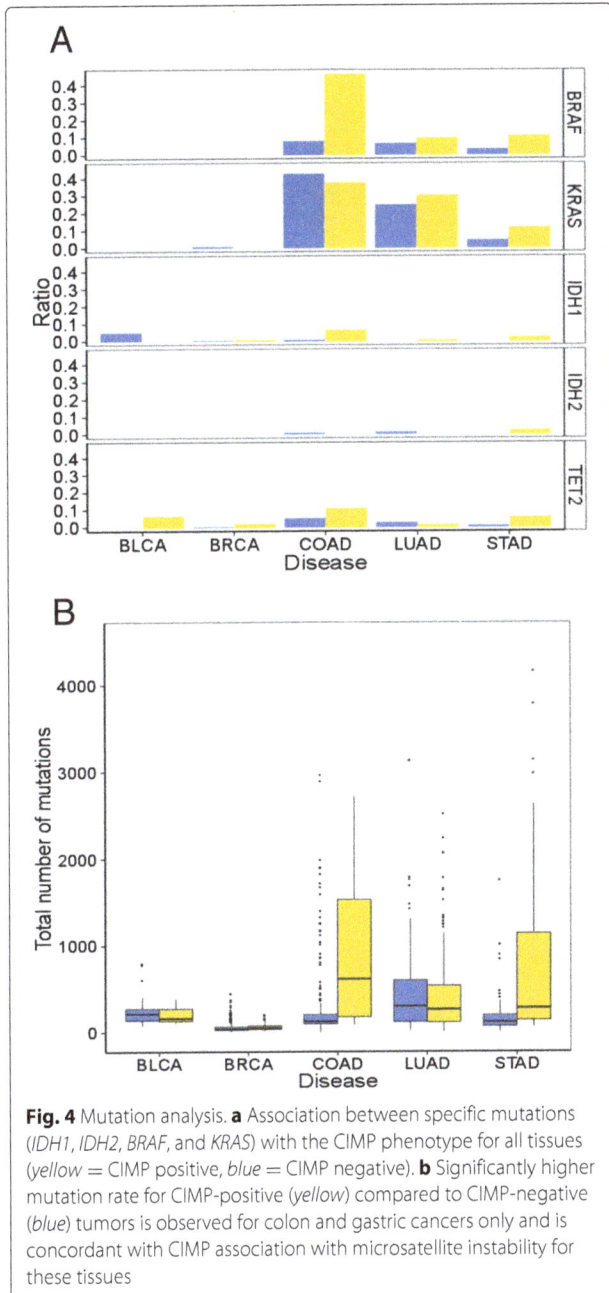

2000 tumor methylation profiles measured with a single technology from 5 different tissue types. We observe a universal epigenetic signature that defines CIMP independently from the tissue of origin, which might suggest a common molecular basis to CIMP across tissues. Genes associated with these CGIs are enriched in several biological pathways linked to organ development and include several interesting genes such as *CDKN2A* coding for p16, a well-characterized tumor suppressor protein [44], which is aberrantly hypermethylated in CIMP-positive tumors and might contribute to tumor development.

Table 7 Clinical impact of CIMP. Overall survival proportion given the CIMP phenotype and the *p* value associated with the survival analysis (logrank test)

Tissue	Event		p value
	CIMP−	CIMP+	
BLCA	47/214	21/96	0.74
BRCA	29/495	9/114	0.20
COAD	28/218	6/54	0.57
LUAD	24/127	67/295	0.49
STAD	26/141	20/193	0.29

usually upregulated in several cancers. Another interesting characteristic of this genetic predictive signature from a clinical point of view is the recurrence of cancer/testis antigens (CTAs) such as *MAGEC2* [49–51], *MAGEA12* [52, 53], *MAGEA2* [54], and *LDHC* [55], which are interesting targets for cancer immunotherapy [56] and are consistently under-expressed in CIMP-positive tumors. Recently, Gevaert et al. [57] also showed a strong association between *MAGEA4* hypomethylation and CIMP-positive tumors which further supports the link between CTAs and the absence of a methylator phenotype.

Mutation analyses are not very conclusive in defining a set of specific somatic mutations significantly associated with CIMP. In particular, lowly mutated cancer sites such as the bladder, breast, or even lung do not show any mutations significantly associated with CIMP. For highly mutated cancer sites such as colon or stomach, our results confirm a strong association between *BRAF* mutation and COAD-CIMP [9] but do not show any particular associations with *IDH1/2*, which have been reported to be causal in gliomas and leukemia [19, 26]. There is a strong association between COAD and STAD-CIMP and the specific mutations of genes related with extracellular matrix and cell adhesion, both reported to be strongly associated with metastasis [58–61]. Interestingly, neuronal developmental processes are highly enriched but affecting different genes from the universal epigenetic signature. Associations with neuronal development were already mentioned in [17].

Studies have often reported a clear distinct clinical prognosis associated with CIMP [9, 14, 17, 32]. This reiterates that a main reason for defining CIMP in each tissue site is its potential use as a prognosis marker. However, CIMP could be associated with a good or bad prognosis depending on the type of tumors. In the current study, we do not observe a significant association with any good nor bad prognosis linked with CIMP.

Conclusion

This meta-analysis of more than 2000 samples sheds new light on CIMP across cancers, its link with gene expression, and its clinical relevance. We found strong evidence that a panel of genes, which we call the pan-cancer CIMP signature, is involved simultaneously in the establishment of the CIMP in various cancer sites, which might be an indicator of a universal biological process behind CIMP. We found that differences in the CIMP status of a sample is associated to differences in the transcriptome, and also found a core set of genes whose expression levels differentiates CIMP-positive and CIMP-negative samples, in all cancers studied. Finally, we found little evidence of association between CIMP and mutations, except for the well-known BRAF mutation in colon cancer and also little association with patient survival.

Materials and methods
Patient selection

All data were retrieved from the TCGA data portal. We selected samples from bladder, breast, colon, lung and gastric adenocarcinomas because large matched datasets were available for methylation, gene expression, and mutation profiles. Moreover, all these tissues were previously reported to exhibit a methylator phenotype. The datasets are detailed in Table 1 and the different institutions that released the data are mentioned in the "Acknowledgements" section.

Methylation profiling

Methylation profiles were retrieved from level 2 TCGA data. They were obtained with the Illumina Human-Methylation450K DNA Analysis BeadChip assay, which is based on genotyping of bisulfite-converted genomic DNA at individual CpG sites to provide a quantitative measure of DNA methylation [62]. Following hybridization, the methylation value for a specific probe was calculated as the ratio $M/(M + U)$ where M is the methylated signal intensity and U is the unmethylated signal intensity. Across the genome, 485,577 CpG methylation levels, associated with 27,176 CGIs and 21,231 genes, were measured as such.

Following [63], we considered not only the CGI methylation profile but also included in the analysis proximal regions in the near vicinity (up to 4 kb), namely the CGI Shores and Shelves regions in a general CGI+SS methylation profile.

Gene expression profiling

Gene expression profiles were retrieved from level 3 TCGA data. They were obtained from the Illumina HiSeq RNASeq technology and processed following [64]. We used the reads per kilobase per million mapped reads (RPKM) to quantify the gene expression level from RNA sequencing data.

Mutation profiling

Mutation profiles were retrieved from somatic mutation profiles from level 2 TCGA data obtained through whole exome sequencing. To compare the rate of mutation given the CIMP status, we performed a hypergeometric test and corrected for multiple testing using Benjamini-Hochberg correction.

CIMP analysis

To assess the existence of CIMP, we performed Ward hierarchical clustering using euclidean distance on the top 5 % most variant CGIs. Variations from 1 to 10 % of the

most variant CGIs had a small impact on the clustering stability (Additional file 3). We then cut the hierarchical clustering tree in two classes namely CIMP-positive and CIMP-negative tumors given their average level of methylation (CIMP-positive = high level of methylation, CIMP-negative = low level of methylation). Robustness of the clustering was obtained through consensus clustering [25].

Predicting CIMP status from gene expression profiles

We performed logistic regression using a lasso penalty [65] to predict CIMP status from gene expression profiles for each tissue separately. Accuracy is calculated through threefold cross-validation averaged over 100 repeats. To combine the different datasets into a single prediction task, we performed group-lasso logistic regression (Additional file 1). Given the imbalanced proportion of CIMP in each datasets, we defined the "random" predictor as a predictor that always predicts the majority class. The statistical significance of a gene expression-based predictor over the "random" predictor was calculated using a Student t test.

To determine the genetic predictive signature, genes were ranked according to the frequency at which they appeared in the optimal lasso estimator signature averaged over the different folds and repeats. Genes with a frequency of at least 50 % were selected.

Survival analysis

Overall survival was estimated using the Kaplan-Meier method [66] to compare the survival between CIMP-positive and CIMP-negative tumors. A multivariate Cox proportional hazards regression model [67] was also fitted to assess the CIMP odd ratio.

Additional files

Additional file 1: Hierarchical clustering and CIMP status of samples in each tissue. Each sample is represented by the methylation levels of the 5 % of the probes that vary most in the tissue considered. Heatmaps range from hypomethylated (*blue*) to hypermethylated (*yellow*). The *column colorbar* represents the resulting assignment of each sample as CIMP positive (*yellow*) or CIMP negative (*blue*). Panel A. bladder; panel B. breast; panel C. colon; panel D. lung; panel E. stomach.

Additional file 2: Stability of CIMP clusters with respect to the size of the CIMP signature. Robustness of cluster assignment for each sample (*columns*) as a function of the proportion of variant CGIs kept to define the CIMP signature, from 1 to 10 % (*rows*) and given the number of CIMP clusters considered (*left panels*: K = 2, *right panels*: K = 3, *yellow* = CIMP-positive, *blue* = CIMP-negative, *black* = CIMP-low) for bladder (panel A/B), breast (panel C/D), colon (panel E/F), lung (panel G/H),

stomach (panel I/J). Panel K. Table summarizing the stability of the cluster assignments for each tissue and different number of CIMP clusters considered.

Additional file 3: Supplementary methods.

Additional file 4: Gene expression profiling on the common genetic predictive signature for each tissue. The *column color bar* represents the CIMP status (*yellow* = CIMP-positive, *blue* = CIMP-negative) while the *row color bar* represents the clustering of genes (*green* = under-expressed in CIMP, *red* = over-expressed in CIMP). Panel A. bladder; panel B. breast; panel C. colon; panel D. lung; panel E. stomach.

Additional file 5: Study of a genetic signature associated with CIMP. Panel A. Venn diagram representing the intersection of the mutations significantly associated with CIMP in colon and gastric cancers. Panel B. Gene ontology analysis of the common genes associated with CIMP.

Additional file 6: Clinical impact of CIMP on the patient surival. The plots show the Kaplan Meier survival curves based on CIMP status for different tissues. Panel A. bladder; panel B. breast; panel C. colon; panel D. lung; panel E. stomach.

Additional file 7: Association between CIMP and clinical annotations. Panel A. Association between CIMP and age: distribution of patients' age given their CIMP phenotype in each tissue. Panel B. Association between CIMP and MSI in colon: ratio of MSI-positive and MSI-negative patient given the CIMP phenotype in the colon. Panel C. Association between CIMP and ER status in breast: ratio of ER-positive patients given the CIMP phenotype in the breast. Panel D. Association between CIMP and PR status in the breast: ratio of PR-positive patients given the CIMP phenotype in the breast. Panel E. Association between CIMP and HER2 status in the breast: ratio of HER2-positive patients given the CIMP phenotype in the breast.

Abbreviations
CIMP: CpG island methylator phenotype; CGI: CpG island; BLCA: Bladder carcinoma; BRCA: Breast carcinoma; COAD: Colon adenocarcinoma; LUAD: Lung adenocarcinoma; STAD: Stomach adenocarcinoma.

Competing interests
The authors declare that they have no competing interests.

Authors' contributions
MM participated in the statistical analyses and the writing of the manuscript. FR and JPV conceived the study and participated in its design and coordination. All authors read and approved the final manuscript.

Acknowledgements
This study was financially supported by "La Ligue Nationale Contre le Cancer" (to MM) and the European Research Council (SMAC-ERC-280032 to JPV and MM).
The authors would like to acknowledge the following: the Cancer Genome Atlas, IUPUI[1,2,4,5], Lahey Clinic[1,4], Research Metrics Pakistan[2], Asterand[1,2,3,4,5], Baylor[1,3,4], Cleveland Clinic[1], UT Southwestern Medical Center at Dallas[1], University of Chicago[1], University of Miami[1], Barretos Cancer Hospital[1,5], Penrose Colorado[1,4], Candler[1,3,4,5], Christiana Healthcare[1,2,3,4,5], Cornell Medical College[1], Cureline[2,3,4,5], Duke[2], Erasmus MC[1], Fox Chase[4,5], Global Bioclinical-Moldova[5], Greater Poland Cancer Center[2,3,5], Gundersen Lutheran Health System[1,4], Hartford Hospital[1], Harvard[3], ILSBio[1,2,3,4,5], Indivumed[1,2,3,4,5], International Genomics Consortium[1,2,3,4,5], Johns Hopkins[4], MD Anderson Cancer Center[1,2,5], Memorial Sloan Kettering Cancer Center[2,3,5], Mayo Clinic[1,2], Medical College of Georgia[1], National Cancer Center Korea[5], Ontario Institute for Cancer Research[1,2,3,4,5], Peter MacCallum Cancer Center[5], Prince Charles Hospital[4], Roswell Park[1,2,3,4], Spectrum Health[1], St Joseph Medical Center[3,4], Tayside Tissue Bank[5], Thoraxklinik at University Hospital Heidelberg[4], UCSF[2], UNC[1,2,3,4,5], University Health Network[5], University of Michigan[3], University of Colorado Denver[1], University of Miami[2,4], University of Minnesota[1], University of Oklahoma[1], University of Pittsburgh[1,2,3,4,5], University of Puerto Rico[1], University of Sheffield[1], University of Southern California[1], Walter Reed[2], Washington University[4] for the distribution of patients data (as specified below).

Author details

[1]CBIO-Centre for Computational Biology, Mines Paristech, PSL-Research University, 35 Rue Saint-Honore, F-77300 Fontainebleau, France. [2]Department of Bioinformatics, Biostatistics and System Biology, Institut Curie, 11-13 Rue Pierre et Marie Curie, F-75248 Paris, France. [3]U900, INSERM, 11-13 Rue Pierre et Marie Curie, F-75248 Paris, France. [4]UMR932, Immunity and Cancer Team, Institut Curie, 26 Rue d'Ulm, 75006 Paris, France. [5]Department of Translational Research, Residual Tumor and Response to Treatment Team, Institut Curie, 26 Rue d'Ulm, 75006 Paris, France. [6]Department of Surgery, Institut Curie, 26 Rue d'Ulm, 75006 Paris, France.

References

1. Jones PA, Baylin SB. The epigenomics of cancer. Cell. 2007;128(4):683–92.
2. Esteller M. Epigenetics in cancer. New Eng J Med. 2008;358(11):1148–59.
3. Rodriguez-Paredes M, Esteller M. Cancer epigenetics reaches mainstream oncology. Nat Med. 2011;17(3):330–339.
4. Jones P. DNA methylation and cancer. Cancer Res. 1986;46(2):461–6.
5. Baylin SB, Herman JG. DNA hypermethylation in tumorigenesis: epigenetics joins genetics. Trends Genet. 2000;16(4):168–74.
6. Esteller M, Corn PG, Baylin SB, Herman JG. A gene hypermethylation profile of human cancer. Cancer Res. 2001;61(8):3225–9.
7. Toyota M, Ahuja N, Ohe-Toyota M, Herman JG, Baylin SB, Issa J-PJ. CpG island methylator phenotype in colorectal cancer. Proc Nat Acad Sci. 1999;96(July):8681–6.
8. Issa J-PJ, Shen L, Toyota M. CIMP, at last. Gastroenterology. 2005;129(3): 1121–4.
9. Weisenberger DJ, Siegmund KD, Campan M, Young J, Long TI, Faasse Ma, et al. CpG island methylator phenotype underlies sporadic microsatellite instability and is tightly associated with BRAF mutation in colorectal cancer. Nat Genet. 2006;38(7):787–93.
10. Estécio MRH, Yan PS, Ibrahim AEK, Tellez CS, Shen L, Huang TH-M, et al. High-throughput methylation profiling by MCA coupled to CpG island microarray. Genome Res. 2007;17(10):1529–36.
11. Curtin K, Slattery ML, Samowitz WS. CpG island methylation in colorectal cancer: past, present and future. Pathol Res Int. 2011;2011:902674.
12. Hinoue T, Weisenberger D, Lange C, Shen H, Byun H, Van Den Berg D, et al. Genome-scale analysis of aberrant DNA methylation in colorectal cancer. Genome Res. 2012;22(2):271–82.
13. Van der Auwera I, Yu W, Suo L, Van Neste L, van Dam P, Van Marck EA, et al. Array-based DNA methylation profiling for breast cancer subtype discrimination. PloS One. 2010;5(9):e12616. doi:10.1371/journal.pone.0012616.
14. Fang F, Turcan S, Rimner A, Kaufman A, Giri D, Morris LGT, et al. Breast cancer methylomes establish an epigenomic foundation for metastasis. Sci Trans Med. 2011;3(75):75–25.
15. Suzuki M, Shigematsu H, Lizasa T, Hiroshima K, Nakatani Y, Minna J, et al, Cancer. Exclusive mutation in epidermal growth factor receptor gene, HER-2, and KRAS, and synchronous methylation of nonsmall cell lung cancer. 2006;106(10):2200–7.
16. Chen HY, Zhu BH, Zhang CH, Yang DJ, Peng JJ, Chen JH, et al. High CpG island methylator phenotype is associated with lymph node metastasis and prognosis in gastric cancer. Cancer Sci. 2012;103(1):73–9.
17. Noushmehr H, Weisenberger DJ, Diefes K, Phillips HS, Pujara K, Berman BP, et al. Identification of a CpG island methylator phenotype that defines a distinct subgroup of glioma. Cancer Cell. 2010;17(5):510–22.
18. Baysan M, Bozdag S, Cam MC, Kotliarova S, Ahn S, Walling J, et al. G-CIMP status prediction of glioblastoma samples using mRNA expression data. PloS One. 2012;7(11):47839.
19. Yilmaz E, Campos C, Fabius AWM, Lu C, Ward PS, Viale A, et al. IDH1 mutation is sufficient to establish the glioma hypermethylator phenotype. Nature. 2012;483(7390):479–83.
20. Bae YK, Brown A, Garrett E, Bornman D, Fackler MJ, Sukumar S, et al. Hypermethylation in histologically distinct classes of breast cancer. Clinical Cancer Res. 2004;10(18):5998–6005.
21. Anacleto C, Leopoldino A, Rossi B, Soares FA, Lopes A, Rocha JC, et al. Colorectal cancer "methylator phenotype": fact or artifact? Neoplasia. 2005;7(4):331–5.
22. Hughes LAE, Melotte V, de Schrijver J, de Maat M, Smit VTHBM, Bovee JVMG, et al. The CpG island methylator phenotype: what's in a name? Cancer research. 2013;73(19):5858–68.
23. Ogino S, Kawasaki T, Kirkner GJ, Loda M, Fuchs CS. CpG island methylator phenotype-low (CIMP-low) in colorectal cancer: possible associations with male sex and KRAS mutations. J Mol Diagn. 2006;8(5):582–8.
24. Shen L, Toyota M, Kondo Y, Lin E, Zhang L, Guo Y, et al. Integrated genetic and epigenetic analysis identifies three different subclasses of colon cancer. Proc Natl Acad Sci USA. 2007;104(47):18654–9.
25. Monti S, Tamayo P, Mesirov J, Golub T. Consensus Clustering : A Resampling-Based Method for Class Discovery and Visualization of Gene. Machine Learning. 2003;52(1):91–118.
26. Figueroa M, Abdel-Wahab O, Lu C, Ward P, Patel J, Shih A, et al. Leukemic IDH1 and IDH2 mutations result in a hypermethylation phenotype, disrupt TET2 function, and impair hematopoietic differentiation. Cancer Cell. 2010;18(6):553–67.
27. Amary M, Damato S, Halai D, Eskandarpour M, Berisha F, Bonar F. Ollier disease and Maffucci syndrome are caused by somatic mosaic mutations of IDH1 and IDH2. Nat Genet. 2011;43(12):1262–5.
28. Pansuriya T, van Eijk R, d'Adamo P, van Ruler M, Kuijjer M, Oosting J, et al. Somatic mosaic IDH1 and IDH2 mutations are associated with enchondroma and spindle cell hemangioma in Ollier disease and Maffucci syndrome. Nat Genet. 2011;43(12):1256–61.
29. Herman J, Umar A, Polyak K, Graff J, Ahuja N, Issa J, et al. Incidence and functional consequences of hMLH1 promoter hypermethylation in colorectal carcinoma. Proc Natl Acad Sci U S A. 1998;95(12):6870–5.
30. Jones S, Li M, Parsons D, Zhang X, Wesseling J, Kristel P, et al. Somatic mutations in the chromatin remodeling gene ARID1A occur in several tumor types. Hum Mutat. 2012;33(1):100–3.
31. Zang Z, Cutcutache I, Poon S, Zhang S, McPherson J, Tao J, et al. Exome sequencing of gastric adenocarcinoma identifies recurrent somatic mutations in cell adhesion and chromatin remodeling genes. Nat Genet. 2012;44(5):570–4.
32. Toyota M, Kopecky K, Toyota M, Jair K, Willman C, Issa J. Methylation profiling in acute myeloid leukemia. Blood. 2001;97(9):2823–9.
33. Garcia-Manero G, Daniel J, Smith T, Kornblau S, Lee M, Kantarjian H, et al. DNA methylation of multiple promoter-associated CpG islands in adult acute lymphocytic leukemia. Clinical Cancer Res. 2002;8(7):2217–24.
34. Roman-Gomez J, Jimenez-Velasco A, Agirre X, Prosper F, Heiniger A, Torres A. Lack of CpG island methylator phenotype defines a clinical subtype of T-cell acute lymphoblastic leukemia associated with good prognosis. J Clin Oncol. 2005;23(28):7043–9.
35. Roman-Gomez J, Jimenez-Velasco A, Agirre X, Castillejo J, Navarro G, Calasanz M, et al. CpG island methylator phenotype redefines the prognostic effect of t(12;21) in childhood acute lymphoblastic leukemia. Clinical Cancer Res. 2006;12(16):4845–50.
36. Maruyama R, Toyooka S, Toyooka K, Harada K, Virmani A, Zochbauer-Muller S, et al. Aberrant promoter methylation profile of bladder cancer and its relationship to clinicopathological features. Cancer Res. 2001;61(24):8659–63.
37. Liu Z, Zhao J, Chen X, Li W, Liu R, Lei Z, et al. CpG island methylator phenotype involving tumor suppressor genes located o chromosome 3p in non-small cell lung cancer. Lung Cancer. 2008;62(1):15–22.
38. Maruyama R, Toyooka S, Toyooka K, Virmani A, Zochbauer-Muller S, Farinas A, et al. Aberrant promoter methylation profile of prostate cancers and its relationship to clinicopathological features. Clinical Cancer Res. 2002;8(2):514–9.
39. Toyota M, Ahuja N, Suzuki H, Itoh F, Ohe-Toyota M, Imai K, et al. Aberrant methylation in gastric cancer associated with the CpG island methylator phenotype. Cancer Res. 1999;59:5438–42.
40. Oue N, Oshimo Y, Nakayama H, Ito R, Yoshida K, Matsusaki K, et al. DNA methylation of multiple genes in gastric carcinoma: association with histological type and CpG island methylator phenotype. Cancer Sci. 2003;94(10):901–5.
41. Kim H, Kim Y, Kim S, Kim N, Noh S. Concerted promoter hypermethylation of hMLH1, p16INK4A, and E-cadherin in gastric carcinomas with microsatellite instability. J Pathol. 2003;200(1):23–31.
42. Etoh T, Kanai Y, Ushijima S, Nakagawa T, Nakanishi Y, Sasako M, et al. Increased DNA methyltransferase 1 (DNMT1) protein expression correlates significantly with poorer tumor differentiation and frequent DNA hypermethylation of multiple CpG islands in gastric cancers. Am J Pathol. 2004;164(2):689–99.
43. Kusano M, Toyota M, Suzuki H, Akino K, Aoki F, Fujita M, et al. Genetic, epigenetic, and clinicopathologic features of gastric carcinomas with the CpG island methylator phenotype and an association with Epstein-Barr virus. Cancer. 2006;106(7):1467–79.
44. Nobori T, Miura K, Wu DJ, Lois A, Takabayashi K, Carson DA. Deletions of

the cyclin-dependent kinase-4 inhibitor gene in multiple human cancers. Nature. 1994;368(April):753–6.

45. Gupta RA, Shah N, Wang KC, Kim J, Horlings HM, Wong DJ, et al. Long noncoding RNA HOTAIR reprograms chromatin state to promote cancer metastasis. Nature. 2010;464(7291):1071–6.

46. Sneddon J, Zhen H, Montgomery K, van de Rijn M, Tward A, West R, et al. Bone morphogenetic protein antagonist gremlin 1 is widely expressed by cancer-associated stromal cells and can promote tumor cell proliferation. Proc Natl Acad Sci USA. 2006;103(40):14842–7.

47. Reddington JP, Sproul D, Meehan RR. DNA methylation reprogramming in cancer: does it act by re-configuring the binding landscape of Polycomb repressive complexes? BioEssays: News Rev Mol Cell Dev Biol. 2014;36(2):134–40.

48. Pourebrahim R, Houtmeyers R, Ghogomu S, Janssens S, Thelie A, Tran H, et al. Transcription factor Zic2 inhibits Wnt/beta-catenin protein signaling. J Biol Chem. 2011;286(43):37732–40.

49. von Boehmer L, Keller L, Mortezavi A, Provenzano M, Sais G, Hermanns T, et al. MAGE-C2/CT10 protein expression is an independent predictor of recurrence in prostate cancer. PLoS ONE. 2011;6(7):1–7.

50. Yang F, Zhou X, Miao X, Zhang T, Hang X, Tie R, et al. MAGEC2, an epithelial-mesenchymal transition inducer, is associated with breast cancer metastasis. Breast Cancer Res Treatment. 2014;145(1):23–32.

51. Reinhard H, Yousef S, Luetkens T, Fehse B, Berdien B, Kröger N, et al. Cancer-testis antigen MAGE-C2/CT10 induces spontaneous CD4+ and CD8+ T-cell responses in multiple myeloma patients. Blood Cancer J. 2014;4:e212. doi:10.1038/bcj.2014.31.

52. Heidecker L, Brasseur F, Probst-Kepper M, Guéguen M, Boon T, Van den Eynde BJ. Cytolytic T lymphocytes raised against a human bladder carcinoma recognize an antigen encoded by gene MAGE-A12. J Immunol (Baltimore, Md. : 1950). 2000;164(11):6041–5.

53. Mollaoglu N, Vairaktaris E, Nkenke E, Neukam FW, Ries J. Expression of MAGE-A12 in oral squamous cell carcinoma. Disease Markers. 2008;24(1): 27–32.

54. Peche LY, Scolz M, Ladelfa MF, Monte M, Schneider C. MageA2 restrains cellular senescence by targeting the function of PMLIV/p53 axis at the PML-NBs. Cell Death Differentiation. 2012;19(6):926–36.

55. Tang H, Goldberg E. Homo sapiens lactate dehydrogenase c (Ldhc) gene expression in cancer cells is regulated by transcription factor Sp1, CREB, and CpG island methylation. J Androl. 2009;30(2):157–67.

56. Scanlan MJ, Gure AO, Jungbluth AA, Old LJ, Chen YT. Cancer/testis antigens: an expanding family of targets for cancer immunotherapy. Immunol Rev. 2002;188(1):22–32.

57. Gevaert O, Tibshirani R, Plevritis SK. Pancancer analysis of DNA methylation-driven genees using MethylMix. Genome Biol. 2015;16(1):17.

58. Gilkes DM, Semenza GL, Wirtz D. Hypoxia and the extracellular matrix: drivers of tumour metastasis. Nat Rev Cancer. 2014;141(6):430–439.

59. Lu P, Weaver VM, Werb Z. The extracellular matrix: a dynamic niche in cancer progression. J Cell Biol. 2012;196(4):395–406.

60. Bendas G, Borsig L. Cancer cell adhesion and metastasis: selectins, integrins, and the inhibitory potential of heparins. Int J Cell Biol. 2012;ID 676731:10. doi:10.1155/2012/676731.

61. Okegawa T, Pong R, Hsieh J. The role of cell adhesion molecule in cancer progression and its application in cancer therapy. Acta Biochim Pol. 2004;51(2):445–57.

62. Bibikova M, Barnes B, Tsan C, Ho V, Klotzle B, Le JM, et al. High density DNA methylation array with single CpG site resolution. Genomics. 2011;98(4):288–95.

63. Irizarry RA, Ladd-Acosta C, Wen B, Wu Z, Montano C, Onyango P, et al. Genome-wide methylation analysis of human colon cancer reveals similar hypo- and hypermethylation at conserved tissue-specific CpG island shores. Nat Genet. 2009;41(2):178–86.

64. Mortazavi A, Williams BA, Mccue K, Schaeffer L, Wold B. Mapping and quantifying mammalian transcriptomes by RNA-Seq. Nat Methods. 2008;5(7):1–8.

65. Tibshirani R. Regression shrinkage and selection via the lasso. J R Stat Soc. 1996;58(1):267–88.

66. Kaplan EL, Meier D. Nonparametric estimation from incomplete observation. J Am Statist. 1958;58:457–81.

67. Cox DR, Oakes D. Analysis of Survival Data. London: Chapman & Hall/CRC Monographs on Statistics & Applied Probability, Taylor & Francis; 1984.

7

Identification of protein complexes from multi-relationship protein interaction networks

Xueyong Li[1,2], Jianxin Wang[1*], Bihai Zhao[2*], Fang-Xiang Wu[3] and Yi Pan[4]

Abstract

Background: Protein complexes play an important role in biological processes. Recent developments in experiments have resulted in the publication of many high-quality, large-scale protein-protein interaction (PPI) datasets, which provide abundant data for computational approaches to the prediction of protein complexes. However, the precision of protein complex prediction still needs to be improved due to the incompletion and noise in PPI networks.

Results: There exist complex and diverse relationships among proteins after integrating multiple sources of biological information. Considering that the influences of different types of interactions are not the same weight for protein complex prediction, we construct a multi-relationship protein interaction network (MPIN) by integrating PPI network topology with gene ontology annotation information. Then, we design a novel algorithm named MINE (identifying protein complexes based on Multi-relationship protein Interaction NEtwork) to predict protein complexes with high cohesion and low coupling from MPIN.

Conclusions: The experiments on yeast data show that MINE outperforms the current methods in terms of both accuracy and statistical significance.

Background

With the completion of the sequencing of the human genome, proteomic research becomes one of the most important areas in the life science. One important task in proteomics is to detect protein complexes based on protein-protein interaction (PPI) data generated by various experimental technologies, e.g., yeast-two-hybrid [1], tandem affinity purification [2], and mass spectrometry [3]. Protein complexes are molecular aggregations of proteins assembled by PPIs, which play critical roles in biological processes. Many proteins are functional only when they are assembled into a protein complex and interact with other proteins in this complex. Protein complexes are key molecular entities to perform cellular functions. Even in the relatively simple model organism *Saccharomyces cerevisiae*, these complexes are comprised of many subunits that work

in a coherent fashion. Besides applications of PPI networks, such as protein function predictions [4] and essential protein discoveries [5–11], prediction of protein complexes is another active topic. Actually, protein complexes are of great importance for understanding the principles of cellular organization and function.

Many computational methods for predicting protein complexes from PPI networks have been developed. Pairwise protein interactions can be modelled as a graph or network, where vertices are proteins and edges are PPIs. Since proteins in the same complex are highly interactive with each other, protein complexes generally correspond to dense subgraphs in the PPI network and many previous studies have been proposed based on this observation, such as MCODE (Molecular Complex detection) [12], MCL (Markov Cluster algorithm) [13], R-MCL (Regularized MCL) [14], CMC (Maximal Clique algorithm) [15], RRW (Repeated Random Walks) [16], SPICi (Speed and Performance in Clustering algorithm) [17], HC-PIN (Hierarchical Clustering based on Protein-Protein Interaction Network) [18], IPC-MCE (Identifying Protein

* Correspondence: jxwang@mail.csu.edu.cn; bihaizhao@163.com
[1]School of Information Science and Engineering, Central South University, Changsha 410083, China
[2]Department of Information and Computing Science, Changsha University, Changsha 410003, China
Full list of author information is available at the end of the article

Complexes based on Maximal Clique Extension) [19], and IPCA (Identification of Protein Complexes Algorithm) [20]. Nepusz et al. [21] proposed an algorithm to find overlapping protein complexes from PPI networks, named ClusterONE (Clustering with Overlapping Neighborhood Expansion). For the convenience of researchers, MCODE, ClusterONE, etc. have been designed as plus-in for protein complex prediction and biological network analysis. ClusterViz [22] is such a Cytoscape APP to complete this work.

However, these abovementioned approaches for extracting dense subgraphs fail to take into account the inherent organization. Recent analysis of experimentally detected protein complexes [23] has revealed that a complex consists of a core component and attachments. Core proteins are highly co-expressed and share high functional similarity, and each attachment protein binds to a subset of core proteins to form a biological complex. Based on the core-attachment concept, some algorithms have been proposed, including COACH (Core-Attachment-based method) [24], CORE [25], MCL-Caw [26], DCU (Detecting Complex based on Uncertain graph model) [27], and WPNCA (a Weighted PageRank-Nibble algorithm with Core-Attachment structure) [28].

In spite of the advances in computational approaches and related fields, accurate identification protein complexes are still a bottleneck. One of the most important reasons is that the PPI network contains a lot of false positives which greatly reduce the complex detection accuracy. To address this problem, biological information other than PPIs has been integrated with network topology to improve the precision of protein complex detection methods. Wu et al. proposed a method called CACHET to discover protein complexes with core-attachment structures from tandem affinity purification (TAP) data [29]. Tang et al. [30] constructed time course PPI networks by incorporating gene expression into PPI networks and applied it successfully to the identification of function modules. Wang et al. [31] proposed a three-sigma method to identify active time points of each protein in a cellular cycle, where three-sigma principle is used to compute an active threshold for each gene according to the characteristics of its expression curve. A dynamic PPI network (DPIN) is constructed for the detection of protein complexes. Li et al. proposed novel algorithms, such as TSN-PCD [32] and DPC [33], to identify dynamic protein complexes by integrating PPI data and dynamic gene expression profiles. Zhao et al. [34] reconstructed a weighted PPI network by using dynamic gene expression data and developed a novel protein complex identification algorithm, named PCIA-GeCo.

There exist complex and diverse relationships among proteins after integrating multiple sources of biological information. However, comparing PPI data is difficult because they are often diverse and play different roles under different conditions. Current existing approaches failed to take into account and combined the interactions with different natures into one interaction effectively. Taking into account the influences of different types of interactions are not the same weight for protein complex prediction, we construct a multi-relationship protein interaction network (MPIN) by integrating PPI network topology with gene ontology (GO) annotation information. Then, a new method named MINE (identify protein complexes based on Multi-relationship protein Interaction NEtwork) is proposed. We have conducted an experiment on yeast data. Experimental results show that MINE outperforms the existing methods in terms of both accuracy and p value.

Methods
Multi-relationship protein interaction network
Complex networks have now been a new research focus because of surging networks in various fields such as engineering, social science, and life science. In reality, connections among nodes in complex networks are diversified. Multi-relationship means that there is more than one connection between two nodes and each of them has its own property. For instance, in social networks [35], persons contact with each other via emails, telephones, MSN, etc. and hence make up a complex multi-relationship network. Similarly, in biological networks, there are diverse links among proteins like physical interaction, co-expression, and co-annotation. However, multi-relationship networks are much more difficult to analyze than single-relationship networks. Multi-relationship networks are also essential in better reflecting the real world.

Definition 1 Multi-relationship network
Consider a PPI network $G = (V, E)$, where $V = \{v_1, v_2,..., v_n\}$ represents a set of proteins and $E = \{e_1, e_2,..., e_m\}$ represents a set of interactions. A multi-relationship network is defined as MG = $(V, E \cup E', T)$, where $T(e_i) = t_i$ $(i = 1, 2...m)$ is the interaction type of e_i. E' is the set of new generated interactions.

In a multi-relationship network, a pair of proteins may be connected by more than one type of links. If there are two or more links between a pair of proteins, they are called parallel interactions. Figure 1 illustrates a typical multi-relationship network. From Fig. 1, we can see that proteins A and B have physical interaction in the PPI network and at the same time, A and B are also co-expression based on gene expression profiles and co-annotations based on gene ontology annotation information. In the multi-relationship network, multiple connections between A and B are kept.

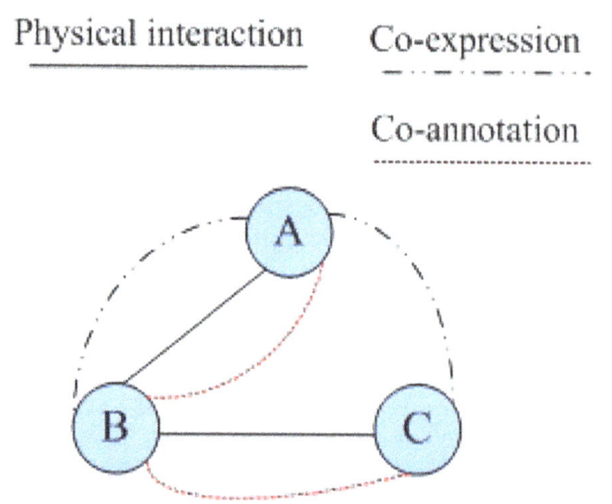

Fig. 1 An example of a typical multi-relationship network. There are three links between A and B, including physical interaction, co-expression, and co-annotation. *Solid line* represents physical interaction of PPI networks, *dotted line* represents co-annotation between two nodes, and the *rest line* indicates co-expression

Researches [27, 36] show that PPI data obtained through high-throughput biological experiments contains relatively high rates of false positives and false negatives. False positives become obstacle to the precision of prediction algorithm. False negatives lead to the loss of interaction data and continue to inhibit the increase of the number of protein complexes correctly matched. To overcome these problems, researches have begun to integrate the PPI network and other biological information, such as gene expression profiles, essential proteins, and GO annotation information. Due to the similar biological properties of protein complexes, GO annotation is a valuable addition to PPI data for protein complex prediction. Therefore, in this study we construct a multi-relationship protein interaction network by integrating PPI network topology and GO annotation information.

The GO database consists of three separate categories of annotations, namely molecular function (MF), biological process (BP), and cellular component (CC). MF describes activities, such as catalytic or binding activities, at the molecular level. BP describes biological goals accomplished by one or more ordered assemblies of molecular functions. CC describes locations, at the levels of subcellular structures and macromolecular complexes. In this study we integrate the PPI network and three categories of GO annotations to construct a multi-relationship protein interaction network. In our constructed multi-relationship network, four kinds of interactions at most can be considered between two proteins, namely the interactions of the PPI network and the interactions of sharing molecular functions, sharing biological processes, and sharing cellular components. Figure 2 describes the process of a multi-relationship network construction.

In the constructed multi-relationship protein interaction network, two proteins are connected if they interact with each other in the PPI network or have common functions, including biological processes, molecular functions, and cellular components. After constructing a multi-relationship protein interaction network, we do some further processing, such as weighting and filtering. Studies [9, 10, 36] show that the performance of prediction algorithms based on weighted networks is generally superior to that based on unweighted networks. The reason is simple: weight stands for the relative reliability/importance of interactions; thus, weighted networks can be more valuable than unweighted networks in the representative of PPI networks. For the first type of interaction in our constructed multi-relationship network, interacting with each other in the PPI network, we weight these interactions through the analysis of topological features of PPI networks. Generally speaking, for a pair of interacting proteins, the strength of an interaction can be reflected by the number of its common neighbors. This study uses ECC to calculate the weight of protein pairs, which is defined as

$$\mathrm{ECC}(v_i, v_j) = \begin{cases} \dfrac{|N_i \cap N_j|^2}{(|N_i|-1) * (|N_j|-1)} & , \ |N_i| > 1 \text{ and } |N_j| > 1 \\ 0 & , \ |N_i| = 1 \text{ or } |N_j| = 1 \end{cases}, \tag{1}$$

where N_i and N_j are the neighborhood sets of v_i and v_j, respectively. To reduce the negative effect of false positive on the protein complex prediction, we remove interactions whose ECC values are zero.

For the rest three types of interaction, we weight interactions according to the number of common functions (including BP, MF, and CC) between two proteins. For a pair of proteins v_i and v_j, BP_i and BP_j are sets of biological processes of v_i and v_j, respectively. W_BP (v_i, v_j) represents the strength of sharing biological processes, which is calculated as follows:

$$\mathrm{W_BP}(v_i, v_j) = \begin{cases} \dfrac{|\mathrm{BP}_i \cap \mathrm{BP}_j|^2}{|\mathrm{BP}_i| * |\mathrm{BP}_j|} & , \ |\mathrm{BP}_i| * |\mathrm{BP}_j| > 0 \\ 0 & , \ |\mathrm{BP}_i| * |\mathrm{BP}_j| = 0 \end{cases} \tag{2}$$

In Eq. (2), $BP_i \cap BP_j$ denotes the set of common biological processes of v_i and v_j. In a similar way, W_MF (v_i, v_j) and W_CC (v_i, v_j) denote the strengths of sharing molecular functions and cellular components of v_i and v_j, respectively. They can be calculated as follows:

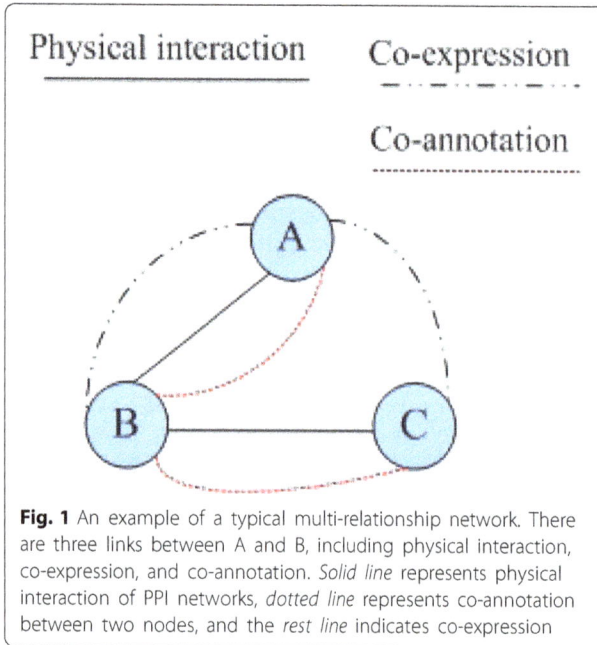

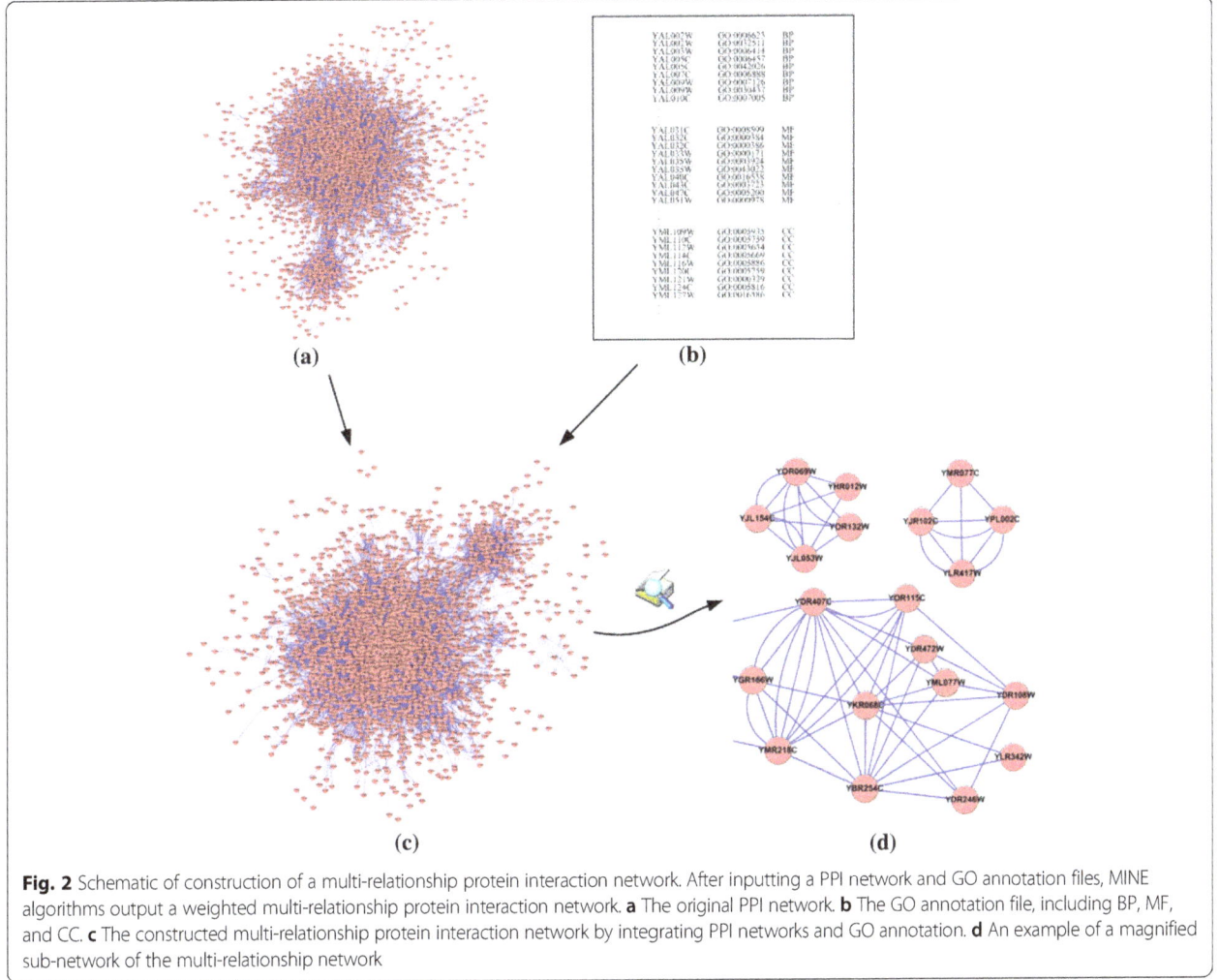

Fig. 2 Schematic of construction of a multi-relationship protein interaction network. After inputting a PPI network and GO annotation files, MINE algorithms output a weighted multi-relationship protein interaction network. **a** The original PPI network. **b** The GO annotation file, including BP, MF, and CC. **c** The constructed multi-relationship protein interaction network by integrating PPI networks and GO annotation. **d** An example of a magnified sub-network of the multi-relationship network

$$W\_MF(v_i, v_j) = \begin{cases} \dfrac{|MF_i \cap MF_j|^2}{|MF_i| * |MF_j|} & , \quad |MF_i| * |MF_j| > 0 \\ 0 & , \quad |MF_i| * |MF_j| = 0 \end{cases} \quad (3)$$

$$W\_CC(v_i, v_j) = \begin{cases} \dfrac{|CC_i \cap CC_j|^2}{|CC_i| * |CC_j|} & , \quad |CC_i| * |CC_j| > 0 \\ 0 & , \quad |CC_i| * |CC_j| = 0 \end{cases} \quad (4)$$

For the three types of interactions, we perform more stringent filter operations than the first type because they are newly generated interactions. For a pair of function-shared proteins, if they have only one common function or no common neighbors in the PPI network, interactions between them are removed. After performing the above operations, a weighted multi-relationship protein interaction network is constructed.

MINE algorithm

Considering the influences of different types of interactions in protein complex prediction are not the same, we construct a multi-relationship protein interaction network by integrating PPI networks and GO annotation information. To test the effectiveness of the multi-relationship network, we design a new method for predicting protein complexes, named MINE (based on Multi-relationship protein Interaction NEtwork). Multi-relationship networks have more complex attributes than single networks. Current protein complex prediction methods are mainly based on single networks. So, converting a multi-relationship network into single networks is key to design the MINE algorithm. A simple way for addressing this problem is to combine interactions with different natures to one interaction effectively. In reality, it is inappropriate for us to combine multiple interactions between two proteins because they are often derived under different conditions and play different roles in protein complex prediction. Considering that different types of interactions play different roles in detecting protein complexes, we decompose the multi-relationship network into several single networks,

including the PPI network, BPN (sharing biological processes), MFN (sharing molecular functions) and CCN (sharing cellular components). Figure 3 displays the framework of multi-relationship decomposition.

And then, we identify protein complexes through mining density subgraphs from the four networks. Intuitively, a subgraph representing a protein complex should satisfy two simple structural properties: it should contain many reliable interactions between its subunits, and it should be well-separated from the rest of the network [21]. Inspired by the notion, we take into account the density of a subgraph and connections between nodes of the subgraph and nodes out of the subgraph. To describe MINE simply and clearly, we provide the following definitions, firstly.

Definition 2 Weighted Density [27]

Given a weighted network $G = (V, E, W)$. $V = \{v_1, v_2, ..., v_n\}$, $E = \{e_1, e_2,..., e_m\}$, $W = \{w(e_1), w(e_2),..., w(e_m)\}$, $w(e_i)$ is the weight of an edge e_i. WD (G) denotes the weighted density of G and is defined as

$$WD(G) = \frac{\sum_{i=1}^{m} p(e_i) \times 2}{\max_{1 \leq i \leq m} (p(e_i)) \times (|V| \times (|V| - 1))} \tag{5}$$

Definition 3 Sub-network Weighted Degree [36]

Given a weighted sub-network $G = (V, E, W)$ and a vertex u, $u \in V$. $V = \{v_1, v_2, ..., v_n\}$, $E = \{e_1, e_2,..., e_m\}$, $W = \{w(e_1), w(e_2),..., w(e_m)\}$, $w(e_i)$ is the weight of an edge e_i. SWD (u, G) denotes the weighted degree of u within G and is defined as

$$SWD(u, G) = \sum_{i=1}^{n} w(u, v_i), (u, v_i) \in E \tag{6}$$

Based on these definitions, we are now ready to describe our proposed MINE algorithm to detect protein complexes. Our method visits the four single networks, respectively, to discover density subgraphs as protein complexes. For a selected network, MINE starts from a randomly chosen protein vertex and add protein vertices via a greedy procedure to form a candidate complex

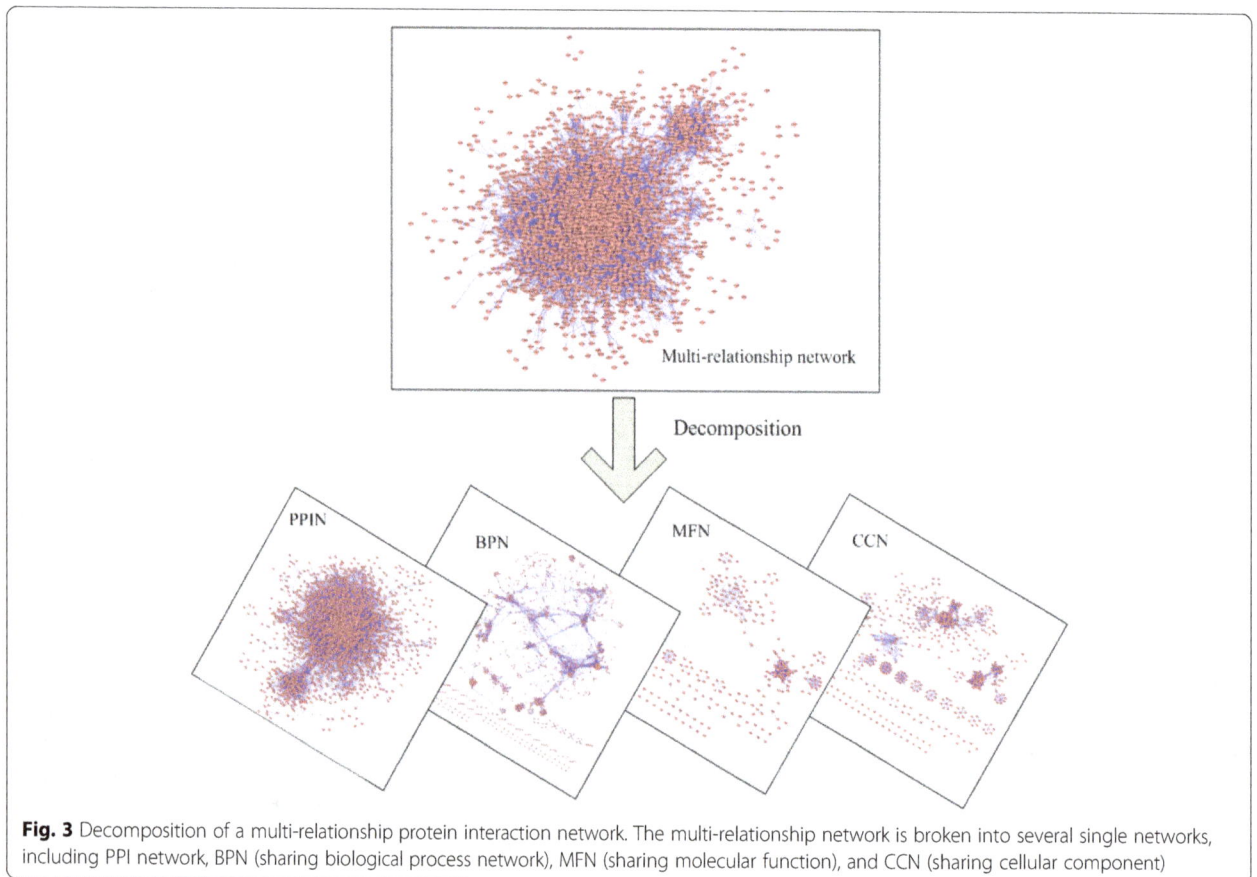

Fig. 3 Decomposition of a multi-relationship protein interaction network. The multi-relationship network is broken into several single networks, including PPI network, BPN (sharing biological process network), MFN (sharing molecular function), and CCN (sharing cellular component)

with high cohesion and low coupling. The growth process is repeated from all vertices to form non-redundant complex sets. Since some vertices have similar neighborhood graphs, the candidate complexes detected from their neighborhood graphs may have large overlaps, which result in high redundancy. Hence, a redundancy-filtering procedure is applied to quantify the extent overlap between each pair of complexes and discard the complexes with low density or small size.

MINE algorithm (Algorithm 1) describes the overall procedure to identify protein complexes. MINE algorithm processes four single networks according to the multi-relationship network, such as PPIN, BPN, MFN, and CCN, in line 1. For a selected network G_k, we first generate candidate complexes according to neighbors of all proteins in the network, in lines 3–8. The seed is inserted into the candidate set CCS, and then all neighbors of the seed are put into CCS one by one. If the weighted density of CCS is less than the threshold WDT, the new added neighbor node is removed from CCS. After this process, a candidate complex with high cohesion is formed. Then, we remove some nodes highly connected with the neighbor subgraph to form a candidate complex with low coupling, in lines 9–12. Figure 4 illustrates an example of removing high-coupling proteins. In Fig. 4, SWD(D, CCS) = 0.2, SWD(D, NS) = 0.3 + 0.4 = 0.7, D is removed from CCS.

Finally, if CCS is not a subset of complex in the set of protein complex SC, CCS is inserted into SC.

The second stage of our method is redundancy-filtering, in lines 15–20. Complexes overlapping to a very high extent should be discarded. With quantifying the extent of overlap between each pair of complexes, a complex with small weighted density or a small number of proteins is discarded for which overlap score of the pair is above the threshold. In our method, the overlap threshold is typically set as 0.8 [21, 27], where the matching score of two complexes A and B is defined as follows [15, 24]:

$$MS(A, B) = \frac{|A \cap B|^2}{|A| \times |B|} \qquad (7)$$

Algorithm 1: Protein complexes identification

Input: multi-relationship network MG= (V, E, W, T);

weighted density threshold WDT; the threshold for overlap T;

Output: SC: the set of protein complexes;

1. GS= {G_1, G_2, G_3, G_4} is a network set generated from MG;

2. for each network $G_k \in GS$ (k=1, 2, 3, 4)

3. for each vertex $v \in V_k$

4. Insert v into CCS;// Candidate complex set

5. for each neighbor q of v

6. insert q into CCS;

7. If WD (CCS)< WDT

8. remove q from CCS

9. NS is a neighbor subgraph of nodes in CCS

10. for each vertex u ∈ CS

11. if SWD(u , CCS) <= SWD (u , NS)

12. remove u from CCS;

13. if CCS is not a subset of element in SC

14. insert CCS into SC;

15. for each element A ∈ SC

16. for each element B ∈ SC and A≠B

17: if NA(A,B)>T

18. if WD(A)≥WD(B) or Size(A)≥Size(B)

19. remove B from SC ;

20. else remove A from SC ;

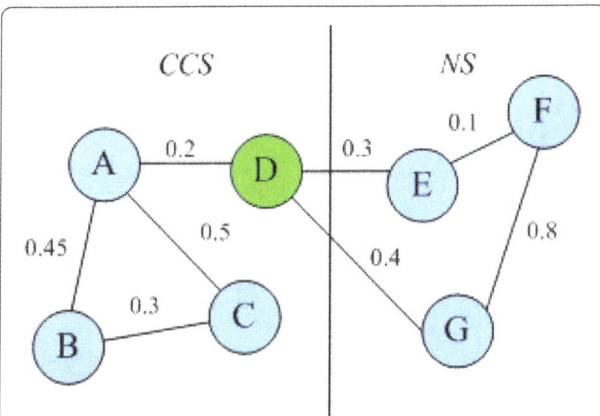

Fig. 4 An example of removing high-coupling proteins. The sum of weighted degree in CCS is 0.2, while that value in NS is 0.7, so D is removed from CCS, due to high coupling with neighbor set NS

Results and discussion

In order to evaluate the performance of our proposed algorithm, we compare it with other five competing algorithms, including CMC [15], RRW [16], COACH [24], SPICi [17], and ClusterONE [21]. For all those competing algorithms, the parameters are set as recommended by their authors. We have applied our MINE method and other methods on two yeast PPI networks, including DIP [37] and Krogan [38]. These PPI datasets are

available online, which varied from each other a lot. In this section, we will first present in details the results on DIP data. The results using Krogan data will also be briefly presented to demonstrate the effectiveness of our proposed method.

The DIP dataset consists of 5023 proteins and 22,570 interactions. The Krogan dataset contains 3672 proteins and 14,317 interactions. Self-interactions and repeated interactions are filtered out in the three PPI networks. To evaluate the protein complexes predicted by our method, a benchmark set is obtained from the reference [39], which consists of 408 complexes.

To assess the quality of predicted complexes, we employed several evaluation measures, including precision, recall, F-measure, and functional enrichment of GO terms.

Precision, recall, and F-measure

We describe how well the predicted protein complexes match with the benchmark complex set, firstly. A predicted protein complex is considered to match with a benchmark complex, if its matching score MS (see Eq. (7)) is no less than a threshold. Typically, the threshold is set as 0.2 [24, 27]. Precision and recall are the commonly used measures to evaluate the performance of protein complex prediction algorithms. Precision measures the percentage of predicted protein complexes that match benchmark complexes in all the predicted protein complexes. Recall is the fraction of benchmark complexes that are retrieved. Mathematically, precision and recall are defined as follows:

$$\text{Precision} = \frac{N_{cp}}{|P|} \tag{8}$$

$$\text{Recall} = \frac{N_{cb}}{|B|} \tag{9}$$

where N_{cp} is the number of predicted complexes matched by benchmark complexes, N_{cb} is the number of benchmark complexes that are matched by predicted complexes, P is the set of predicted protein complexes and B is the benchmark complex set.

F-measure, as the harmonic mean of precision and recall, can be used to evaluate the overall performance of the different techniques [21, 24]. Table 1 shows the basic information about predicted complexes by various methods on DIP data, where the best values are italized.

In Table 1, PC represents the total number of predicted complexes, while N_{pcp} is the number of complexes perfectly matching the benchmark complexes. In other words, the matching score between a predicted complex and a benchmark complex is 1. From Table 1, we can see that MINE produces the largest number of correctly predicted complexes and the second-largest number

Table 1 The matching results of various algorithms

Algorithms	PC	N_{cp}	N_{cb}	N_{pcp}
MINE	606	345	218	19
CMC	235	119	124	8
COACH	902	319	219	15
RRW	250	118	136	4
SPICi	574	118	143	7
ClusterONE	371	155	136	6

of benchmark complexes after COACH, respectively, while PC of our method (606) is far less than COACH's (902). The fifth column of Table 1 shows that MINE has the absolute advantage to obtain the largest number of perfectly matched complexes. N_{pcp} of MINE is 137.5, 26.67, 375, 171.43, and 216.67 % higher than that of CMC, COACH, RRW, SPICi, and ClusterONE, respectively. Figure 5 shows the overall comparison in terms of precision, recall, and F-measure.

On DIP data, F-measure of MINE is 0.551, which is 45.05, 29.23, 41.02, 112.62, and 48.59 % higher than that of CMC, COACH, RRW, SPICi, and ClusterONE, respectively. Our MINE method can achieve the highest F-measure by providing the highest precision and the same highest recall as COACH, which shows that our method can predict protein complexes very good.

Functional enrichment analysis

Another evaluation measure is the function enrichment which measures the biological significance of predicted protein complexes by various algorithms. To substantiate the biological significance of our predicted complexes, we calculate their p values, which represent the probability of co-occurrence of proteins with common functions [27]. In this wok, we employ the tool BiNGO

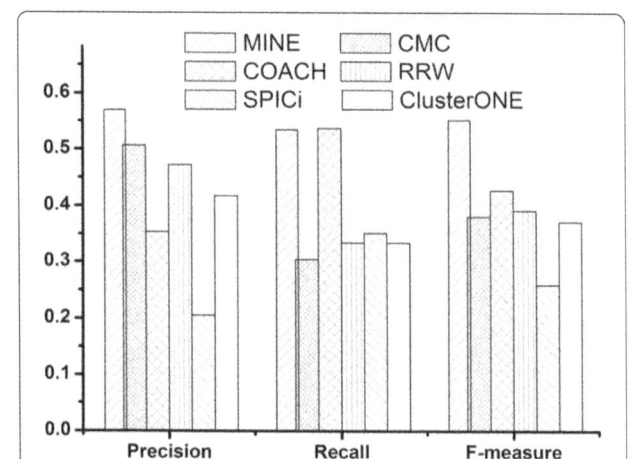

Fig. 5 The performance comparison for various algorithms on DIP data. MINE achieves the highest precision, recall, and F-measure among all the six methods

[40] to calculate p values for predicted complexes. BiNGO is a Java-based tool to determine which GO categories are statistically overrepresented in a set of genes or a subgraph of a biological network. BiNGO is implemented as a plug-in for Cytoscape [41], which is an open-source bioinformatics software platform for visualizing and integrating molecular interaction networks. A low p value of a predicted complex indicates that those proteins in the complex do not happen merely by chance, so the complex has high statistical significance. Generally, a complex is considered to be significant with p value <0.01. In addition, the p-score is also used as an effective evaluation measure, which is defined as

$$p\text{-score} = \frac{1}{n}\sum_{i=1}^{n} -\lg(p \text{ value}_i)|p \text{ value}_i < 0.01 \quad (10)$$

Table 2 lists comparative results of various algorithms based on GO annotation, where the best values are italized. In Table 2, SC is the number of significant predicted complexes. That is, their p values are less than 0.01. Our MINE method achieves the highest proportion of significantly predicted complexes and p-score values among all algorithms. The p-score of MINE is 12.16, 18.41, 32.08, 48.38, and 20.20 % higher than that of CMC, COACH, RRW, SPICi, and ClusterONE, respectively. In addition, Table 2 indicates that RRW gets the highest proportion of significant complexes, while achieves a lower p-score values than ClusterONE because the p value of significant complexes predicted by ClusterONE are lower than RRW's. These results suggest that the complexes predicted by MINE had the most biological significance.

Effect of parameters on prediction performance

In MINE, we introduce a user-defined parameter WDT (weighted density threshold) to discover density subgraphs with high cohesion to form candidate complexes. To investigate the effect of parameter WDT on performance of MINE, we evaluate the prediction accuracy in terms of precision, recall, and F-measure by setting different values of WDT, ranging from 0 to 1. Figure 6 shows that the performance of our method fluctuates

Table 2 The comparison of various methods in terms of function enrichment

Algorithms	PC	SC	Proportion (%)	p-score
MINE	*606*	*499*	*82.34*	*11.9*
CMC	235	187	79.57	10.61
COACH	902	676	74.94	10.05
RRW	250	191	76.40	9.01
SPICi	574	262	45.64	8.02
ClusterONE	371	235	63.34	9.9

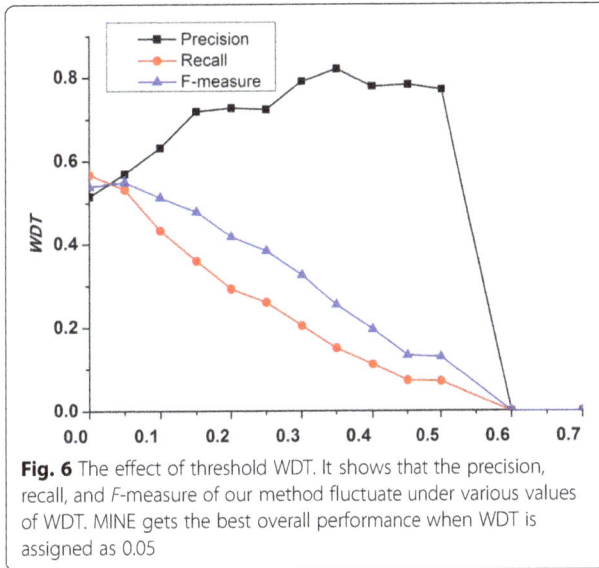

Fig. 6 The effect of threshold WDT. It shows that the precision, recall, and F-measure of our method fluctuate under various values of WDT. MINE gets the best overall performance when WDT is assigned as 0.05

under various values of WDT. Figure 6 clearly indicates that MINE gets the best performance when WDT is assigned as 0.05.

Results using Krogan data

We also performed MINE method on the Krogan PPI network. The precision, recall, and F-measure of each algorithm based on Krogan data are shown in Fig. 7.

Figure 7 indicates that our method gets the best performance among all these methods in terms of precision, recall, and F-measure. The F-measure of our method is 0.5, which is 68.63, 33.52, 45.53, 69.71, and 47.73 % higher than that of CMC, COACH, RRW, SPICi, and ClusterONE, respectively.

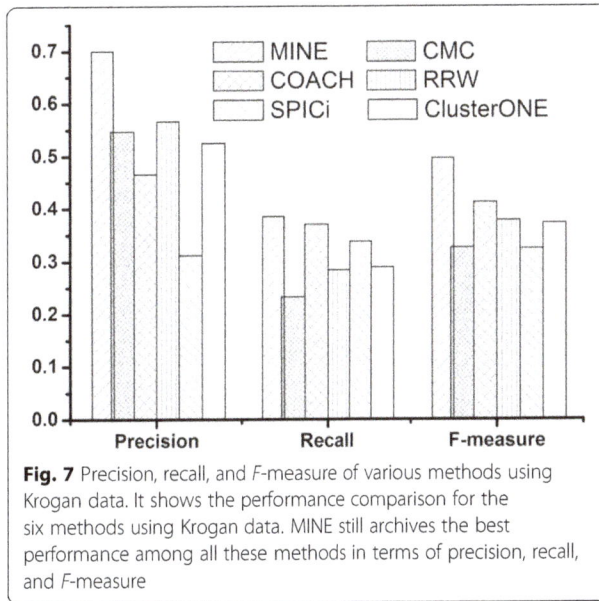

Fig. 7 Precision, recall, and F-measure of various methods using Krogan data. It shows the performance comparison for the six methods using Krogan data. MINE still archives the best performance among all these methods in terms of precision, recall, and F-measure

Conclusions

In this paper, we have constructed a multi-relationship protein interaction network (MPIN) by integrating PPI network topology with GO annotation information. For a pair of proteins in the MPIN, there exists more than one kind of interactions between them. To test the effectiveness of the MPIN, we have developed a novel method named MINE to predict protein complexes. MINE first decomposes the MPIN into four single relationship networks. Then, MINE visits four networks in turn for predicting protein complexes with high cohesion and low coupling. The results of experiments based on yeast PPI networks show that not only MINE achieves higher prediction accuracy than other existing methods but also majority of complexes predicted by MINE possess high biological significance. All results have proved that the constructed MPIN is useful for predicting protein complexes.

Competing interests
The authors declare that they have no competing interests.

Authors' contributions
XYL and BHZ obtained the protein-protein interaction data, gene ontology annotation data, and protein complex data. XYL and BHZ designed the new method MINE and analyzed the results. XYL and BHZ drafted the manuscript together. JW, FXW, and YP participated in revising the draft. All authors have read and approved the manuscript.

Acknowledgements
This work is supported in part by the National Natural Science Foundation of China under Grant No. 61472133, No. 31560317, and No. 61428209 and the Program for New Century Excellent Talents in University under Grant NCET-12-0547.

Declarations
Publication of this article has been funded by the National Natural Science Foundation of China (No. 61472133).
This article has been published as part of *Human Genomics* Volume 10 Supplement 2, 2016: From genes to systems genomics: human genomics. The full contents of the supplement are available online at http://humgenomics.biomedcentral.com/articles/supplements/volume-10-supplement-2.

Author details
[1]School of Information Science and Engineering, Central South University, Changsha 410083, China. [2]Department of Information and Computing Science, Changsha University, Changsha 410003, China. [3]Department of Mechanical Engineering and Division of Biomedical Engineering, University of Saskatchewan, Saskatoon, SK S7N 5A9, Canada. [4]Department of Computer Science, Georgia State University, Atlanta, GA 30302-4110, USA.

References
1. Ito T, Chiba T, Ozawa R, Yoshida M, Hattori M, Sakaki Y. A comprehensive two-hybrid analysis to explore the yeast protein interactome. Proc Natl Acad Sci. 2001;98:4569–74.
2. Rigaut G, Shevchenko A, Rutz B, Wilm M, Mann M, Séraphin B. A generic protein purification method for protein complex characterization and proteome exploration. Nat Biotechnol. 1999;17:1030–2.
3. Ho Y, Gruhler A, Heilbut A, et al. Systematic identification of protein complexes in Saccharomyces cerevisiae by mass spectrometry. Nature. 2002;415:180–3.
4. Peng W, Li M, Chen L, and Wang L S. Predicting protein functions by using unbalanced random walk algorithm on three biological networks. IEEE/ACM Transactions on Computational Biology and Bioinformatics. doi 10.1109/TCBB.2015.2394314.
5. Li M, Lu Y, Niu Z B, Wu F X. United complex centrality for identification of essential proteins from PPI networks. IEEE/ACM Transactions on Computational Biology and Bioinformatics. doi: 10.1109/TCBB.2015.2394487.
6. Li M, Lu Y, Wang JX, Wu FX, Pan Y. A topology potential-based method for identifying essential proteins from PPI networks. IEEE/ACM Trans Comput Biol Bioinform. 2015;12(2):372–83.
7. Ren J, Wang JX, Li M, Wu FX. Discovering essential proteins based on PPI network and protein complex. Int J Data Min Bioinform. 2015;12(1):24–43.
8. Li M, Zheng RQ, Zhang HH, Wang JX, Pan Y. Effective identification of essential proteins based on priori knowledge, network topology and gene expressions. Methods. 2014;67(3):325–33.
9. Li M, Wang JX, Wang H, Pan Y. Identification of essential proteins from weighted protein interaction networks. J Bioinform Comput Biol. 2013;11(3):1341002.
10. Wang JX, Li M, Wang H, Pan Y. Identification of essential proteins based on edge clustering coefficient. IEEE/ACM Trans Comput Biol Bioinform. 2012;9(4):1070–80.
11. Tang Y, Li M, Wang JX, Pan Y, Wu FX. CytoNCA: a cytoscape plugin for centrality analysis and evaluation of biological networks. BioSystems. 2015;127:67–72.
12. Bader G, Hogue C. An automated method for finding molecular complexes in large protein interaction networks. BMC Bioinformatics. 2003;4:2.
13. Enright AJ, Dongen SV, Ouzounis CA. An efficient algorithm for large-scale detection of protein families. Nucleic Acids Res. 2002;30:1575–84.
14. Shih YK, Parthasarathy S. Identifying functional modules in interaction networks through overlapping Markov clustering. Bioinformatics. 2012;28:i473–9.
15. Liu G, Wong L, Chua HN. Complex discovery from weighted PPI networks. Bioinformatics. 2009;25:1891–7.
16. Macropol K, Can T, Singh AK. RRW: repeated random walks on genome-scale protein networks for local cluster discovery. BMC Bioinformatics. 2009;10:283.
17. Jiang P, Singh M. SPICi: a fast clustering algorithm for large biological networks. Bioinformatics. 2010;26:1105–11.
18. Wang JX, Li M, Chen JE, Pan Y. A fast hierarchical clustering algorithm for functional modules discovery in protein interaction networks. IEEE/ACM Trans Comput Biol Bioinform. 2011;8(3):607–20.
19. Li M, Wang JX, Chen JE, Cai Z, Chen G. Identifying the overlapping complexes in protein interaction networks. Int J Data Min Bioinform (IJDMB). 2010;4(1):91–108.
20. Li M, Chen J, Wang J, et al. Modifying the DPClus algorithm for identifying protein complexes based on new topological structures. BMC Bioinformatics. 2008;9(1):398.
21. Nepusz T, Yu H, Paccanaro A. Detecting overlapping protein complexes in protein-protein interaction networks. Nature Methods. 2012;9:471–5.
22. Wang JX, Zhong JC, Chen G, et al. ClusterViz: a Cytoscape APP for clustering analysis of biological network. IEEE/ACM Trans Comput Biol Bioinform. 2015;12(4):815–22.
23. Gavin A, Aloy P, Grandi P, et al. Proteome survey reveals modularity of the yeast cell machinery. Nature. 2006;440:631–6.
24. Wu M, Li X, Kwoh CK, Ng S. A core-attachment based method to detect protein complexes in PPI networks. BMC Bioinformatics. 2009;10:169.
25. Leung HC, Xiang Q, Yiu SM, Y CF. Predicting protein complexes from PPI data: a core-attachment approach. J Comput Biol. 2009;16:133–44.
26. Srihari S, Ning K, Leong HW. MCL-CAw: a refinement of MCL for detecting yeast complexes from weighted PPI networks by incorporating core-attachment structure. BMC Bioinformatics. 2010;11:504.
27. Zhao B, Wang J, Li M, et al. Detecting protein complexes based on uncertain graph model. IEEE/ACM Trans Comput Biol Bioinform. 2014;11:486–97.
28. Peng W, Wang J, Zhao B, et al. Identification of protein complexes using weighted PageRank-Nibble algorithm and core-attachment structure. IEEE/ACM Trans Comput Biol Bioinform. 2014;12:179–92.
29. Wu M, Li X, Kwoh C, et al. Discovery of protein complexes with core-attachment structures from Tandem Affinity Purification (TAP) data. J Comput Biol. 2012;19:1027–42.

30. Tang X, Wang J, Liu B, et al. A comparison of the functional modules identified from time course and static PPI network data. BMC Bioinformatics. 2011;12(1):339.
31. Wang J, Peng X, Li M, et al. Construction and application of dynamic protein interaction network based on time course gene expression data. Proteomics. 2013;13:301–12.
32. Li M, Wu X, Wang J, et al. Towards the identification of protein complexes and functional modules by integrating PPI network and gene expression data. BMC Bioinformatics. 2012;13:109.
33. Li M, Chen W, Wang J, et al. Identifying dynamic protein complexes based on gene expression profiles and PPI networks. BioMed Res Int. 2014;2014 (375262):10.
34. Zhao J, Hu X, He T, et al. An edge-based protein complex identification algorithm with gene co-expression data. IEEE Trans Nanobioscience. 2014; 13:80–8.
35. Fan W, Yeung KH. Similarity between community structures of different online social networks and its impact on underlying community detection. Commun Nonlinear Sci Numer Simul. 2015;20:1015–25.
36. Zhao BH, Wang JX, Li M, et al. Prediction of essential proteins based on overlapping essential modules. IEEE Transactions on NanoBioscience. 2014;13:415–24.
37. Xenarios X et al. DIP: the database of interacting proteins. Nucleic Acids Res. 2000;28:289–91.
38. Krogan N et al. Global landscape of protein complexes in the yeast Saccharomyces cerevisiae. Nature. 2006;440:637–43.
39. Pu S, Wong J, Turner B, et al. Up-to-date catalogues of yeast protein complexes. Nucleic Acids Res. 2009;37:825–31.
40. Maere S, Heymans K, Kuiper M. BiNGO: a Cytoscape plugin to assess overrepresentation ontology categories in biological network. Bioinformatics. 2005;21:3448–9.
41. Shannon P et al. Cytoscape: a software environment for integrated models of biomolecular interaction networks. Genome Res. 2003;13:2498–504.

8

Multiplex SNaPshot—a new simple and efficient CYP2D6 and ADRB1 genotyping method

Songtao Ben[1], Rhonda M. Cooper-DeHoff[2], Hanna K. Flaten[3], Oghenero Evero[3], Tracey M. Ferrara[1], Richard A. Spritz[1] and Andrew A. Monte[3,4*]

Abstract

Background: Reliable, inexpensive, high-throughput genotyping methods are required for clinical trials. Traditional assays require numerous enzyme digestions or are too expensive for large sample volumes. Our objective was to develop an inexpensive, efficient, and reliable assay for *CYP2D6* and *ADRB1* accounting for numerous polymorphisms including gene duplications.

Materials and methods: We utilized the multiplex SNaPshot® custom genotype method to genotype *CYP2D6* and *ADRB1*. We compared the method to reference standards genotyped using the Taqman Copy Number Variant Assay followed by pyrosequencing quantification and determined assigned genotype concordance.

Results: We genotyped 119 subjects. Seven (5.9 %) were found to be *CYP2D6* poor metabolizers (PMs), 18 (15.1 %) intermediate metabolizers (IMs), 89 (74.8 %) extensive metabolizers (EMs), and 5 (4.2 %) ultra-rapid metabolizers (UMs). We genotyped two variants in the β1-adrenoreceptor, rs1801253 (Gly389Arg) and rs1801252 (Ser49Gly). The Gly389Arg genotype is Gly/Gly 18 (15.1 %), Gly/Arg 58 (48.7 %), and Arg/Arg 43 (36.1 %). The Ser49Gly genotype is Ser/Ser 82 (68.9 %), Ser/Gly 32 (26.9), and Gly/Gly 5 (4.2 %). The multiplex SNaPshot method was concordant with genotypes in reference samples.

Conclusions: The multiplex SNaPshot method allows for specific and accurate detection of *CYP2D6* genotypes and *ADRB1* genotypes and haplotypes. This platform is simple and efficient and suited for high throughput.

Keywords: *ADRB1*, *CYP2D6*, Copy number variation, Gene deletion, Genotyping, Personalized medicine, Precision medicine

Background

Cytochrome P450 family 2 subfamily D member 6 (*CYP2D6*) is one of the most important drug-metabolizing enzymes expressed in humans. The enzyme metabolizes 20–30 % of all xenobiotics including numerous antidepressants, antipsychotics, anti-emetics, analgesics, and cardiovascular medications. The expression of *CYP2D6* is highly polymorphic; more than 70 allelic variants have been identified, and deletion or duplication of the gene leads to variable enzyme function in individual patients [1, 2]. The *CYP2D6* gene has been mapped to chromosome 22q13.1 and consists of nine exons with an open reading frame of 1491 base pairs coding for 497 amino acids [2–5]. Variability in enzyme expression has been associated with altered drug effectiveness and safety [2]. In fact, there are four functional genotype groups identified for the enzyme, poor metabolizers (PMs), intermediate metabolizers (IMs), extensive metabolizers (EMs), and ultra-rapid metabolizers (UMs) based upon the number and activity of the gene copies the patients express (Table 1) [1, 6]. Therefore, many researchers and clinicians have targeted genotyping *CYP2D6* in efforts to improve drug therapy.

Expanding on previously published work [7–9], we sought to determine the effectiveness and safety of metoprolol succinate utilizing a systems biology approach, whereby we genotyped the drug-metabolizing enzyme

* Correspondence: andrew.monte@ucdenver.edu
[3]Department of Emergency Medicine, School of Medicine, University of Colorado, 12401 E. 17th Ave, B215, Aurora, CO 80045, USA
[4]Rocky Mountain Poison & Drug Center, Denver Health and Hospital Authority, Denver, CO 80203, USA
Full list of author information is available at the end of the article

Table 1 Predicted *CYP2D6* enzyme function based upon the activity score derived from SNV identification

Predicted CYP2D6 enzyme function	Activity score
Poor metabolizer (PM)	0
Intermediate metabolizer (IM)	0.5
Extensive metabolizer (EM)	1.0–2.0
Ultra-rapid metabolizer (UM)	>2.0

Some genes have partial functionality based upon in vitro enzyme activity for specific substrates. This table is adapted from Crews et al. and Gaedigk et al. [1, 6]

CYP2D6 as well as the drug target, β1-adrenoreceptor (*ADRB1*). Additionally, we captured demographic and clinical factors that are necessary to understand the potential for individualized therapy in patients taking metoprolol succinate. Metoprolol is a B1 selective antagonist, known as a beta-blocker. Beta-blockers are first-line treatment for heart failure, hypertension (HTN), angina, and myocardial infarction [10–14]. In 2011, 34.5 million prescriptions for metoprolol were written in the USA [15]. Thus, metoprolol is a first-line therapy for several of the most common chronic diseases and is one of the most commonly prescribed drugs in the USA. CYP2D6 is the only clinically pertinent pathway of metoprolol metabolism, and polymorphisms have been associated with altered levels of metoprolol [16, 17]. *ADRB1* is the drug target, and polymorphisms in this receptor have been associated with variable drug response [7, 18]. Prediction of drug response with knowledge of one without knowledge of the other could be incomplete since both contribute to the ultimate clinical effect [19]. This highlights the importance of accounting for both factors. Prior investigators have identified imperfect associations with *CYP2D6* [16, 17] and *ADRB1* [7, 18] genotypes, but no study has been appropriately powered to account for variability in both genotypes [19]. No single assay accounts for variants in both genes. Given that multiple single-nucleotide variants (SNVs) and the presence of gene duplication can affect *CYP2D6* and *ADRB1*, a simple, efficient, and inexpensive method that identifies both SNVs and gene duplication is required to allow for an adequately powered systems biology approach.

CYP2D6 and *ADRB1* variants have been identified using long-range polymerase chain reaction (XL-PCR) with PCR restriction fragment length polymorphism (PCR-RFLP) or microarray analysis. Microarray analysis remains expensive for large-scale genotyping. Commercially available tests cost a minimum of $500 per patient. XL-PCR with PCR-RFLP accounts for allele variants, and multiplication is inexpensive, easy, and reliable in patients regardless of ethnicity or race [6]. However, XL-PCR with PCR-RFLP requires numerous enzymatic digestions and multiple amplification steps to identify more than one polymorphism in the *CYP2D6* gene. In order to conduct a large-scale clinical trial to determine

the effectiveness and safety of metoprolol in treating hypertension, we sought to develop a simple, rapid, yet less expensive assay for *CYP2D6* and *ADRB1* that accounts for numerous polymorphisms including gene duplications. Drug effectiveness data will be presented at a later date when the trial closes.

Material and methods
Study design and setting
This was a prospective observational clinical trial (NCT02293096) that enrolled patients with uncontrolled HTN from clinics, the emergency department, and across the community at the University of Colorado Hospital in Aurora, Colorado.

Subjects
In accordance with the University of Colorado IRB approval, we enrolled subjects with uncontrolled HTN between 30 and 80 years of age. Exclusion criteria included end-stage liver disease, glomerular filtration rate < 60 ml/min/1.73 m^2, pregnancy, American Society of Anesthesiologists (ASA) classification of >3, prisoners or wards of the state, decisionally challenged, heart rate < 60 beats per minute, AV block > 240 ms, active reactive airway disease, illicit drug use in the preceding 30 days (excluding marijuana), allergy to metoprolol succinate, or severe peripheral arterial circulatory disorders.

Drug intervention
Patients were initially managed according to the Eighth Joint National Committee guidelines for management of HTN [20]. Patients were started on angiotensin-converting enzyme inhibitor or angiotensin receptor blockers as first-line therapy. If blood pressure remained uncontrolled, then metoprolol succinate was added. Patients were then followed up and metoprolol titrated weekly for 4 weeks. Drug effectiveness data will be presented at a later date.

DNA isolation
Genomic DNA was extracted from whole blood via the Puregene® Blood Core kit B (Qiagen) according to the manufacturer's instructions.

Long-range PCR analyses to determine CYP2D6 gene multiplication
We analyzed *CYP2D6* duplication using long-range PCR with the primers described by Lovlie et al. [21]. This primer combination amplifies a 5.2-kb PCR fragment from the *CYP2D7-CYP2D6* intergenic regions in all individuals and a 3.6-kb PCR fragment from the *CYP2D6-CYP2D6* region in individuals with a duplication of the gene [22]. The amplification was done using the Phusion High-Fidelity DNA Polymerase kit (New England

Biolabs). The reaction components included 2 μl of 5×
Phusion HF buffer, 0.2 μl of 25 mM dNTPs, 0.5 μl of
10 μM/μl forward primer, 0.5 μl of 10 μM/μl reverse primer, 1 μl of template DNA, 0.2 μl of Phusion DNA polymerase, 0.3 μl DMSO, 2.5 μl of BETAIN, and 2.8 μl of
nuclease-free water. The total reaction volume was 10 μl.
Thermal cycling conditions were as follows: 98 °C for 30 s
followed by 35 cycles of 98 °C for 10 s, 67 °C for 30 s, and
72 °C for 2 min 30 s. After cycling, samples were stored at
4 °C. PCR products were analyzed by electrophoresis.

Long-range PCR analyses to determine CYP2D6 gene deletion

To detect deletion of the entire CYP2D6 gene, (CYP2D6 *5)
we performed a multiplex longer-PCR reaction with the
primers described by Okubo et al. [23]. The reaction components included 2 μl of 5× Phusion HF buffer, 0.2 μl of
25 mM dNTPs, 0.2 μl of 10 μM F1 primer, 0.2 μl of 10 μM
R2 primer, 2 μl of 10 μM F3 primer, 1 μl of template DNA,
0.2 μl of Phusion DNA polymerase, 2.5 μl of BETAIN, and
1.7 μl of nuclease-free water. The PCR conditions were as
follows: initiation at 98 °C for 3 min, 35 cycles of 98 °C for
10 s and 70 °C for 30 s, termination at 72 °C for 2 min 30 s,
and a final elongation at 72 °C for 10 min. The PCR products were examined by electrophoresis and include the
amplification of a 3.5-kb fragment, which indicates CYP2D6
*5, and a 4.7-kb fragment, which indicates CYP2D6 wild
type.

SNV selection

We detected 20 variants associated with altered CYP2D6
enzyme activity. SNVs were chosen because they are
representative of major haplotypes associated with altered enzyme function without providing redundancy of
rsIDs leading to the same functional activity within the
same allele variant (Table 2). If none of these SNVs were
identified, the allele designation was defaulted to the reference allele, CYP2D6*1. Haplotype analysis for CYP2D6
was based upon predicted enzyme activity of the SNVs
identified in the genotyping stage.

Additionally, two ADRB1 SNVs were genotyped (Table 2)
because haplotypes of these alleles are known to be associated with altered clinical response to metoprolol treatment
(Table 3). If neither of these SNVs were identified, the allele was defaulted to Ser49 and Gly389, the reference
allele.

Genotyping

Genotyping was performed using two separate multiplex
reactions. First, we used PCR to amplify a purified template DNA fragment that included the target nucleotide.
We designed a total of eight pairs of PCR primers (Table 4)
that were divided into two separate pools. The PCR
was performed with the AmpliTaq® Gold kit (Applied

Table 2 CYP2D6 and ADRB1 variant alleles with rsIDs and subsequent effect on the gene sequence

Allele	Major nucleotide variation	SNV	Effect
*1	Presumed	NA	Wild type
*2	2850C>G	rs16947	Arg296Cys
*3	2549delA	rs35742686	Frameshift
*4	100C>T	rs1065852	Pro34Ser
	1846G>A	s3892097	Splicing defect
*6	1707delT	rs5030655	Frameshift
*7	2935A>C	rs5030867	His324Pro
*9	2615_2617delAAG	rs5030656	Lys281del
*10	100C>T	rs1065852	Pro34Ser
*12	124G>A	rs5030862	Gly42Arg
*14	1758G>A	rs5030865	Gly169Arg
*17	1023C>T	rs28371706	Thr107Ile
	2850C>T	rs16947	Arg296Cys
*19	2539_2542delAACT	rs72549353	255Frameshift
*20	1973_1974insG	rs72549354	211Frameshift
*38	2587_2590delGACT	rs72549351	271Frameshift
*40	1863_1864insTTTCGCCCCX2	rs72549356	174_175insFRP × 2
*41	2850C>T	rs16947	Arg296Cys
	2988G>A	rs38371725	Splicing defect
*42	3259_3260insGT	rs72549346	363Frameshift
*49	100C>T	rs1065852	Pro34Ser
	1611T>A	rs1135822	Phe120Ile
*69	100C>T	rs1065852	Pro34Ser
	2850C>T	rs16947	Arg296Cys
	2988G>A	rs38371725	Splicing defect
ADRB1 Ser49Gly	A>G	rs1801252	Ser49Gly
ADRB1 Gly389Arg	G>C	rs1801253	Gly389Arg

Biosystems) using a hot start/touchdown PCR assay. The
10 μl reaction included 1 μl of template DNA, 1.6 μl of
10× GeneAmp® PCR Buffer, 0.15 μl of 25 mM/each
dNTPs, 0.4 μl of 1 μM/each primer mix, 2.15 μl of 50 mM
MgCl$_2$, 0.15 μl of 5 U/μl AmpliTaq Gold, and 4.55 μl of
nuclease-free water. Thermal cycling conditions were as
follows: 94 °C for 5 min followed by 15 cycles of 94 °C for
30 s; annealing temperature steps down every cycle by

Table 3 ADRB1 haplotype and associated metoprolol clinical effect

ADRB1 haplotypes	Metoprolol clinical effect
49Ser389Arg/49Ser389Arg	Threefold greater diastolic blood pressure reduction [7, 12]
49Ser389Arg/49Gly389Arg	Good responder [12]
49Ser389Gly/49Gly389Arg and 49Ser389Gly/49Ser389Gly	Non-responders [7, 12]

Table 4 The oligonucleotide PCR primers used for amplification of genomic DNA to obtain template

Multiplex pool	Fragment name	Sequence 5′ → 3′	Length (bases)	Product size (bp)	SNV
1	P1PCR1	agcccggtaacctgtcgt	18	162	rs1801252
		ccatcagcagacccatgc	18		
	P1PCR2	tggaggaggtcaggcttaca	20	342	rs16947
		ggtgcagaattggaggtcat	20		rs5030867
					rs28371725
	P1PCR3	gtgtggtggcattgaggact	20	332	rs72549346
		gtggggacgcatgtctgt	18		
	P1PCR4	gatgcactggtccaaccttt	20	223	rs35742686
		ctggtgtaggtgctgaatgc	20		rs5030656
					rs72549353
					rs72549351
2	P2PCR1	gccttcaaccccatcatcta	20	328	rs1801253
		ggccctacaccttggattc	19		
	P2PCR2	ctcacctggtcgaagcagta	20	145	rs1065852
		ccatcttcctgctcctggt	19		rs5030862
	P2PCR3	cagctcggactacggtcatc	20	272	rs28371706
		cttgacaagaggccctgacc	20		
	P2PCR4	gtccttcccaaacccatct	20	562	rs3892097
		gtggggctaatgccttcat	19		rs5030655
					rs5030865
					rs72549354
					rs72549356
					rs1135822

0.5 °C (from 63 to 56.5 °C) every 30 s and then 72 °C for 1 min. The annealing temperature for the final 25 cycles was 56 °C with a denaturation temperature of 94 °C for 30 s and extension temperature of 72 °C for 1 min. After PCR, the products were first treated with shrimp alkaline phosphatase (SAP) and Exonuclease I (Exo I) to remove excess primers and dNTPs. Two units of SAP and 1 unit of Exo I were added to 5 μl of PCR product and incubated at 37 °C for 30 min and then 80 °C for 15 min. Second, we performed multiplex single-base extraction (SBE) reactions. We designed 19 SBE primers and divided the reactions into two pools. See Table 5. Each SBE reaction was carried out in a 10 μl final volume containing 5 μl of SNaPshot Multiplex Ready Reaction Mix, 2 μl of pooled PCR products, 1 μl of pooled SNaPshot primers (0.3 μmol/ each μl), and 2 μl of deionized water. Extension was performed for 25 cycles under the following conditions: 96 °C for 10 s, 50 °C for 5 s, and 60 °C for 30 s. To remove ddNTPs, the SBE reactions were then treated with 1.0 unit of calf-intestinal phosphatase (CIP) and incubated at 37 °C for 30 min. The enzyme was deactivated by incubating at 80 °C for 15 min. The samples were run by electrophoresis on the 3130 Genetic Analyzer (Applied Biosystems). The subsequent data were analyzed with GeneScan software and GeneScan-120 LIZ size standard.

CYP2D6 multiple-copy allele determination

The SNaPshot Multiplex System (Applied Biosystems) is a primer extension-based method developed for the analysis of SNVs. Theoretically, with the same SNV peak in the same reacting system with the same reaction conditions, a comparatively higher density means more DNA copies, and the ratios of any two peak densities of the same sample are relatively inflexible. We identified the copy numbers and duplicated/multiplicated allele by the ratio of the peak densities. In each pool, we designed one primer to identify an *ADRB1* gene's SNV. *ADRB1* is a single-copy gene which allowed us to use the peaks of ADRB1 gene's SNVs as the inner standard peaks. We calculated the ratios of the peak density of *CYP2D6* SNVs to the peak density of the *ADRB1* SNV. We set the reliability value range of the ratios of each SNV based on the values of the ratio of single-copy control samples with the same SNVs. If a ratio of a SNV of a known duplicated sample was greater than the maximum of the reference reliability range, this SNV was

Table 5 The oligonucleotide primers for SNaPshot primer extension reactions

Multiplex pool	SNV	Primer direction	Peak to SNV correspondence	Primer length	Primer sequence (5' → 3')
1	rs28371725C/T	F(C/T)	C=C T=T	16	CCCCGCCTGTACCCTT
	rs1801252A/G	F(A/G)	A=A G=G	18	GACTCTCCGCCAGCGAA
	rs72549346-/AC	R(T/G)	T=AC C=-	30	GACTGACTGCCGTGATTCATGAGGTG
	rs16947A/G	F(A/G)	A=A G=G	40	GACTGACTGACTGACTGACTGAGGTCAGCCACCACTATGC
	rs5030656-/CTT	F(T/C)	T=CTT C=-	40	GACTGACTGACTGACTGACTGATGGCAGCCACTCTCACCT
	rs72549351-/AGTC	R(C/G)	C=AGTC G=-	46	GACTGACTGACTGACTGACTGACTGACTGACCCCCGAGACCTGA
	rs72549353-/AGTT	F(T/A)	T=AGTT A=-	46	GACTGACTGACTGACTGACTGACTGACTGACCAGGTCATCCTGTCTCAG
	rs35742686-/T	F(T/G)	T=T G=-	52	GACTGACTGACTGACTGACTGACTGACTGACTGACTGGGTCCCAGGTCATCC
	rs5030867T/G	R(A/C)	A=T C=G	52	GACTGACTGACTGACTGACTGACTGACTGACTGACTCCTCGCTCATGATCCTAC
	rs1801253C/G	F(C/G)	C=C G=G	15	CGCAAGGCCTTCCAG
	rs72549356-/GGGGCGAAAGGGGCGAAA	R(T/A)	T=GGGGCGAAAGGGGCGAAA A=-	18	GACTGCCCCTTTCGCCCC
2	rs1065852G/A	R(C/T)	C=G T=A	36	GACTGACTGACTGACTGACTGGGCTGCACGCTAC
	rs3892097T/C	R(A/G)	A=T G=C	36	GACTGACTGACTGACTGACTGCATCTCCCACCCCA
	rs28371706G/A	R(C/T)	C=G T=A	43	GACTGACTGACTGACTGACTGACTGCCTGTGCCCATCA
	rs5030862T/C	R(A/G)	A=T G=C	43	GACTGACTGACTGACTGACTGACTGACTCCCTGCCACTGCCC
	rs72549354-/C	F(T/C)	T=- C=C	47	GACTGACTGACTGACTGACTGACTGACTGACTCCTTCAGTCCC
	rs5030655-/A	R(T/G)	T=T G=-	51	GACTGACTGACTGACTGACTGACTGACTGACTCAAGAAGTCGCTGGAGCAG
	rs1135822A/T	F(A/T)	A=A T=T	55	GACTGACTGACTGACTGACTGACTGACTGACTGACTGACTCATAGCGCGCCAGGA
	rs5030865A/C/T	F(A/C/T)	A=A C=C T=T	59	TGACTGACTGACTGACTGACTGACTGACTGACTGACTGACTCTTCGCCCATCACCCAC

designated a duplication. We identified the copy number based upon that value. Based on published results and the values of the ratios, we assume all duplicated samples have the total copy number of three since this was the highest number of copy number variants (CNVs) in our controls.

Activity scoring and predicted phenotype assignment

Each identified *CYP2D6* SNV was assigned a predicted enzyme activity score [1, 6]. Gene deletions were designated as an activity score of zero. The predicted enzyme phenotype was determined by addition of the individual gene activity scores, accounting for gene copies yielding decreased enzyme activity and gene duplications in each patient. A score of 0 was predicted to be a PM, 0.5 was predicted to be IM, 1–2 was predicted to be an EM, and 2.5 or greater was predicted to have a UM phenotype.

Assay verification

Genotypes were confirmed with known reference genotype samples from 5 PMs, 4 IMs, and 24 EMs [8, 24]. Copy number variations were determined by Taqman Copy Number Assay (Life Technologies, CA) and then by pyrosequencing allele quantification in the known samples [24].

Results

Subjects

The demographics on the initial 79 subjects with unknown haplotypes in this cohort were as follows: the median age was 52 (IQR 45, 60), 46 (58.2 %) were males, 14 (19.9 %) were Hispanic/Latino, 37 (46.8 %) were Black or African American, 3 (3.8 %) were American Indian or Alaskan Native, 3 (3.8 %) were of Asian decent, 1 (1.3 %) was Native Hawaiian or Pacific Islander, and 38 (48.1 %) were Caucasian.

Genotypes

Our genotyping method demonstrated consistent results with all 30 reference standards. The *CYP2D6* haplotype analysis revealed 7 (5.88 %) *CYP2D6* PMs, 18 (15.1 %) IMs, 89 (74.78 %) EMs, and 5 (4.2 %) UMs (Table 6). We also genotyped two variants of *ADRB1*, rs1801253 (Gly389Arg) and rs1801252 (Ser49Gly). The Gly389Arg genotype is Gly/Gly 18 (15.1 %), Gly/Arg 58 (48.7 %), and Arg/Arg 43 (36.1 %). The Ser49Gly genotype is Ser/Ser 82 (68.9 %), Ser/Gly 32 (26.9), and Gly/Gly 5 (4.2 %). See Table 7. *CYP2D6* allele frequencies are shown in Table 8. All genotypes were in the Hardy-Weinberg equilibrium (HWE) after allele designation by SNV identification and CNV determination.

Table 6 Distribution of *CYP2D6* genotypes and phenotypes (*n* = 119, including 79 unknown subjects and 30 reference subjects)

Genotype	Number of subjects	Frequency (%)	Active score	Predicted phenotype	Phenotype frequency (%)
*1/*2xN	1	0.84	3.0	UM	4.2
*1xN/*2	1	0.84	3.0	UM	4.2
*1/*1xN	1	0.84	3.0	UM	4.2
*2/*2xN	1	0.84	3.0	UM	4.2
*1xN/*2	1	0.84	3.0	UM	4.2
*1/*1	15	12.61	2.0	EM	74.8
*1/*2	15	12.61	2.0	EM	74.8
*1/*3	4	3.36	1.0	EM	74.8
*1/*4	7	5.88	1.0	EM	74.8
*1/*6	2	1.68	1.0	EM	74.8
*1/*10	4	3.36	1.5	EM	74.8
*1/*17	8	6.72	1.5	EM	74.8
*1/*41	7	5.88	1.5	EM	74.8
*2/*2	4	3.36	2.0	EM	74.8
*2/*3	1	0.84	1.0	EM	74.8
*2/*4	2	1.68	1.0	EM	74.8
*2/*5	2	1.68	1.0	EM	74.8
*2/*10	2	1.68	1.5	EM	74.8
*2/*17	6	5.04	1.5	EM	74.8
*2/*40	1	0.84	1.0	EM	74.8
*2/*41	3	3.52	1.5	EM	74.8
*17/*17	1	0.84	1.0	EM	74.8
*1/*4xN	1	0.84	1.0	EM	74.8
*2xN/*12	1	0.84	2.0	EM	74.8
*2/*17xN	2	1.68	2.0	EM	74.8
*1/*5	1	0.84	1.0	EM	74.8
*4/*9	1	0.84	0.5	IM	16
*4/*17	2	1.68	0.5	IM	16
*5/*17	2	1.68	0.5	IM	16
*4/*41	4	3.36	0.5	IM	16
*5/*41	2	1.68	0.5	IM	16
*6/*17	1	0.84	0.5	IM	16
*6/*41	1	0.84	0.5	IM	16
*10/*40	1	0.84	0.5	IM	16
*17/*40	2	1.68	0.5	IM	16
*4xN/*10	1	0.84	0.5	IM	16
*4xN/*41	1	0.84	0.5	IM	16
*3/*4	1	0.84	0.0	PM	5.9
*4/*4	4	3.36	0.0	PM	5.9
*4xN/*5	2	1.68	0.0	PM	5.9

PM poor metabolizer, *IM* intermediate metabolizer, *EM* extensive metabolizer, *UM* ultra-rapid metabolizer

Table 7 *ADRB1* genotype

ADRB1 haplotype	Number of subjects	Frequencies (%)	Expected metoprolol clinical effect
49Ser389Arg/49Ser389Arg	30	25.2	Threefold greater diastolic blood pressure reduction [7, 12]
49Ser389Arg/49Gly389Arg	8	6.7	Good responder [12]
49Ser389Gly/49Gly389Arg	22	18.5	Non-responders [7, 12]
49Ser389Gly/49Ser389Gly	16	13.4	Non-responders [7, 12]
49Ser389Gly/49Gly389Gly	2	1.7	Unknown, designated good responders by prior investigators [7, 12]
49Ser389Gly/49Ser389Arg	36	30.3	Good responder [12]
49Gly389Arg/49Gly389Arg	5	4.2	Unknown, designated good responders by prior investigators [7, 12]

Copy number variants

We identified 13 subjects with CNVs with predicted enzyme activities scores ranging from 0 to 3 (Table 9). An example CNV determination is as follows. One sample had a genotype of *4/*41 and the total gene copy number of three. In this sample, we detected four heterozygous SNVs: rs1065852, rs3892097, rs28371725, and rs16947. The ratio of rs1065852 was 0.92, higher than the maximum of the reference reliability range (0.7–0.85). The ratio of rs3892097 was 0.98, higher than the maximum of the reference reliability range (0.6–0.9). The ratio of rs28371725 was 1.97, lower than the minimum of the reference reliability range (2.4–2.9). The ratio of rs16947 was 1.09, lower than minimum of the reference reliability range (1.28–1.38). All these ratios suggest that the variant allele, *4 was duplicated in this sample. Thus, the genotype is *4xN/*41. Since *4 is a non-functional allele, the activity score is assigned 0, and *41 is a reduced function allele, assigned an activity score of 0.5, even if the *4 is duplicated, the activity

score remains 0.5. Hence, the genotype for this sample was designated IM (Table 6).

Assay verification

Genotyping results for the UM alleles showed a high degree of concordance between the Taqman Copy Number Assay paired with pyrosequencing quantification [24] and our SNaPshot methods. In fact, the only difference between the two assays was the identification of an additional *CYP2D6*5* gene deletion with our multiplex longer-PCR method. Therefore, this assay is reliable for *CYP2D6* genotyping dependent upon the polymorphisms listed.

Limitations

This method should be validated in an additional cohort with known genotypes to ensure concordance with other haplotype designations. This assay has not been validated to determine more than three CNVs because the control samples did not contain subjects with more than three. However, more than three CNVs of the identified SNVs are universally considered UMs. If additional CNVs of SNVs with associated lower activity scores are found, this may become important for haplotype distinction, depending upon the genotype at the second loci. Additional SNVs with altered predicted enzyme activity will not be captured unless the additional primers are added to the reaction pool. This flexibility of the method is an advantage though the assay is limited by 10 SNVs per pool and identification of allele duplications requires samples to be tested in batches in order to establish reference data. Oversaturation of the assay with additional primers in each pool may lead inconsistent detection of SNVs. Therefore, significantly increasing the number of SNVs identified will require larger volumes of DNA. With the presented assay, less than 60 ng of DNA was necessary to genotype 10 SNVs in these two genes in each sample.

Discussion

We have demonstrated that the SNaPshot method of genotyping *CYP2D6* and *ADRB1* is reliable and efficient

Table 8 *CYP2D6* Allele frequencies

CYP2D6 allele	Number of subjects	Frequency (%)
*1	81	34
*1xN	3	1.3
*2	45	18.9
*2xN	3	1.3
*3	6	2.5
*4	25	10.5
*4xN	5	2.1
*5	9	3.8
*6	4	1.7
*9	1	0.4
*10	8	3.4
*12	1	0.4
*17	23	9.7
*17xN	2	0.8
*41	18	7.6 %

Table 9 Genotypes, activity score and predicted phenotypes for samples with gene duplications

Genotype (xN) before revision	Activity score	Predicted phenotype before revision	Genotype after revision	Activity score	Predicted phenotype after revision
*1/*2	3	UM	*1/*2xN	3	UM
*4/*41	0.5–1.0	IM OR EM	*4xN/*41	0.5	IM
*1/*2	3	UM	*1xN/*2	3	UM
*1/*1	3	UM	*1/*1xN[a]	3	UM
*1/*4	1.0–2.0	EM	*1/*4xN	1	EM
*4/*5	0	PM	*4xN/*5	0	PM
*2/*2	3	UM	*2/*2xN[a]	3	UM
*2/*17	2.0–2.5	UM OR EM	*2/*17xN	2	EM
*4/*10	0.5–1.0	IM	*4xN/*10[b]	0.5	IM
*1/*2	3	UM	*1xN/*2	3	UM
*2/*17	2.0–2.5	UM OR EM	*2/*17xN	2	EM
*4/*5	0	PM	*4xN/*5	0	PM
*2/*12	1.0–2.0	EM	*2xN/*12	2	EM

[a]The number of duplication alleles is far less than single-copy alleles. We assumed the subject was heterozygous for a single-copy allele and the duplicated/multiplicated allele
[b]No reference
*4 had a higher distribution or copy number in comparison with *10

for rapidly identifying numerous SNVs. We demonstrate concordance with known reference standards using this method that requires no enzymatic digestion and can be performed at high volumes. The only difference was an additional identification of a CYP2D6*5 gene deletion not identified by the Taqman method. Less than 60-ng genomic DNA was used for each sample in our assay. The assay is flexible; it would be easy to add additional primers to cover more CYP2D6 SNVs should this be necessary. This customizable assay has advantages given the speed of discovery of CYP SNV identification. As demonstrated by the success with CYP2D6, the multiplex SNaPshot method can be used for designing genotype assays for complex genotyping circumstances rapidly and inexpensively.

This method allows genotyping patients for two genes at only $40 per sample compared to commercially available microarray assays that cost in excess of $500 per sample. While the assay is limited by 10 SNVs per pool, this provided flexibility of the assay to add additional SNVs should they be clinically important. Additional SNVs would require additional DNA, but only 30 ng is required per pool allowing for significant up-scaling of the assay. This method is well suited for pharmacogenes with well-established clinical associations and a finite number of SNVs.

Subjects in our cohort had predicted phenotype frequencies similar to European populations.

Genotypes for CYP2D6 and ADRB1 demonstrate that 25 % may have a gene-gene interaction affecting their metoprolol therapy. Results in our cohort demonstrate a non-responder haplotype in 31.9 % of subjects. Liu et al.

demonstrated a non-responder phenotype based upon these same haplotypes in 45.9 % of subjects in that trial [18]. Variability in ethnic proportions may explain this discrepancy though the frequencies are similar. These genes do not co-segregate, requiring knowledge of both genotypes to determine this interaction potential. Paired with clinical outcomes, these gene-gene interactions can be further clarified as the cohort matures. Overall, the multiplex SNaPshot represents a valuable tool for systems biology studies in need of flexible genotyping methods.

Conclusion

Our multiplex SNaPshot protocol allows for specific and accurate detection of CYP2D6 and ADRB1 genotypes. This platform is flexible, simple, efficient, and suited for high throughput.

Competing interests
The authors declare that they have no competing interests.

Authors' contributions
SB and AAM designed the assay. SB performed the multiplex assay. RMC provided the standards for the assay. HKF and OE enrolled the subjects and processed the samples. TF and RS coordinated the laboratory where the assay was performed and worked on troubleshooting the assay. AAM is the principle investigator of this study. All authors contributed to the development and critical review of the manuscript. All authors read and approved the final manuscript.

Acknowledgements
None

Disclosures and funding
Dr. Monte receives support from NIH 1 K23 GM110516 and NIH CTSI UL1 TR001082. Dr. Cooper-DeHoff receives funding from NIH U01 GM074492 and U01 HG007269.

Author details
[1]Human Medical Genetics Program, University of Colorado, Aurora, CO 80045, USA. [2]Center for Pharmacogenomics, College of Pharmacy, University of Florida, Gainesville, FL 32610, USA. [3]Department of Emergency Medicine, School of Medicine, University of Colorado, 12401 E. 17th Ave, B215, Aurora, CO 80045, USA. [4]Rocky Mountain Poison & Drug Center, Denver Health and Hospital Authority, Denver, CO 80203, USA.

References

1. Crews KR, Gaedigk A, Dunnenberger HM, Klein TE, Shen DD, Callaghan JT, Kharasch ED, Skaar TC, Clinical Pharmacogenetics Implementation C. Clinical Pharmacogenetics Implementation Consortium (CPIC) guidelines for codeine therapy in the context of cytochrome P450 2D6 (CYP2D6) genotype. Clin Pharmacol Ther. 2012;91:321–6.

2. Zhou SF. Polymorphism of human cytochrome P450 2D6 and its clinical significance: part I. Clin Pharmacokinet. 2009;48:689–723.

3. Gough AC, Smith CA, Howell SM, Wolf CR, Bryant SP, Spurr NK. Localization of the CYP2D gene locus to human chromosome 22q13.1 by polymerase chain reaction, in situ hybridization, and linkage analysis. Genomics. 1993;15:430–2.

4. Kimura S, Umeno M, Skoda RC, Meyer UA, Gonzalez FJ. The human debrisoquine 4-hydroxylase (CYP2D) locus: sequence and identification of the polymorphic CYP2D6 gene, a related gene, and a pseudogene. Am J Hum Genet. 1989;45:889–904.

5. Eichelbaum M, Baur MP, Dengler HJ, Osikowska-Evers BO, Tieves G, Zekorn C, Rittner C. Chromosomal assignment of human cytochrome P-450 (debrisoquine/sparteine type) to chromosome 22. Br J Clin Pharmacol. 1987;23:455–8.

6. Gaedigk A, Simon SD, Pearce RE, Bradford LD, Kennedy MJ, Leeder JS. The CYP2D6 activity score: translating genotype information into a qualitative measure of phenotype. Clin Pharmacol Ther. 2008;83:234–42.

7. Johnson JA, Zineh I, Puckett BJ, McGorray SP, Yarandi HN, Pauly DF. Beta 1-adrenergic receptor polymorphisms and antihypertensive response to metoprolol. Clin Pharmacol Ther. 2003;74:44–52.

8. Hamadeh IS, Langaee TY, Dwivedi R, Garcia S, Burkley BM, Skaar TC, Chapman AB, Gums JG, Turner ST, Gong Y, Cooper-DeHoff RM, Johnson JA. Impact of CYP2D6 polymorphisms on clinical efficacy and tolerability of metoprolol tartrate. Clin Pharmacol Ther. 2014;96:175–81.

9. Terra SG, Pauly DF, Lee CR, Patterson JH, Adams KF, Schofield RS, Belgado BS, Hamilton KK, Aranda JM, Hill JA, Yarandi HN, Walker JR, Phillips MS, Gelfand CA, Johnson JA. beta-Adrenergic receptor polymorphisms and responses during titration of metoprolol controlled release/extended release in heart failure. Clin Pharmacol Ther. 2005;77:127–37.

10. Chobanian AV, Bakris GL, Black HR, Cushman WC, Green LA, Izzo Jr JL, Jones DW, Materson BJ, Oparil S, Wright Jr JT, Roccella EJ. The Seventh Report of the Joint National Committee on Prevention, Detection, Evaluation, and Treatment of High Blood Pressure: the JNC 7 report. Jama. 2003;289:2560–72.

11. Pfeffer MA, Braunwald E, Moye LA, Basta L, Brown Jr EJ, Cuddy TE, Davis BR, Geltman EM, Goldman S, Flaker GC, et al. Effect of captopril on mortality and morbidity in patients with left ventricular dysfunction after myocardial infarction. Results of the survival and ventricular enlargement trial. The SAVE Investigators. N Engl J Med. 1992;327:669–77.

12. Hunt SA, Baker DW, Chin MH, Cinquegrani MP, Feldman AM, Francis GS, Ganiats TG, Goldstein S, Gregoratos G, Jessup ML, Noble RJ, Packer M, Silver MA, Stevenson LW, Gibbons RJ, Antman EM, Alpert JS, Faxon DP, Fuster V, Jacobs AK, Hiratzka LF, Russell RO, Smith Jr SC. ACC/AHA guidelines for the evaluation and management of chronic heart failure in the adult: executive summary: a report of the American College of Cardiology/American Heart Association Task Force on practice guidelines (Committee to Revise the 1995 Guidelines for the Evaluation and Management of Heart Failure): developed in collaboration with the International Society for Heart and Lung Transplantation; endorsed by the Heart Failure Society of America. Circulation. 2001;104:2996–3007.

13. Gibbons RJ, Abrams J, Chatterjee K, Daley J, Deedwania PC, Douglas JS, Ferguson Jr TB, Fihn SD, Fraker Jr TD, Gardin JM, O'Rourke RA, Pasternak RC, Williams SV. ACC/AHA 2002 guideline update for the management of patients with chronic stable angina—summary article: a report of the American College of Cardiology/American Heart Association Task Force on practice guidelines (Committee on the Management of Patients with Chronic Stable Angina). J Am Coll Cardiol. 2003;41:159–68.

14. Shin J, Johnson JA. Pharmacogenetics of beta-blockers. Pharmacotherapy. 2007;27:874–87.

15. The use of medicines in the United States: review of 2011. IMS Institute for Healthcare Informatics. [https://www.imshealth.com/files/web/IMSH%20Institute/Reports/The%20Use%20of%20Medicines%20in%20the%20United%20States%202011/IHII_Medicines_in_U.S_Report_2011].

16. Goryachkina K, Burbello A, Boldueva S, Babak S, Bergman U, Bertilsson L. CYP2D6 is a major determinant of metoprolol disposition and effects in hospitalized Russian patients treated for acute myocardial infarction. Eur J Clin Pharmacol. 2008;64:1163–73.

17. Ismail R, Teh LK. The relevance of CYP2D6 genetic polymorphism on chronic metoprolol therapy in cardiovascular patients. J Clin Pharm Ther. 2006;31:99–109.

18. Liu J, Liu ZQ, Yu BN, Xu FH, Mo W, Zhou G, Liu YZ, Li Q, Zhou HH. beta1-Adrenergic receptor polymorphisms influence the response to metoprolol monotherapy in patients with essential hypertension. Clin Pharmacol Ther. 2006;80:23–32.

19. Beitelshees AL, Zineh I, Yarandi HN, Pauly DF, Johnson JA. Influence of phenotype and pharmacokinetics on beta-blocker drug target pharmacogenetics. Pharmacogenomics J. 2006;6:174–8.

20. James PA, Oparil S, Carter BL, Cushman WC, Dennison-Himmelfarb C, Handler J, Lackland DT, LeFevre ML, MacKenzie TD, Ogedegbe O, Smith Jr SC, Svetkey LP, Taler SJ, Townsend RR, Wright Jr JT, Narva AS, Ortiz E. 2014 evidence-based guideline for the management of high blood pressure in adults: report from the panel members appointed to the Eighth Joint National Committee (JNC 8). Jama. 2014;311:507–20.

21. Lovlie R, Daly AK, Molven A, Idle JR, Steen VM. Ultrarapid metabolizers of debrisoquine: characterization and PCR-based detection of alleles with duplication of the CYP2D6 gene. FEBS Lett. 1996;392:30–4.

22. Bathum L, Johansson I, Ingelman-Sundberg M, Horder M, Brosen K. Ultrarapid metabolism of sparteine: frequency of alleles with duplicated CYP2D6 genes in a Danish population as determined by restriction fragment length polymorphism and long polymerase chain reaction. Pharmacogenetics. 1998;8:119–23.

23. Okubo M, Murayama N, Miura J, Shimizu M, Yamazaki H. A rapid multiplex PCR assay that can reliably discriminate the cytochrome P450 2D6 whole-gene deletion allele from 2D6*10 alleles. Clin Chim Acta. 2012;413:1675–7.

24. Langaee T, Hamadeh I, Chapman AB, Gums JG, Johnson JA. A novel simple method for determining CYP2D6 gene copy number and identifying allele(s) with duplication/multiplication. PLoS One. 2015;10:e0113808.

9

Systematic analysis of the molecular mechanism underlying atherosclerosis using a text mining approach

Dan Xi[1], Jinzhen Zhao[1], Wenyan Lai[2*] and Zhigang Guo[1*]

Abstract

Background: Atherosclerosis is one of the common health threats all over the world. It is a complex heritable disease that affects arterial blood vessels. Chronic inflammatory response plays an important role in atherogenesis. There has been little success in fully identifying functionally important genes in the pathogenesis of atherosclerosis.

Results: In the present study, we performed a systematic analysis of atherosclerosis-related genes using text mining. We identified a total of 1312 genes. Gene ontology (GO) analysis revealed that a total of 35 terms exhibited significance ($p < 0.05$) as overrepresented terms, indicating that atherosclerosis invokes many genes with a wide range of different functions. Pathway analysis demonstrated that the most highly enriched pathway is the Toll-like receptor signaling pathway. Finally, through gene network analysis, we prioritized 48 genes using the hub gene method.

Conclusions: Our study provides a valuable resource for the in-depth understanding of the mechanism underlying atherosclerosis.

Keywords: Atherosclerosis, Pathogenesis, Text mining

Background

Atherosclerosis is a complex heritable disease involving multiple cell types and the interactions of many different molecular pathways [1]. Atherosclerosis is therefore a syndrome affecting arterial blood vessels due to a chronic inflammatory response [2, 3]. Atherosclerosis is at the core of cardiovascular diseases, often leading to myocardial infarctions, stroke, and peripheral vascular diseases.

Recent genome-wide association studies (GWAS), involving hundreds of thousands of individuals, have identified numerous loci contributing to atherosclerotic traits and to risk factors such as blood lipoprotein levels and blood pressure [4]. Plasma lipids are primarily of importance for driving early atherosclerosis development, consistent with the notion that loci identified by GWAS will be more useful for primary prevention and with the experimental finding that atherosclerosis regression in response to LDL lowering is much greater for early lesions than for mature and advanced lesions [5]. The extensive ongoing studies into the molecular mechanisms of the 153 confirmed GWA-defined CAD loci will shed light on this issue, as these mechanisms will likely be traceable to early versus late events in the pathogenesis of atherosclerosis [6]. Despite a large number of genes are identified, there has been little success in fully identifying functionally important genes in the pathogenesis of atherosclerosis.

Recently, the text mining methodology has been implemented, providing a necessary means to retrieve disease-related genes in an automated way [7]. Here, we reported on a systematic analysis of atherosclerosis-related genes using text mining. Our study provides in-depth insights into the molecular mechanisms underlying atherosclerosis.

* Correspondence: frame007@163.com; guozhigang126@126.com
[2]Laboratory of Department of Cardiology, Nanfang Hospital, Southern Medical University, Guangzhou 510515, Guangdong, People's Republic of China
[1]Division of Cardiology, Huiqiao Medical Center, Nanfang Hospital, Southern Medical University, 1838 North Guangzhou Avenue, Guangzhou 510515, Guangdong, People's Republic of China

Results

Identification of atherosclerosis-related genes by using text mining

We ran a key word search in the PubMed database for articles related to atherosclerosis and obtained 45,304 entries as a result (from January 1980 to April 2016). Abstracts of these articles were downloaded and processed through a text mining pipeline shown in Fig. 1a. Cumulative distribution analysis indicated that the number of articles published on atherosclerosis is growing linearly in recent years (Fig. 1b). From these articles, we extracted atherosclerosis-associated genes via text mining. We compiled a list of 1312 atherosclerosis-related genes (Fig. 1c; Additional file 1: Table S1).

Functional clustering analysis

All 1312 unique genes were functionally categorized based on gene ontology (GO) annotation terms using the BiNGO program package. Enrichment analysis revealed that a total of 35 terms exhibited significance ($p < 0.05$) as overrepresented terms. In the biological process category, response to stimulus, cell communication, regulation of biological process, cellular process, behavior, multicellular organismal development, cell motility, cell death, metabolic process, cell differentiation, enzyme regulator activity, transcription regulator activity, electron carrier activity, secretion, catabolic process,

transport, macromolecule metabolic process, and unspecific monooxygenase activity were found to be significantly enriched. GO terms related to extracellular region, extracellular space, cell surface, cytoplasm, membrane, proteinaceous extracellular matrix, and cell were overrepresented under the cellular component category. The overrepresented GO terms in the molecular function category were protein binding, binding, signal transducer activity, receptor activity, antioxidant activity, oxidoreductase activity, catalytic activity, hydrolase activity, kinase activity, and transferase activity (Additional file 2: Table S2). The hierarchical organization of these GO terms is shown in Fig. 2, together with the significance of enrichment indicated by different colors.

Pathway analysis

In addition to the GO analysis, we also performed pathway analysis by using the DAVID tools. Unlike GO, which only contains lists of functional gene groups, the pathway database also stores the information of gene dependencies in each pathway. In the present study, all atherosclerosis-related genes were linked to a total of 50 pathways. Among these pathways, 20 pathways, namely Toll-like receptor signaling pathway, complement and coagulation cascades, hematopoietic cell lineage, NOD-like receptor signaling pathway, adipocytokine signaling pathway, focal adhesion, Jak-STAT signaling pathway,

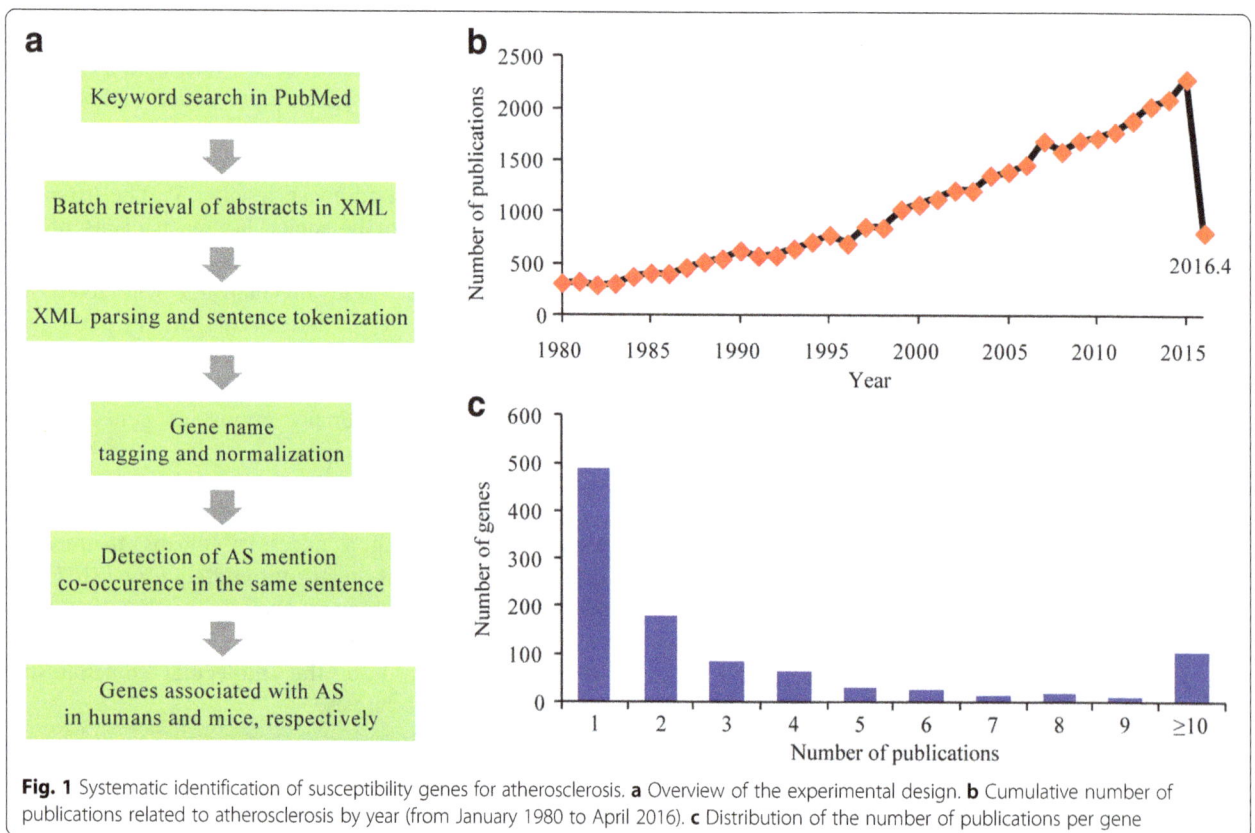

Fig. 1 Systematic identification of susceptibility genes for atherosclerosis. **a** Overview of the experimental design. **b** Cumulative number of publications related to atherosclerosis by year (from January 1980 to April 2016). **c** Distribution of the number of publications per gene

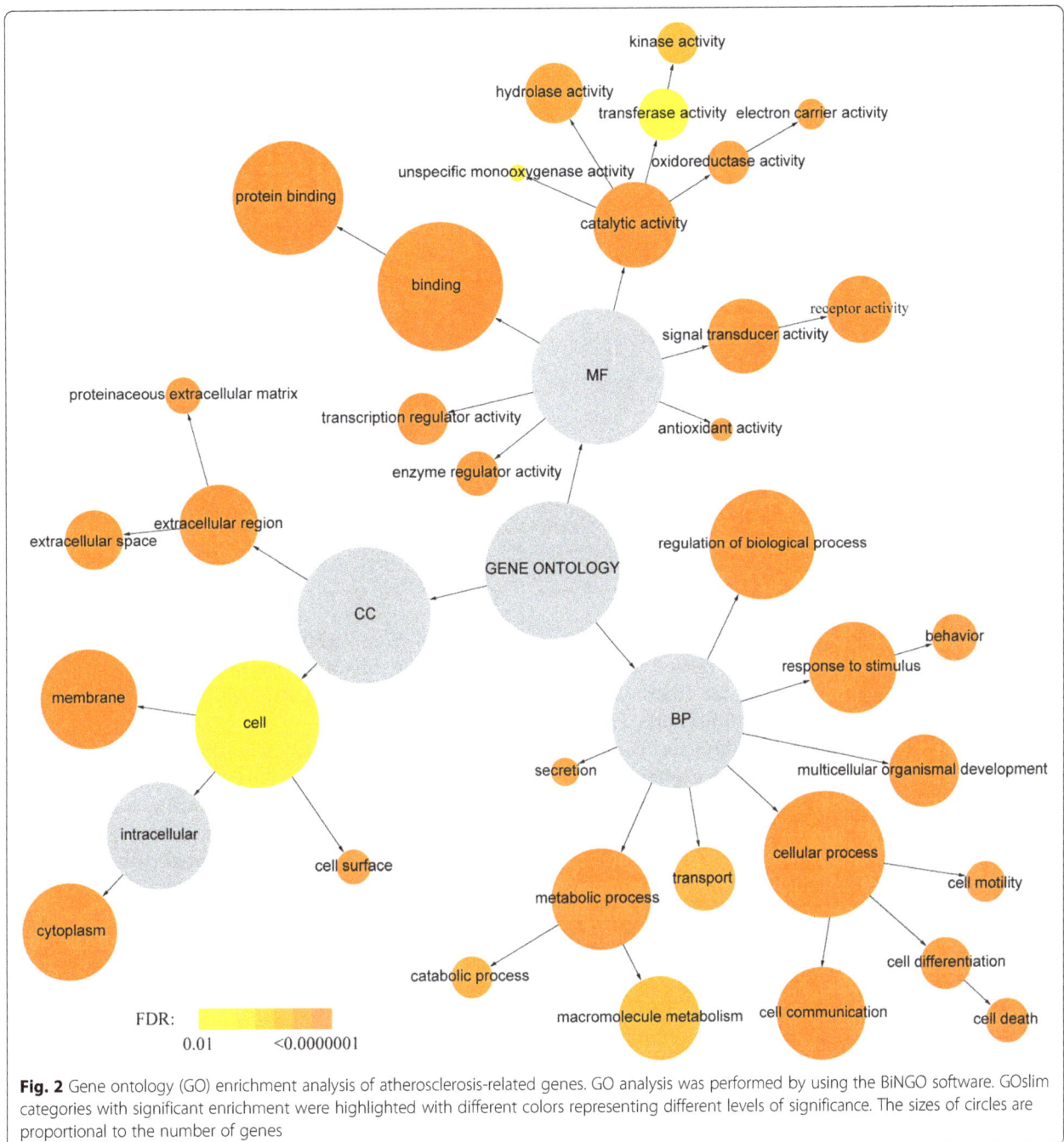

Fig. 2 Gene ontology (GO) enrichment analysis of atherosclerosis-related genes. GO analysis was performed by using the BiNGO software. GOslim categories with significant enrichment were highlighted with different colors representing different levels of significance. The sizes of circles are proportional to the number of genes

apoptosis, T cell receptor signaling pathway, neurotrophin signaling pathway, Fc epsilon RI signaling pathway, PPAR signaling pathway, VEGF signaling pathway, B cell receptor signaling pathway, renin-angiotensin system, leukocyte transendothelial migration, ErbB signaling pathway, TGF-beta signaling pathway, MAPK signaling pathway, and natural killer cell mediated cytotoxicity were significantly enriched ($p < 0.05$) (Fig. 3a; Additional file 3: Table S3). Based on enrichment p value, the most highly

overrepresented pathway went to the Toll-like receptor signaling pathway (Fig. 3b). The Toll-like receptor signaling pathway is known to play an important role during atherosclerosis in both immune and inflammatory response.

Network analysis

In the present study, a genome-wide protein-protein interaction (PPI) network was constructed by merging up-to-date protein-protein interactions available in

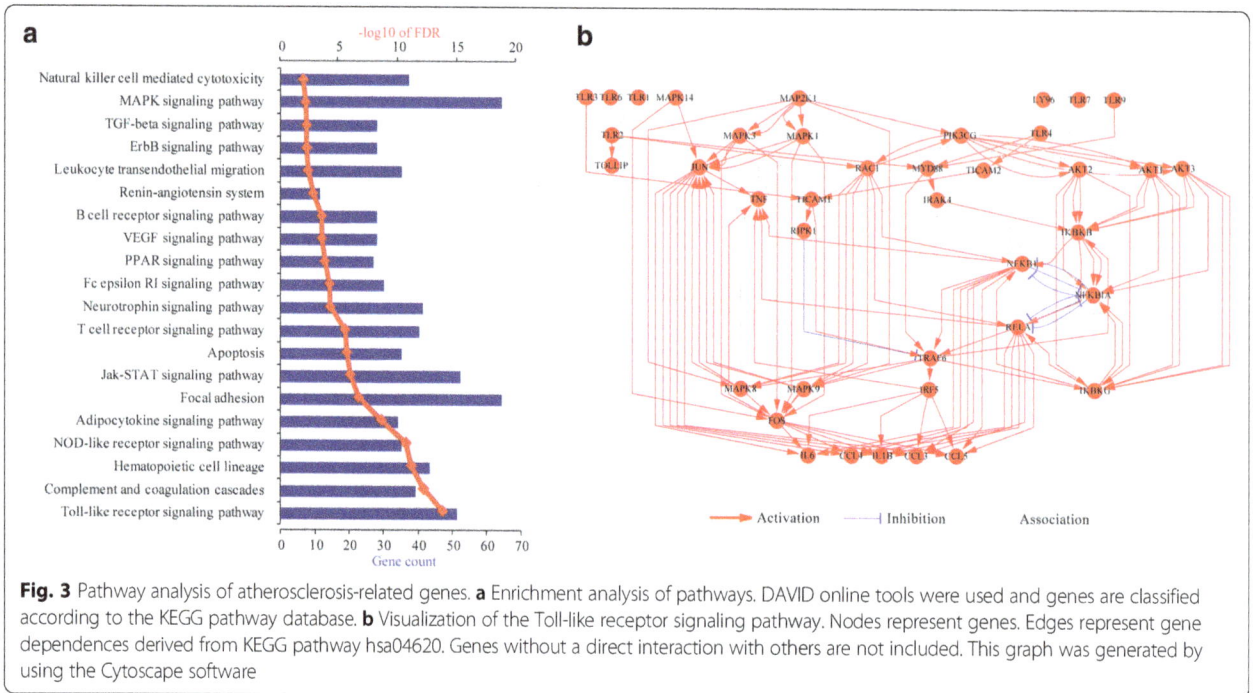

Fig. 3 Pathway analysis of atherosclerosis-related genes. **a** Enrichment analysis of pathways. DAVID online tools were used and genes are classified according to the KEGG pathway database. **b** Visualization of the Toll-like receptor signaling pathway. Nodes represent genes. Edges represent gene dependences derived from KEGG pathway hsa04620. Genes without a direct interaction with others are not included. This graph was generated by using the Cytoscape software

IntAct [8], BioGRID [9], MINT [10], DIP [11], HPRD [12, 13], and MIPS [13]. The network related to atherosclerosis was generated by mapping the atherosclerosis-related genes to the genome-wide PPI network. The atherosclerosis network consisted of 1079 nodes connected via 6089 edges (Fig. 4a). Topological analysis showed that the network follows a power-law distribution (Fig. 4b) and therefore is a scale-free, small-world network [14]. This type of networks has the particular feature that some nodes are highly connected compared with others within the network. These highly connected nodes, also known as hub genes, represent important genes in the network and therefore are treated with special attention. Using a defined cut-off value, we identified 48 hub genes. These hub genes and their connections were extracted from the whole network and rendered as a simplified sub-network (Fig. 4c).

Discussion

In the present study, we attempted to compile a complete list of genes involved in atherosclerosis. In recent years, high-throughput transcriptomic and proteomic approaches make it possible for studying the expression levels of thousands of genes and proteins simultaneously. However, these data suffer from high technical variability and high dimension size [15, 16]. On the contrary, there is a large body of research using conventional gene-by-gene methods. Text mining provides the necessary means to retrieve these data through automated processing of texts [7]. Here, we performed a text mining analysis of atherosclerosis-associated genes. We identified

1312 genes from 45,304 publications. Considering the large body of literature we analyzed, our result may have reasonably good coverage of all atherosclerosis-associated genes.

We found that 1312 genes were associated with atherosclerosis. Based on GO analysis, 35 GO terms were significantly enriched. Additionally, our study also revealed 20 enriched pathways. Based on enrichment p value, the most highly overrepresented pathway went to the Toll-like receptor (TLR) signaling pathway. The Toll-like receptor signaling pathway is known to play an important role during atherosclerosis in both immune and inflammatory response. The disruptions of cellular or organismal cholesterol homeostasis that occur as a risk factor of atherosclerosis may lead to an augmentation of inflammatory responses via enhanced TLR signaling or inflammasome activation [17]. TLR activation leads to the expression of pro-inflammatory cytokines and also induces the expression of many negative regulators, acting to limit signal transduction, messenger RNA (mRNA) transcription, or translation [18].

A genome-wide gene network was constructed by using up-to-date interaction data available in the PINA2 database [19]. We obtained a gene network consisting of 1079 nodes connected via 6089 edges. So far, several studies have been conducted to incorporate the topology of gene network in prioritization of disease candidate genes [20–22]. The main concern for these studies is that the incompleteness and noisiness of interaction data may affect the accuracy of prioritization result. By merging up-to-date protein-protein interactions available in

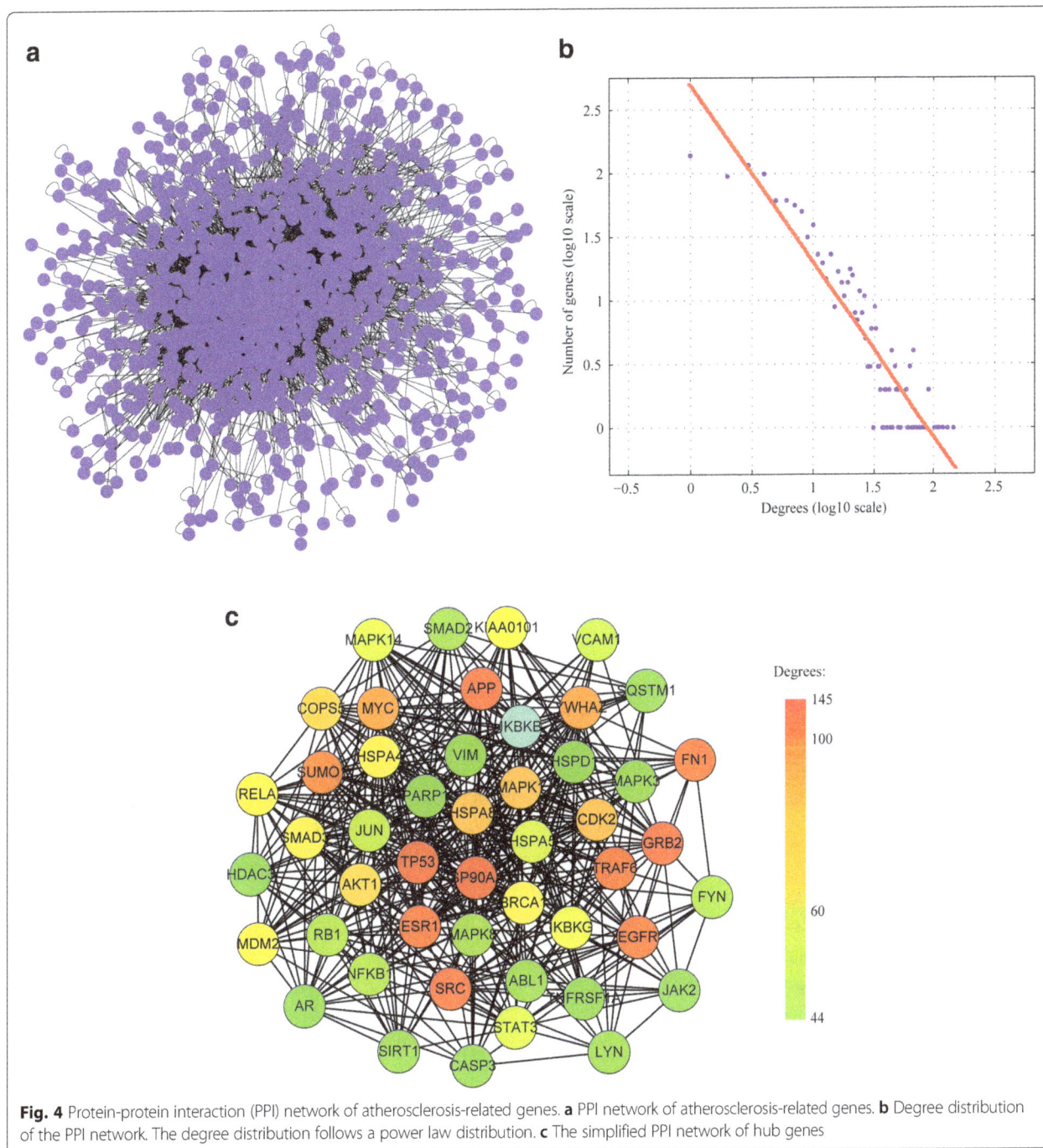

Fig. 4 Protein-protein interaction (PPI) network of atherosclerosis-related genes. **a** PPI network of atherosclerosis-related genes. **b** Degree distribution of the PPI network. The degree distribution follows a power law distribution. **c** The simplified PPI network of hub genes

IntAct [8], BioGRID [9], MINT [10], DIP [11], HPRD [12], and MIPS [13], the PINA2 database provides a comprehensive gene network at genome-wide scale. We expected that the use of the PINA2 database may alleviate this problem to a certain extent. Using a defined threshold value for degree, we identified a total of 48 hub genes in this network.

The top 20 hub genes are the following: APP (amyloid beta A4 precursor protein), HSP90AA1 (heat shock

protein 90 kDa alpha class A member 1), GRB2 (growth factor receptor-bound protein 2), SRC (v-src sarcoma viral oncogene homolog), TP53 (tumor protein p53), ESR1 (estrogen receptor 1), FN1 (fibronectin 1), TRAF6 (TNF receptor-associated factor 6), EGFR (epidermal growth factor receptor), SUMO1 (SMT3 suppressor of mif two 3 homolog 1), YWHAZ (14-3-3 zeta), MYC (v-myc myelocytomatosis viral oncogene homolog), CDK2 (cyclin-dependent kinase 2), HSPA8 (heat shock 70-kDa

protein 8), MAPK1 (mitogen-activated protein kinase 1), AKT1 (v-akt murine thymoma viral oncogene homolog 1), COPS5 (COP9 constitutive photomorphogenic homolog subunit 5), MDM2 (Mdm2 p53 binding protein homolog), RELA (v-rel reticuloendotheliosis viral oncogene homolog A, NFKB3, p65), and HSPA4 (heat shock 70-kDa protein 4). APP is present in advanced human carotid plaques, in proximity to activated macrophages and platelets [23], and lack of APP attenuates atherogenesis and leads to plaque stability [24]. HSP90 is a candidate autoantigen, target of cellular and humoral immune reactions in patients with carotid atherosclerosis [25]. HSP90 expression is associated with features of plaque instability in advanced human lesions [26]. GRB2 is required for atherosclerotic lesion formation and uptake of oxidized LDL by macrophages [27]. In endothelial cell, SRC contributes to atherosclerotic lesion development by disrupting adherence junction integrity and promoting monocyte transmigration [28]. Increasing P53 activity protects against atherosclerosis by causing proliferation arrest of lesional macrophages [29]. The product of the MDM2 gene is a nuclear protein which forms a complex with P53, thereby inhibiting the negative regulatory effects of wild-type P53 on cell cycle progression. P53 and MDM2 are expressed in human atherosclerotic lesions; P53 and MDM2 may therefore play an important role in regulating cellularity and inflammatory activity in human atherosclerotic plaques [30, 31]. SUMOylation of P53 by SUMO1 contributes to the atherosclerotic plaque formation [32]. ESR1 is expressed in macrophages and other immune cells known to exert dramatic effects on glucose homeostasis. A study suggests that diminished ESR1 expression in hematopoietic/myeloid cells promotes aspects of the metabolic syndrome and accelerates atherosclerosis in female mice [33]. FN is one of the earliest extracellular matrix (ECM) proteins deposited at atherosclerosis-prone sites and was suggested to promote atherosclerotic lesion formation [34]. TRAF6 is expressed in atherosclerotic aortic tissue of low-density lipoprotein-null mice [35]. Endothelial-specific TRAF6 deficiency in females was associated with diminished atherosclerosis and decreased plaque macrophage burden [36]. EGFR mRNA was detected in atherosclerotic plaques but not in morphologically normal aortae and EGFR receptor staining co-localized with macrophage staining in these plaques [37]. Secreted from activated platelets, YWHAZ is present at the atherosclerotic plaques [38]. In cholesterol-fed roosters, MYC was seen in lipid-rich thickened intimal lesions of the entire aorta [39]. CDK2 negatively regulates neointimal thickening in animal models of restenosis and atherosclerosis, and its expression in human neointimal lesions is consistent with a protective role [40]. HSP70 (HSPA4 and HSPA8) is present in human atherosclerotic lesions [41]. Treatment with platelet-

derived growth factor which caused vascular smooth muscle cell migration in an MAPK1 activation-dependent manner suggests a role for MAPK1 in the pathogenesis and/or progression of atherosclerosis [42]. AKT1 expression in vascular smooth muscle cells influences early and late stages of atherosclerosis. The absence of AKT1 in VSMCs induces features of plaque vulnerability including fibrous cap thinning and extensive necrotic core areas [43]. Macrophage migration inhibitory factor (MIF), a cytokine with potent inflammatory functions, was thus considered to be important in atherosclerotic lesion evolution. COPS5 is able to form complexes with MIF and serves critical regulatory functions in atherosclerotic lesion evolution [44]. The transcription factor NF-κB p65 is a key regulator in the regulation of an inflammatory response and in the pathology of atherosclerosis [45].

A limitation for text mining-based strategies is that there is no chance to discover new genes. In order to solve this problem, we enlarged the network by inducing new genes that are not reported to be involved in atherosclerosis. According to the rule of "guilty-by-association," these new genes may be potential susceptibility genes. Finally, we made a list of 50 new genes, all of which have more than 44 connections with known genes (Additional file 4: Table S4).

Conclusions

In summary, we have reported here the first systematic analysis of the molecular mechanism underlying atherosclerosis using a text mining approach. Our study provides a valuable resource for the in-depth understanding of the mechanism underlying atherosclerosis.

Methods

Identification of atherosclerosis-related genes by using text mining

The PubMed database was used as a source of literature for text mining. We conducted a search with the following combinations of query key words: "atherosclerosis" OR "atherogenesis" OR "atheroma" OR "atherosclerotic." The search tag "[Title/Abstract]" was added after each keyword. The relevant articles were retrieved in XML format, which makes information extraction more precise due to the presence of content enclosed within XML tag pairs. For each article, titles and abstract texts were fetched using the dom4j XML parser class in JAVA. Abstract texts were further divided into sentences through sentence tokenizer implemented in LingPipe (Alias-I, Inc.). Text mining was performed at sentence level.

Gene mention recognition was performed using two different gene mention taggers, the hidden Markov model (HMM) tagger implemented in LingPipe and the ABNER tagger based on a machine learning system of conditional random fields (CRF) [46]. Gene mentions

from both taggers were merged. Because researchers mention genes in a highly variable manner, we built a gene synonym dictionary from entrez gene database [47]. The dictionary was used for the gene name normalization process during which gene mentions were linked to entrez genes using exact string match. If multiple entrez genes share the same gene mention, the ambiguity was resolved manually. In order to minimize the false positive rate, we required the co-occurrence of atherosclerosis mention and gene mention within a single sentence. Finally, we compiled a complete list of atherosclerosis-related genes.

Enrichment test of gene ontology (GO) terms

GO enrichment analysis was performed by using BiNGO 2.3 with GOslim dataset [48]. To test for enrichment, a hypergeometric test was conducted followed by Benjamini and Hochberg multiple test correction. The adjusted p value <0.01 was used as significance threshold to identify enriched categories.

Pathway analysis

To rank overall importance of pathways involved in atherosclerosis, we calculated Fisher's exact test p values and Benjamini-Hochberg adjusted p values through the DAVID bioinformatics resource 6.7 [49]. The significance threshold was set at 0.01. After enrichment tests, gene sets were collected for each pathway. Gene dependencies in a certain pathway were determined using the R package KEGGSOAP and visualized in Cytoscape [50, 51].

Construction of protein-protein interaction (PPI) network

The genes associated with atherosclerosis were cross-referenced with the PINA2 database to create the PPI network [19]. The PINA2 database provides integrated and up-to-date protein-protein interactions available in IntAct, BioGRID, MINT, DIP, HPRD, and MIPS, which simplifies the task of inter-database mapping [8–13]. To query the PINA2 database, interaction was restricted to human and mouse and all kinds of experimental procedures were included. Cytoscape software was applied for visualization and analysis of PPI network. The topological parameters of PPI network were analyzed by NetworkAnalyzer [52]. The edges in the network were treated as undirected. The degree of a node was the number of its directly connecting neighbors in the network. The threshold degree value for hub genes was the mean plus two standard deviations. As a result, genes with a degree value of larger than 26 were considered hub genes. Hub genes and their connections were extracted from the whole network and rendered as a simplified sub-network.

Acknowledgements

We would like to thank Jilong Liu for his contributions to database support and statistical analysis. This project was supported by the National Natural Science Foundation of China (81370380), the Natural Science Foundation of Guangdong Province of China (S2013010014739), and the Science and Technology Foundation of Guangdong Province of China (2012B091100155).

Authors' contributions

GZG conceived and designed the study. LWY and ZJZ collected the data. XD completed the study and preformed the statistical tests. XD wrote the manuscript. All authors read and approved the final manuscript.

Competing interests

The authors declare that they have no competing interests.

References

1. Stylianou IM, Bauer RC, Reilly MP, Rader DJ. Genetic basis of atherosclerosis: insights from mice and humans. Circ Res. 2012;110:337–55.
2. Glass CK, Witztum JL. Atherosclerosis. The road ahead. Cell. 2001;104:503–16.
3. Andersson J, Libby P, Hansson GK. Adaptive immunity and atherosclerosis. Clin Immunol. 2010;134:33–46.
4. Bennett BJ, Davis RC, Civelek M, Orozco L, Wu J, Qi H, et al. Genetic architecture of atherosclerosis in mice: a systems genetics analysis of common inbred strains. PLoS Genet. 2015;11, e1005711.
5. Bjorkegren JL, Hagg S, Talukdar HA, Foroughi Asl H, Jain RK, Cedergren C, et al. Plasma cholesterol-induced lesion networks activated before regression of early, mature, and advanced atherosclerosis. PLoS Genet. 2014;10, e1004201.
6. Bjorkegren JL, Kovacic JC, Dudley JT, Schadt EE. Genome-wide significant loci: how important are they? Systems genetics to understand heritability of coronary artery disease and other common complex disorders. J Am Coll Cardiol. 2015;65:830–45.
7. Rebholz-Schuhmann D, Oellrich A, Hoehndorf R. Text-mining solutions for biomedical research: enabling integrative biology. Nat Rev Genet. 2012;13:829–39.
8. Aranda B, Achuthan P, Alam-Faruque Y, Armean I, Bridge A, Derow C, et al. The IntAct molecular interaction database in 2010. Nucleic Acids Res. 2010;38:D525–31.
9. Stark C, Breitkreutz BJ, Chatr-Aryamontri A, Boucher L, Oughtred R, Livstone MS, et al. The BioGRID interaction database: 2011 update. Nucleic Acids Res. 2011; 39:D698–704.
10. Ceol A, Chatr Aryamontri A, Licata L, Peluso D, Briganti L, Perfetto L, et al. MINT, the molecular interaction database: 2009 update. Nucleic Acids Res. 2010;38:D532–9.
11. Salwinski L, Miller CS, Smith AJ, Pettit FK, Bowie JU, Eisenberg D. The database of interacting proteins: 2004 update. Nucleic Acids Res. 2004;32:D449–51.
12. Keshava Prasad TS, Goel R, Kandasamy K, Keerthikumar S, Kumar S, Mathivanan S, et al. Human protein reference database—2009 update. Nucleic Acids Res. 2009;37:D767–72.
13. Mewes HW, Dietmann S, Frishman D, Gregory R, Mannhaupt G, Mayer KF, et al. MIPS: analysis and annotation of genome information in 2007. Nucleic Acids Res. 2008;36:D196–201.
14. Barabasi AL, Oltvai ZN. Network biology: understanding the cell's functional organization. Nat Rev Genet. 2004;5:101–13.
15. Frantz S. An array of problems. Nat Rev Drug Discov. 2005;4:362–3.
16. Chandramouli K, Qian PY. Proteomics: challenges, techniques and possibilities to overcome biological sample complexity. Hum Genomics Proteomics. 2009;2009.
17. Tall AR, Yvan-Charvet L. Cholesterol, inflammation and innate immunity. Nat Rev Immunol. 2015;15:104–16.
18. Medzhitov R, Horng T. Transcriptional control of the inflammatory response. Nat Rev Immunol. 2009;9:692–703.
19. Cowley MJ, Pinese M, Kassahn KS, Waddell N, Pearson JV, Grimmond SM, et al. PINA v2.0: mining interactome modules. Nucleic Acids Res. 2012;40:D862–5.

20. Chen J, Aronow BJ, Jegga AG. Disease candidate gene identification and prioritization using protein interaction networks. BMC Bioinformatics. 2009;10:73.

21. Guo H, Dong J, Hu S, Cai X, Tang G, Dou J, et al. Biased random walk model for the prioritization of drug resistance associated proteins. Sci Rep. 2015;5:10857.

22. Morrison JL, Breitling R, Higham DJ, Gilbert DR. GeneRank: using search engine technology for the analysis of microarray experiments. BMC Bioinformatics. 2005;6:233.

23. De Meyer GR, De Cleen DM, Cooper S, Knaapen MW, Jans DM, Martinet W, et al. Platelet phagocytosis and processing of beta-amyloid precursor protein as a mechanism of macrophage activation in atherosclerosis. Circ Res. 2002;90:1197–204.

24. Van De Parre TJ, Guns PJ, Fransen P, Martinet W, Bult H, Herman AG, et al. Attenuated atherogenesis in apolipoprotein E-deficient mice lacking amyloid precursor protein. Atherosclerosis. 2011;216:54–8.

25. Businaro R, Profumo E, Tagliani A, Buttari B, Leone S, D'Amati G, et al. Heat-shock protein 90: a novel autoantigen in human carotid atherosclerosis. Atherosclerosis. 2009;207:74–83.

26. Madrigal-Matute J, Lopez-Franco O, Blanco-Colio LM, Munoz-Garcia B, Ramos-Mozo P, Ortega L, et al. Heat shock protein 90 inhibitors attenuate inflammatory responses in atherosclerosis. Cardiovasc Res. 2010;86:330–7.

27. Proctor BM, Ren J, Chen Z, Schneider JG, Coleman T, Lupu TS, et al. Grb2 is required for atherosclerotic lesion formation. Arterioscler Thromb Vasc Biol. 2007;27:1361–7.

28. Sun C, Wu MH, Lee ES, Yuan SY. A disintegrin and metalloproteinase 15 contributes to atherosclerosis by mediating endothelial barrier dysfunction via Src family kinase activity. Arterioscler Thromb Vasc Biol. 2012;32:2444–51.

29. Sayin VI, Khan OM, Pehlivanoglu LE, Staffas A, Ibrahim MX, Asplund A, et al. Loss of one copy of Zfp148 reduces lesional macrophage proliferation and atherosclerosis in mice by activating p53. Circ Res. 2014;115:781–9.

30. Ihling C, Haendeler J, Menzel G, Hess RD, Fraedrich G, Schaefer HE, et al. Co-expression of p53 and MDM2 in human atherosclerosis: implications for the regulation of cellularity of atherosclerotic lesions. J Pathol. 1998;185:303–12.

31. Barillari G, Iovane A, Bonuglia M, Alboinci L, Garofano P, Di Campli E, et al. Fibroblast growth factor-2 transiently activates the p53 oncosuppressor protein in human primary vascular smooth muscle cells: implications for atherogenesis. Atherosclerosis. 2010;210:400–6.

32. Heo KS, Chang E, Le NT, Cushman H, Yeh ET, Fujiwara K, et al. De-SUMOylation enzyme of sentrin/SUMO-specific protease 2 regulates disturbed flow-induced SUMOylation of ERK5 and p53 that leads to endothelial dysfunction and atherosclerosis. Circ Res. 2013;112:911–23.

33. Ribas V, Drew BG, Le JA, Soleymani T, Daraei P, Sitz D, et al. Myeloid-specific estrogen receptor alpha deficiency impairs metabolic homeostasis and accelerates atherosclerotic lesion development. Proc Natl Acad Sci U S A. 2011;108:16457–62.

34. Rohwedder I, Montanez E, Beckmann K, Bengtsson E, Duner P, Nilsson J, et al. Plasma fibronectin deficiency impedes atherosclerosis progression and fibrous cap formation. EMBO Mol Med. 2012;4:564–76.

35. Zirlik A, Bavendiek U, Libby P, MacFarlane L, Gerdes N, Jagielska J, et al. TRAF-1, -2, -3, -5, and -6 are induced in atherosclerotic plaques and differentially mediate proinflammatory functions of CD40L in endothelial cells. Arterioscler Thromb Vasc Biol. 2007;27:1101–7.

36. Polykratis A, van Loo G, Xanthoulea S, Hellmich M, Pasparakis M. Conditional targeting of tumor necrosis factor receptor-associated factor 6 reveals opposing functions of Toll-like receptor signaling in endothelial and myeloid cells in a mouse model of atherosclerosis. Circulation. 2012;126:1739–51.

37. Lamb DJ, Modjtahedi H, Plant NJ, Ferns GA. EGF mediates monocyte chemotaxis and macrophage proliferation and EGF receptor is expressed in atherosclerotic plaques. Atherosclerosis. 2004;176:21–6.

38. Hernandez-Ruiz L, Valverde F, Jimenez-Nunez MD, Ocana E, Saez-Benito A, Rodriguez-Martorell J, et al. Organellar proteomics of human platelet dense granules reveals that 14-3-3zeta is a granule protein related to atherosclerosis. J Proteome Res. 2007;6:4449–57.

39. Toda T, Tamamoto T, Shimajiri S, Sadi AM, Nakashima Y, Takei H. Expression of PDGF and C-myc in atherosclerotic lesions in cholesterol-fed chicken. Immunohistochemical and in situ hybridization study. Ann N Y Acad Sci. 1995;748:514–6.

40. Sanz-Gonzalez SM, Melero-Fernandez de Mera R, Malek NP, Andres V.

41. Johnson AD, Berberian PA, Tytell M, Bond MG. Atherosclerosis alters the localization of HSP70 in human and macaque aortas. Exp Mol Pathol. 1993;58:155–68.

42. Yoshizumi M, Kyotani Y, Zhao J, Nagayama K, Ito S, Tsuji Y, et al. Role of big mitogen-activated protein kinase 1 (BMK1)/extracellular signal-regulated kinase 5 (ERK5) in the pathogenesis and progression of atherosclerosis. J Pharmacol Sci. 2012;120:259–63.

43. Rotllan N, Wanschel AC, Fernandez-Hernando A, Salerno AG, Offermanns S, Sessa WC, et al. Genetic evidence supports a major role for Akt1 in VSMCs during atherogenesis. Circ Res. 2015;116:1744–52.

44. Burger-Kentischer A, Goebel H, Seiler R, Fraedrich G, Schaefer HE, Dimmeler S, et al. Expression of macrophage migration inhibitory factor in different stages of human atherosclerosis. Circulation. 2002;105:1561–6.

45. Ye X, Jiang X, Guo W, Clark K, Gao Z. Overexpression of NF-kappaB p65 in macrophages ameliorates atherosclerosis in apoE-knockout mice. Am J Physiol Endocrinol Metab. 2013;305:E1375–83.

46. Settles B. ABNER: an open source tool for automatically tagging genes, proteins and other entity names in text. Bioinformatics. 2005;21:3191–2.

47. Brown GR, Hem V, Katz KS, Ovetsky M, Wallin C, Ermolaeva O, et al. Gene: a gene-centered information resource at NCBI. Nucleic Acids Res. 2015;43:D36–42.

48. Maere S, Heymans K, Kuiper M. BiNGO: a Cytoscape plugin to assess overrepresentation of gene ontology categories in biological networks. Bioinformatics. 2005;21:3448–9.

49. Huang DW, Sherman BT, Tan Q, Kir J, Liu D, Bryant D, et al. DAVID Bioinformatics Resources: expanded annotation database and novel algorithms to better extract biology from large gene lists. Nucleic Acids Res. 2007;35:W169–75.

50. Gentleman RC, Carey VJ, Bates DM, Bolstad B, Dettling M, Dudoit S, et al. Bioconductor: open software development for computational biology and bioinformatics. Genome Biol. 2004;5:R80.

51. Shannon P, Markiel A, Ozier O, Baliga NS, Wang JT, Ramage D, et al. Cytoscape: a software environment for integrated models of biomolecular interaction networks. Genome Res. 2003;13:2498–504.

52. Assenov Y, Ramirez F, Schelhorn SE, Lengauer T, Albrecht M. Computing topological parameters of biological networks. Bioinformatics. 2008;24:282–4.

Atheroma development in apolipoprotein E-null mice is not regulated by phosphorylation of p27(Kip1) on threonine 187. J Cell Biochem. 2006;97:735–43.

10

Transcriptome sequencing of gingival biopsies from chronic periodontitis patients reveals novel gene expression and splicing patterns

Yong-Gun Kim[1,2†], Minjung Kim[3†], Ji Hyun Kang[4], Hyo Jeong Kim[4], Jin-Woo Park[1], Jae-Mok Lee[1], Jo-Young Suh[1], Jae-Young Kim[2,4], Jae-Hyung Lee[3,5*] and Youngkyun Lee[2,4*] (iD)

Abstract

Background: Periodontitis is the most common chronic inflammatory disease caused by complex interaction between the microbial biofilm and host immune responses. In the present study, high-throughput RNA sequencing was utilized to systemically and precisely identify gene expression profiles and alternative splicing.

Methods: The pooled RNAs of 10 gingival tissues from both healthy and periodontitis patients were analyzed by deep sequencing followed by computational annotation and quantification of mRNA structures.

Results: The differential expression analysis designated 400 up-regulated genes in periodontitis tissues especially in the pathways of defense/immunity protein, receptor, protease, and signaling molecules. The top 10 most up-regulated genes were *CSF3, MAFA, CR2, GLDC, SAA1, LBP, MME, MMP3, MME-AS1,* and *SAA4*. The 62 down-regulated genes in periodontitis were mainly cytoskeletal and structural proteins. The top 10 most down-regulated genes were *SERPINA12, MT4, H19, KRT2, DSC1, PSORS1C2, KRT27, LCE3C, AQ5,* and *LCE6A*. The differential alternative splicing analysis revealed unique transcription variants in periodontitis tissues. The EDB exon was predominantly included in *FN1*, while exon 2 was mostly skipped in *BCL2A1*.

Conclusions: These findings using RNA sequencing provide novel insights into the pathogenesis mechanism of periodontitis in terms of gene expression and alternative splicing.

Keywords: Periodontitis, Transcriptome sequencing, Alternative splicing, Gene expression profile

Introduction

Periodontitis is a chronic inflammatory disease of periodontium, characterized by massive destruction of both soft and hard tissues surrounding the teeth [1]. The current concept for the periodontal diseases involve complex interaction between the microbial biofilm and host immune responses that leads to the alteration of bone and connective tissue homeostasis [2, 3]. Understanding the

molecular mechanisms underlying the pathogenesis as well as development of efficient therapeutics is furthermore important since periodontitis is linked to other metabolic and/or systemic diseases including diabetes, cardiovascular diseases, and rheumatoid arthritis [4–6].

The analysis of transcriptome by microarrays has been a valuable tool to study the changes in gene expression profiles in gingival tissues of periodontitis patients [7–9]. However, recent advances in the high-throughput RNA sequencing technology revolutionarily enhanced our understanding on the complexity of eukaryotic transcriptome [10, 11]. RNA sequencing has several key advantages over the hybridization-based microarray techniques. First of all, direct sequencing enables an unbiased approach compared

* Correspondence: jaehlee@khu.ac.kr; ylee@knu.ac.kr
†Equal contributors
3Department of Life and Nanopharmaceutical Sciences, Kyung Hee University, Seoul 02447, Korea
2Institute for Hard Tissue and Bone Regeneration, Kyungpook National University, Daegu 41940, Korea
Full list of author information is available at the end of the article

with the microarrays that depends on the predetermined genome sequences. Secondly, RNA sequencing is highly accurate in detecting gene expression with very wide dynamic detection ranges with low background. Thus, RNA sequencing is not only useful to precisely determine gene expression profiles but also particularly powerful to detect novel transcription variants via alternative splicing [10].

In the present study, we analyzed the pooled transcriptome from gingival tissues of periodontitis patients and compared with that of healthy patients. The large sum of novel information on the gene expression profiles as well as novel transcripts through alternative splicing would provide not only insights into the pathogenesis of periodontitis but also basis for the development of biomarkers and therapeutic targets.

Materials and methods
Periodontitis patient characteristics and gingival tissue samples
Gingival tissue samples were collected from chronic periodontitis patients or healthy individuals. On the basis of clinical and radiographic criteria, the periodontitis-affected site had a probing depth of ≥4 mm, clinical attachment level of ≥4 mm, and bleeding on probing. A total of 10 gingival samples were collected from 9 periodontal healthy patients who visited Kyungpook National University Hospital. Similarly, a total of 10 periodontitis tissue samples were obtained from 4 periodontitis patients with pocket depth of 4~6 mm and 3 severe periodontitis patients with pocket depth of 7 mm or deeper. The patient characteristics are given in Additional file 1: Table S1. All patients were non-smoking and did not have untreated metabolic/systemic diseases nor associated with infection/autoimmune diseases at the time of tissue collection. The size of 3-mm^2 gingival biopsies were obtained from the marginal gingiva during periodontal flap surgery and immediately stored in RNAlater solution (Thermo Fisher Scientific, Waltham, MA) at −70 °C after removal of blood by brief washing in phosphate-buffered saline. The study was approved by the institutional review board of the Kyungpook National University Hospital with informed consent from all patients.

Isolation of RNA and RNA sequencing
Frozen tissues were disrupted in the lysis solution of mirVana RNA isolation kit (Thermo Fisher Scientific) using disposable pestle grinder system (Thermo Fisher Scientific). After RNA extraction, the same amount of total RNA isolated from each individual sample (1 μg) was pooled into 2 groups (healthy and periodontitis) and used for further analysis. The integrity of pooled total RNA was analyzed by Agilent 2100 Bioanalyzer (Agilent Technologies, Santa Clara, CA). After purification of mRNA molecules by poly-T oligo-attached magnetic

beads followed by fragmentation, the RNA of approximately 300-bp size was isolated using gel electrophoresis. The cDNA synthesis and library construction was performed using the Illumina Truseq RNA sample preparation kit (Illumina, San Diego, CA), following the manufacturer's protocol. The PCR-amplified cDNA templates on a flow cell was loaded and sequenced in the HiSeq 2000 sequencing system (Illumina) in the paired-end sequencing mode (2 × 101 bp reads).

Sequencing data analysis
All sequencing raw reads were aligned to the human genome reference hg19 using the GSNAP alignment tool (2013-11-27) [12]. Only uniquely and properly mapped read pairs were used for further analysis. The differentially expressed genes between gingival tissues from periodontal healthy patients and periodontitis patients were identified using the DESeq R package [13]. Differentially expressed genes were defined as those with changes of at least 2-fold between samples and at a false discovery rate (FDR) cutoff of 5 % based on DESeq adjusted p values. The analysis of alternative splicing events was performed using MATS software [14]. The differences in the alternative splicing in genes were considered significant when the inclusion difference between samples was equal or greater than 5 % at a 10 % FDR. Each alternative splicing change of the skipped exon vent was manually inspected in UCSC genome browser using the sequencing data. The functional classification analysis of differentially expressed genes was performed using the PANTHER tools (http://www.pantherdb.org). The GO term and KEGG pathway enrichment analysis was performed as described previously [15]. Briefly, the fraction of genes in a test set associated with each GO category was calculated and compared with that of control set comprised of randomly chosen genes of the same number and length of the test genes. The random sampling was repeated 100,000 times for the calculation of empirical p value. The significance of enriched GO terms or KEGG pathways were determined by the p value cutoff, which was 1/total number of GO terms considered.

Validation of differentially expressed genes and alternative splicing events
From the pooled RNA samples, 1 μg of RNA was reversed transcribed using the Superscript II Reverse Transcriptase (Thermo Fisher Scientific). Quantitative real-time PCR analysis was performed by the addition of 1 μg of cDNA and SYBR green master mix in MicroAMP optical tubes using the AB 7500 system (Thermo Fisher Scientific). The expression of genes relative to that of HPRT1 was determined by the $2^{-\Delta\Delta C_T}$ method [16]. The differential alternative splicing events were confirmed via RT-PCR analysis with the addition of 1 μg of cDNA and Takara premix Taq polymerase (Takara Bio Inc, Shiga, Japan) for 33 cycles of

10 s at 98 °C, 30 s at 60 °C, and 1 min at 72 °C. The primers for the detection of alternative splicing were designed by the PrimerSeq software [17] in order that the PCR product to span the region of exon inclusion/skipping, enabling the differentiation of alternative splicing events by product size. The primer sequences for the real-time RT-PCR analysis of selected genes and those for the RT-PCR detection of alternative splicing events of *FN1* and *BCL2A1* gene were provided in the supplemental tables (Additional file 2: Table S2 and Additional file 3: Table S3).

Results

RNA sequencing results

Total RNA was extracted from 10 healthy gingival tissue samples and 10 chronic periodontitis-affected gingival tissues as described above. Then, cDNAs synthesized from the pooled RNA samples of both groups were sequenced using the Illumina HiSeq 2000 system, which generated approximately 80 million pairs of reads of 101 bp in size. When compared with the reference sequence of Genome Reference Consortium GRCh37 (hg19), more than 90 % of read pairs were uniquely mapped on the human genome (Table 1). Gene annotation using the Ensembl (release 75) identified that a total of 36,814 genes have at least 1 read mapped on the exonic regions. Among these, 4800 genes were unique to the periodontitis tissue sample, while 2811 transcripts were detected only in healthy gingival sample.

Identification and classification of differentially expressed genes between periodontitis and healthy gingiva

The differential expression of genes between periodontitis and healthy gingival samples was analyzed by DESeq package [13]. By applying the cutoff of at least twofold change in the number of reads with 5 % FDR, we found a total of 462 genes differentially expressed between the samples (Fig. 1a, volcano plot). While 400 genes were up-regulated in the periodontitis tissue sample, 62 genes were down-regulated compared with the healthy control (Additional file 4: Table S4). Previously, Davanian et al. reported the discovery of 381 genes up-regulated in the periodontitis-affected gingival tissues by RNA sequencing [18]. Notably, 182 genes among them were also found to be up-regulated in the present study (Additional file 5: Figure S1), demonstrating an overlap between the two sets of gene lists when analyzed by a hypergeometric test ($p < 2.2e^{-16}$) [19].

The top 20 up-regulated genes listed in Table 2 included cytokines and immune response-related genes (*CSF3*, *CR2*, *LBP*, *CXCL1*, and *IL19*), serum amyloid proteins (*SAA1*, *SAA4*, and *SAA2*), and proteases (*MME*, *MMP3*, *MME-AS1*, and *MMP7*). The 20 most down-regulated genes (Table 3) included peptidase inhibitors (*SERPINA12* and *SPINK9*) and structural proteins (*KRT2*, *KRT27*, *LCE3C*, *LCE6A*, *LCE1B*, *LCE2D*, and *KRT1*).

To classify the differentially expressed genes into functionally related subgroups, we utilized the PANTHER classification system (http://pantherdb.org). As a result, the 462 differentially expressed genes between periodontitis and healthy gingival tissues were segregated into 20 different classes of proteins. When we compared the composition of these protein classes, there was a significant difference in the number of genes between periodontitis and healthy gingival samples in 6 protein classes. In the periodontitis tissue, genes classified as defense/immunity protein, receptor, protease, and signaling molecules were significantly enriched (Fig. 1b). On the other hand, genes in the categories of cytoskeletal protein and structural protein were predominant in healthy tissue sample compared with periodontitis. Furthermore, functional annotation of GO and KEGG pathway enrichment analyses as previously described [15] revealed enhanced immune responses in the periodontal tissues, including NOD-like receptor signaling, cytokine and chemokine activities, response to lipopolysaccharide, Jak-STAT signaling pathway, and B cell receptor signaling pathway (Additional file 6: Table S5 and Additional file 7: Table S6).

Validation of differentially expressed genes between periodontitis and healthy gingiva by quantitative real-time PCR analysis

To validate the differential gene expression results by RNA sequencing analysis, we selected 10 up-regulated or down-regulated genes in periodontal tissue and assessed their expression by quantitative real-time RT-PCR analysis. Figure 1c shows that the examination of differential gene expression by both methods is significantly concordant, with the Pearson's correlation coefficient (*R*) value of 0.81 ($p = 0.005$). Since the current study design employed pooling of samples, we further validated the variations in gene expression in individual samples of healthy and periodontitis patients. The real-time RT-PCR analyses for selected genes (Additional file 8: Figure S2) mostly repeated the RNA sequencing results, showing significant reduction in *NOS1*, *CHP2*, *CDON*, and *MT4*. Similarly, significant

Table 1 Summary of RNA sequencing read mapping results

	Number of total sequencing pairs	Number of unique pairs	Number of unmapped pairs	Percentage of uniquely mapped pairs
Periodontitis tissue	87,118,086	80,778,080	6,340,006	92.72
Healthy tissue	72,014,202	67,035,158	4,979,044	93.09

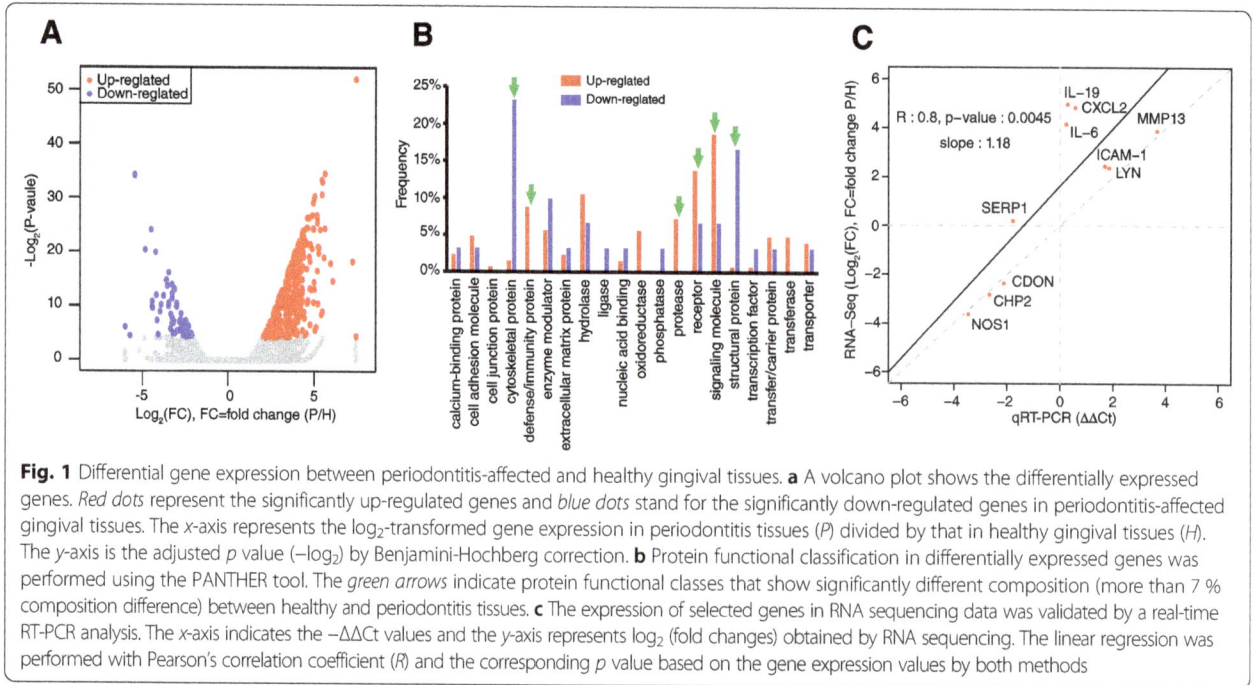

Fig. 1 Differential gene expression between periodontitis-affected and healthy gingival tissues. **a** A volcano plot shows the differentially expressed genes. *Red dots* represent the significantly up-regulated genes and *blue dots* stand for the significantly down-regulated genes in periodontitis-affected gingival tissues. The *x*-axis represents the log$_2$-transformed gene expression in periodontitis tissues (*P*) divided by that in healthy gingival tissues (*H*). The *y*-axis is the adjusted *p* value (−log$_2$) by Benjamini-Hochberg correction. **b** Protein functional classification in differentially expressed genes was performed using the PANTHER tool. The *green arrows* indicate protein functional classes that show significantly different composition (more than 7 % composition difference) between healthy and periodontitis tissues. **c** The expression of selected genes in RNA sequencing data was validated by a real-time RT-PCR analysis. The *x*-axis indicates the −ΔΔCt values and the *y*-axis represents log$_2$ (fold changes) obtained by RNA sequencing. The linear regression was performed with Pearson's correlation coefficient (*R*) and the corresponding *p* value based on the gene expression values by both methods

elevation was observed in *ICAM1*, *MMP13*, *LYN*, *CSF3*, *MMP3*, *LBP*, and CXCL2 while the expression of IL6 and IL19 only slightly increased. However, a large individual variation was observed in *SERP1* and *KRT2* expression.

Alternative splicing events in periodontitis and healthy gingival tissues

More than 90 % of human genes are alternatively spliced through different types of splicing [20, 21]. To identify the differential splicing events between the healthy and

Table 2 Top 20 up-regulated genes in periodontitis tissues

Ensemble ID	Gene symbol	Fold change	*q* value	Description
ENSG00000108342	*CSF3*	181.6	5.9E-21	Colony stimulating factor 3 (granulocyte)
ENSG00000182759	*MAFA*	157.5	8.2E-09	V-MAF avian musculoaponeurotic fibrosarcoma oncogene homolog A
ENSG00000117322	*CR2*	69.6	1.5E-07	Complement component (3D/Epstein Barr virus) receptor 2
ENSG00000178445	*GLDC*	50.8	2.6E-11	Glycine dehydrogenase (decarboxylating)
ENSG00000173432	*SAA1*	46.4	1.8E-14	Serum amyloid A1
ENSG00000129988	*LBP*	45.1	1.4E-05	Lipopolysaccharide binding protein
ENSG00000196549	*MME*	45.0	1.2E-14	Membrane metallo-endopeptidase
ENSG00000149968	*MMP3*	39.6	7.1E-10	Matrix metallopeptidase 3 (stromelysin 1, progelatinase)
ENSG00000240666	*MME-AS1*	38.8	2.1E-09	MME antisense RNA 1
ENSG00000148965	*SAA4*	37.3	4.7E-06	Serum amyloid A4, constitutive
ENSG00000137673	*MMP7*	37.1	4.5E-12	Matrix metallopeptidase 7 (matrilysin, uterine)
ENSG00000130513	*GDF15*	36.6	3.5E-08	Growth differentiation factor 15
ENSG00000134339	*SAA2*	36.4	2.6E-12	Serum amyloid A2
ENSG00000163739	*CXCL1*	33.3	1.6E-13	Chemokine (C-X-C motif) ligand 1 (melanoma growth stimulating activity, alpha)
ENSG00000117215	*PLA2G2D*	32.8	2.1E-04	Phospholipase A2, group IID
ENSG00000142224	*IL19*	32.4	5.3E-07	Interleukin 19
ENSG00000134873	*CLDN10*	31.8	1.8E-07	Claudin 10
ENSG00000255071	*SAA2-SAA4*	31.7	3.0E-11	SAA2-SAA4 readthrough
ENSG00000145113	*MUC4*	31.3	2.6E-11	Mucin 4, cell surface associated

Table 3 Top 20 down-regulated genes in periodontitis tissues

Ensemble ID	Gene symbol	Fold change	q value	Description
ENSG00000165953	SERPINA12	0.015	1.3E-04	Serpin peptidase inhibitor, clade A (alpha-1 antiproteinase, antitrypsin), member 12
ENSG00000102891	MT4	0.018	5.5E-04	Metallothionein 4
ENSG00000130600	H19	0.023	4.0E-15	H19, imprinted maternally expressed transcript (non-protein coding)
ENSG00000172867	KRT2	0.023	5.4E-15	Keratin 2
ENSG00000134765	DSC1	0.045	3.6E-11	Desmocollin 1
ENSG00000204538	PSORS1C2	0.046	7.2E-06	Psoriasis susceptibility 1 candidate 2
ENSG00000171446	KRT27	0.047	9.1E-06	Keratin 27
ENSG00000244057	LCE3C	0.053	1.8E-09	Late cornified envelope 3C
ENSG00000161798	AQP5	0.055	1.3E-06	Aquaporin 5
ENSG00000235942	LCE6A	0.057	5.3E-05	Late cornified envelope 6A
ENSG00000188959	C9orf152	0.059	4.2E-04	Chromosome 9 open reading frame 152
ENSG00000196734	LCE1B	0.067	5.4E-06	Late cornified envelope 1B
ENSG00000187223	LCE2D	0.073	1.5E-06	Late cornified envelope 2D
ENSG00000089250	NOS1	0.083	1.4E-07	Nitric oxide synthase 1 (neuronal)
ENSG00000204909	SPINK9	0.085	3.1E-07	Serine peptidase inhibitor, Kazal type 9
ENSG00000130595	TNNT3	0.091	6.7E-06	Troponin T type 3 (skeletal, fast)
ENSG00000110675	ELMOD1	0.091	4.1E-06	ELMO/CED-12 domain containing 1
ENSG00000167768	KRT1	0.091	4.5E-08	Keratin 1
ENSG00000237515	SHISA9	0.097	1.8E-04	Shisa family member 9

periodontitis gingival tissues, the inclusion level of alternative spliced exons was compared using the MATS tool [14] based on a statistical model that calculates the difference in the isoform ratio of a gene. The MATS analysis of RNA sequencing data revealed 183 significantly differential alternative splicing events in 155 genes with a cutoff of 5 % inclusion difference and 10 % FDR (Table 4 and Additional file 9: Table S7). The GO and KEGG pathway enrichment analyses for the determination of the biological relevance of those differentially spliced genes showed significant difference in the pathways including RNA splicing regulation, substrate adhesion-dependent cell spreading, response to wound healing, and positive regulation of cell migration (Additional file 10: Table S8 and Additional file 11: Table S9).

Among the genes that exhibited prominently novel included exons was *FN1* that encodes one of the major extracellular matrix protein fibronectin [22]. Fibronectin structure consists of 2 nearly identical ~250-kDa glycoprotein subunits with each monomer composed of repetitive units of type I, II, and III domains [23]. The type III domains contain 2 exons called extra domain A (EDA) and extra domain B (EDB), the latter showed significantly increased inclusion in periodontitis gingival tissues compared with healthy samples (Fig. 2a; left panel). The preferential formation of EDB-containing isoform in periodontitis was further corroborated by the RT-PCR analysis designed to amplify the included EDB exon regions (Fig. 2a; right panel). The analysis of alternative splicing events also indicated that *BCL2A1* (BCL2-related protein A1) exhibited prominently skipped exon 2 (Fig. 2b; left panel). RT-PCR analysis designed to amplify the skipped region revealed significantly increased shorter isoform (Fig. 2b; right panel). The individual variation between healthy and periodontitis tissues for these differences in the alternative slicing events was further confirmed by RT-PCR analyses (Additional file 12: Figure S3). For *FN1*, the inclusion of EDB exon was

Table 4 Summary of the differential alternative splicing event analysis

	Alternative 3' splicing sites	Alternative 5' splicing sites	Mutual exclusive exon	Retained intron	Skipped exon
Number of total alternative splicing events (genes)	3125 (2177)	2124 (1622)	4424 (2562)	2272 (1800)	32,824 (10,026)
% of total alternative splicing events (45,259)	6.9	4.7	9.8	6.1	72.5
Number of differential alternative splicing events (genes)	10 (10)	4 (4)	34 (32)	82 (77)	53 (42)
% of total differential alternative splicing events (183)	5.5	2.2	18.6	44.8	28.9

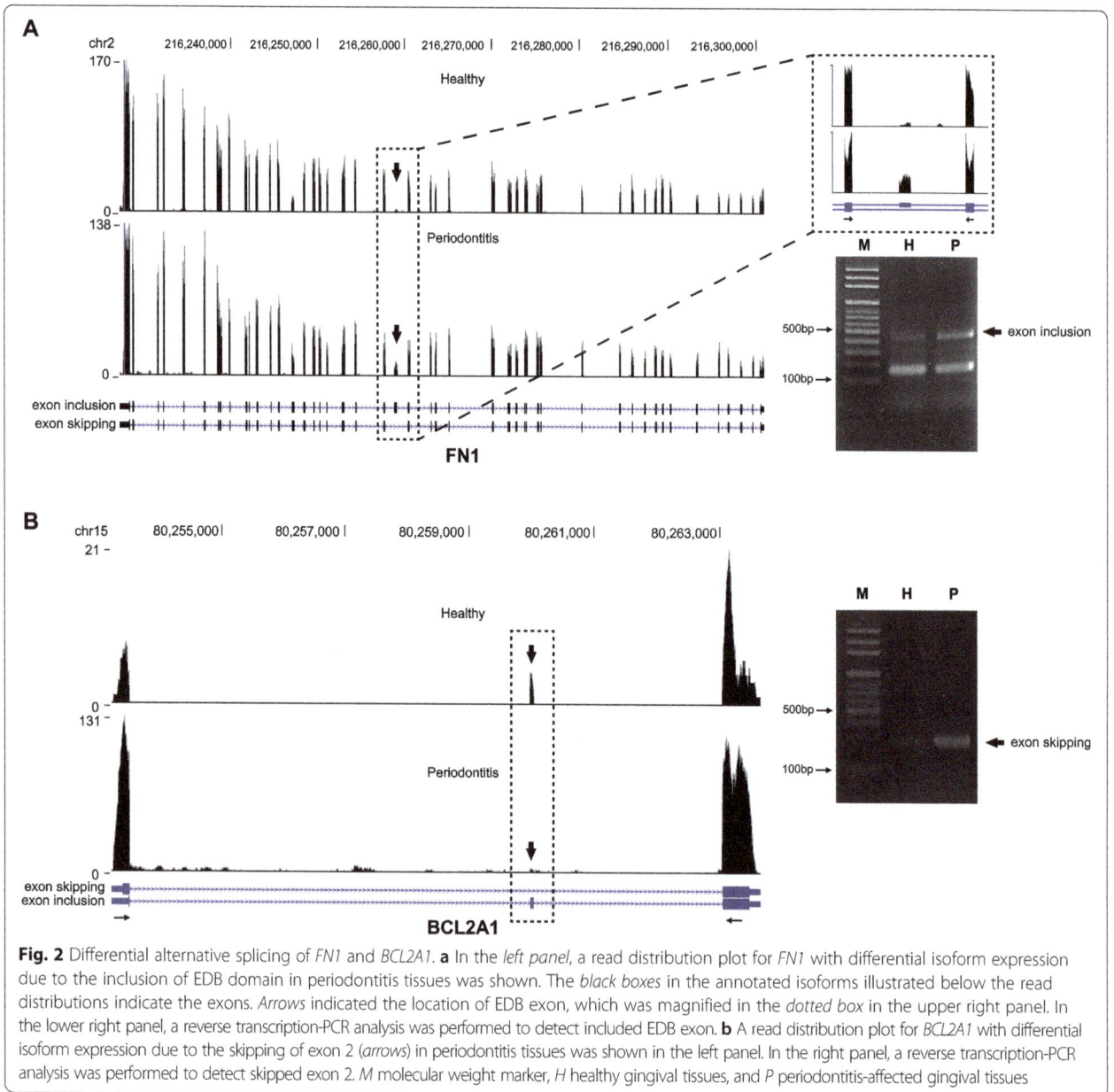

Fig. 2 Differential alternative splicing of *FN1* and *BCL2A1*. **a** In the *left panel*, a read distribution plot for *FN1* with differential isoform expression due to the inclusion of EDB domain in periodontitis tissues was shown. The *black boxes* in the annotated isoforms illustrated below the read distributions indicate the exons. *Arrows* indicated the location of EDB exon, which was magnified in the *dotted box* in the upper right panel. In the lower right panel, a reverse transcription-PCR analysis was performed to detect included EDB exon. **b** A read distribution plot for *BCL2A1* with differential isoform expression due to the skipping of exon 2 (*arrows*) in periodontitis tissues was shown in the left panel. In the right panel, a reverse transcription-PCR analysis was performed to detect skipped exon 2. *M* molecular weight marker, *H* healthy gingival tissues, and *P* periodontitis-affected gingival tissues

preferentially observed in periodontitis tissues (7/10) compared with healthy tissues (3/8) tested. Similarly, the skipping of exon 2 in *BCL2A1* was predominant in periodontitis tissues (9/10), compared with healthy tissues (2/8).

Discussion

Recent developments in the RNA sequencing technology and bioinformatics tools enabled elaborate analysis of gene expression in numerous human diseases. However in periodontitis research, most RNA sequencing studies have focused on the identification of microbiome that constitutes periodontal biofilm, with little attention to the host responses against such microbial challenge. The

current study provides extensive information on gene expression as well as alternative splicing in periodontitis gingival tissues, which is crucial for the understanding the pathogenesis and development of biomarkers and therapeutic targets. The gene expression analysis revealed 62 down-regulated and 400 up-regulated genes in periodontitis tissues, suggesting the effectiveness of mRNA sequencing as a tool to scrutinize the differential gene expression during the development of periodontitis. Davanian et al. previously reported a series of up-regulated genes as well as enriched biological pathways in periodontitis [18]. When we compared these results with ours, the current results only partially overlap in terms of differential gene expression, possibly originated from the difference in the

ethnic group of the subjects as well as in the methods to eliminate individual fluctuations in the gene expression. For example, Davanian et al. used healthy gingival tissue of the same periodontitis-affected individual as healthy control tissue. However, in the current study, the healthy and periodontitis tissues were pooled, allowing the dilution of individual differences in the gene expression. Indeed, the RNA sequencing analysis of pooled samples proved effective, since the expression levels of genes (except *IL6* and *IL19*) identified as differentially expressed by RNA sequencing were also significantly different between healthy and periodontitis samples, when we confirmed by real-time PCR analysis of individual samples (Additional file 8: Figure S2). Most of the top 20 up-regulated genes in periodontitis tissue (Table 2) were associated with inflammation and tissue degradation. Notably, serum amyloid A isoforms consisted 3 of 20 most up-regulated mRNAs, supporting the notion that these can serve as biomarkers for periodontitis-associated acute as well as chronic inflammation [24].

Until recently, gene expression analyses mostly focused on the genes whose expression was significantly increased in periodontitis. In line with this, 18 of top 20 up-regulated genes were associated with periodontal disease at least once by previous studies. The current study revealed 2 novel genes highly overexpressed in periodontitis tissues compared with healthy control. *MAFA* is a subgroup member of the basic leucine-zipper family transcription factor prominently known for its role in glucose-responsive insulin secretion [25]. *CLDN10* is an ion channel-forming member of claudin family, which is a constituent of tight junction [26]. The role of these genes in periodontitis is of great interest and requires further investigation.

In contrast to the highly expressed genes in periodontitis, fewer highlights have been drawn on the genes down-regulated in periodontal diseases. In accordance, most of the top 20 down-regulated genes (Table 3) have not been studied with regard to periodontitis, although investigating the role of those genes in periodontitis compared with that in normal tissues would greatly enhance our knowledge regarding the pathogenesis of periodontal diseases. Notably, keratin (*KRT2*, *KRT27*, and *KRT1*) and late cornified envelope (*LCE3C*, *LCE6A*, *LCE1B*, and *LCE2D*) genes constituted significant part of the down-regulated genes, suggesting the loss of epithelial barrier [27]. The causal relationship between the loss of these genes and the development of periodontal diseases requires further investigation.

It has long been suggested that different sites in the same individual exhibit different patterns of disease progression, morphology, and often response to therapy [28]. In addition, the oral microbiota responsible for the induction of periodontal diseases is distinct from site-to-site in the same individual [29, 30]. Accordingly, it is

recommended to design clinical studies based on individual sites rather than individual person [31]. In agreement of this notion, the analysis of gene expression in individual sites by real-time RT-PCR (Additional file 8: Figure S2) revealed site-specific variation. In different sites from the same periodontitis patients (P2: P3, P7: P8, and P9: P10), it was clearly noticeable that *MMP3*, *MMP13*, and *LBP* expressions differ in a site-specific manner. An individual RNA sequencing study with larger number of patients is ongoing, which will further provide detailed information on the site specificity of periodontitis.

The gene ontology and KEGG pathway enrichment analyses revealed both innate and adaptive immune responses in the periodontal tissues, including NOD-like receptor signaling, response to lipopolysaccharide, cytokine and chemokine activities, and B cell receptor signaling pathways (Additional file 6: Table S5 and Additional file 7: Table S6). The NOD1 and NOD2 have been suggested to mediate the sensing of periodontal bacteria [32]. In addition, NOD2 has been linked to the *P. gingivalis*-induced bone resorption, since *NOD2* knockout mice were protected from bone loss in a periodontitis model [33]. Bellibasakis and Johansson showed that a periodontal pathogen *A. actinomyceptemcomitans* regulated NLRP3 and NLRP6 expression in human mononuclear cells [34]. Considering the existence of 22 human NOD-like receptor protein members and their crucial functions in immune diseases, it will be of great interest and importance to elucidate the involvement of these receptors in the pathogenesis of periodontitis.

In the periodontitis lesions, it has been estimated that more than 75 % of infiltrating immune cells are plasma cells and B cells, suggesting the importance of these cells in adaptive immunity during the development of periodontitis [35]. In accordance, molecules involved in B cell activation including *CD79*, *CD19*, *Lyn*, and *CR2* were significantly increased in periodontitis tissue. An increasing body of evidence indicates that B cells with autoreactive propensities might be linked to tissue destruction in periodontitis [36, 37]. Indeed, recent reports demonstrated that B cell-deficient mice were protected from alveolar bone loss in experimental periodontitis [38, 39].

Numerous studies attempted to delineate the role of T helper (Th) cell subsets in human periodontitis by examining the cytokine mRNA levels by RT-PCR, flow cytometry, and immunohistochemistry. However, those studies are incoherent in terms of Th1 and Th2 cytokine expression, although the Th17 cytokines are consistently increased [37]. The current study revealed that the levels of Th1 cytokines *IFNG* and *IL12* did not change between healthy and periodontitis-affected gingival samples while that of *TNF* slightly increased in periodontitis (Additional file 13: Table S10). The Th2 cytokines *IL10* and *IL33*

remained unaltered in periodontitis patients. Interestingly, Th17 cytokines *IL6*, *IL23A*, and *IL17C* significantly increased in gingival tissues from periodontitis patients compared with those of healthy control, supporting the concept of Th17 cells as crucial mediators of inflammation, although it is still controversial whether these cells contribute to tissue destruction or protection in periodontitis [40, 41].

Alternative splicing of genes contribute to the diversity of proteome as well as genome evolution, control of developmental processes, and physiological regulation of various biological systems [42]. Not surprisingly, dysregulation of alternative splicing is often linked to various human diseases such as cancer, metabolic, neurological, and skeletal diseases [43–47]. However, alternative splicing events in the context of periodontitis has rarely been investigated. The current study uncovered significant differential alternative splicing events in *BCL2A1* and *FN1*. *BCL2A1* is a target gene of NF-kB, implicated in the survival of leukocytes thereby inflammation [48]. However, the role of alternative splicing on the activity of the protein has not been suggested until the present. Interestingly, recent discovery showed that *BCL2A1* was increased not only in periodontitis but also in systemic diseases such as cardiovascular diseases and ulcerative colitis [49]. Therefore, research regarding the multiple layers of regulatory mechanisms including mRNA expression and alternative splicing of *BCL2A1* are required to fully understand the role of this gene during the pathogenesis of periodontitis.

Parkar et al. previously suggested that *FN1* is differentially spliced in periodontitis [50]. Interestingly, the authors reported exon skipping of both EDA and EDB domain in periodontitis, while the current study showed conspicuously increased inclusion of EDB domain. Although whether these differences originated from the use of periodontal ligament [51] versus gingival tissues (the present study) yet to be cleared, it would be of great interest to fully identify the role of fibronectin isoforms in the pathogenesis of periodontitis considering the suggested role of EDA- and EDB-containing isoforms of fibronectin during embryonic development and tissue repair [23, 51].

In conclusion, the current study presented novel gene expression profiles as well as alternative splicing in gingival tissues from periodontitis patients by RNA sequencing experiments. Considering its effectiveness for whole transcriptome analysis, the use of RNA sequencing in periodontitis research would facilitate the elucidation of pathogenesis.

Additional files

Additional file 1: Table S1. The characteristics of patients involved in the current study. The information on age, gender, and disease severity is given in this table.

Additional file 2: Table S2. Primer sequences used for the real-time RT-PCR validation of RNA sequencing differential gene expression results. The primer sequences are given in this table.

Additional file 3: Table S3. Primer sequences used for the RT-PCR validation of RNA sequencing alternative splicing results. The primer sequences are given in this table.

Additional file 4: Table S4. The full list of differentially expressed genes in healthy and periodontitis tissues. This excel file contains the full list of deferentially expressed genes.

Additional file 5: Figure S1. Comparison of up-regulated genes in periodontitis with those of the previous study by Davanian et al. The Venn diagram shows the number of genes unique for each study and that of commonly detected genes.

Additional file 6: Table S5. The GO term analysis of genes of up- and down-regulated genes in periodontitis. This excel file contains the list of the differentially expressed genes, categorized according to the GO terms.

Additional file 7: Table S6. The KEGG pathway analysis of genes of up- and down-regulated genes in periodontitis. This excel file contains the list of the differentially expressed genes, categorized according to the KEGG pathway enrichment analysis.

Additional file 8: Figure S2. The expression levels of selected genes in individual samples. The individual variation in gene expression was examined by real-time RT-PCR analysis of individual healthy and periodontitis samples. The *p* values of the Wilcoxon rank-sum test between healthy and periodontitis groups are given in each graph.

Additional file 9: Table S7. The full list of differential alternative splicing events in healthy and periodontitis tissues. This excel file contains the full list of exons which were included or excluded in periodontitis.

Additional file 10: Table S8. The GO term analysis of genes with differential alternative splicing in periodontitis. This excel file contains the list of the genes with differential alternative splicing, categorized according to the GO terms.

Additional file 11: Table S9. The KEGG pathway analysis of genes with differential alternative splicing in periodontitis. This excel file contains the list of the genes with differential alternative splicing, categorized according to the KEGG pathway enrichment analysis.

Additional file 12: Figure S3. The alternative splicing events in individual samples. The individual variation in alternative splicing events in *FN1* and *BCL2A1* was examined by RT-PCR analysis of individual healthy and periodontitis samples.

Additional file 13: Table S10. The expression of Th1, Th2, and Th17 cytokines in healthy and periodontitis tissues. This excel file contains the selected list of the Th1, Th2, and Th17 cytokine genes with their expression levels in healthy and periodontitis tissues.

Abbreviations
EDA, extra domain A; EDB, extra domain B; FDR, false discovery rate; GO, gene ontology; KEGG, kyoto encyclopedia of genes and genomes; PANTHER, protein analysis through evolutionary relationships

Acknowledgements
Not applicable.

Funding
This work was supported by grants from the National Research Foundation of Korea (NRF) funded by the Ministry of Science, ICT, and Future Planning (NRF-2012M3A9B6055415, NRF-2014R1A2A2A01004161, and NRF-2008-0062282 to YL). This work was also supported by grants from the Korea Health Technology R&D Project through the KHIDI, funded by the Ministry of Health & Welfare (HI14C0175 to J-HL).

Authors' contributions

Y-GK designed the experiments, collected the tissue samples, analyzed the data, and wrote the paper. MK analyzed the bioinformatics data and wrote the paper. JHK, HJK, J-YK, J-WP, J-ML, and J-YS performed the experiments and analyzed the data. J-HL designed and performed the experiments, analyzed the bioinformatics data, and wrote the paper. YL designed and performed the experiments, analyzed the data, and wrote the paper. All authors read and approved the final manuscript.

Competing interests

The authors declare that they have no competing interests.

Consent for publication

Not applicable.

Author details

[1]Department of Periodontology, School of Dentistry, Kyungpook National University, Daegu 41940, Korea. [2]Institute for Hard Tissue and Bone Regeneration, Kyungpook National University, Daegu 41940, Korea. [3]Department of Life and Nanopharmaceutical Sciences, Kyung Hee University, Seoul 02447, Korea. [4]Department of Biochemistry, School of Dentistry, Kyungpook National University, 2177 Dalgubeol-daero, Joong-gu, Daegu 41940, Korea. [5]Department of Maxillofacial Biomedical Engineering, School of Dentistry, Kyung Hee University, 26 Kyunghee-daero, Dongdaemun-gu, Seoul 02447, Korea.

References

1. Pihlstrom BL, Michalowicz BS, Johnson NW. Periodontal diseases. Lancet. 2005;366:1809–20.
2. Offenbacher S, Barros SP, Beck JD. Rethinking periodontal inflammation. J Periodontol. 2008;79:1577–84.
3. Garlet GP. Destructive and protective roles of cytokines in periodontitis: a re-appraisal from host defense and tissue destruction viewpoints. J Dent Res. 2010;89:1349–63.
4. Bullon P, Morillo JM, Ramirez-Tortosa MC, Quiles JL, Newman HN, Battino M. Metabolic syndrome and periodontitis: is oxidative stress a common link? J Dent Res. 2009;88:503–18.
5. Pischon N, Heng N, Bernimoulin JP, Kleber BM, Willich SN, Pischon T. Obesity, inflammation, and periodontal disease. J Dent Res. 2007;86:400–9.
6. Linden GJ, Lyons A, Scannapieco FA. Periodontal systemic associations: review of the evidence. J Periodontol. 2013;84:S8–19.
7. Abe D, Kubota T, Morozumi T, Shimizu T, Nakasone N, Itagaki M, Yoshie H. Altered gene expression in leukocyte transendothelial migration and cell communication pathways in periodontitis-affected gingival tissues. J Periodontal Res. 2011;46:345–53.
8. Beikler T, Peters U, Prior K, Eisenacher M, Flemmig TF. Gene expression in periodontal tissues following treatment. BMC Med Genomics. 2008;1:30.
9. Kim DM, Ramoni MF, Nevins M, Fiorellini JP. The gene expression profile in refractory periodontitis patients. J Periodontol. 2006;77:1043–50.
10. Wang Z, Gerstein M, Snyder M. RNA-Seq: a revolutionary tool for transcriptomics. Nat Rev Genet. 2009;10:57–63.
11. Garber M, Grabherr MG, Guttman M, Trapnell C. Computational methods for transcriptome annotation and quantification using RNA-seq. Nat Methods. 2011;8:469–77.
12. Wu TD, Nacu S. Fast and SNP-tolerant detection of complex variants and splicing in short reads. Bioinformatics. 2010;26:873–81.
13. Anders S, Huber W. Differential expression analysis for sequence count data. Genome Biol. 2010;11:R106.
14. Shen S, Park JW, Huang J, Dittmar KA, Lu ZX, Zhou Q, Carstens RP, Xing Y. MATS: a Bayesian framework for flexible detection of differential alternative splicing from RNA-Seq data. Nucleic Acids Res. 2012;40:e61.
15. Lee JH, Gao C, Peng G, Greer C, Ren S, Wang Y, Xiao X. Analysis of transcriptome complexity through RNA sequencing in normal and failing murine hearts. Circ Res. 2011;109:1332–41.
16. Livak KJ, Schmittgen TD. Analysis of relative gene expression data using real-time quantitative PCR and the $2^{(-\Delta\Delta C(T))}$ method. Methods. 2001;25:402–8.
17. Tokheim C, Park JW, Xing Y. PrimerSeq: design and visualization of RT-PCR primers for alternative splicing using RNA-seq data. Genomics Proteomics Bioinformatics. 2014;12:105–9.
18. Davanian H, Stranneheim H, Bage T, Lagervall M, Jansson L, Lundeberg J, Yucel-Lindberg T. Gene expression profiles in paired gingival biopsies from periodontitis-affected and healthy tissues revealed by massively parallel sequencing. PLoS One. 2012;7:e46440.
19. Fury W, Batliwalla F, Gregersen PK, Li W. Overlapping probabilities of top ranking gene lists, hypergeometric distribution, and stringency of gene selection criterion. Conf Proc IEEE Eng Med Biol Soc. 2006;1:5531–4.
20. Wang ET, Sandberg R, Luo S, Khrebtukova I, Zhang L, Mayr C, Kingsmore SF, Schroth GP, Burge CB. Alternative isoform regulation in human tissue transcriptomes. Nature. 2008;456:470–6.
21. Xiao X, Lee JH. Systems analysis of alternative splicing and its regulation. Wiley Interdiscip Rev Syst Biol Med. 2010;2:550–65.
22. Geiger B, Spatz JP, Bershadsky AD. Environmental sensing through focal adhesions. Nat Rev Mol Cell Biol. 2009;10:21–33.
23. White ES, Muro AF. Fibronectin splice variants: understanding their multiple roles in health and disease using engineered mouse models. IUBMB Life. 2011;63:538–46.
24. D'Aiuto F, Orlandi M, Gunsolley JC. Evidence that periodontal treatment improves biomarkers and CVD outcomes. J Periodontol. 2013;84:S85–105.
25. Hang Y, Stein R. MafA and MafB activity in pancreatic beta cells. Trends Endocrinol Metab. 2011;22:364–73.
26. Krug SM, Schulzke JD, Fromm M. Tight junction, selective permeability, and related diseases. Semin Cell Dev Biol. 2014;36:166–76.
27. Presland RB, Jurevic RJ. Making sense of the epithelial barrier: what molecular biology and genetics tell us about the functions of oral mucosal and epidermal tissues. J Dent Educ. 2002;66:564–74.
28. Socransky SS, Haffajee AD, Goodson JM, Lindhe J. New concepts of destructive periodontal disease. J Clin Periodontol. 1984;11:21–32.
29. Preza D, Olsen I, Willumsen T, Grinde B, Paster BJ. Diversity and site-specificity of the oral microflora in the elderly. Eur J Clin Microbiol Infect Dis. 2009;28:1033–40.
30. Aas JA, Paster BJ, Stokes LN, Olsen I, Dewhirst FE. Defining the normal bacterial flora of the oral cavity. J Clin Microbiol. 2005;43:5721–32.
31. Lindhe J, Socransky S, Wennstrom J. Design of clinical trials of traditional therapies of periodontitis. J Clin Periodontol. 1986;13:488–99.
32. Okugawa T, Kaneko T, Yoshimura A, Silverman N, Hara Y. NOD1 and NOD2 mediate sensing of periodontal pathogens. J Dent Res. 2010;89:186–91.
33. Prates TP, Taira TM, Holanda MC, Bignardi LA, Salvador SL, Zamboni DS, Cunha FQ, Fukada SY. NOD2 contributes to Porphyromonas gingivalis-induced bone resorption. J Dent Res. 2014;93:1155–62.
34. Belibasakis GN, Johansson A. Aggregatibacter actinomycetemcomitans targets NLRP3 and NLRP6 inflammasome expression in human mononuclear leukocytes. Cytokine. 2012;59:124–30.
35. Berglundh T, Donati M. Aspects of adaptive host response in periodontitis. J Clin Periodontol. 2005;32:87–107.
36. Berglundh T, Donati M, Zitzmann N. B cells in periodontitis: friends or enemies? Periodontol 2000. 2007;45:51–66.
37. Gonzales JR. T- and B-cell subsets in periodontitis. Periodontol 2000. 2015;69:181–200.
38. Abe T, AlSarhan M, Benakanakere MR, Maekawa T, Kinane DF, Cancro MP, Korostoff JM, Hajishengallis G. The B cell-stimulatory cytokines BLyS and APRIL are elevated in human periodontitis and are required for B cell-dependent bone loss in experimental murine periodontitis. J Immunol. 2015;195:1427–35.
39. Oliver-Bell J, Butcher JP, Malcolm J, MacLeod MK, Adrados Planell A, Campbell L, Nibbs RJ, Garside P, McInnes IB, Culshaw S. Periodontitis in the absence of B cells and specific anti-bacterial antibody. Mol Oral Microbiol. 2015;30:160–9.
40. Gaffen SL, Hajishengallis G. A new inflammatory cytokine on the block: re-thinking periodontal disease and the Th1/Th2 paradigm in the context of Th17 cells and IL-17. J Dent Res. 2008;87:817–28.
41. Cheng WC, Hughes FJ, Taams LS. The presence, function and regulation of IL-17 and Th17 cells in periodontitis. J Clin Periodontol. 2014;41:541–9.
42. Gamazon ER, Stranger BE. Genomics of alternative splicing: evolution, development and pathophysiology. Hum Genet. 2014;133:679–87.
43. Biamonti G, Catillo M, Pignataro D, Montecucco A, Ghigna C. The alternative splicing side of cancer. Semin Cell Dev Biol. 2014;32:30–6.

44. Tazi J, Bakkour N, Stamm S. Alternative splicing and disease. Biochim Biophys Acta. 2009;1792:14–26.

45. Juan-Mateu J, Villate O, Eizirik DL. Mechanisms in endocrinology: alternative splicing: the new frontier in diabetes research. Eur J Endocrinol. 2016;174:R225.

46. Raj B, Blencowe BJ. Alternative splicing in the mammalian nervous system: recent insights into mechanisms and functional roles. Neuron. 2015;87:14–27.

47. Fan X, Tang L. Aberrant and alternative splicing in skeletal system disease. Gene. 2013;528:21–6.

48. Vogler M. BCL2A1: the underdog in the BCL2 family. Cell Death Differ. 2012; 19:67–74.

49. Lundmark A, Davanian H, Bage T, Johannsen G, Koro C, Lundeberg J, Yucel-Lindberg T. Transcriptome analysis reveals mucin 4 to be highly associated with periodontitis and identifies pleckstrin as a link to systemic diseases. Sci Rep. 2015;5:18475.

50. Parkar MH, Bakalios P, Newman HN, Olsen I. Expression and splicing of the fibronectin gene in healthy and diseased periodontal tissue. Eur J Oral Sci. 1997;105:264–70.

51. White ES, Baralle FE, Muro AF. New insights into form and function of fibronectin splice variants. J Pathol. 2008;216:1–14.

11

An efficient method for protein function annotation based on multilayer protein networks

Bihai Zhao, Sai Hu[*], Xueyong Li, Fan Zhang, Qinglong Tian and Wenyin Ni[*]

Abstract

Background: Accurate annotation of protein functions is still a big challenge for understanding life in the post-genomic era. Many computational methods based on protein-protein interaction (PPI) networks have been proposed to predict the function of proteins. However, the precision of these predictions still needs to be improved, due to the incompletion and noise in PPI networks. Integrating network topology and biological information could improve the accuracy of protein function prediction and may also lead to the discovery of multiple interaction types between proteins. Current algorithms generate a single network, which is archived using a weighted sum of all types of protein interactions.

Method: The influences of different types of interactions on the prediction of protein functions are not the same. To address this, we construct multilayer protein networks (MPN) by integrating PPI networks, the domain of proteins, and information on protein complexes. In the MPN, there is more than one type of connections between pairwise proteins. Different types of connections reflect different roles and importance in protein function prediction. Based on the MPN, we propose a new protein function prediction method, named function prediction based on multilayer protein networks (FP-MPN). Given an un-annotated protein, the FP-MPN method visits each layer of the MPN in turn and generates a set of candidate neighbors with known functions. A set of predicted functions for the testing protein is then formed and all of these functions are scored and sorted. Each layer plays different importance on the prediction of protein functions. A number of top-ranking functions are selected to annotate the unknown protein.

Conclusions: The method proposed in this paper was a better predictor when used on *Saccharomyces cerevisiae* protein data than other function prediction methods previously used. The proposed FP-MPN method takes different roles of connections in protein function prediction into account to reduce the artificial noise by introducing biological information.

Background

The accurate annotation of protein functions is the key to understanding life at the molecular level and has great biomedical and pharmaceutical implications. Due to high-throughput biological technologies, a large number of protein sequences [1] are available, while majority of their functions are still unknown. With its inherent difficulty and expense, experimental characterization of protein functions cannot accommodate the ever-increasing number of sequences and structures produced by Genomics Centers. Recent developments in experiments such as

yeast two-hybrid [2], tandem affinity purification [3] and mass spectrometry [4] have resulted in the publications of many high-quality, large-scale protein-protein interaction (PPI) data, which make it possible and feasible to use computational methods to predict functions for un-annotated proteins [5].

The past decade has witnessed a rapid development of computational methods for predicting protein functions from PPI datasets. A neighbor counting (NC) method proposed by Schwikowski et al. [6] predicted an un-annotated protein with the functions that occurred most frequently among its neighbor proteins. However, this method ignored the background frequency of different function annotations. Hishigaki et al. [7] improved the

---
* Correspondence: husaiccsu@163.com; wenyinccsu@163.com
Department of Mathematics and Computing Science, Changsha University, Changsha, Hunan 410022, China

neighbor counting method by using the Chi-Square statistics instead of frequency as a scoring function. Besides direct neighbors, Chua et al. [8] inferred the functional information within both direct (level 1) and indirect (level 2) neighbors by giving them different weights. Prior methods typically measured proximity as the shortest-path distance in the network, while most proteins are close to each other. Cao et al. [9] introduced diffusion state distance (DSD), a new metric based on a graph diffusion property, designed to capture finer-grained distinctions in proximity for transferring functional annotation in PPI networks. Other methods have been introduced to make functional prediction by getting the most consistent agreement throughout the whole PPI networks [10]. Chi et al. [11] proposed an approach that predicted protein functions iteratively. This iterative approach incorporated the local and global semantic influence of protein functions into the prediction. Some kind of network-based methods partitioned proteins in PPI networks into several function modules [12], and the proteins in the same modules are assigned with the same functions. Lee et al. [13] applied a novel method that generated improved modularity solutions, and developed a better method to use this community information to predict protein's functions.

Taking both high noise in PPI data and insufficient number of available annotated proteins into account, some researchers have tried to improve the prediction performance by incorporating other heterogeneous data sources. Cozzetto et al. [14] proposed an integrative approach for addressing annotation challenge, which combines into a wide variety of biological information sources encompassing sequence, gene expression, and PPI data. Zhang et al. [15] presented a novel protein function prediction method that combined protein domain composition information and PPI networks. Domain combination similarity (DCS) [16] was applied to predict protein function by integrating PPI networks and proteins' domain information. Different from Zhang's, DCS changed the method to calculate domain context similarity and combined the domain compositions of both proteins and their neighbors. Liang et al. [17] built a network model called protein overlap network (PON) using domain co-occurrence information. In a PON, each node represented a protein and two nodes were connected with an edge if they share a common domain. The function of a protein can be predicted by counting the occurrence frequency of gene ontology (GO) terms associated with domains of direct neighbors in the PON. Recently, some new algorithms are proposed to predict protein function from PPI networks. Gong et al. [18] developed a method named GoFDR for predicting GO-based protein functions. The input for GoFDR is simply a query sequence-based multiple sequence alignment (MSA) produced by PSI-BLAST (Position-Specific Iterated BLAST). Kumar et al.

[19] proposed an improved approach for protein function prediction by exploiting the connectivity properties of prominent proteins. Yu et al. [20] proposed a method called Predicting Protein Function using Multiple Kernels (ProMK). ProMK iteratively optimizes the phases of learning optimal weights and reduces the empirical loss of multi-label classifier for each of the labels simultaneously.

In conclusion, many computational methods that integrate heterogeneous data for predicting protein (or gene) functions have been suggested. Most of these techniques follow the same basic paradigm: firstly, they generate various functional association networks by analyzing implicit information of shared functions of proteins from different data sources. Then these individual networks are combined into a composite and highly reliable network through a weighted sum. The weight of each individual network represents the contribution of the corresponding data source to the function prediction. A correct setting of these weights is thought to be the key to designing an effective function prediction method. In general, the weights adjustment of individual networks is mainly influenced by human experience and statistical analysis. The major drawback of how each network is weighted is that it varies between different datasets. Furthermore, functions of proteins are diverse and some of them only occur under specific conditions. Different functional association networks play different roles and have varying importance in function prediction. Combining a heterogeneous data source into a single weighted network could obscure the inherent nature of the protein function.

To address these difficulties, we construct a multilayer protein network which integrates PPI network topology, domain information, and protein complexes. Additionally, we propose an efficient protein function annotation method, named FP-MPN (function prediction based on multilayer protein networks). FP-MPN takes into account the varying influences by multiple connections in the prediction of protein function. Given an un-annotated protein, FP-MPN generates candidate functions by examining multilayer networks systematically in turn. The performance of FP-MPN was tested on the well-studied species of *Saccharomyces cerevisiae*. Compared to several previously reported protein function prediction algorithms, FP-MPN achieved a greater degree of accuracy in predicting protein function. The experimental results demonstrate that this method, which distinguishes different types of connections in function prediction, is more robust and effective than those methods combining multiple interactions, and that FP-MPN is a good example of this.

Materials and methods
Assessment criteria

Cross-validation is a widely used method to evaluate the performance of protein function prediction algorithms.

The proteins in the PPI network are partitioned into two subsets, the training set and the testing set. Functions are removed from the part of proteins in the PPI network artificially. These proteins consist of the testing set and the rest proteins form the training set. Functions of proteins in the testing set are predicted, using functional information of proteins in the training set. Finally, the comparing results of predicted functions with actual functions are used to evaluate the performance of protein function prediction algorithms. The cross-validation methods can be classified into two categories: leave-one-out cross-validation and leave-percent-out cross-validation. The leave-one-out cross-validation method puts one protein into the testing set and the remaining proteins into the training set, while the leave-percent-out cross-validation method randomly selects a percentage of proteins as the testing set and then puts other proteins into the training set. Each function of proteins in the testing set is assigned with a probability, according to the functions of proteins in the training set. Then a number of top-ranking functions are selected to annotate the protein with unknown functions. The quality of prediction depends on the matching results of predicted functions with actual ones. There are two widely used criteria to measure the predicted results. The one is Precision which measures the percentage of predicted functions that match the known functions. The other is Recall which measures the fraction of known functions that are matched by the predicted ones. They can be calculated as follows:

$$\text{Precision} = \frac{TP}{TP + FP} \quad (1)$$

$$\text{Recall} = \frac{TP}{TP + FN} \quad (2)$$

where TP (true positive) is the number of predicted functions matched by known functions. FP (false positive) is the number of predicted functions that are not matched by known functions. FN (false negative) is the number of known functions that are not matched by predicted functions. Selecting more functions can improve the recall, but it may lead to the reduction of precision. F-measure, as the harmonic mean of precision and recall, is another measure to evaluate the performance of a method synthetically, which is calculated as follows:

$$F\text{-measure} = \frac{2 \times \text{Precision} \times \text{Recall}}{\text{Precision} + \text{Recall}} \quad (3)$$

At the same time, the coverage rate (CR) [21] is also used to evaluate a function prediction algorithm, which shows how many functions of proteins in the testing set can be covered by predicted functions. Given a testing protein set TP = {tp_1, tp_2, ..., tp_n}, KF = {kf_{11}, kf_{12},..., kf_{ij},

..., kf_{nm}} is a list of known function sets of TP, KF_i = {kf_{i1}, kf_{i2},..., kf_{il}} is a known function set of the protein tp_i. PF = {pf_{11}, pf_{12},..., pf_{ij}, ..., pf_{nm}} is a list of predicted function sets of TP, PF_i = {pf_{i1}, pf_{i2},..., pf_{il}} is a predicted function set of the protein tp_i. The coverage rate is then defined as

$$CR = \sum_{i=1}^{n} |KF_i \cap PF_i| \Big/ \sum_{i=1}^{n} |KF_i| \quad (4)$$

Motivation

Some methods try to reconstruct more reliable networks by integrating PPI networks and biological information, in order to reduce the impact of random noise on predicting performance. There exist complex and diverse relationships between proteins as demonstrated after integrating biological information. For example, proteins can interact with each other through physical interactions which can be identified by biological experiments, co-expression based on time course gene expression data [22, 23], or co-annotation based on gene ontology [24, 25], etc. Most of these methods generate various functional association networks, such as co-expression networks and co-annotation networks. Then a single network can be constructed through a weighted sum of these individual networks. The weight assigned to each individual network reflects its contribution towards protein function annotation, which is computed by a specific similar metric for the related biological data.

Figure 1 describes an example of constructed networks by integrating the PPI network and heterogeneous data. Figure 1a shows an original physical PPI network, which was derived from experimental methods. In the co-annotation network, as shown in Fig. 1b, there exists a connection between a pair of proteins if they perform the same functions. As for the co-expressed network, it is based on time course gene expression data. For a protein v, its gene expression at n different times is denoted as a variate: $Gen(v) = \{T(v, 1), T(v, 2), ..., T(v, n)\}$, $T(v, i)$ denotes the expression level of gene v at the time point i. Generally, the Pearson correlation coefficient [26] is used to assess the probability of whether two particular proteins are co-expressed. If the Pearson correlation coefficient of two proteins over all time points is greater than 0.8, then they are considered to be co-expressed and are connected in the co-expressed network. The network shown in Fig. 1d is a reconstructed network based on three networks currently used. This network shows that proteins could have a diversity of functions when exposed to different conditions or at different time points. Therefore, the importance and roles of different types of interactions between proteins are not the same for the protein function prediction. When functions are predicted for the unknown protein YJL115W using the

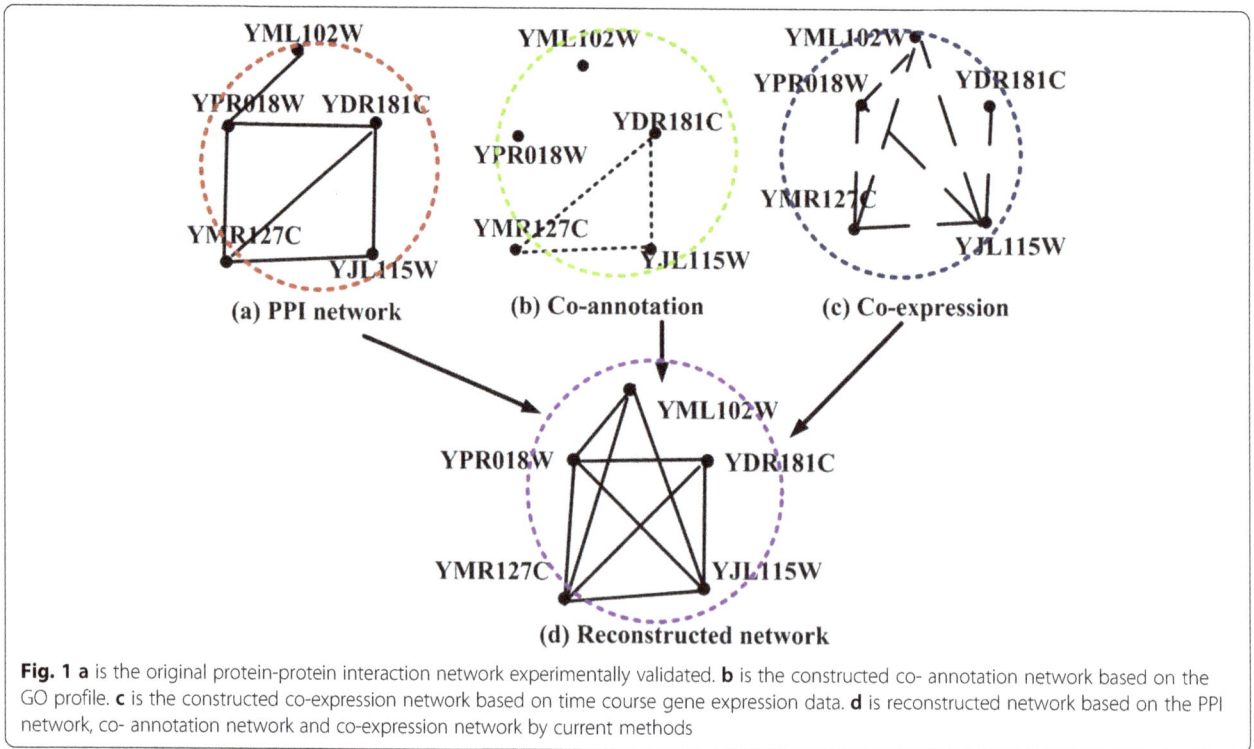

Fig. 1 a is the original protein-protein interaction network experimentally validated. **b** is the constructed co- annotation network based on the GO profile. **c** is the constructed co-expression network based on time course gene expression data. **d** is reconstructed network based on the PPI network, co- annotation network and co-expression network by current methods

constructed network in Fig. 1d, YPR018W and YDR181C are treated in the same way. The connection (YPR018W, YJL115W) and (YDR181C, YJL115W) has the same status and reliability (they both have an edge clustering coefficient [27] of one). After analyzing the original PPI network, co-annotation network, and co-expression network as shown in Fig. 1, it is demonstrated that the connection (YDR181C, YJL115W) is more reliable than (YPR018W, YJL115W), due to its occurrence in all three networks. YPR018W and YJL115W are only co-expressed at the gene expression level, based on gene expression data. Therefore, YDR181C should contribute more to the function prediction of YJL115W, than the protein YPR018W. Connections between YDR181C and YJL115W overlap in the reconstructed network; therefore, it is difficult to determine their relationship. The information mentioned above was obtained from the reconstructed network.

The analysis of this experiment suggests that existing methods have two deficiencies. Different biological data sources (i.e., PPI networks, protein domains, and subcellular information) often describe protein properties in different ways and have different correlations with different GO terms. Combining multiple biological data into a single network can not only enhance the matching accuracy (i.e., recall, which measures the fraction of known functions that are matched to the predicted ones) to a certain extent but also introduce a lot of noise functions and reduce predicting accuracy (i.e., precision, which measures the percentage of predicted functions which match the known

functions). As a result, the comprehensive performance improvement is not apparent. Current methods set different weights for heterogeneous data based on the quality of data sources in order to integrate them into a single network. Setting the weighting system for multiple biological data is the key to ensuring the accuracy of protein function prediction. These optimal weighting methods rely on empirical analysis and have differences between datasets. Furthermore, these weighting methods may also lead to the inconsistency of these prediction algorithms.

In conclusion, it is inappropriate to combine multiple interactions or connections between two proteins, as they often occur under different conditions and play different roles in protein function prediction. In this paper, we describe a multilayer protein network developed by integrating PPI network topology and heterogeneous data. In the constructed network, a pair of proteins has more than one connection which is connected through multiple links. Based on the multilayer protein network, we propose a new method for predicting protein functions, named FP-MPN.

Multilayer protein networks construction

The complex network is a hot, new research area as a result of the increased use of networks in various fields, such as mathematics, social science, and life science. The features of many real-life complex networks are that they are small-world (i.e., high clustering coefficient and small average path length) and scale-free (i.e., follow the power-law distributions in node degree and display the growth and

preferential attachment). In reality, connections among nodes in complex networks are diversified. For instance, in social networks, people can contact each other via emails, telephones or MSN, etc., and hence make up a complex network with multi-links. Similarly, in biological networks there are diverse links among proteins via co-expression or co-annotation of the proteins. Multilayer networks are more complex than those with single link.

We consider a multilayer network $G = (V, E)$, where $V = \{v_1, v_2,..., v_n\}$ represents a set of proteins, the edge set $E = \{Me_1, Me_2,..., Me_m\}$ consists of edges of L different types representing different relations. That is, $Me_i = \{e_{i1}, e_{i2},..., e_{iL}\}$ $(0 < i < =m)$, e_{ij} $(0 < j < =L)$ represents the ith connection in the jth layer of G. We can view the multilayer network as a graph with vector valued edge information, i.e., the adjacency matrix A consists of elements A_{ij}, who are themselves L dimensional vectors: $A_{ij} = \{A_{ij}^{(1)}, A_{ij}^{(2)},..., A_{ij}^{(L)}\}$. An alternative way to approach the problem is to view the multi-graph as a collection of L, $N \times N$ adjacency matrices $\{A^{(1)}, A^{(2)},..., A^{(L)}\}$, each corresponding to one type of relation. Figure 2 describes an example of a multilayer network according to Fig. 1. The multilayer network consists of five nodes and three

layers. Each layer represents a different level of connection or relationship between nodes.

Functions are often performed by proteins physically interacting with each other, located within the same complex, or by having similar structures. A protein consists of one or more domains which have independent functions. There may be discrepancies within domain combinations among different proteins and it is of great significance to recognize these. In this paper, we develop a multilayer network by integrating the PPI network, protein domain information, and protein complexes. The multilayer network consists of three layers, which include the physical interaction layer (PIL), sharing domain layer (SDL), and sharing complex layer (SCL). The physical interaction layer is derived from original PPI networks. On the SDL, two proteins are physically connected if there is at least one domain common to both of them. On the SCL, each node represents a protein and two nodes are physically connected if they are contained in a common complex. Our previous research on protein complex prediction [28] and essential protein identification [26] suggests that the performance of the prediction algorithm based on weighted networks is superior to that based on un-weighted networks. An

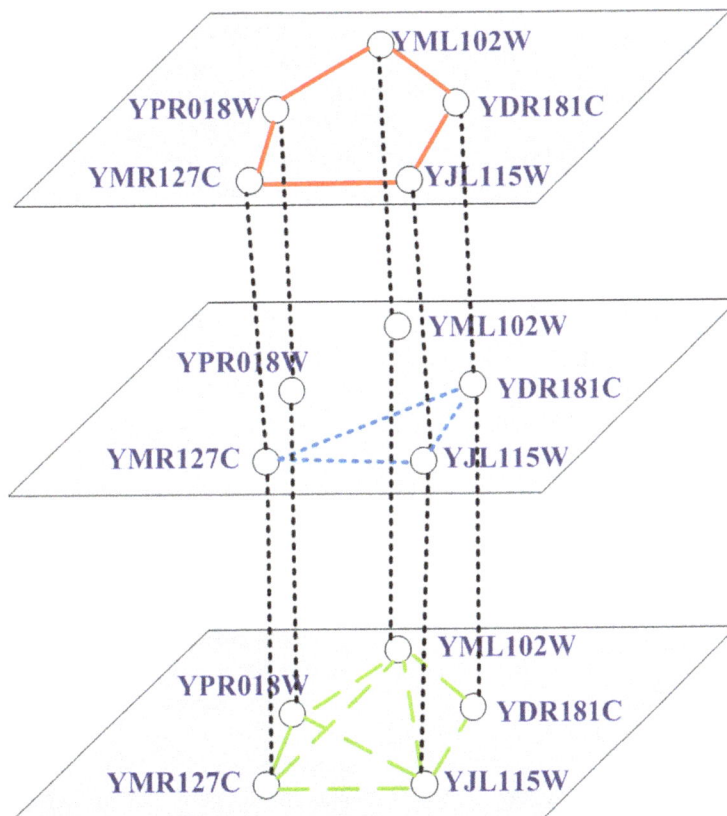

Fig. 2 Example of multilayer protein networks

explanation for this could be that the weight stands for the reliability of interactions and therefore, weighted networks can be more useful than un-weighted networks in the representative of PPI networks. In this work, appropriate weighting methods for the three types of connections are developed for the multilayer network.

Methods of Zhang and DCS successfully integrated domain information and PPI networks, improving the performance of protein function prediction. The two methods rely on the same principle, which is to implement function prediction by way of computing similarities between the two proteins. The two methods differ in that the method described by Zhang only computes similarity through the domain information of the protein itself, while the DCS method expands on the extra domain information of the neighbors surrounding it. The two methods are all based on the computing similarity of the combination formula. However, they have the problem of being highly complex to program. To balance the pros and cons of the two methods, this study has set up the weighting computational formula aiming at the interaction of shared domain as follows:

$$W(v_i, v_j) = \begin{cases} \dfrac{|D_i \cap D_j|^2}{|D_i| \times |D_j|} & , \quad D_i \neq \varnothing \text{ and } D_j \neq \varnothing \\ 0 & , \quad \text{otherwise} \end{cases} \quad (5)$$

where D_i and D_j are sets of distinct domain types of v_i and v_j, respectively.

In a similar way, the weight of sharing complexes between v_i and v_j on the SCL can be calculated as follow:

$$W(v_i, v_j) = \begin{cases} \dfrac{|C_i \cap C_j|^2}{|C_i| \times |C_j|} & , \quad C_i \neq \varnothing \text{ and } C_j \neq \varnothing \\ 0 & , \quad \text{otherwise} \end{cases} \quad (6)$$

where C_i and C_j are the sets of protein complexes that contained v_i and v_j, respectively, and $C_i \cap C_j$ denotes the set of common protein complexes.

As for the weight of connections on the PIL, we suggest that the weight of an interaction can be reflected by the number of common neighbors between the proteins. Here we use a variant of edge clustering coefficient (ECC) [27] to calculate the weight of protein pairs. Given a pair of proteins v_i and v_j, the weight of edge (v_i, v_j) on the PIL is defined as follows:

$$W(v_i, v_j) = \begin{cases} \dfrac{|N_i \cap N_j|^2}{(|N_i|-1) * (|N_j|-1)} & , \quad |N_i| > 1 \text{ and } |N_j| > 1, \text{otherwise} \end{cases}$$

$$(7)$$

where N_i and N_j are sets consisting of all neighbors of v_i and v_j, respectively.

Figure 3 is the visualization of our constructed multilayer protein network. The network consists of three layers, i.e., PIL, SDL, and SCL. There are the same set of proteins and different connections sets on these three layers. The multilayer protein network can be modeled as $G = (V, E)$, where $V = \{v_1, v_2,..., v_n\}$, $E = \{Me_1, Me_2,..., Me_m\}$. $Me_i = \{e_{i1}, e_{i2}, e_{i3}\}$ $(0 < i <= m)$, e_{ij} $(0 < j < =3)$ represents the ith connection in the jth layer of G.

FP-MPN algorithm

Based on the weighted multilayer protein network, we propose a new method for protein functional prediction, named FP-MPN. How to deal with the multilayer networks is the first problem to be addressed. Current algorithms combine different connections into a single connection when dealing with these complex biological networks. In reality, it is inappropriate to combine multiple connections between two proteins, as they often occur under different conditions and play different roles in protein function prediction. The influences of different types of interactions in protein function prediction are not the same. Combining different interactions into a single event can lead to false positive results. So, it is necessary to deal with multilayer networks in another way.

The different connections among proteins may have different impacts on function prediction. To address this, FP-MPN visits each layer of the multilayer network in turn to generate candidate functions. Each layer has different contribution to predict ion of functions for an unannotated protein. The FP-MPN algorithm operates in two stages, pre-processing data and predicting functions.

To assign functions of proteins in the testing of a set of probabilities, pre-processing of the multilayer protein network is required. The constructed multilayer protein network can be represented as a tensor $A = (a_{i,j,k})_{n \times n \times m}$, where n is the number of proteins and m is the number of types of interconnections. If node i is connected to node j by the kth type link, $a_{i,j,k}$ is equal to 1; otherwise, it equals 0. Figure 4 depicts the tensor representation of the multilayer network as shown in Fig. 2. Given a tensor A, we can get a new tensor $A^{(1)}$, which is calculated as follows:

$$a_{i,j,k}^{(1)} = \begin{cases} a_{i,j,k} / \sum_{j=1}^{n} a_{i,j,k}, & \sum_{j=1}^{n} a_{i,j,k} > 0, \text{otherwise} \end{cases} \quad (8)$$

Therefore, for each row i of the tensor $A^{(1)}$, $\sum_{j=1}^{n} a_{i,j,k}^{(1)} = 1$ or $\sum_{j=1}^{n} a_{i,j,k}^{(1)} = 0$.

The second stage of FP-MPN is predicting functions for un-annotated proteins. The FP-MPN method visits each layer of the corresponding multilayer network of the tensor $A^{(1)}$, Given that the proteins interact with each other under different conditions or stimuli in order to perform different functions, FP-MPN generates predicted functions across all layers. While the importance of each layer to the prediction is not the same. We assign different importance

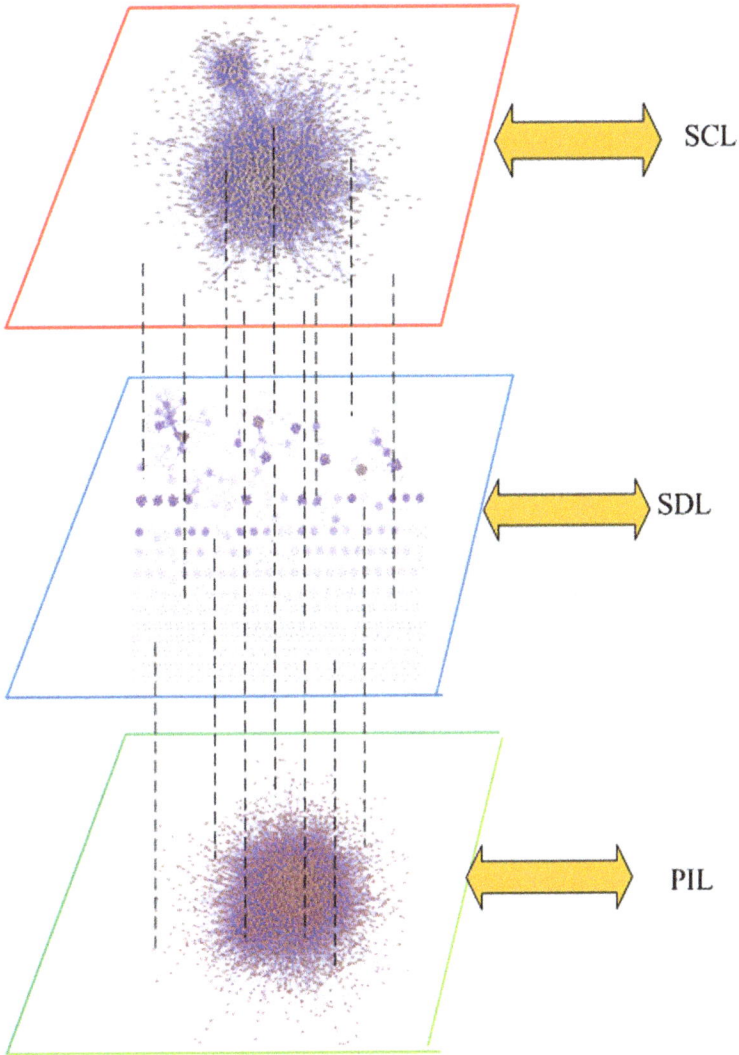

Fig. 3 Visualization of constructed multilayer protein networks

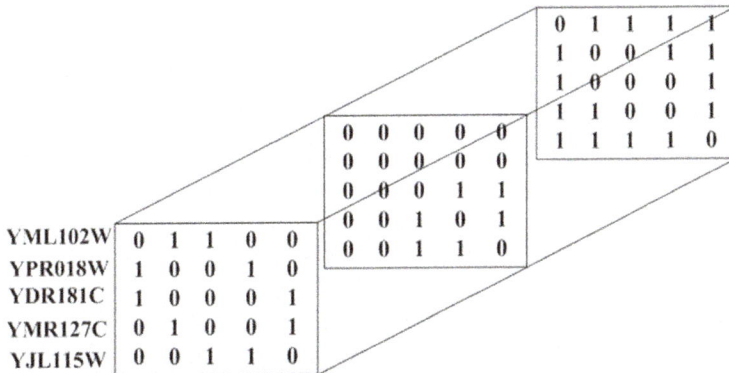

Fig. 4 The tensor representation of a multilayer protein network

coefficient (IC) for each layer of the MPN. For the ith layer, its IC value can be calculated as follow:

$$IC(i) = \frac{1}{2^i} \quad (9)$$

The final score of a predicted function is the weighed sum of scores achieved from all layers. The IC value of a layer is used to present the weight. The layer accessed firstly has higher IC value than that rest of the layers. For this reason, the set up access sequence of each layer in the MPN is critical for the FP-MPN method. This paper addresses the problem of the impact of each layer on the accuracy of function predictions using statistical analysis. More detailed statistical results can be found in Table 1.

In this experiment, we used the NC [6] method on the SDL, SCL, and PIL to annotate all unknown proteins, using leave-one-out cross-validation. Then, we calculate the average Precision, Recall, and F-measure to evaluate the significance of each layer for function prediction. The original PPI network consisted of 5093 proteins with 24,743 interactions. For the PIL, SDL, and SCL, there are 13,871, 23,749, and 7337 connections, respectively. Using PIL, there are 2388 proteins, which had at least one neighbor. The number of nodes with neighbors on the SDL and SCL is 2972 and 1494, respectively. From Table 1, it can be seen that SCL archives the highest F-measure among the three layers. In addition, 73.83 % (1103/1494 = 73.83 %) of proteins with neighbors on the SCL have been annotated as at least one function. While the proportion of PIL and SDL is 53.35 % (1274/2388 = 53.35 %) and 40.88 % (1215/2972 = 40.88 %), respectively. The SDL gets the second highest F-measure and Recall after SCL among all the layers. Thus, we assigned the highest access sequence to SCL, the second highest priority to SCL, and the lowest order to PIL.

The second stage of FP-MPN consists of two major steps. The first step is to search its neighbors in the MPN for a particular protein u with unknown function, to generate candidate functions. Starting from the layer in MPN which has the highest access sequence, the FP-MPN method creates a functions list PF. These lists of functions are derived from neighbors of the testing protein u. Assume that $P = \{p_1, p_2, ..., p_n\}$ is a set of neighbors of the protein u on the first layer, $F = \{f_1, f_2, ..., f_m\}$ is a set of functions of all these proteins in P. The score of

a certain function f_j in F can be calculated by the following formula:

$$S\left(f_j\right) = \sum_{i=1}^{n} W(u, p_i) \times t_{ij}, \quad (j \in [1, m]) \quad (10)$$

where $W(u, p_i)$ represents the weight of the connection between u and p_i. If p_i contains function f_j, then $t_{ij} = 1$, otherwise $t_{ij} = 0$. Then, the FP-MPN enters the next layer of MPN and continues to predict functions. If a function has been predicted on previous layers, its score is accumulated. This process is repeated for the next layer etc., until all the layers are traversed. For a predicted function f, its final score is the weighed sum of scores on all layers and can be calculated as follow:

$$Score(f) = \sum_{i=1}^{L} IC(i) * S(f_i) \quad (11)$$

where L is the number of layers, $IC(i)$ is the IC value of the ith layer, and $S(f_i)$ is the score of function f on the ith layer calculated using Equation (10). From Equation (9), it is not difficult to deduce the formula $\sum_{i=1}^{m} IC(i) < 1$, thus ensuring that $Score(f)$ is less than 1 and can be used as a probability of the function f. Figure 5 illustrates how the FP-MPN method gets the predicted functions list. Figure 5a depicts the constructed multilayer protein network. Numbers on the edges of each layer in the MPN represent their corresponding weights. Figure 5b is the tensor representation of MPN after pre-processing, using Equation (8). Figure 5c shows the predicted functions list for the unknown protein A generated by the FP-MPN method. In this example, FP-MPN predicts functions $f3$ and $f4$ according to its neighbors on the SCL. FP-MPN computes the scores of $f3$ and $f4$ on the SCL by Equation (10), which is 1 and 1, respectively. Then, FP-MPN enters the SDL and continues to generate functions. The candidate function set of A's neighbors on SDL consists of $\{f1, f2, f3, f4\}$. The score of $f1, f2, f3, f4$ on the SDL is 0.28, 0.28, 0.72, and 0.72, respectively. In a similar way, FP-MPN records the functions $\{f1, f2, f3, f4, f5\}$ on the PIL. Scores of the five functions are the same that is 0.5. According to Equation (11), the final score of $f3$ can be calculated as follow:

$$Score(f_3) = 1*\frac{1}{2} + 0.72*\frac{1}{2^2} + 0.5*\frac{1}{2^3} = 0.7425$$

The final score of $f1, f2, f4, f5$ is 0.1325, 0.1325, 0.7425, and 0.0625, respectively.

The last step of the second stage is to rank functions according to their scores and select a top N of the ranked functions for the protein with unknown function. This is a key factor which influences the performance of the function prediction algorithm. Existing methods for function selection

Table 1 Statistical analysis of the influence of three layers

Layers	Annotated proteins	Precision	Recall	F-measure
PIL	1274	0.3791	0.1094	0.1697
SDL	1215	0.3595	0.1538	0.2154
SCL	1103	0.3404	0.1829	0.238

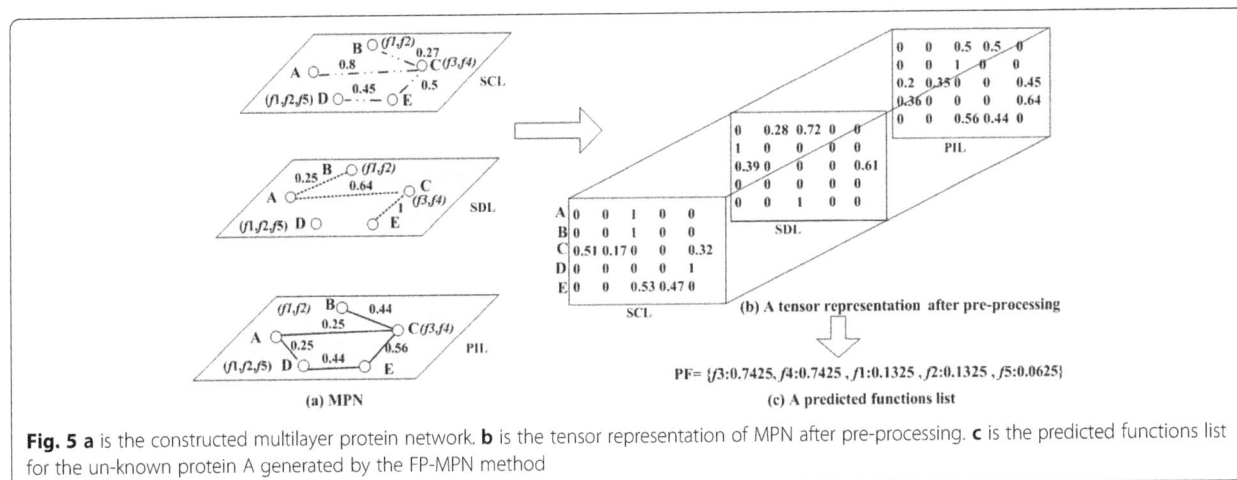

Fig. 5 a is the constructed multilayer protein network. **b** is the tensor representation of MPN after pre-processing. **c** is the predicted functions list for the un-known protein A generated by the FP-MPN method

are mainly implemented in two ways: one is represented by the methods of Zhang [15] and DCS [16], which computes the similarity between proteins and endow all functions of the protein with the highest similarity to the protein with unknown function. Another is represented by the method of NC, which forms candidate functions set by all the functions of the neighbors, then grades and ranks these functions according to a strategy. We have performed statistical analysis for the overlap of functions between the annotated proteins, in order to determine a solution to function selection, as shown in Table 2.

The first column in Table 2 refers to the function overlap between each pair of proteins. The function overlap score of two proteins u and v is defined as follows [28]:

$$OS(u,v) = \frac{|F_u \cap F_v|^2}{|F_u| \times |F_v|} \tag{12}$$

where F_u and F_v is the function set of proteins u and v, respectively. The second column in Table 2 has shown statistical results of overlaps of all pairs of proteins with shared functions, among which the overlap score of 54.22 % protein pairs has exceeded 0.8. As many proteins have only one function, we made statistics again after excluding those with only one function (the result is shown in the third column). It turned out that the overlap score of more than half of the protein pairs falls in (0.4, 0.6], and

the protein pairs with overlap score over 0.6 accounts for only 11.99 %. Based on these statistical results, the FP-MPN method adopts the second strategy of function selection mentioned above.

All functions are sorted in descending order according to their scores. The top N of these functions can be selected to annotate the testing protein u, where N is the number of functions of the protein most closely associated with u. In this paper, we used the highest weight of a pair of proteins to evaluate the close degree of all their layers. We limited the number of predicted functions to be less than or equal to that of the annotated GO terms in the protein with highest weight to u. Algorithm FP-MPN illustrates the overall framework to predict protein functions based on multilayer protein networks.

Table 2 Statistical analysis of overlaps of functions

OS	Proportion (all proteins)	Proportion (proteins with more than one function)
(0, 0.2]	2.81 %	5.64 %
(0.2, 0.4]	13.90 %	27.95 %
(0.4, 0.6]	27.05 %	54.41 %
(0.6, 0.8]	2.02 %	4.06 %
(0.8, 1]	54.22 %	7.93 %

FP-MPN Algorithm

Input: A PPI network $G = (V, E)$

Output: The set of predicted functions PF

1. Generate a weighted multilayer protein network WG by Equation (5-7);
2. FOR each un- annotated protein u DO
3. $PF = \Phi$; // initialization;
4.   FOR J=1 TO 3 DO
5. $CP = \{v_i \mid dis(v_i, u) = 1\} \cup \{u\}$
6. $CF = \{f_i \mid \exists v_i, v_i \in CP, f_i \text{ is a function of } v_i\}$
7.   FOR each function f_i in CF DO
8.   IF $f_i \notin PF$ THEN
9. $Score(f_j) = \frac{1}{2^J} \times \sum_{i=1}^{k} W(u, p_i) \times t_{ij}, \ (p_i \in CP)$;
10.   insert f_i into PF;
11.   ELSE
                $Score(f_j) = Score(f_j) + \frac{1}{2^J} \times \sum_{i=1}^{k} W(u, p_i) \times t_{ij}, \ (p_i \in CP)$
12.   END IF
13.   END FOR
14.   END FOR
15.   Sort functions in PF descendant by their score
16.   Select Top N functions from PF; ///N is the number of functions of the protein, which has close degree with u on all layers.
17.   Output PF
18. END FOR

Results and discussion

Experimental data

The *S. cerevisiae* (yeast) PPI networks are widely used in the research of network-based function prediction methods, because the species of yeast has been well characterized by knockout experiments and is the most complete and convincible. Here, we also adopt the yeast PPI network to test our method. We have applied our method and four other competing algorithms by integrating network topological features, domain information, and protein complexes data: Zhang [15], DCS [16], domain combination similarity in context of protein complexes (DSCP) [16], and PON [17] on DIP data [29]. DSCP is a variant of DSC, which combines protein complex information. The DIP dataset, updated to Oct. 1, 2014, consists of 5017 proteins and 23,115 interactions among the proteins. The self-interactions and the repeated interactions are filtered out in DIP data. The annotation data of proteins used for method validation is the latest version (2012.3.3) downloaded from GO official website [30]. The GO system consists of three separate categories of annotations, namely molecular function (MF), biological process (BP), and cellular component (CC). The predictions are validated separately for each of the three GO categories. To avoid too special or too general, only those GO terms that annotate at least 10 and at most 200 proteins will be kept in the experiments. After processing by this step, the number of GO terms is 267. The domain data is derived from Pfam database [31], including 1107 different types of domains among 3056 proteins. As for the protein complex information, we used the dataset CYC2008 [32], which consists of 408 protein complexes involving 1492 proteins in the yeast PPI network. The GO data and Pfam domain data are transformed to use the ensemble genome protein entries because the original PPI network uses such a labeling system.

Effect of access sequence of each layer

The access sequence of each layer in the MPN plays an important role in the performance of the proposed FP-MPN method. In this paper, the priority of each layer was determined using statistical analysis. Different schemes were used to sequence layers of the MPN and then compare these results to verify the effectiveness of the FP-MPN method. Table 3 depicts the results of FP-MPN when different schemes were adopted. Table 3 demonstrates that the first scheme (SCL → SDL → PIL), in which SCL was visited first and the SDL was visited second, performed the highest in terms of BP (biological process), MF (molecular function), and CC (cellular component). The comparison of these results with the statistical results show they are in agreement. Experimental results also verify the method used to access the sequence of each layer in the FP-MPN.

Table 3 The influence of access sequence

Categories	Schemes	Precision	Recall	*F*-measure	CR
BP	SCL → SDL → PIL	0.444	0.427	0.435	0.426
	SCL → PIL → SDL	0.462	0.401	0.429	0.374
	SDL → PIL → SCL	0.452	0.404	0.426	0.396
	SDL → SCL → PIL	0.442	0.424	0.433	0.422
	PIL → SDL → SCL	0.453	0.404	0.427	0.397
	PIL → SCL → SDL	0.459	0.398	0.426	0.372
MF	SCL → SDL → PIL	0.569	0.544	0.556	0.508
	SCL → PIL → SDL	0.566	0.535	0.55	0.495
	SDL → PIL → SCL	0.585	0.54	0.561	0.505
	SDL → SCL → PIL	0.568	0.543	0.555	0.507
	PIL → SDL → SCL	0.584	0.539	0.561	0.504
	PIL → SCL → SDL	0.573	0.541	0.557	0.5
CC	SCL → SDL → PIL	0.463	0.439	0.451	0.415
	SCL → PIL → SDL	0.468	0.43	0.448	0.4
	SDL → PIL → SCL	0.473	0.424	0.447	0.402
	SDL → SCL → PIL	0.461	0.439	0.45	0.413
	PIL → SDL → SCL	0.473	0.424	0.448	0.403
	PIL → SCL → SDL	0.467	0.429	0.447	0.4

Leave-one-out cross-validation

A representative set of function prediction algorithms was run: FP-MPN, Zhang, DCS, DSCP, and PON, and their performance was examined using the leave-one-out cross-validation method. In the DIP PPI network, 2870, 1592, and 2427 proteins from a total of 5017 proteins were annotated by BP, MF, and CC, respectively. We analyzed the overall prediction performance of FP-MPN on these annotated proteins, as well as four other

Table 4 Overall comparisons of various methods

Categories	Methods	MP	Precision	Recall	*F*-measure	CR
BP	FP-MPN	1595	0.444	0.427	0.435	0.426
	Zhang	810	0.225	0.220	0.222	0.216
	DCS	1148	0.312	0.314	0.313	0.327
	DSCP	1298	0.357	0.359	0.358	0.363
	PON	572	0.150	0.140	0.145	0.161
MF	FP-MPN	995	0.569	0.544	0.556	0.508
	Zhang	608	0.332	0.332	0.332	0.316
	DCS	839	0.461	0.462	0.461	0.441
	DSCP	927	0.518	0.515	0.516	0.489
	PON	413	0.223	0.216	0.22	0.228
CC	FP-MPN	1265	0.463	0.439	0.451	0.415
	Zhang	561	0.197	0.196	0.197	0.198
	DCS	876	0.306	0.309	0.307	0.315
	DSCP	1014	0.364	0.363	0.364	0.356
	PON	440	0.148	0.138	0.143	0.158

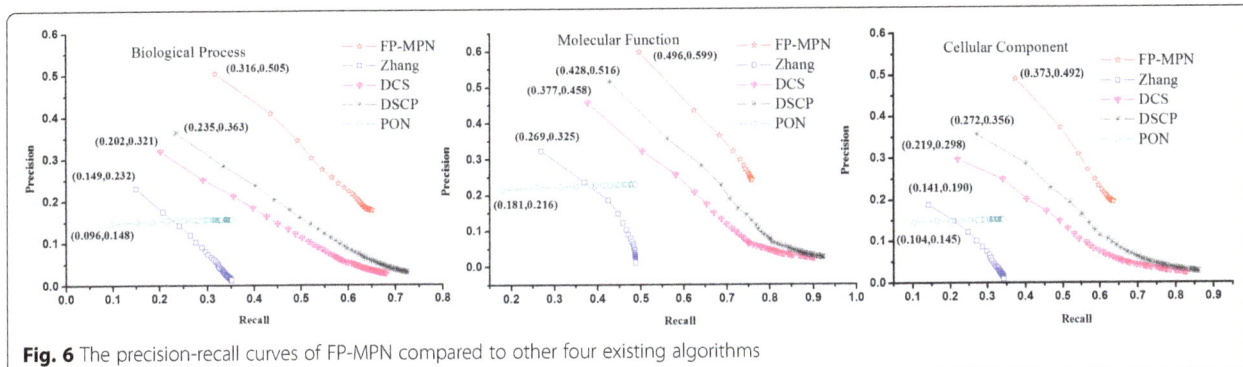

Fig. 6 The precision-recall curves of FP-MPN compared to other four existing algorithms

methods. The results are shown in Table 4, which include the average Precision, Recall, and F-measure and coverage rate (CR) of the various algorithms.

In Table 4, MP is the number of proteins which have been matched to at least one function with known function. Among the five methods, FP-MPN and PON are two methods of selecting top-ranking functions from the set of candidate functions, whereas the methods of Zhang, DCS, and DSCP are three methods of endowing un-annotated proteins with all functions of proteins with the highest similarity values. From Table 4, we can see that FP-MPN can predict functions for more proteins and archive higher performance than the other four methods, with respect to BP, MF, and CC. For BP, the F-measure of FP-MPN is 95.95, 38.98, 21.51, and 200 % higher than Zhang, DCS, DSCP, and PON, respectively. After integrating protein complexes and domains, DSCP improves the performance compared to DCS. FP-MPN outperforms DSCP, including the F-measure and coverage rate. When looking at MF, the performances of these five methods are better. The F-measure of FP-MPN is 67.47, 20.61, 7.75, and 152.73 % higher than the results using the methods of Zhang, DCS, DSCP, and PON, respectively. As for CC, the F-measure of FP-MPN is 128.93, 46.91, 23.9, and 215.38 % higher than the results using the methods of Zhang, DCS, DSCP, and PON, respectively. Compared to BP and MF, FP-MPN had a higher F-measure growth rate compared to other methods.

A comprehensive comparison of the performances of these five methods was undertaken using a Precision-Recall (PR) curve to evaluate the global performance of every method in terms of the different strategies of function selection adopted by the five prediction methods. The same number of functions was chosen for each method, i.e., the top K functions of each prediction method. When examining the methods of Zhang, DCS, and DSCP, the top M ($M < =K$) proteins which had the highest similarity value were selected and the top K functions from the function list as a predictor of functions was listed in descending order according to the maximum value of protein similarity (e.g., given a certain function F_i found in more than one protein, the score of F_i is the similarity value of this protein when compared to the tested proteins). As for the FP-MPN and PON methods, the top K GO terms are chosen to assign functional properties to the unknown proteins (K ranges from 1 to 50). The areas under the curve (AUC) for FP-MPN and other methods are used to compare their performance. AUC is considered to be a standard method to assess the accuracy of predictive distribution models. From Fig. 6, we can see that FP-MPN outperforms other methods in terms of BP, MF, and CC. For example, on the BP, the AUC of FP-MPN is 347.67, 53.76, 31.76, and 195.46 % higher than Zhang, DCS, DSCP, and PON, respectively.

The number of incorrect predicted functions when matching a function correctly using these methods was

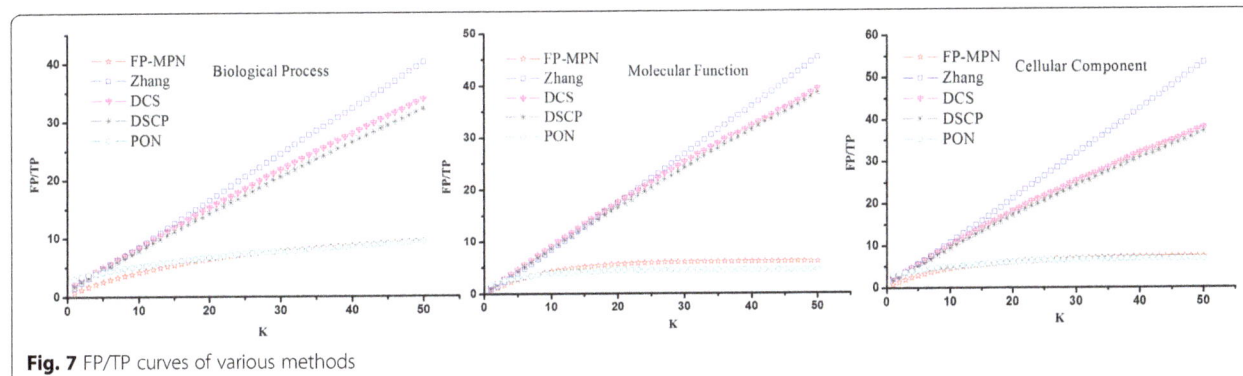

Fig. 7 FP/TP curves of various methods

Table 5 Statistical analysis of FP/TP of various methods

Categories	Methods	Maximum	Minimum	Average	Middle
BP	FP-MPN	9.44	0.72	6.48	7.18
	Zhang	40.29	1.59	20.96	21.04
	DCS	33.94	2.12	18.64	18.94
	DSCP	32.14	1.75	17.49	17.75
	PON	9.39	3.07	6.98	7.41
MF	FP-MPN	6.19	0.53	5.23	5.99
	Zhang	45.5	0.9	22.81	22.71
	DCS	39.41	1.18	21.28	21.88
	DSCP	38.54	0.94	20.4	20.73
	PON	4.57	1.85	4.2	4.57
CC	FP-MPN	7.39	0.72	5.88	6.59
	Zhang	53.51	2.12	27.29	27.09
	DCS	38.15	2.36	21.49	22.25
	DSCP	37.02	1.81	20.45	21.04
	PON	6.88	3.07	6.02	6.57

methods of FP-MPN, Zhang, DCS, DSCP, and PON is 7, 21, 19, 18, and 7, respectively. The results illustrate that FP-MPN has the high prediction efficiency and accuracy.

Tenfold cross-validation

The performance of FP-MPN was tested using leave-one-out validation. Experimental results demonstrate improvements when predicting protein functions by the FP-MPN method compared to competing methods. However, in practical applications, there are much more proteins without annotations, instead of one unknown protein. In this section, we will use the leave-percent-out cross-validation method to verify the effectiveness of FP-MPN on PPI networks that have less functional information. Tenfold cross-validation is a widely used leave-percent-out cross-validation, which is used in this paper. The tenfold cross-validation requires the entire set of examples to be divided into ten equal sets randomly. Nine of the ten parts are used for training, and one part is used for testing. This is repeated ten times, each time using another testing set. We evaluate the performance of each method using area under precision-recall (PR) curve. Figure 8 illustrates the PR curve using tenfold cross-validation, in terms of biological processes, molecular functions, and cellular components. When compared to the results of leave-one-out cross-validation, the performance of all methods using tenfold cross-validation decrease slightly, due to the decrease of the number of training proteins. It appears that Fig. 8 is very similar to Fig. 6, except for the coordinate values of the various methods. Figure 8 demonstrates that FP-MPN still outperforms other methods when tenfold cross-validation is used to test all methods.

Analysis of the overlaps and differences between FP-MPN and other methods

To further analyze the differences between the FP-MPN and other methods, we selected 12 testing proteins and predicted their functions using the five methods. Table 6 lists the functions of these selected proteins predicted by

determined. For each testing protein, the top *K* functions are selected as its predicted ones, and TP and FP values are calculated according to its known functions. The TP and FP values of all testing proteins are added to calculated TP and FP pairs. Selecting different values of *K* (ranging from 1 to 50), a FP/TP curve can be generated with different TP and FP pairs, as shown in Fig. 7. Figure 7 clearly shows that the curvature of FP-MPN curve is the lowest as compared to others, which means that, if matched functions are the same, the number of functions incorrectly matched by FP-MPN is the least. Table 5 lists the statistical results of the various FP/TP curves, including maximum value, the minimum value, the average value, and the middle value. These results indicates that to match a protein function correctly, the number of average noise functions (i.e., predicted function incorrectly matched) produced by FP-MPN is smaller compared to the Zhang, DCS, and DSCP methods. FP-MPN has comparable results with PON's. For example, on the BP, the number of average noise functions of the

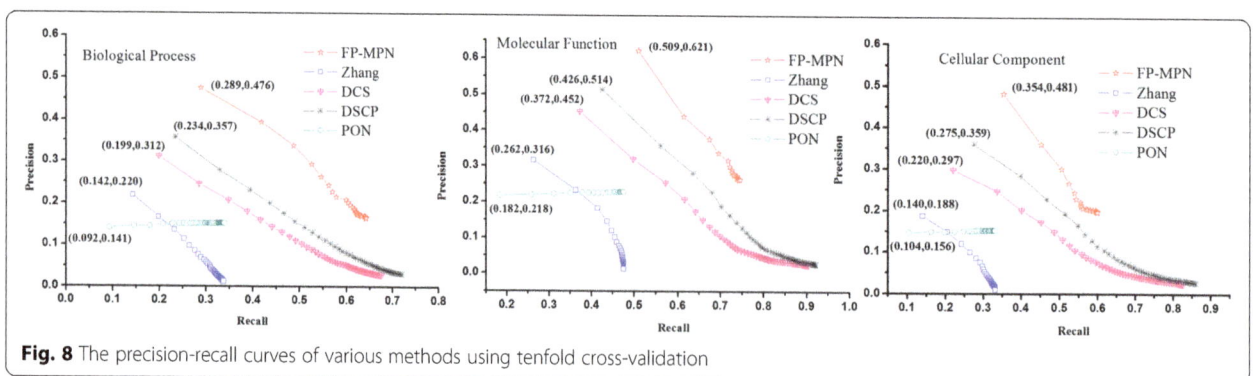

Fig. 8 The precision-recall curves of various methods using tenfold cross-validation

Table 6 Selected functions predicted by various methods

Categories	Proteins	FP-MPN	Zhang	DCS	DSCP	PON
BP	YGL100W (8 GO terms)	*GO:0006409* *GO:0006607* *GO:0006913* *GO:0006999* *GO:0006406* *GO:0006609* *GO:0006611* *GO:0006407* *GO:0000973* *GO:0000055*	GO:0000723 GO:0006348 GO:0006355 GO:0051568	GO:0043161	GO:0043161	GO:0000001 GO:0000002 GO:0000027 GO:0000055 GO:0000082 GO:0000086 GO:0000122 GO:0000209
	YNL262W (7 GO terms)	*GO:0006272* *GO:0006273* *GO:0006289* *GO:0006298* *GO:0000084*	*GO:0006273* GO:0000084 GO:0006270	*GO:0006273* GO:0000084 GO:0006270	*GO:0006273* GO:0000084 GO:0006270	*GO:0006272* *GO:0006273* *GO:0006289* GO:0000084 GO:0006260 GO:0006270 GO:0006284
	YLR321C (6 GO terms)	*GO:0006337* *GO:0006368* *GO:0043044* *GO:0000086*	*GO:0006302* *GO:0043044* GO:0006338 GO:0042766 GO:0045944	*GO:0006302* GO:0043044 GO:0006338 GO:0042766 GO:0045944	*GO:0006302* GO:0043044 GO:0006338 GO:0042766 GO:0045944	*GO:0006302* GO:0043044 GO:0006338 GO:0042766 GO:0045944
	YBR278W (5 GO terms)	*GO:0006272* *GO:0006273* *GO:0006289* *GO:0006298* *GO:0006348* GO:0006303 GO:0007064		*GO:0006348* GO:0000723 GO:0006281 GO:0007064 GO:0030466	*GO:0006348* GO:0000723 GO:0006281 GO:0007064 GO:0030466	
MF	YBR114W (3 GO terms)	*GO:0004842* *GO:0003684* *GO:0008094*	*GO:0008094*	*GO:0008094*	*GO:0008094*	GO:0000386 GO:0000990 GO:0001102
	YJR052W (3 GO terms)	*GO:0004842* *GO:0003684* *GO:0008094*		GO:0008134	GO:0043130	
	YJR140C (3 GO terms)	*GO:0003677* *GO:0031491* *GO:0003714*		GO:0046933 GO:0046961	*GO:0003677* *GO:0031491*	
	YBL021C (2 GO terms)	*GO:0001077* *GO:0000978*	GO:0003713 GO:0003714	GO:0003713 GO:0003714	GO:0003713 GO:0003714	GO:0003713 GO:0003714
CC	YNL161W (6 GO terms)	*GO:0005933* *GO:0005934* *GO:0005935* *GO:0043332*	*GO:0005935* GO:0005816	*GO:0005935* GO:0005816	*GO:0005935* GO:0005816	*GO:0000131* GO:0000139 GO:0000142 GO:0000307 GO:0000324 GO:0000329
	YBR198C (3 GO terms)	*GO:0000124* *GO:0046695* *GO:0005669*	GO:0070210	GO:0070210	GO:0070210	*GO:0000124* GO:0000139 GO:0000228
	YDR167W (3 GO terms)	*GO:0000124* *GO:0046695* *GO:0005669*		GO:0005666	*GO:0000124* *GO:0046695*	
	YNL273W (3 GO terms)	*GO:0031298* *GO:0000228* *GO:0043596*		GO:0005751	GO:0005751	

various methods. The third to the seventh column of Table 6 lists functions predicted by the FP-MPN, Zhang, DCS, DSCP, and PON methods, respectively. In this table, functions in italics represent the matched functions of the testing proteins, the rest are mismatched functions. In Table 6, we can see that FP-MPN can record more correct functions and fewer error functions compared to the other competing methods.

In addition, we continued to look for sources of functions predicted by various methods. For the protein YGL100W, the functions set predicted by the method of Zhang consists of GO:0000723, GO:0006348, GO:0006355, and GO:0051568, which were derived from the protein YAR003W. In this study, YAR003W is regarded as having the most similar domain to YGL100W among all the proteins. Unfortunately, these predicted functions are mismatched by the real functions of YGL100W. As for DCS and DSCP, the protein YCL039W is considered to be the most similar in domain to YGL100W than the other known proteins. Similarly, the predicted functions of GO:0043161, which were derived from YCL039W, created errors in predicted functions for YAR003W. Predicted functions by PON were GO:0000001, GO:0000002, GO:0000027, GO:0000055, GO:0000082, GO:0000086, GO:0000122, and GO:0000209, which were derived from YBR234C, YJL112W, YKL021C, YDR267C, YDR364C, YFL009W, YLR055C, and YIL046W, respectively. All of these proteins have at least one domain with YGL100W. So, we can draw a conclusion that we cannot predict functions for the protein YGL100W based on domain information only. Our FP-MPN predicts ten functions, in which eight are matched and two are mismatched. These matched functions were derived from protein YDL116W, which is located in the transcription factor TFIID complex with the YGL100W protein. FP-MPN successfully matched eight functions for the protein YGL100W, with the help of protein complexes information. The results suggest that complexes information improves the accuracy of protein function prediction. However, protein complexes data is also used in the DSCP methods, which has a different predictor results compared to that of FP-MPN. This could be due to the difference in how the data is used between the two methods. For the protein YNL262W, the methods of Zhang, DCS, and DSCP created the same function lists, consisting of GO:0006273, GO:0000084, and GO:0006270. These three functions are derived from the protein YNL102W, which has common domains with the protein YNL262W. In the predicted functions list, only GO:0006273 is correct as a function for the protein YNL102W. Compared to the methods of Zhang, DCS, and DSCP, PON can identify two other correct functions GO:0006273 and GO:0006289 from another protein YDL102W, which shares domains with the protein YNL262W. The result suggests that annotating proteins according to multiple known proteins is more reliable than predicting functions from a single protein. Besides the three matched functions identified by other methods, FP-MPN identifies a new correct function GO:0006298. In this example, FP-MPN predicts more matched functions compared to other methods, due to the domain and complexes information being used. This phenomenon suggests that proper use of multiple heterogeneous biological data can effectively improve the performance of function prediction algorithms. The analysis for the rest of the ten proteins described above is consistent with that of YGL100W and YNL262W.

Efficiency analysis

To compare the efficiency of these methods, we ran FP-MPN and competing methods under the same conditions and looked at their running time. All methods in this paper were run on a notebook computer with Inter(R) Core(TM) i5-4300M 2.6 GHz CPU and 4 GB RAM. Figure 9 illustrates a comparison of the running time of FP-MPN and the other four methods used for predicting protein functions. The methods of Zhang, DCS, and DSCP are all based on combined number computation. So, they have the disadvantage of being time consuming. From Fig. 9, it can be seen that FP-MPN is extremely fast, 25, 52, 55, and 0.8 times faster than the methods of Zhang, DCS, DSCP, and PON, respectively. As protein-protein interactions are accumulating, FP-MPN can be used in larger scale PPI networks.

Conclusions

Different types of interactions or connections play different roles in protein function prediction. Combining multiple interactions or connections between two proteins could reduce the impact of false negatives and increase the number of correct predicted functions. However, there appears to be more false functions identified compared to positive functions, thus the overall performance of function prediction would not be improved greatly. In this paper, multilayer protein networks (MPN) are constructed based on topological characteristics, protein domain information, and protein complex information, with each layer given various priorities. Based on the constructed networks, we proposed a new method, named FP-MPN, to predict the functions of a particular protein. The proposed

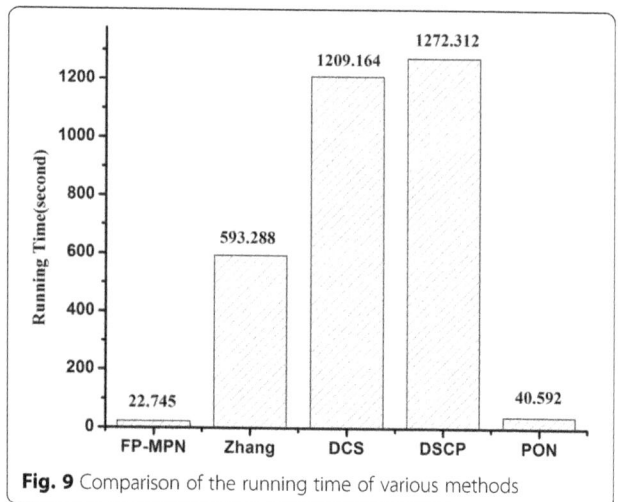

Fig. 9 Comparison of the running time of various methods

method is based around visiting each layer of the MPN in turn and forming a set of candidate neighbors with known functions. The set of predicted functions is then formed and all of these functions are scored and sorted. Each layer contributes differently to the predicted functions in the un-annotated protein. The experimental results indicate that it is an effective method to predict protein functions.

Additional file

Additional file 1: Supplementary Data are available, including Complexes.txt: The CYC2008 dataset, which consists of 408 protein complexes involving 1492 proteins in the yeast PPI network. DIP141001.txt: The DIP dataset, updated to Oct.1, 2014, consists of 5017 proteins and 23115 interactions among the proteins. Domain.txt: The domain data derived from Pfam database, including 1107 different types of domains among 3056 proteins. FP-MPN.exe: The FP-MPN algorithm. FPMPN.txt: The predicted results by FP-MPN. GO_C.txt, GO_F.txt and GO_P.txt: Represents cellular component (CC), molecular function (MF) and biological process (BP), respectively.

Acknowledgements
Not applicable

Funding
This work is supported in part by the National Natural Science Foundation of China under Grant No. 11501054; the Science and Technology Plan Project of Hunan Province, China No. 2015GK3072; the Natural Science Foundation of Hunan Province, China No. 2016JJ3016; the National Scientific Research Foundation of Hunan Province Education Department, China No. 16A020, 16B028, No. 16C0133, No. 16C0137; and the Education Scientific Planning Project of Hunan Province, China No. XJK016BGD078.

Authors' contributions
BHZ and SH obtained the protein-protein interaction data, gene ontology annotation data, protein complex data, and domain data. BHZ and SH designed the new method, FP-MPN, and analyzed the results. BHZ, XYL, and SH drafted the manuscript together. FZ, QLT, and WYN participated in revising the draft. All authors have read and approved the manuscript.

Competing interests
The authors declare that they have no competing interests.

Consent for publication
Not applicable.

References

1. Liu B, Fang L, Liu F, Wang X, Chen J, Chou K-C. Identification of real MicroRNA precursors with a pseudo structure status composition approach. PLoS One. 2015;10(3):e0121501.
2. Ito T, Chiba T, Ozawa R, et al. A comprehensive two-hybrid analysis to explore the yeast protein interactome. Proc Natl Acad Sci. 2001;98(8):4569–74.
3. Uetz P, Giot L, Cagney G, et al. A comprehensive analysis of protein–protein interactions in Saccharomyces cerevisiae. Nature. 2000;403(6770):623–7.
4. Gavin AC, et al. Functional organization of the yeast proteome by systematic analysis of protein complexes. Nature. 2002;415(6868):141–7.
5. Enright AJ, Van Dongen S, Ouzounis CA. An efficient algorithm for large-scale detection of protein families. Nucleic Acids Res. 2002;30(7):1575–84.
6. Schwikowski B, Uetz P, Fields S. A network of protein–protein interactions in yeast. Nat Biotechnol. 2000;18(12):1257–61.
7. Hishigaki H, Nakai K, Ono T, et al. Assessment of prediction accuracy of protein function from protein–protein interaction data. Yeast. 2001;18(6):523–31.
8. Chua HN, Sung WK, Wong L. Exploiting indirect neighbours and topological weight to predict protein function from protein–protein interactions. Bioinformatics. 2006;22(13):1623–30.
9. Cao M, Zhang H, Park J, et al. Going the distance for protein function prediction: a new distance metric for protein interaction networks. PLoS One. 2013;8(10):e76339.
10. Nabieva E, Jim K, Agarwal A, et al. Whole-proteome prediction of protein function via graph-theoretic analysis of interaction maps. Bioinformatics. 2005;21 suppl 1:i302–10.
11. Chi X, Hou J. An iterative approach of protein function prediction. BMC Bioinformatics. 2011;12(1):437.
12. Wu Z, Zhao X, Chen L. Identifying responsive functional modules from protein-protein interaction network. Mol Cells. 2009;27(3):271–7.
13. Lee J, Gross SP, Lee J. Improved network community structure improves function prediction. Sci Rep. 2013;3:2197.
14. Cozzetto D, Buchan DWA, Bryson K, et al. Protein function prediction by massive integration of evolutionary analyses and multiple data sources. BMC Bioinformatics. 2013;14 Suppl 3:S1.
15. Zhang S, Chen H, Liu K, et al. Inferring protein function by domain context similarities in protein-protein interaction networks. BMC Bioinformatics. 2009;10(1):395.
16. Peng W, Wang J, Cai J, et al. Improving protein function prediction using domain and protein complexes in PPI networks. BMC Syst Biol. 2014;8(1):35.
17. Liang S, Zheng D, Standley DM, et al. A novel function prediction approach using protein overlap networks. BMC Syst Biol. 2013;7(1):61.
18. Gong Q, Ning W, Tian W. GoFDR: a sequence alignment based method for predicting protein functions. Methods. 2016;93:3-14.
19. Kumar DS, Reddy PK. Improved approach for protein function prediction by exploiting prominent proteins: IEEE International Conference on Data Science and Advanced Analytics (DSAA). 2015. p. 1–7.
20. Yu G, Rangwala H, Domeniconi C, et al. Predicting Protein Function Using Multiple Kernels. IEEE/ACM Trans Comput Biol Bioinform. 2015;12(1):219–33.
21. Wu M, Li X, Kwoh CK, et al. A core-attachment based method to detect protein complexes in PPI networks. BMC Bioinformatics. 2009;10(1):169.
22. Wang J, Peng X, Li M, et al. Construction and application of dynamic protein interaction network based on time course gene expression data. Proteomics. 2013;13(2):301–12.
23. Zhao BH, Wang JX, Li XY, et al. Essential Protein Discovery based on a Combination of modularity and conservatism. Methods. 2016. doi:10.1016/j.ymeth.2016.07.005.
24. Liu W, Li D, Zhu Y, et al. Reconstruction of signalling network from protein interactions based on function annotations. IEEE/ACM Trans Comput Biol Bioinform. 2013;10(2):514–21.
25. Zhao BH, Wang JX, Li M, et al. A new method for predicting protein functions from dynamic weighted interactome networks. IEEE Trans NanoBioscience. 2016;15(2):415–24.
26. Zhao BH, Wang JX, Li M, et al. Prediction of essential proteins based on overlapping essential modules. IEEE Trans NanoBioscience. 2014;13(4):415–24.
27. Peng W, Wang J, Zhao B, et al. Identification of protein complexes using weighted PageRank-Nibble algorithm and core-attachment structure. IEEE/ACM Trans Comput Biol Bioinform. 2015;12(1):179–92.
28. Zhao B, Wang J, Li M, et al. Detecting protein complexes based on uncertain graph model. IEEE/ACM Trans Comput Biol Bioinform. 2014;11:486–97.
29. Xenarios X, et al. DIP: the database of interacting proteins. Nucleic Acids Res. 2000;28:289–91.
30. Ashburner M, Ball CA, Blake JA, et al. Gene ontology: tool for the unification of biology. Nat Genet. 2000;25(1):25–9.
31. Bateman A, Coin L, Durbin R, et al. The Pfam protein families database. Nucleic Acids Res. 2004;32 suppl 1:D138–41.
32. Pu S, Wong J, Turner B, et al. Up-to-date catalogues of yeast protein complexes. Nucleic Acids Res. 2009;37:825–31.

12

Falling giants and the rise of gene editing: ethics, private interests and the public good

Benjamin Capps[1*], Ruth Chadwick[2], Yann Joly[3], John J. Mulvihill[4], Tamra Lysaght[5] and Hub Zwart[6]

Abstract

This paper considers the tensions created in genomic research by public and private for-profit ideals. Our intent is to strengthen the public good at a time when doing science is strongly motivated by market possibilities and opportunities. Focusing on the emergence of gene editing, and in particular CRISPR, we consider how commercialisation encourages hype and hope—a sense that only promise and idealism can achieve progress. At this rate, genomic research reinforces structures that promote, above all else, private interests, but that may attenuate conditions for the public good of science. In the first part, we situate genomics using the aphorism that 'on the shoulders of giants we see farther'; these giants are infrastructures and research cultures rather than individual 'heroes' of science. In this respect, private initiatives are not the only pivot for successful discovery, and indeed, fascination in those could impinge upon the fundamental role of public-supported discovery. To redress these circumstances, we define the extent to which progress presupposes research strategies that are for the public good. In the second part, we use a 'falling giant' narrative to illustrate the risks of over-indulging for-profit initiatives. We therefore offer a counterpoint to commercialised science, using three identifiable 'giants'—scientists, publics and cultures—to illustrate how the public good contributes to genomic discovery.

Keywords: Public good, Public interest, CRISPR, Solidarity, Benefit sharing, Ethics in genomic research, Human genome project, Biobank

Introduction

Cutting-edge bioscience is a public good: in addition to economic benefits, it can generate social value in healthcare, agriculture and industry. Sometimes, however, preoccupation with a pecuniary imperative encourages 'hype and hope': predictions that beget idealism and claims which promise too much. Hyperbole has been a feature of genomics since its inception, and high hopes continue to shape perceptions of private interests and the public good. In this paper, we present an egalitarian-type response to the entrenchment of commercialisation in genomics research. Using the concept of genomic solidarity, we endorse undertaking research for the public good and question the current commercial speculation in genomics.

From the Human Genome Project (HGP) [1] as the flagship project of the 'Genomic Era' [2] up to the new wave of post-genomics research, there has been an overarching narrative about the value of high-profile discoveries. Most recently, this has been highlighted by gene editing—a platform of converging scientific expertise organised around similar methods—and specifically, CRISPR-Cas9. As we discuss later, these discoveries are often promoted as the exclusive achievement of processes of commercialisation. This gratification bias, which creates pathways for exclusion and validates outlandish (and sometimes unjustified) rewards for innovators, is at least partly responsible for the devaluing of the public infrastructure. On closer inspection of the sophisticated pathways of scientific discovery, it becomes clear that in various ways, the quest for significant returns potentially jeopardises the ways that the public good contributes to the production and implementation of scientific knowledge. It is, therefore, essential that responsible research in

* Correspondence: benjamin.capps@dal.ca
[1]Department of Bioethics, Faculty of Medicine, Dalhousie University, Halifax, Canada
Full list of author information is available at the end of the article

genomics and post-genomics include the management of promises (or 'promisomics') [3] which we argue requires the reification of the public good. We contend that society-driven research anchored in the public good should be acknowledged as essential to progress. Refocusing upon the public good could, to some degree, challenge the culture of hype and hope [4].

The paper is structured in two parts. In the first part, we situate genomics within a 'giants' narrative. In making our case for the public good, we draw on the work of Robert Merton, who recounts the Newtonian idea of giants in science upon whose shoulders others stand [5]. Applying this to genomics, we argue that the giants are collective infrastructures and broad research cultures. With these in mind, we then offer a challenge to market ideologies as the pivot for successful discovery by emphasising the extent to which progress *presupposes* research strategies that are 'for the public good'.

In the second part, we build a 'falling giant' narrative to illustrate the risks of over-indulging for-profit initiatives that come about because of the predominant ideology that is imposed upon research. That trend has devalued the public good. Thus, we offer a counterpoint to marketisation using three 'giants'—scientists, publics and cultures—to illustrate how the public good contributes to genomic discovery.

Giants and genomic technologies

CRISPR-Cas9 (hereafter CRISPR) is the latest highly prized biotechnology. It is a gene editing tool developed from bacterial adaptive immunity, based on Clustered Regularly Interspaced Short Palindromic Repeats and 'CRISPR-associated' enzymes. It is a precise, rapid and cheap tool for editing DNA that by far outperforms previous genetic engineering capabilities; it has become essential in laboratories across the globe. Like the HGP before, CRISPR promises to revolutionise genetics and genomics as a quantum advance, and much like the sequencers who laid the groundwork for next-generation technologies, it will allow superior analytics to become available to diverse laboratories [6].[1] The emergent gene editing platform is a switch from the slow but widely available genetic engineering tools of yesterday, to new, sharp and shiny ones.

Understanding CRISPR's place in scientific creativity and its implications for society can suggest ways in which technologies are defined by vested interests, policy goals and public imaginations. In the *emerging landscape* of gene editing technology, a number of themes are re-emerging from past innovations. One of these themes is the anticipation that technology brings vast clinical benefits. Before gene editing, the promise of stem cell science predicted sudden and immediate prospects—a technology that still envisages immense progress in areas like regenerative medicine but has yet to meet expectations. Of ethical concern is the repeated use of hype and hope to attract funding, promote more permissive regulations and mislead vulnerable patients [7]. We do not see this simply as malicious action by greedy scientists or institutions promoting their investments (but witness the recent court battles between scientists and institutions in respect to the CRISPR patents [8]), nor do we dismiss it as the work of media hunting for headlines. Rather, it is a feature of the profit-driven constellation whose basic premises we question—their 'catallactic bias' [9][2] towards promoting markets as podiums for progress without also questioning their unfairness and their failures, as well as their accomplishments.

A second theme is access to technologies [8]. We may more generally question the pervasive idea that profit is not an afterthought to doing worthwhile science, but the *raison d'être*. While patents are important in biosciences, at least according to the socio-economic argument that they stimulate innovation and investment, critics question the effectiveness (and desirability) of monopolies as incentive models for innovation [10]. To this end, whoever successfully receives the rights to CRISPR could assert to a large extent the still unspecified terms under which that technology is adopted in laboratories in clinical, animal and agricultural areas [11]. The consequences of these legal trials could redefine traditional *genetic engineering*—that has become an inclusive platform over the years—in terms of the exclusive context of modernistic *gene editing*. Critically, patents may also encourage the kinds of unreasonable dominance that elevates inventors and privileges investors, while subordinating public goods. The current landscape of marketisation as means to discovery and value, might push valuations of CRISPR-based therapeutics into the stratosphere of reasonable cost in order to satiate returns. There, they become out of reach of most including insured patients and those dependent on national health systems. In such circumstances, only the wealthiest can be optimistic of benefiting from CRISPR technologies.

However, rather than focusing on legal and clinical access conundrums, we want to recast this debate by using Merton's narrative about the words famously uttered by Isaac Newton: *Without the giants we would see nothing; on their shoulders we see farther.*[3] The first giants are the individual innovators and inventors and, in this respect, Newton's well-known aphorism is an expression for the dependencies of scientific researchers on predecessors. In other words, however original the present endeavour, its success can be traced back to many prior discoveries. The development of CRISPR involved many incremental steps, including the discovery of DNA itself and many contributions since [12, 13].

Second, there are the giants in contemporary biomedical research. Scientists increasingly rely on vast networks and infrastructures, such as big international research consortia, big machines and big data. It is here that one finds further significance in the aphorism in respect to the biases of collaboration (who works with whom and why), particularly in a culture in which being first brings global fame and (not just monetary) fortune [14].

Third among the giants is the enormous influence of culture, pegged to the zeitgeist of any particular time; these might be categorised as neoliberalism, the Anthropocene, populism, post-truth and so on. Gaining cultural ascendency is significant for scientific discovery: it can determine what, where and by whom science is done, and who is acknowledged, compensated and rewarded. In these times, arguably, science is dominated by neoliberalism, and that involves planning scientific research ultimately to translate discoveries into consumer products and industrial technology; progress, in this respect, is possible only because of the ascendency of corporations, competition and 'degovernmentalization' [15]; innovations and discoveries are celebrated primarily because of their exceptional contributions to the vast biomedical-market. As a result, the current CRISPR debate is dominated by the clinical prospects rather than the undoubted contributions it will make in many other areas such as animal and agricultural engineering. However, it is our conjecture that within all three giants, the real contributions of the public good are distorted to make the case for marketisation. To understand how this came about, we need to go back to the ideologies that grounded the genomic revolution.

Genomics and emerging giants

In 2010, the journal *Nature* asked whether the 'genomic revolution' had arrived. Contributors to the issue included the key architects of the HGP, Francis Collins [16] and Craig Venter [17], whose answers, and those of other contributors, were essentially 'Not yet'. The reference human genome dramatically changed the capabilities of genomic research, yet so far (in 2017), the benefits for individuals and society have been limited. There remain to date three grand challenges in genomics: genomics to biology (elucidating the structure and function of genomes), genomics to health (translating genome-based knowledge into health benefits) and genomics to society (promoting the use of genomics to maximise benefits and minimise harms in populations) [2]. The revolution is progressing more slowly than many first envisioned; in particular, there is still some way to go in the translation of genomic science into widespread clinical applications. It is difficult to pinpoint any one single reason for this [18], but perhaps it is the right time to consider conceivable flaws in the ideologies that inform the industry-research complex responsible for undertaking genomic sciences.

The HGP was a moment of high visibility for science that attracted vast public financing and private entrepreneurship[4]; now, standing on the shoulders of *this* giant, we can appreciate discoveries such as CRISPR. Maintaining momentum in genomics has become the hard sell to investors and funders, both public and private, so that waves of hype (and some hope) continue to fluctuate [3]. While technologies become more effective, our dexterity in managing expectations hardly seems to improve at all; for genomics, prospects are being transferred to new initiatives, such as personal and precision genomics [19], and now, gene editing can be added to that list.

Reflecting on the HGP, Maynard Olson writes:

> There are two stories of the Human Genome Project. One describes a century of scientific progress that began with the rediscovery of Mendel's laws in 1900 and ended in a frenzy of genomic sequencing. The other is a story about contemporary societal values—particularly those that framed the project's endgame and continue to shape public perceptions towards this defining event in time ([20], p. 931).

The first story alludes to the many giants that enabled progress in genetics and genomics—all of which surely contributed, in various ways, to the post-genomic era; that must include many other confluent technologies such as computing and data storage. The idea to sequence the human genome, then, was as much about historical socio-political events as about the technological feasibility that would lead to an opportune 'time to sequence' [21].

The second story is about translating genomics into society. The HGP was characterised by some as a race between two competing parties—the International Human Genome Sequencing Consortium (IHGSC) [22] and Celera Genomics [23]. It is a story that is multi-layered, involving partisan politics and indiscriminating press coverage documented in the public-private competition between the 'players' [20]. At the time, two key players claimed a special connection to the public interest or good—the public project of IHGSC wanted to publish their sequences so that it was freely and therefore widely accessible; Celera argued that it could get the job done more quickly and save countless human lives by using intellectual property to generate exclusive rights and revenue from the human genome [20]. Taking liberty to distill that rivalry to its most basic point, two ideologies surface: on the one hand, Venter and Celera's interest in sequencing the human genome was billed as a way to accelerate the laboured efforts of the public initiative. On the other hand, Collins, praising the public investment as 'arguably one of the more impressive success stories... of all time', recognised the

implications if the Consortium 'dropped the ball' ([24], p. 60, 80); the only way to assure unrestricted access to the sequence was to continue with the public project, perhaps in partnership with other private entities [24]. Thus, it was either a story of mavericks challenging the slow-witted establishment or a lament about how private interests seemed about to *capture* public goods [25].[5] These competing ideologies persist in bioscience today, often because of promises and pitfalls of scientific research are created, sustained and leveraged via ethical and social norms expressed by the leaders in the field. These opinions echo within complex social and political networks and are sustained by immense public and private infrastructures.

The public good

The response of Collins to Celera's strategy was to reaffirm the significance of the public good. What, then, is meant by the public good?

The Human Genome Organisation (HUGO) has a long tradition of advocacy for 'benefit sharing' to realise societal *as well as* economic opportunities [26]. In a HUGO statement from 2000, it was stated:

A benefit is a good that contributes to the well-being of an individual and/or a given community. ... Thus, a benefit is not identical with profit in the monetary or economic sense. Determining a benefit depends on needs, values, priorities and cultural expectations...
The HUGO Ethics Committee recommends ... that all humanity share in, and have access to, the benefits of genetic research [27].

HUGO's statement, we believe, reflects the public-private intellectual climate of that time.[6] At the outset of the HGP, it was proposed, and then codified in the 1997 Bermuda Principles, that human DNA sequences 'should be freely available and in the public domain in order to encourage research and development and to maximise its benefit to society.'

It was agreed that these principles should apply for all human genomic sequence generated by large-scale sequencing centres, funded for *the public good*, in order to prevent such centres establishing a privileged position in the exploitation and control of human sequence information [our emphasis.] [28]

The HUGO Committee on Ethics, Law and Society has more recently stated that 'genomic solidarity' ideally supports collaborations between individuals, communities and populations, with research communities and industry [29].[7] Significantly, benefit sharing and genomic solidarity work together through an idea of the public good [30]. In the most rigorous terms, benefit sharing suggests that research must be preceded by engagement with all stakeholders rather than allowing for exclusion and domination and, consequently, disunity. Relatedly, a notion of solidarity requires collective agreement upon common ends to be achieved and how to do so, and thereby, differentiates between public goods and public bads. A public good is valued distributively, i.e., to each and every person that value is secured through equality of rights. In rights talk, that idea recalls the right to claim a good (such as food, water or shelter) and affirms the justified protection of persons' important interests (to claim a right is to claim access, protection or provision of a good). That claim is not limited to the goods that are traded but includes all goods that establish a basic level of healthy living and contributes to opportunities, within egalitarian societies. Public bads do the opposite: they exist in a way that affect people distributively (such as pollution spewing into a river from a factory upstream of a village) and are expressed in terms of those affected having their rights infringed. Within a solidary framework, the scope for public goods to do good and public bads to do the opposite is understood; institutions and cultures thereby adapt to priorities that most likely support public goods. In the context of genomics, the public good means that everyone is entitled to access to the fruits of research because that meets HUGO's ethical conditions for benefit sharing and solidarity.

The idea of genomic solidarity is likely to be challenged, for it confronts the engrained idea of public goods as something which hinders the benefits of economies of exclusion and rivalry. In that classical estimation, goods are 'public' depending on whether private investment has any interest in them; in other words, if a good is profitable, then it is economically wasteful to consider it a public one. This illustrates what Samuelson originally called *collective consumption goods* (what later became known as 'public goods') [31]. It leads to a particular view of goods that can be applied to the human genome: in the course of human history, every human being, living or dead, has been part of the genome's conception (for instance, by adding variants) and contributed to its continuation. Although it is our legacy, no one person has written the chapters, and sequencers are now 'reading' the book and genomists are 'translating' it. This process of curiosity, understanding and innovation converts the genome from a status of *public legacy* to one of *value* (it is now a chapter or verse that deserves a price), and ownership becomes a significant factor in that conversion. In short, particles of, or even the entirety of the genome, have become someone's property [26].

And now, gene editing is likely to extend the interest in ownership of DNA in the same way that economics

shaped the claims of ownership of other human materials [32]. These rewritten or novel sequences will exist outside of any normal or representative human genome. Thus, it is more likely than ever that human genes will become commodities [33], and society will have to decide how tolerable such claims are to be in light of the alleged benefits of a flourishing gene market. It is in this context that we find HUGO's benefit sharing model and genomic solidarity as a challenge to uncritical characterisations of human genomic commodities. In the next section, we use the narrative of giants to explain the public's role in genomics, the pitfalls of profit-driven science, and to thereby strengthen the conception of the public good.

Giants and the public good

How do CRISPR and other gene editing tools become an opportunity for the public good? We now explore in greater detail the 'giants' metaphor; a term which not only means 'greater than normal' but also refers to ways that people who are exceptional in talents and abilities contribute to ideal conditions for discovery. There are three giants: (1) individuals with great creativity and insight involved in the development of the technology (not just the inventors but also the policy makers, politicians and administrators who will create the regulatory conditions in which gene editing occurs); (2) institutions of great size and reach, where research is housed and applications transpire; and (3) the prevailing zeitgeist, namely the cultures that exert influence in this area of research.

Individuals as giants

The prevailing social narrative of CRISPR concerns the first of our giants: inventors pitted against one another in conjunction with their lawyers and administrators [34]. As a result of this perspective, there is a tendency to think about cutting-edge technologies only in terms of economics, thus venerating scientists for their endeavours within systems that primarily promote profit [35].

However, from the *observation* that clustered repeats might be significant, to CRISPR's sensational harnessing and refining, it involved, as with nearly all other discoveries, many scientists, working for many years on many topics [36]; CRISPR's discovery, therefore, is contentious in respect to the 'giants' metaphor. On the one hand, a legal narrative prompts us to focus excessively on isolated contributions attributable to individuals; on the other hand, that approach disavows the extent to which novelty builds on vast networks of knowledge and technology that are already in place. This understanding of discovery is also relevant in respect to technology's translation into applications and useful products [36]. In this respect, we might ask whether the gene editing

platform should be a public resource by acknowledging multiple contributions.

Our premise is that the links of discovery are much wider that is currently appreciated by the legal narrative. Rather than standing in isolation, scientists, their affiliates and institutions rely on publics who volunteer their time, bodies and experiences for clinical trials, become patient-participants in research through their giving of data and tissue samples, and have interests by way of meeting their tax obligations (that are spent on industry partnerships and subsidies). Science, therefore, consumes enormous amounts of public time and resources; its progress is felt through the flow of capital, user products, and necessary oversight and regulation. It is because of these factors that scientists are accountable to publics: the public good, therefore, refocuses progress upon what the public needs, or expects, from investments in bioscience. These arguments become more pressing when the technology is as significant as potentially gene editing is. If science is answerable to the public, then there might be an expectation that there are good reasons for commodification. In that case, arguments for the exclusivity for CRISPR might be contentious because of the public interest in public goods and the ways in which markets cause mischief in this respect: the patterns of hype and hope and limited access contribute in ways that are public bads. The public good requires a revaluation of progress so that science justifies investment and rewards, by maximising social progress via promoting pathways in which better medicines lead to better health, and those benefits are reasonably accessible. For instance, perhaps by acknowledging the interdependent paths of thought and discovery, we would become more prudent when rewarding serendipitous discoveries, and, moreover, question industries that often require secrecy and delay dissemination [14]. In these respects, benefit sharing and genomic solidarity acknowledge that discoveries happen, not just because society venerates and rewards innovators, but because their discoveries stand on the shoulders of those that contribute to worthwhile aspects of society.

Institutions as giants

There are case examples that can usefully show how marketisation effects social progress. We have already seen how the architects of the IHGSC believed approaching the project as a public good was the most effective way of deciphering the human genome and making sure it reached as many as possible. Their efforts are unambiguous in underlining the importance of the capacity for public innovation, and yet, in the hubbub of entrepreneurship, the public contribution is easily overlooked. In fact, because of the HGP, individual innovators stand to benefit from these kinds of gigantic and

collective knowledge-producing institutions. Thus, we might think of public research as a broad partnership in which information is shared between institutions, researchers, participants and publics, and this framework signifies the importance of the public good in biosciences.

Examining CRISPR as a broad social phenomenon draws attention to the kinds of institutions that contributed: education (high schools, universities), research and training (research facilities and supervision, as well as public funders), and security and stability (from sophisticated enabling infrastructures up to legal systems). These signify the public infrastructures' role in innovation. Pierre Teilhard de Chardin once referred to these giant, global, intelligent networks as the 'noosphere' (derived from the Greek term vo ς: i.e. 'mind' or 'intellect') [37]: the world-wide network of research facilities, discourses, devices, circuits and repositories. He describes a collective and distributed web of collaborators, working together in order to co-create the technologies and insights needed to address global challenges. In other words, collaboration and technical and resource dependency is necessary, and individual achievement is only possible because of these giant techno-scientific networks. It is apt, then, to recall these giants as part of the gene editing narrative about the public good, benefit sharing and genomic solidarity.

When we shift our focus from the innovators to the research participants, for instance, one perhaps recognises the importance of other active and 'passive' contributors to science. In this respect, there have been some notable developments since the outset of the Genomic Era, especially under the moniker of *big data*, which segues logically from the vastness of the human genome. Big data applies to the creation of extremely large data sets for computational analysis to generate value [38]; these data are sourced from vast, indiscriminate methods of trawling random information for patterns and opportunities. Others create data within the public commons, namely, a data repository or resource that is 'of the people' who voluntarily contribute. In respect to the latter, biobanks have become significant in terms of activating public collaborations in ways that are characterised as 'for the public good' [25]. This public good distinction has considerable impact on governance and the norms that define collection methods and processes to use resources. A strong sense of the public good contributes to 'open science'; conversely, institutions that are pursuing big data for commercial reasons often consolidate and conceal their collections. The latter are the traditional giants of private enterprise, such as the pharmaceutical industry. Their practices for accumulating and sharing these data are much different from the aforementioned public good practices, instead using private business models rather than public engagement

to appropriate and withhold data [39]. Sometimes, that business acumen amounts to capture.[5]

An example of public good capture is illustrated by the Icelandic deCODE health sector database (*qua* biobank). The rise and fall of that biobank is a complicated story of political and scientific intrigue that has been widely documented [40]. In essence, proponents for the deCODE biobank claimed there was a public good in aggregating health records to be used by affiliates of the biobank and those purchasing licences. deCODE had to make the biobank attractive to venture capitalists; to do so, they realised that ownership of the data would be necessary. That business strategy was defended by arguing for a public interest for economic growth and national revitalisation (i.e. investment in scientific infrastructure). Many of the data were amassed legally (after a much debated and enacted law), but without having to obtain the express consent of individuals in Iceland (who could only opt out).

That strategy was successfully challenged in court [41], and this proved ultimately a pivotal moment that exposed the inadequacy of deCODE's 'public good-public interest' rhetoric. From the outset, the argument about the public good was doubted by many in the scientific and medical communities [40]. In defence of the strategy, 'The theme of solidarity, through the idea that deCODE could help keep families together, was invoked to outweigh abstract notions of autonomy, patient–doctor confidentiality, and erosion of scientific integrity' ([40], p. 89). What the 'public good' really meant to deCODE was the embrace of 'naive scientific hype, commercial dominance, and the privatization of common cultural and scientific resources' ([40], p. 100). During the days of the HGP, Celera used similar rhetoric about the 'importance of this information to the entire biomedical research community'; ([20], p. 934) but had no intention of depositing its sequence data into the public GenBank database. The company still intended to restrict public access to their sequence, suggesting that, as Collins' predicted, (paraphrasing) perhaps it is not a sound market strategy to be giving data away for free ([20], p. 935) (also see [24]).

Why is it important to challenge private data acquisition? Firstly, Collins argued that the fruits of the HGP should be kept in the public realm because he believed in facilitating access as widely as possible: the bottom line was progress through collaboration, rather than progress by the bottom dollar. He felt discovery would come from collaboration, and not privatisation and capture. It is worth pointing out in light of the assumption that public-based research is antiquated, that the public IHGSC sequencers not only competed on time and under budget but also published a more comprehensive genome (Celera's speed had come with gaps in its

sequence that had to be filled using the publicly available data) [1], and, most of all, that success ensured that the human genome remained in the public sphere. Secondly, the public good model means that *all* researchers—not just those allied to the IHGSC, Celera, deCODE affiliates or those willing (or able) to afford licencing costs—have access. That becomes an inclusive path to progress rather than the narrow trajectory of consumerism: it opens up exploration beyond profit motivations. Doing so does not morally shut any doors to profit: venture capitalists would still be welcome and encouraged to develop products and derivative technologies, but without being allowed a monopoly on the tools or data.

Consider an exemplar of this model: the UK Biobank. Its participants are not paid, and they receive few direct benefits. Research is not prioritised simply because it likely leads to profit, but it must contribute to the public good. Half a million people voluntarily and enthusiastically took part. Why? Perhaps they understood the purpose of the biobank to be about the advantages of creating a sustainable public resource, and endorsed its intention to provide inclusive access *for the good of all*. In the case of UK Biobank, the Ethics and Governance Council acts as a 'steward' of participants' data and samples, and therefore takes a direct responsibility for their interests [30]. This role is only possible because of a governance framework that incorporates participants' rights as conceived by the public interest and public good, and creates a broad steering role for participants through a vision that is informed by ethics rather than business. Even so, UK Biobank does recognise 'reasonable' patents, which refers to inventions that are 'not used to restrict health-related research and/or access to healthcare anywhere in the world.'[8] Compare that to deCODE's strategy to exploit the enthusiasm of the Icelandic people or Celera's mission. Both had begun with overt economic ideals to capture the public good.

The examples of IHGSC and UK Biobank, we believe, counter the presumptive folly of public ineptitude, and indeed illustrate ethically and economically reasonable arguments to support public goods [42].

The culture of giants

The US National Research Council saw the opportunity for creating a framework to create, manage and coordinate access to the vast amounts of information generated from genomics research, but did not state how this was to happen. Instead, they urged the key players, agencies and institutions to grow into their roles to avoid stifling innovation and adaptation [43]. However, developments in biobanking, where there are clear signs that the public good can be enhanced for inclusive benefit [44], and the tempering of patent claims over human genes,[9] suggest

the possibilities for securing genomics for the public good. Capture, in this respect, has a more general danger: 'There is a significant risk that if certain commercial deals are struck or if public access is somehow limited, there may be a real or perceived sense in which managers have reneged on an implied promise to advance the "public good"' ([45], p. 449). The problem resides in 'the corporate skew of the research agenda' ([45], p. 448). Capture is also an issue of trustworthiness. Holding institutions to be trustworthy is far more significant for those in the public sphere, where publics expect their interests to be respected (such as privacy), than those clear about the priorities of their own private pursuits (like commercialisation) [46, 47]. This can be explained by different sets of values or cultures on display by public and private institutions [30], just as the public good (as we have argued) does not always mean the same thing to an ethicist as it does to an economist.

As we have defined it, the public good means that stakeholders may not agree ultimately on the goal (or fate) of research in terms of shared benefit and solidarity. In this vein, it was written about biobanks that 'competing, but ultimately compatible, interests' of multiple agents often find they share values ([48] p. 9). We would disagree: it might be better said that conflicts between public and private are not inexorable but are culturally engrained; compatibility should certainly not be assumed, and, in fact, scepticism seems warranted.[6] The chief danger is a misconstrued narrative about the public good that conceals a reluctance to be critical of business approaches (or to be less than appreciative of public ones). This is, of course, also a comment on the role of private and state interests in innovation, and the controversial aspects of expanding or shrinking the role of government in innovation (e.g. basic science, research and development) that go beyond the focus of this paper [49].

Instead, our message is short: one should challenge superficial explanations that hold to the benefits of markets without being critical of the processes—even injustices—that are involved [9]. In a culture that does not question such assumptions, there are clear trade-offs, as Olson speculates, 'Perhaps science has assimilated the mores of the "new economy" a bit too readily' ([20], p. 941). We have given examples of the achievements possible when not driven by financial gain; these accomplishments challenge the assumptions about public inefficiencies. Regardless, these assumptions continue to be carefully and advantageously communicated to convince others that innovation is driven by profit. In reality, the public infrastructure is not only essential, but ultimately able to compete, accelerate, and achieve.

In this regard, the rhetoric of hype or hope may be driven by questionable short-term gains (such as a patient's vulnerablity for a last chance cure), but one

should also question the significance of vested interests when a new, expensive drug is marketed to a deliberately impoverished health system. These must be challenged in the context of the long-term damages to socio-political elements rooted in the public sphere.

We caution, therefore, the erosion of the public good that produces, conserves, and preserves resources for current and future generations. The public good may substantiate ideas such as 'open science'—one scheme in this regard is presented by Maynard Olson in this issue of *Human Genomics*, in which 'investigators and small laboratories tap directly into a truly communal resource' as an alternative to the tendency to build (and the connotations of) research behemoths. Our model of genomic solidarity supports this sentiment, although giant infrastructures can be welcomed, and to a degree, are inescapable when studying large publics because 'communal' is defined as providing information to publics but also inviting them to take part *en masse* as *individually informed* participants. A public bad leaves scope for secrecy or hype-oriented misinformation; and is bound to discourage participation, and instead the public become *subjects*. The public good, then, creates space for engagement and information dissemination; and requires an obligation for truthful claims, honest brokering and research integrity. Researchers in both big and more modest institutions equally are bound by the same solidarity. In this respect, if these giants fall—the active structures that support public science—then so would its contributions to the public commons and scientific method. Instead, researchers focused on the shared ideals of Merton's principles of transparency, objectivity, disinterest and scepticism, will be forced into the private vision of 'interested enquiry' and 'secret knowledge' [14]. Merton's principles are still fundamental to the way in which the dialogue between the science complex and society occurs, and they are necessary to ensure that research programmes evolve in a way that societal needs, expectations and concerns can be addressed, and benefits can be produced. It is likely that in market-obsessed environs, scientific integrity becomes eroded. Moreover, market priorities particularly affect the ways in which research endeavours are disseminated to colleagues and publics [50]. We therefore challenge prevailing neoliberal ideas and suggest that continuation of these strategies may evermore rely on hype and hope. Ultimately, that likely will undermines public infrastructure and, inevitably, the cost will be public trust.

Conclusion

While legalisation has tended to dominate social commentaries about CRISPR, we have explored the role of other 'giants' as key players in the innovation of gene editing. In so doing, we have examined the biomedical research complex comprising individuals, institutions and cultures. Although multiple disassociated visions of social, legal and fiscal reality thrive within this complex, the question for us has been whether these separate visions of discovery evoke one ethical paradigm that outclasses another. We do not conceive of our world as one in which markets should be entrusted with making important decisions. Rather, we acknowledge the extent to which market players remain fundamentally dependent on public infrastructures and previous efforts made by (often anonymous, often public) others. An enquiry is urgently required to see whether forgoing these contributions through marketisation could in fact stifle or already has stifled progress.

The current world is one in which both public and private players have their place, and their supporters and detractors. If there is such a thing as an ideal community, there is unlikely to be a single guiding vision of innovation. However, careful study of the specific incremental contributions of individuals, institutions and networks within science, and how social and economic ideals affect them, will allow us to articulate the values that lead to truly prosperous science—not just economic gains, but also the augmentation of discoveries that may fundamentally change peoples' lives. CRISPR is already a technical 'disruptor' [51]. We should consider now how its potential should be turned into a 'health disruptor'. Otherwise, it may become another an unrealised promise, another invention encumbered by hype and hope. In this respect, public interest rhetoric is entangled with market agendas and, arguably, still dominates the ways in which many parties think about doing successful research—yet, the examples of UK Biobank and the IHGSC, used here, should be a warning to those acquiescent to that stilted dogma.

In the end, the single, major giant that supports biomedical sciences, including genomics, is 'the public' or, realistically, a constellation of all publics. We have articulated the concept of the public good as a solidary, a community that finds it worthy to protect the shared interests of research. This is what HUGO meant when it articulated 'genomic solidarity', with the public and scientists as joint owners in discovery and opportunity [29]. If genomic research becomes infused with the public good—one in which participants and scientists stand together for a common purpose that benefits humanity—then that is something to nurture to provide a viable and sustainable alternative to purely commercial research.

Here, our aim has been to open up avenues for deliberation for progress in genomic research. What is required now is a differential investigation into the roles of public and private efforts to create and translate basic science into public benefits via a ethical framework that does not instigate hype and hope as a 'tried and trusted'

mechanism to drive innovation. There are a number of challenges for translating genomics for the public good, which can be brought together under three headings:

1. Conceptual—How can these challenges be practically framed within a conception of the public good?
2. Scientific—How can this framework create trust, promote progress and encourage investment in science?
3. Political and social—How can different agents (private and public) work within, and promote the goals of, this framework?

It is important, however, not only to investigate these issues and how they can be addressed but also to examine the indicators of scientific success (metrics) and the mechanisms that best reward fruitful research. The concept of genomic solidarity allows us to measure the extent to which genomics is or can be harnessed for the public good, so that both the public and the scientists share the benefits and opportunities.

Endnotes

[1]The first human draft genome cost at 'least $500 million', and a genome sequence in 2016 was below $1000); see https://www.genome.gov/sequencingcosts/

[2]Heath's reference is to the term 'catallaxy' used by Friedrich Hayek (via Ludwig von Mises) to describe spontaneous market order that would come about between different economies. In effect, Hayek believed that trade is fundamentally a pattern of mutually beneficial interactions

[3]Paraphrasing from Umberto Eco's 'Dicebat Bernardus Carnotensis', foreword in: [5] p. pxiii

[4]See https://www.genome.gov/sequencingcosts/

[5]Capture is the tendency for 'private interests' to appropriate public goods for their market value. It explains how market disinterest in worthless goods becomes a newfound interest, often by way of a spark that ignites a flurry of profiteering hubbub. A public good suddenly becomes a marketisation imperative. It might be as simple as turning a public road into a toll road; or as complex as sequencing a (very long) reference genome and claiming that being the first to do so rightly transfers ownership of the 'human genome' [25]

[6]Knoppers, the then Chair of the HUGO Ethics Committee, suggested that patents would not be incompatible with 'common heritage' as long as consent and contracts were valid. Below, we discuss one of the examples used by Knoppers to strengthen our difference to her claim, noting that in fact deCODE's efforts, rather than justifying 'moving beyond historical distrust' of industry [26], became an example of why scepticism and suspicion are well placed

[7]*UNESCO Universal Declaration on Bioethics and Human Rights* (2005): Article 13—solidarity and cooperation: 'Solidarity among human beings and international cooperation towards that end are to be encouraged'

[8]Section 3.8. The Biobank's Access Procedures also offer up a remedy: if 'UK Biobank considers that an Unreasonable Restriction exists or is likely to exist, then it shall promptly notify the Applicant, and automatically, on receipt of such notification, the Applicant shall be deemed to grant a perpetual, irrevocable, worldwide, fully paid-up, royalty-free, fully sub-licensable licence to UK Biobank to use such Applicant-Generated Invention in order to remove or mitigate the Unreasonable Restriction'; (s. B8) http://www.ukbiobank.ac.uk/wp-content/uploads/2011/11/Access_Procedures_Nov_2011.pdf

[9]See Myriad case: cDNA could be patented, but not human genomic DNA in its natural form; *Assoc. for Molecular Pathology v. Myriad Genetics, Inc.* 569 U.S. (2013).

Acknowledgements

All authors are members of the Human Genome Organisation's Committee on Ethics, Law and Society. The opinions herein do not necessarily represent those of HUGO.

Funding

None.

Authors' contributions

BC conceived of the idea for this paper and led its drafting. All authors took part in drafting, revising and reviewing the content. All authors read and approved the final manuscript.

Consent for publication

Not applicable.

Competing interests

The authors declare that they have no competing interests.

Author details

[1]Department of Bioethics, Faculty of Medicine, Dalhousie University, Halifax, Canada. [2]School of Law, University of Manchester, Manchester, UK. [3]Department of Human Genetics, Centre of Genomics and Policy, McGill University, Québec, Canada. [4]Department of Pediatrics, University of Oklahoma Health Sciences Center, Oklahoma, USA. [5]Centre for Biomedical Ethics, Yong Loo Lin School of Medicine, National University of Singapore, Singapore, Singapore. [6]Faculty of Science, Department of Philosophy and Science Studies, Radboud University Nijmegen, Nijmegen, The Netherlands.

References

1. Shreeve J. The genome war: how Craig Venter tried to capture the code of life and save the world. New York: Random House; 2005.
2. Collins F, Green E, Guttmacher A, Guye M. A vision for the future of genomics research. Nature. 2003;422:835–47.
3. Chadwick R. Zwart H. Editorial: from ELSA to responsible research and promisomics. Life sciences, society and policy 2013; doi:https://doi.org/10.1186/2195-7819-9-3.

4. Braude P, Minger S, Warwick R. Stem cell therapy: hope or hype? BMJ. 2005; 330:1159–60.

5. Merton R. On the shoulders of giants: a Shandean postscript: the post-Italianate edition. Chicago: University of Chicago Press; 1965. (1985; 1993).

6. Fox E, Reid-Bayliss K, Emond M, Loeb L. Accuracy of next generation sequencing platforms. Next Gener Seq Appl. 2014;1:1000106.

7. Lysaght T, Kerridge I, Sipp D, Porter G, Capps B. Ethical and regulatory challenges with autologous adult stem cells: a comparative review of international regulations. J Bioeth Inq. 2017;14:261–73.

8. Egelie K, Graff G, Strand S, Johansen B. The emerging landscape patent landscape of CRISPR-Cas gene editing technology. Nat Biotechnol. 2016; 34:1025–31.

9. Heath J. The benefits of cooperation. Philos Public Aff. 2016;34:313–51.

10. Heller M, Eisenberg R. Can patents deter innovation? The anticommons in biomedical research. Science. 1998;280:698–701.

11. Begley S. Broad institute prevails in heated dispute over CRISPR patents. STAT 2017. https://www.statnews.com/2017/02/15/crispr-patent-ruling/. Accessed 25 Aug 2017.

12. Ratner H, Sampson T, Weiss D. Overview of CRISPR-Cas9 biology. Cold Spring Harb Protoc. 2016. doi:10.1101/pdb.top088849.

13. Brinegar K, Yetisen A, Choi S, Vallillo E, Ruiz-Esparza G, et al. The commercialization of genome-editing technologies. Crit Rev Biotechnol. 2017; https://doi.org/10.1080/07388551.2016.1271768.

14. Capps B. The funding of medical research by industry: can a good tree bring forth evil fruit? Br Med Bull. 2016;118:5–15.

15. Mirowski P. Science-Mart: privatizing American science. Harvard: Harvard University Press; 2011.

16. Collins F. Has the revolution arrived? Nature. 2010;464:674–5.

17. Venter C. Multiple personal genomes await. Nature. 2010;464:676–7.

18. Hayden E. Human genome at ten: life is complicated. Nature. 2010;464:664–7.

19. Zwart H. Francis Collins: the language of life. Book review. Genomics. Soc Policy. 2011;6:67–76.

20. Olson M. The human genome project: a player's perspective. J Mol Biol. 2002;319(931-942):931.

21. Olson M. A time to sequence. Science. 1995;270:394–6.

22. International Human Genome Sequencing Consortium, Lander E, et al. Initial sequencing and analysis of the human genome. Nature. 2001;409:860–921.

23. Venter C, et al. The sequence of the human genome. Science. 2001;291:1304–51.

24. Subcommittee on Energy and Environment of the Committee on Science. The human genome project: how private sector developments affect the government program: hearing before the subcommittee… U.S. House of Representatives, One Hundred Fifth Congress, second session, June 17. Washington: U.S. G.P.O; 1998.

25. Capps B. Public goods in the ethical reconsideration of research innovation. In: Capps P, Pattinson S, editors. Ethical rationalism and the law. Oxford: Hart Publishing; 2016. p. 149–69.

26. Knoppers B. Sovereignty and sharing. In: Caulfield T, Williams-Jones B, editors. The commercialisation of genetic research: ethical, legal and policy issues. New York: Kluwer Academic/Plenum Publishers; 1999. p. 1–11.

27. HUGO Ethics Committee. 2000. Statement on benefit-sharing April 9. http://www.hugo-international.org/Resources/Documents/CELS_Statement-BenefitSharing_2000.pdf.

28. Summary of Principles Agreed Upon at the First International Strategy Meeting on Human Genome Sequencing (Bermuda, 25-28 February 1996) as reported by HUGO: http://web.ornl.gov/sci/techresources/Human_Genome/research/bermuda.shtml.

29. Mulvihill J, et al. Ethical issues of CRISPR technology and gene editing. Br Med Bull. 2017;122:109–122.

30. Capps B. Defining variables of access to UK Biobank: the public interest and the public good. Law Innov Technol. 2013;5:113–39.

31. Samuelson P. The pure theory of public expenditure. Rev Econ Stat. 1954;36:387–9.

32. Capps B. Redefining property in human body parts: an ethical enquiry in the stem cell era. In: Akabayashi A, editor. The future of bioethics: international dialogues. Oxford: Oxford University Press; 2014. p. 235–63.

33. Kaebnick G, Murray T, editors. Synthetic biology and morality: artificial life and the bounds of nature. Cambridge: MIT Press; 2013.

34. Comfort N. Genome editing: that's the way the CRISPR crumbles. Nature. 2017;546:30–1.

35. Lander E. The Heroes of CRISPR. Cell. 2016;164:18–28.

36. Zwart H. The nobel prize as a reward mechanism in the genomics era: anonymous researchers, visible managers and the ethics of excellence. J Bioeth Inq. 2010;7:299–312.

37. Teilhard de Chardin P. The human phenomenon (transl. Sarah Appleton-Weber). Eastbourne: Sussex Academic Press. [Le Phénomène humain. Œuvres 1. Paris: Editions du Seuil]; 1955; 2003.

38. Zwart H. The obliteration of life: depersonalisation and disembodiment in the terabyte age. New Genet Soc. 2016;35:69–89.

39. Sterckx S, Cockbain J, Howard H, Huys I, Borry P. 'Trust is not something you can reclaim easily': patenting in the field of direct-to-consumer genetic testing. Genet Med. 2013;15:382–7.

40. Winickoff D. Genome and nation: Iceland's health sector database and its legacy. Innovations. 2006;1:80–105.

41. Meyer M. Icelandic supreme court holds that inclusion of an individual's genetic information in national database infringes on the privacy interests of his child. Guðmundsdóttir v. Iceland, no 151/2003 (Nov. 27, 2003) (ice.). Harvard Law Rev. 2004;118:810–7.

42. Caulfield T, Gold E, Cho M. Patenting human genetic material: refocusing the debate. Nat Rev Genet. 2000;1:227–31.

43. Committee on a Framework for Developing a New Taxonomy of Disease. Board on life sciences, division on earth and life studies. Toward precision medicine: building a knowledge network for biomedical research and a new taxonomy of disease. National Academies Press: Washington, DC; 2011.

44. Capps B. Models of biobanks and implications for reproductive health innovation. Monash Bioeth Rev. 2015;33:238–57.

45. Winickoff D. Partnership in UK Biobank: a third way for genomic property. J Law Med Ethics. 2007;35:440–56.

46. Hoeyer K, Olofsson B, Mjörndal T. The ethics of research using biobanks: reason to question the importance attributed to informed consent. Arch Intern Med. 2005;165:97–100.

47. Lipworth W, Morrell B, Irvine R, Kerridge I. An empirical reappraisal of public trust in biobanking research: rethinking restrictive consent requirements. J Law Med. 2009;17:119–32.

48. Chalmers D, Nicol D, Kaye J, Bell J, Campbell A, et al. Has the biobank bubble burst? Withstanding the challenges for sustainable biobanking in the digital era. BMC Med Ethics. 2016;17:39.

49. Musgrave R. The theory of public finances: a study in public economy. New York: McGraw-Will; 1959.

50. Lundh A, Lexchin J, Mintzes B, Schroll J, Bero L. Industry sponsorship and research outcome. Cochrane Database Syst Rev. 2017; https://doi.org/10.1002/14651858.MR000033.pub3.

51. Ledford H. CRISPR, the disruptor. Nature. 2015;522:20–4.

13

Beyond genomics: understanding exposotypes through metabolomics

Nicholas J. W. Rattray[1], Nicole C. Deziel[1], Joshua D. Wallach[2,3], Sajid A. Khan[4,5], Vasilis Vasiliou[1,5], John P. A. Ioannidis[6,7,8,9,10] and Caroline H. Johnson[1,5*] (iD)

Abstract

Background: Over the past 20 years, advances in genomic technology have enabled unparalleled access to the information contained within the human genome. However, the multiple genetic variants associated with various diseases typically account for only a small fraction of the disease risk. This may be due to the multifactorial nature of disease mechanisms, the strong impact of the environment, and the complexity of gene-environment interactions. Metabolomics is the quantification of small molecules produced by metabolic processes within a biological sample. Metabolomics datasets contain a wealth of information that reflect the disease state and are consequent to both genetic variation and environment. Thus, metabolomics is being widely adopted for epidemiologic research to identify disease risk traits. In this review, we discuss the evolution and challenges of metabolomics in epidemiologic research, particularly for assessing environmental exposures and providing insights into gene-environment interactions, and mechanism of biological impact.

Main text: Metabolomics can be used to measure the complex global modulating effect that an exposure event has on an individual phenotype. Combining information derived from all levels of protein synthesis and subsequent enzymatic action on metabolite production can reveal the individual exposotype. We discuss some of the methodological and statistical challenges in dealing with this type of high-dimensional data, such as the impact of study design, analytical biases, and biological variance. We show examples of disease risk inference from metabolic traits using metabolome-wide association studies. We also evaluate how these studies may drive precision medicine approaches, and pharmacogenomics, which have up to now been inefficient. Finally, we discuss how to promote transparency and open science to improve reproducibility and credibility in metabolomics.

Conclusions: Comparison of exposotypes at the human population level may help understanding how environmental exposures affect biology at the systems level to determine cause, effect, and susceptibilities. Juxtaposition and integration of genomics and metabolomics information may offer additional insights. Clinical utility of this information for single individuals and populations has yet to be routinely demonstrated, but hopefully, recent advances to improve the robustness of large-scale metabolomics will facilitate clinical translation.

Keywords: Chemometrics, Exposome, Exposotype, Genomics, Genetic epidemiology, Metabolomics

Background

The main concepts underpinning genetic epidemiology developed rapidly after the delineation of the structure of DNA. Neel and Schull provided the first description of these concepts in 1954 [1, 2]. While the original goal of genetic epidemiology was to understand the nature of population and familial genetic inheritance, it soon became evident that environmental factors and gene-environment interactions were important to consider simultaneously [3].

Currently, the study of the whole genome (genomics) has evolved into a multidisciplinary area of science with highly diverse applications [4, 5]. Improved efficiency of genome technology combined with a sharp decrease in cost has enabled genomic assessments in large study populations [6, 7] using genotyping and next-generation-sequencing (NGS) approaches [8]. Thousands of genome-wide association

* Correspondence: caroline.johnson@yale.edu
[1]Department of Environmental Health Sciences, Yale School of Public Health, Yale University, New Haven, CT, USA
[5]Yale Cancer Center, Yale University School of Medicine, New Haven, CT, USA
Full list of author information is available at the end of the article

studies (GWAS) have tracked relationships between base-pair/gene patterns in genomic loci and hundreds of diseases or exposures [9]. However, the discovered loci from these large-scale studies still explain only the minority of presumed heritability for most phenotypes of interest [10]. Moreover, it has been established that genes alone account for the minority of disease etiology for many important illnesses such as cancer, and environmental and lifestyle influences play a critical role [11]. However, quantifying the myriad of environmental and lifestyle risk factors including diet, smoking, exposure to hazardous chemicals, and pathogenic microorganisms is challenging [12, 13]. An individual can be exposed to a complex mix of chemical and biological contaminants, with multiple sources, for varying durations across their life course. This concept has been termed the "exposome," a framework for the collective analysis, and measurement of an individual's exposures over their lifetime [14]. Moreover, different environmental exposures may be heavily correlated with each other or may act in concert to produce adverse effects, which makes studying them one at a time challenging for assigning causality [15]. Therefore, it is essential to find tools that can measure

the cumulative impact of multiple exposures alongside their interactions with the genetic background of individuals. Several multidimensional analytical approaches have been developed, beyond genomics, that try to capture different aspects of this complexity, and their integration into environmental health is discussed in this review.

Application of high-dimensional biology to the environmental health paradigm

Referred to as high-dimensional biology, or a multi-omics/systems-level approach, the combined analysis of data from the genome (genomics), RNA transcription (transcriptomics), proteins/peptides (proteomics), and metabolites (metabolomics) enables researchers to overlay gene information onto complementary datasets towards a more systemic understanding of diseases or other phenotypes of interest [16]. The complexity of high-dimensional datasets becomes even more convoluted when the interaction of environmental exposures is added to the system.

The environmental health paradigm (Fig. 1) integrates the knowledge of exposures and environmental health

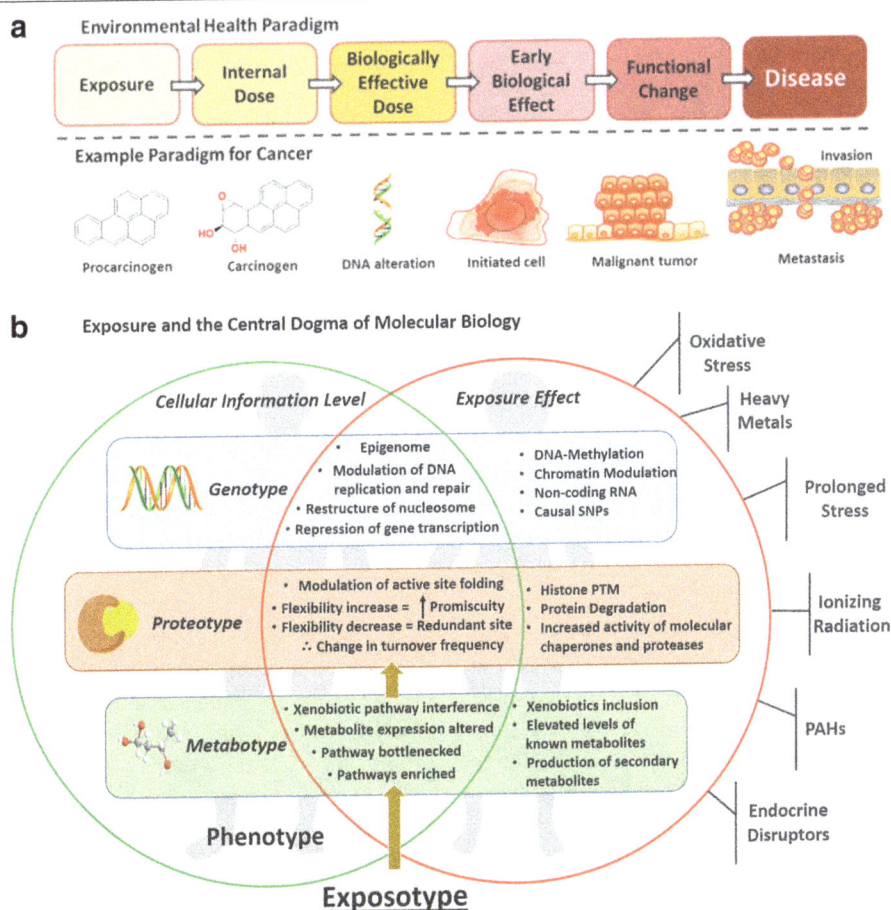

Fig. 1 a Environmental health paradigm. b Exposure and the central dogma of molecular biology

sciences to gain a deeper understanding of the consequences of exposure towards expression of a disease phenotype [17]. Exposures can elicit subtle effects at different stages of gene-encoding, protein synthesis, and on circulating metabolites. Multi-omics approaches using combined data from genomics, proteomics, and metabolomics techniques can identify downstream chemical alterations contributing to the development of an exposotype, the exposure phenotype (Fig. 1), that describes the accrued biological changes within a system that has undergone a specific exposure event [18]. Combining information from all levels of protein synthesis and subsequent enzymatic action on metabolite production is an essential step to start comprehending the complex global modulating effect that an exposure event has on an individual phenotype. This may allow for a greater direct understanding of molecular mechanisms that underpin the route of exposure, and the effect of molecular transit on different areas of metabolism, cellular reproduction, and ultimately the resulting exposotype.

Metabolites are the substrates and products of metabolism that drive essential cellular processes such as energy production, and signal transduction [19]. Of all the molecular entities (genes, transcripts, proteins, metabolites), metabolites have the closest relationship to expressed phenotype as they are the final end-points of upstream biochemical processing. Quantitative readouts of metabolite abundance reflect both this cellular processing and xenobiotics (foreign substances such as environmental chemicals, pollutants, drugs, food additives, dyes) that are physico-chemically distinct from molecular entities that originate in the host. Xenobiotics can be processed by enzymatic machinery, and metabolomics also allows quantification of these metabolites. Therefore, metabolomics can simultaneously analyze both exogenous chemicals and their metabolites, and changes to the endogenous metabolome, to allow assessment of broadly defined exposures and their biological impact [20–23]. One such example was a recent study of occupational exposure to trichloroethylene (TCE) [24]. TCE metabolites were identified in human plasma and associated with changes to endogenous metabolites that were known to be involved in immunosuppression, hepatotoxicity, and nephrotoxicity. This allowed the investigation into how the toxic effects of TCE exposure were manifested [24]. Another study, from the EXPOsOMICS project (http://www.exposomicsproject.eu/), examined human biofluids and exhaled breath for exposure to swimming pool disinfection by-products (DBPs) and for concomitant changes to endogenous metabolites. The study revealed a possible association between DBPs and perturbations to metabolites in the tryptophan pathway [25]. However, these studies and others which have measured exposures in relation to the metabolome highlight the challenge of attempting to unravel the effect of one circumscribed exposure versus combinations of different environmental exposures on the metabolome [26, 27].

One of the major bottlenecks of metabolomics is metabolite identification. However, the expansion and development of metabolite databases have eased this issue. Tens of thousands of metabolites have been identified and uploaded onto metabolite databases such as The Human Metabolome Database (HMDB) (http://www.hmdb.ca/metabolites), which to date houses 114,113 metabolites with associated chemical, clinical, and biochemical information. HMDB also hosts four additional databases including the Toxic Exposome Database (T3DB) (http://www.t3db.ca/) which contains information on 3763 toxins [28, 29]. METLIN (https://metlin.scripps.edu), another large database containing 961,829 metabolites, recently expanded due to the integration of xenobiotics from the United States Environmental Protection Agency's "Distributed Structure-Searchable Toxicity (DSSTox)" database [30, 31]. The Exposome-Explorer database was recently designed to contain information on biomarkers of exposure to environmental risk factors for diseases. This database has information on 692 dietary and pollutant biomarkers, and importantly concentration values measured in biospecimens, with correlation values to assess quality of the biomarkers [32]. These databases, and others that house both xenobiotics and endogenous metabolites, appear in Table 1 [33–38]. With the recent expansion of these databases to include xenobiotics, metabolomics can facilitate both biomonitoring of exposures, assessment of biological impact, and identification of exposotypes [39]. However, one potential gap in these databases still exists, the prediction of phase I and phase II biotransformed metabolites of xenobiotics which can be used as proxy biomarkers for the chemical exposure. Metabolomics has revealed numerous novel metabolites of previously well-characterized pharmaceutical drugs such as acetaminophen [40], dietary supplements [41], and the genotoxic heterocyclic amine 2-amino-1-methyl-6-phenylimidazo[4,5-b]pyridine (PhIP) [42], present in meats cooked at high temperatures. Metabolomics provides a window to identifying these new metabolites, as the biotransformed metabolite will only be present in a sample from an exposed individual. Secondly, there is typically more than one biotransformation metabolite present for each xenobiotic, which will have a similar covariance and correlation within the biological sample examined, thus making it possible to easily map out the related metabolites. One way to overcome this gap in the metabolite databases would be to have a tool housed on these databases that could automatically predict any potential biotransformations, and display the resultant

Table 1 Mass spectrometry metabolite databases for identification of environmental exposures

Database name	Description	URL
Human metabolome database (HMDB)	114,113 xenobiotic and endogenous metabolites with chemical, biochemical, and clinical information.	http://www.hmdb.ca/ [33]
Toxic exposome database (T3DB)	3767 toxic compounds, targets and gene expression data, part of the HMDB suite.	http://www.t3db.ca/ [28]
METLIN	961,829 xenobiotic and endogenous metabolites with chemical information. Contains information from DSSTox.	https://metlin.scripps.edu/ [34]
Exposome-Explorer	692 dietary and pollutant biomarkers, with concentration values measured from biospecimens with intra class correlation coefficients.	http://exposome-explorer.iarc.fr/ [32]
Madison-Qingdao Metabolomics Consortium Database	20,300 xenobiotics and endogenous metabolites, with chemical information	http://mmcd.nmrfam.wisc.edu/ [35]
Drugbank	10,513 drug entries with drug target information, part of the HMDB suite	https://www.drugbank.ca/ [36]
PubChem	93,977,784 compounds, xenobiotic and endogenous metabolites but also peptides, and chemically altered macromolecules. Data is derived from hundreds of sources.	https://pubchem.ncbi.nlm.nih.gov/ [37]
CompTox Chemistry Dashboard	758,000 xenobiotics with chemical information compiled from multiple sources; PubChem, and US EPA's DSSTox, ACToR, ToxCast, EDSP21, and CPCat.	https://comptox.epa.gov/dashboard [38]

important chemical information for identification. A few tools currently available for predicting phase I and II drug metabolism have been recently reviewed, along with the development of "DrugBug" which can predict xenobiotic metabolism by human gut microbiota enzymes [43]. Integration of such tools would facilitate exposome analysis.

The broad range of chemical classes that exist among the thousands of endogenous and environmentally derived metabolites contained within a biological sample has given rise to the need for analytical strategies that can separate and detect as much chemical diversity as possible from within the biological system under examination. The assessment of all metabolites present in a sample, untargeted metabolomics, is typically carried out using chromatography-based mass spectrometry and/or nuclear magnetic resonance spectroscopy, alongside bioinformatics that help understand the complex data generated [44]. Metabolomics research has undergone significant refocus over the past few years due to the improvements made in bioanalytical protocols and an evident shift towards the development of new chemoinformatic and bioinformatic tools [45]. These tools are designed to improve metabolite identification, particularly for microbial metabolites, and biological interpretation, which remain a major challenge for the field. For example, the mass spectrometry data generated in a metabolomics study have a high degree of degeneracy where the same metabolite can be represented as multiple signals [46]. Tools such as CAMERA [47], RAMClust [48], and "Credentialing" [49] have helped overcome this problem and improve peak annotation. Other notable tools include CSI:FingerID [50] which predicts the fragmentation of metabolites using an in silico

method, thus aiding in metabolite identification, and "integrated-omics" housed on XCMSOnline [51] (http://xcmsonline.scripps.edu/) which aids in both metabolite identification and biological interpretation. Excellent reviews on the technological advancements in this area can be found elsewhere [52–54]; in addition, an extensive list of all current metabolomics software and data analysis resources is available [55, 56]. For population-level studies, the application of metabolomics for the analysis of thousands of samples has been optimized and demonstrated [57, 58], but the field could still benefit from decades' worth of research and lessons learning in genetic epidemiology related to study design, statistical analyses, and reproducibility in large-scale population consortia.

Methodological challenges and considerations

Relevant and a priori formulated research questions and rigorous study designs and methods lay the foundation to perform a potentially successful piece of population-based research, after which replication is essential to confirm any associations, and to avoid the dissemination of potentially false research claims [59–61]. Prospective cohort studies follow a predefined population over time, capturing exposure information prior to occurrence of health events. This study design accommodates the appropriate temporal relationship between exposure and outcome, allows for testing of multiple risk factors and health outcomes, and permits collection of multiple pre-clinical biological specimens throughout the follow-up period. Although this is ideal from a metabolomics perspective, this study design often requires long follow-up durations and great expense. Case-control studies can be more efficient, and less expensive ways to test associations, but they

lack the temporality criterion for causality, and metabolic profiles may be influenced by disease status. The use of nested case-control studies offers an efficient approach with the appropriate temporality between exposure and outcome. "Meet-in-the-middle" approaches, which involve linking intermediate biomarkers to both the exposure and outcome within cohort and nested case-control studies, are gaining popularity for their ability to reveal important linkages along the exposure-outcome pathway [62, 63].

While systems-level approaches hold great promise, they also pose challenges in the analysis of high-dimensional, complex data structure. The use of appropriate statistical tests within genomics, metabolomics, and epidemiology is dictated by the study design and the number of dimensions of data under investigation, with the application of univariate or multivariate techniques being applied to low-dimensional and high-dimensional datasets, respectively. Incorrect analytical decisions and interpretations that are made when conducting a study are a direct threat to reproducibility [64]. Table 2 [65–87] provides a list of some of the most commonly used statistical methods and tests in the interface of epidemiology, genetics, and metabolomics.

Many analyses in metabolomics involve the use of null hypothesis significance testing (NHST) and the reporting of p values. The p value, one of the most misused statistics in science [88], has not escaped the focus of members of the fields of epidemiology [89], metabolomics [90], and general biomedicine [91]. Poor application has contributed to the irreproducible nature of many studies, so much that the American Statistical Association felt moved to release a statement highlighting six underlying principles to dictate the proper use and interpretation of the p value [92, 93]. One should examine in each application whether NHST is best suited as an inferential tool or whether alternative approaches, such as the use of Bayesian methods or false discovery rates (FDR), are preferable [90, 94–96]. If p values are still used in multidimensional experiments, proper correction for multiplicity is important. There are numerous methods for accommodating family-wise error rates [90]. There are also some standard thresholds that can be used in specific settings, e.g., genome-wide significance $p < 5 \times 10^{-8}$ for genome-wide analyses. Some multiplicity corrections are more conservative than others; for instance, the Bonferroni correction (dividing the p value threshold required for significance by the number of tests performed) may be too conservative [97]. FDR and variants of FDR may be better suited [96] and can accommodate correlation structures between the multiple tested variables [98, 99].

Several methods are available that can help reduce complexity, detect trends, and generate predictive models within multidimensional datasets (Table 2) such as those generated by NGS and mass spectrometry when target genes or metabolites are not known. Unsupervised

methods such as principal component analysis (PCA) provide an initial step to help reduce the complexity and indicate variables of interest by determining discriminant features linked to the "loadings" of different clusters. These loadings can be considered as the impact that a certain variable has on measured variance, so a high-level loading value displays a strong influence on clustered groups [100]. There also exist several extensions of the PCA architecture such as multiblock PCA, consensus PCA, or ANOVA-PCA that enable the user to control for underlying influential factors within datasets such as the intra-patient variability or other experimental confounders [65]. These approaches have been used for metabolomics and genetics analyses and also lend themselves to other cross-validation methods [66]. Supervised methods apply grouping stratification to the data based on some already known outcome variable(s). They aim to develop models that can accurately predict the correct grouping based on the input and identify genes, metabolites, or other statistical associations that underlie the grouping. The most commonly used methods are variants of regression tools (Table 2). Regression modeling can identify associations relevant to the disease [101], can predict association within gene expression patterns [102], and in metabolomics [103] can generate sample classification. However, as these tests are supervised, one of the issues with multivariate regression is that it tends to over-fit the data. Therefore, cross-validation (in the same dataset) and external validation (in additional datasets) are essential.

Perhaps, the biggest challenge yet for exposome researchers is integration of the multiple types of data generated from systems-level analyses and assessing the role of one versus multiple exposures on the phenotype. Currently, there are platforms that enable biochemical pathway analysis and integration of systems-level data, and these platforms can identify pathways and networks that are related to a known exposure or health outcome (such as disease). Dissection of pathways may help direct mechanistic studies into causality. The most useful to date for untargeted metabolomics data is "mummichog," which uses computational algorithms to predict metabolic pathway effects directly from spectral feature tables without prior identification of metabolites [104]. Mummichog was recently integrated onto the XCMSOnline platform, with an added function to upload transcriptomic and proteomic data, for integrated pathway analysis [51]. Other notable software includes MarVis-Pathway [105], InCroMAP [106], GAM [107], and MetaCore™ (Thomson Reuters Corporation, Toronto, Canada) that can integrate multiple types of systems-level data for pathway interrogation. Combining this type of data with multiple measurements of xenobiotics has not yet been demonstrated, but tools are under development. Up to now, studies have

Table 2 Common statistical methods and tests used in epidemiology, genetics, and metabolomics, with reference link to descriptive articles on appropriate general use

Class of test	Type of test	Application/description	Refs
Descriptive	Mean Median Mode	The simplest of tests used to describe basic features within data.	Covered in all general statistical textbooks and used in most if not all scientific disciplines. [67–69]
	Range, variance, SD	Describe spreads of data within a population	
Inferential	z test, t test, chi-square	Predicts/infers an observed mean, frequency, or proportion to a predetermined value, respectively.	
	ANOVA	Parametric method that tests the hypothesis that the means of two or more populations are equal. Frequently used to compare variance among groups relative to variance within groups	
	Kruskal-Wallis	Non-parametric method to rank statistical significant differences between two or more groups of an independent variable on a continuous/ordinal variable	
Scaling	Centering, auto, pareto, log, MD	Data pretreatment methods aim at reducing biological and analytical bias	[70, 71]
Principal component	PCA	Unsupervised dimensional reduction procedure used to explain the maximum variance within complex datasets.	[72–74]
	Multiblock PCA	PCA extension designed to find the underlying relationships between sets of related data	[65, 66, 75]
	ANOVA-PCA	Uses PC dimensional reduction to determines the effect of the experimental factors on multiple dependent variables	[65, 76]
	PC-DFA	Supervised test that summarizes the differentiation between groups while overlooking within-group variation.	[65, 77, 78]
Regression	Linear	Summarizes and quantifies the relationship between two continuous variables	[72, 79]
	PLS	Used to predict a set of dependent variables from a large set of independent variables	[73, 77, 80–82]
	O-PLS	orthogonal signal correction on PLS that maximizes the explained covariance on the first latent variable	[77, 81, 83]
	PLS-R	Combination of the predictive power of regression alongside the ability to deal with high dimensionality and multicollinearity of variables.	[77, 84]
	PLS-DA	Supervised approach to prediction on discrete variables	[77, 79, 83]
	LASSO	Parsimonious approach to variable selection and regularization in order to enhance interpretability and reduce noise	[79, 80, 85–87]
	Elastic net	Variable reduction approach where strongly correlated predictors coalesce in or out of the model together	[79, 80, 85, 87, 167]

Definitions: *SD* standard deviation, *MD* median, *PCA* principal component analysis, *ANOVA* analysis of variance, *PC-DFA* principal component discriminant function analysis, *PLS* partial least squares (also known as projection of latent structures), *O-PLS* orthogonal PLS, *PLS-R* PLS regression, *LASSO* least absolute shrinkage and selection operator

primarily assessed the effect of individual exposures and have combined multiple systems-level approaches to assess biological response (i.e., benzene exposure and toxicity, susceptibility genes, mRNA and DNA methylation) [108]. Phenome data has also been integrated into studies to account for population variability and reduce false positives [22]. A recent example, from the analysis of preterm birth in the Rhea mother-child cohort study, selected those metabolites that had significant association with birth outcomes in logistic regression models and significant correlation coefficients with metabolic syndrome traits to construct odds ratios (BMI, blood pressure, blood

glucose) [109]. Moreover, new tools are being specifically designed with the exposome in mind; xMWAS can integrate metabolomics data with that derived from the transcriptome [110], microbiome [111], and cytokine [112] and can be used for genome, epigenome, proteome, and other integrated omics analyses. However, modeling the effect of combined exposures is extremely complex. Co-exposures can be linked and cause an additive effect on the biological outcome, but it is not possible to know beforehand which combinations of exposures may have the largest biological effect. A recent novel method was developed that first estimates the correlation between

pairs of exposures, then groups the highly correlated exposures by unsupervised machine learning [26], and identifies co-occurring exposure networks. This technique reduces the total number of combinations of exposures to "prevalent co-occurring combinations"; however, integration with other systems-level data still remains very complex. The additional challenges associated with integrating exposome data with metabolomics, genomics, and proteomics have been recently reviewed [27] and were also highlighted in a recent symposium report [113].

Analytical bias and biological variance in metabolomics analyses for epidemiologic studies

Metabolomics analyses in epidemiologic studies require additional consideration of sources of variability beyond traditional epidemiologic studies. There are a very large number of chemical features that can be detected by current highly sensitive mass spectrometers, and differences in metabolite recovery may arise from biological samples that are not collected under identical protocols. Additional batch variation can be introduced when handling large sample numbers [114], due to contaminant build-up and sample degradation [115].

Analytical bias in genomics and metabolomics can arise from practical laboratory aspects that, by their nature, favor the preselection of one type of variable (single nucleotide polymorphism (SNP) or chemical) over another. This is particularly evident when performing "untargeted" analyses in which the researcher is looking to maximize chemical coverage with a technology that cannot cover the full chemical space. With currently over 24 million SNPs having been documented within the human genome [116], the technology within SNP microarray chips has yet to catch up to this depth of coverage. The same issues are also

present within metabolomics as no single technology can analyze the thousands of different metabolites within a sample. Therefore, pre-selecting approaches are commonly applied, be it using a gene-expression chip predefined for a subset of SNPs [117–120] or untargeted chromatography methods for metabolomics with a restricted spectrum of which metabolites can be captured [121]. These analytical biases are described in Fig. 2, but include the type of metabolite extraction method and column chemistry, which can enhance the analysis of some chemical functional groups and classes over others. For example, reversed-phase liquid chromatography (RPLC) can effectively analyze non-polar compounds such as lipids, carnitines, and bile acids, whereas hydrophilic interaction liquid chromatography (HILIC) is more suitable for the analysis of polar metabolites such as nucleotides, sugars, and amino acids. The two column chemistries have an analytical overlap of only 34%; thus, both column chemistries are needed if one wishes to obtain a relative quantification of the broadest chemical classes from a sample [122]. All types of study design need to consider inherent biological intra-individual variability as a potential source of variation (Fig. 2) as well as a source of discriminatory features. In addition to understanding and addressing potential methodological challenges and various sources of biases, open science practices are necessary to support the subsequent verification of research and use of the obtained data and results in subsequent secondary analyses and meta-analyses.

Moving from genome-wide association studies (GWAS) to metabolome-wide association studies (MWAS)

One of the most-used study approaches in big data genome research, first demonstrated in 2005, is GWAS

Biological Factors

Accounting for the high biological background arising from inter-subject variability is an essential part of experimental design in metabolomics. Other sources of variability can include:

- Dietary effects
- Body mass index
- Age
- Sex
- Subjects at high risk of disease
- Subjects at high risk of exposure
- Observer selection
- Timing of subject sampling
- Gene/Environmental effects
- Circadian/Temporal effects

Analytical Factors

Sample preparation and analytical procedures used within metabolomics experiments vary greatly across labs. Sources of variability that should be considered for study design include:

- Collection and storage procedures
- Between sample and batch variance
- Between instrumental comparisons
- Lack of global 'Gold Standard' methodology driven by constantly evolving instrumentation (annually) from multiple vendors.
- Selection of extraction solvents
- Selection of column chemistry
- Selection of eluent composition
- Ionization technique and source geometry
- Signal processing algorithms
- Compound library

Fig. 2 The biological and analytical aspects of bias and variance that can lead to a tendency towards erroneous results in both untargeted and targeted metabolomics

[123]. This technique examines genome-wide sets of genetic variants in samples of individuals to determine if any variants are associated with a trait and help pinpoint genes that may contribute to a person's risk for a certain disease or other phenotype of interest. GWAS can be described as an untargeted and sometimes a hypothesis-generating approach to associate genetic variants with specific phenotypes. GWAS and consortia-based meta-analyses have been conducted with increasing sample size [124], allowing for improved power [125] to detect genome-wide significant signals for what are typically very small effect sizes. Due to the analytical uniformity of sequencing, this is one area where genomic research has advanced more quickly than metabolomics.

Most of the early untargeted metabolomics experiments have had limited sample sizes ($n = 10$–100) often a result of technological, run-time, and statistical limitations. Given the large number of metabolic features that are typically generated by untargeted metabolomics (typically 1000s for liquid chromatography mass spectrometry), using such small sample sizes has led to overfitting of data and spurious results [100]. Moreover, the highly collinear nature of metabolomics multivariate data [67] have not generally been properly factored in performing a priori power and sample size calculations, and there is no widely accepted method for sample size determination in metabolomics. In the absence of specific metabolic target hypothesis, one can use a data driven sample size determination (DSD) algorithm [126] where sample size estimation depends on the purpose of the study: whether it aims to find at least one statistically significant variation (biomarker discovery) or a maximum of statistically significant variations (metabolic exploration). Alternatively, one may adapt methods that have been developed for use with microarray gene expression(s) [127–129]. One common problem is that there is often high correlation between variables in one dataset, and in addition, not all variables have the same power. However, new more promising approaches have been generated using multivariate simulation to deal with this type of data structure [130].

Predictive power increases with sample size, and the current application of metabolomics to larger longitudinal cohort studies ($n > 1000$) is helping to give access to broader population data that can be linked to specific exposure such as alcohol [131, 132]. These types of studies are needed to improve biomarker discovery and inference of molecular mechanisms. Key issues continuously arise in the application of metabolomics to human subjects which can be overcome by putting metabolomics into epidemiological context. Common problems include causal and mechanistic claims based on differences between groups that have low numbers of individuals, lack of longitudinal data to avoid the possibility of

reverse causation (a health outcome influencing pharmacokinetics and metabolite concentrations), limited information on lifestyle, socioeconomic and other influences, and the lack of multiple statistical tests and biological replication [133]. As metabolomics is incorporated into more population-level studies, it may be possible to more reliably model potential associations of metabolic profiles with phenotypes. The goal is to stratify metabolic data over exposure event data and ultimately determine the related disease risk. Confounding associations may still distort results and lead to erroneous conclusions. Yet it is more readily possible, with larger study numbers, and longitudinal testing, to control confounding by matching samples in to related sub-groups such as age, sex, or level-of-exposure.

Metabolome-wide association studies (MWAS) were first described in 2008 as the capture of "environmental and genomic influences to investigate the connections between phenotype variation and disease risk factors" [134, 135], thus helping reveal the complex gene-environment interactions on disease outcome. The method differs from conventional metabolomics in that high-throughput metabolomics is applied to large-scale epidemiologic studies at the population level and uses specialized algorithms to maximize the identification of biomarkers of disease risk [57]; for example, a recent algorithm was developed to correct for multiple testing using a permutation-based method to derive a metabolome-wide significance level controlling the family-wise error rate [136]. Initial studies showed that using high-throughput metabolomics, MWAS can be carried out on large population cohorts to provide individual metabolic phenotypes (metabotypes), and metabolic biomarkers correlated to exposures [137], and/or biological outcomes [138]. The proof-of-principle study used to coin the term MWAS identified discriminatory biomarkers of blood pressure and cardiovascular risk in 4630 individuals [138]. These types of studies may point to otherwise unknown features of the disease etiology or pathophysiology, which may be used to lead further mechanistic studies and potentially new avenues for therapeutic design, although the complexity of mechanisms makes such translation to therapeutic discovery very difficult. Comparison of metabotypes at the human population level can identify a signature of metabolites statistically correlated to disease risk and/or an exposure. Recent studies have shown the application of MWAS to identify metabolites correlated with cardiovascular events in a dietary intervention trial [139]. In another study, trimethylamine N-oxide (TMAO) was identified as a biomarker predictive of cardiovascular disease risk [140, 141] and was also shown to be involved in the production of atherosclerotic plaques. This discovery has resulted in a clinical test for TMAO, Cleveland Heart-Lab, and is the first to provide this blood test, and therapeutics are currently being designed to inhibit TMAO

production as well as recommendations for dietary changes. Another application is to identify the enrichment of metabolites within specific biochemical pathways [142] to aid in the identification of genes and proteins/enzymes that may be related to the mechanism of disease. This method has gained traction within drug evaluation studies [143] trying to obtain more comprehensive understanding of individual responses to drug therapy [144, 145]. This application may be particularly useful for the design of immunotherapeutics where metabolites have been shown to modulate autoimmunity and can be targeted to improve the efficacy of these drugs [146, 147]. However, it should be acknowledged that therapeutic discovery or improvement in therapeutic management with known interventions has not yet been accomplished using metabolomics data; however, recent development in metabolomics technologies in both the bioanalytical and chemometric components is markedly improving, and thus, there is optimism for clinical translation as well.

Transparency, reproducibility, and open science

There is growing recognition of the need for improved transparency, reproducibility, and replication in the biomedical literature [64, 91, 148, 149]. With respect to multidimensional, big data analyses, transparency can be improved with the sharing of data, protocols, and analytical codes. Furthermore, the number of metabolomics studies that investigate reproducibility across multiple research centers are few in number, and ongoing interlaboratory efforts have struggled to generate metabolite data that is both accurate and reproducible across different labs [150]. Replication has been accepted as a sine qua non in certain disciplines, such as human genome epidemiology [149], and the same should apply across all multidimensional fields using big data. However, the research community is aware of this issue, and groups are convening to provide solutions to address this problem. For example, the European Centre for Ecotoxicology and Toxicology of Chemicals have provided a framework to facilitate the regulatory applicability and use of big data in chemical risk assessment [151, 152].

It is also important to protect inferences from data dredging/p-hacking (mining datasets prior to specifying a causal hypothesis), and unaccounted multiple comparisons in complex datasets that can lead to the inflation of false-positive rates. Therefore, to improve the reproducibility of metabolomics, it is necessary to understand certain methodological and statistical challenges, to protect against analytical biases and biological variance, and to promote transparency and open science. These open science practices, which include "the process of making the content and process of producing evidence and claims transparent and accessible to other researchers" [64], can increase the credibility of research. For metabolomics in particular, both raw and metadata are essential to facilitate reproducibility, secondary analyses, and the synthesis of evidence by external metabolomics researchers [153]. Several measures can support the transparency and reproducibility of metabolomics. For maximal impact, the whole metabolomics research community should adopt and adhere to standards that promote the uniform preparation of study results. The metabolomics standards initiative (MSI), which was conceived in 2005 by the Metabolomics Society, highlights a range of minimum reporting standards covering biological [154], chemical [155], analytical, and data reporting methods [156] within the metabolomics experimental pipeline. However, ideally, metabolomics funders, reviewers, editors, and journals should require researchers to share their protocols, raw data, and analytical code. Broadly speaking, this does not happen (the Springer Journal *Metabolomics* (https://link.springer.com/journal/11306) and MDPI journal *Metabolites* (http://www.mdpi.com/journal/metabolites) being notable exceptions in which MSI compliance is asked for from authors and assessed by reviewers). Currently, most journals leave the suitability of metabolite submission data to reviewer and editor discretion.

Support is also beginning to appear from some funding bodies to help improve the reliability and efficiency of metabolomics. For example, the Data Repository and Coordination Center, which is part of the United States National Institutes of Health (NIH) Common Fund's Metabolomics Program, has created the Metabolomics Data Repository. All NIH Common Fund Metabolomics Program supported research projects which create metabolomics data as part of the funded research are required to submit all raw data (e.g., spectrometric, spectrographic, and chromatographic data) and metadata (e.g., details on how samples were obtained and the analytical methods that were used) to the repository [157]. In addition, the European Union funded data repository MetaboLights (http://www.ebi.ac.uk/metabolights/) has already assembled data from 317 metabolomics studies as of December 2017. Common data submission formats, such as *mzML/mzXML* for mass spectrometry, *nmrML* for NMR data, and ISA-Tab format for metadata, have helped to unify this process [158, 159]. But the research community must be careful to not generate an excess of unconnected data repositories. Multiple and potentially overlapping repositories could confuse researchers as to where they should submit their data and therefor limit the chance of uniform acceptance and adoption of standards. To this end, the COSMOS project (COordination of Standards in MetabOlomicS—http://www.cosmos-fp7.eu/) has been designed to address the challenges of e-infrastructure diversity in metabolomics by developing an interface that globally links community projects and output.

The predominant reason behind the lack of data sharing in metabolomics is the complexity and lack of standardization in the data generated. For research areas such as genomics, transcriptomics, and, to a lesser extent, proteomics, the chemistry of the molecules under detection is highly symmetrical. Regardless of nucleobase-pair connectivity, DNA and RNA constructs can be detected and typed using highly reproducible sequencing chips that can work in a high-throughput manner. The sheer range of molecular chemistries available within the human metabolome demand a multitude of separation strategies when mass spectrometry is used as the detection technology. Consequently, different research groups align their experimental pipelines to one of the many instrument vendors (often dictated by geography and cost) leading to a multitude of protocols that cover all aspects of experimentation. Just within the confines of liquid chromatography mass spectrometry-based metabolomics, 84% use open source software and/or commercial software from instrument vendors, and within the open source software group, the majority use XCMS, and a smaller percentage use MZmine and MZmine 2. Therefore, variability in just the data processing limits integration of the MSI. One way to enable standardized data processing and biostatistics is to encourage the use of a universal workflow platform such as Galaxy (https://galaxyproject.org) [160]. In addition, the use of a standard reference material that can normalize and compare the detection levels from different instruments would be of value. A concerted effort is still needed by the community to enable broader reproducibility [161]. The lack of standardization and reporting is preventing the validation of metabolomics research [162].

Conclusions

Human populations are exposed to a complex mix of chemicals and toxicants, from multiple sources, for varying durations. These exposures are affecting the health of the global population dramatically, for example, over seven million premature deaths annually linked to air pollution exposure alone [163]. It is vital that a more comprehensive understanding of how these environmental exposures affect biology at the systems level to determine cause, effect, and susceptibility. In doing so, a compound specific "exposotype" can be developed that accounts for the totality of the multileveled downstream biological changes that an individual exposure event produces [18]. To better understand these effects, metabolomics can be used to develop not only metabolic biomarkers of exposure but can also be used to build metabolic models that identify upstream genetic and enzymatic changes. This may complement GWAS studies as knowledge of a potential enzymatic mutation can narrows down the DNA search space needed to identify relevant SNPs linked to the exposure [144, 145].

In-depth biological data generated by metabolomics can be used to enhance exposure studies by supplying information not only on directly affected metabolic pathways but also on off-target metabolic effects. The value of metabolomics to identify gene-environment interactions lends itself to the study of the exposome and will be the most complex and important integration of metabolomics to date. Further characterization of gene variants associated with those metabolic pathways could help forecast disease prevalence by either using pre-diagnostic metabolic signatures (collections of metabolites that change prior to disease onset) and genetic risk data. Therefore, preventive measures may be tailored specifically for those individuals. The combination of metabolomics with genomics offers one tool that may prove helpful towards materializing precision medicine. Success in precision medicine has been difficult to achieve [164], but the recent US Food and Drug Administration approval of pembrolizumab, a "tumor-agnostic" therapeutic which targets any solid tumor with a specific genetic feature, shows that the field is starting to head in that direction [165]. Given recent evidence that non-genomic influences such as the microbiome can influence therapeutic response, metabolomics may be used in this context to identify factors that are related to non-responders and responders [166].

However, some of the caveats that still exist within conventional metabolomics and population studies are still present, such as accurate identification of new metabolites, controlling for multiple levels of confounders, and the integration of different forms of data from different analytical platforms. Further advancement can be made by routine application of appropriate statistical tools to metabolomics as well as the adoption and promotion of transparent and reproducible research practices. Reproducible, transparent advances may then be examined for their impact in changing outcomes in single patients and at the population level to judge their utility.

Abbreviations
FDR: False discovery rate; GWAS: Genome-wide association studies; MSI: Metabolomics standards initiative; MWAS: Metabolome-wide association studies; NGS: Next-generation sequencing; NHST: Null hypothesis significance testing; PCA: Principal component analysis; SNP: Single nucleotide polymorphism

Acknowledgements
Not applicable.

Funding
This work is supported in part by NIH grants EY17963 (VV), AA021724 (VV), and AA022057 (VV) and American Cancer Society (ACS) grant MRSG-15-147-01-CNE (ND).

Authors' contributions
All authors were involved in writing and contributing to the manuscript. All authors read and approved the final manuscript.

Consent for publication
Not applicable.

Competing interests
JDW receives research support through Yale University from the Laura and Arnold Foundation to support the Collaboration for Research Integrity and Transparency.

Author details
[1]Department of Environmental Health Sciences, Yale School of Public Health, Yale University, New Haven, CT, USA. [2]Collaboration for Research Integrity and Transparency (CRIT), Yale Law School, New Haven, CT, USA. [3]Center for Outcomes Research and Evaluation (CORE), Yale-New Haven Health System, New Haven, CT, USA. [4]Department of Surgery, Section of Surgical Oncology, Yale University School of Medicine, New Haven, CT, USA. [5]Yale Cancer Center, Yale University School of Medicine, New Haven, CT, USA. [6]Stanford Prevention Research Center, Department of Medicine, Stanford University, Stanford, CA, USA. [7]Department of Health Research and Policy, Stanford University, Stanford, CA, USA. [8]Department of Biomedical Data Science, Stanford University, Stanford, CA, USA. [9]Department of Statistics, Stanford University, Stanford, CA, USA. [10]Meta-Research Innovation Center at Stanford, Stanford University, Stanford, CA, USA.

References
1. Neel JV, Schull WJ. Human heredity. Chicago: Chicago Press; 1954.
2. DeWan AT. Five classic articles in genetic epidemiology. Yale J Biol Med. 2010;83:87–90.
3. Beaty TH, Khoury MJ. Interface of genetics and epidemiology. EpidemiolRev. 2000;22:120–5.
4. Sanger F, Nicklen S, Coulson AR. DNA sequencing with chain-terminating inhibitors. Proc Natl Acad Sci U S A. 1977;74:5463–7.
5. National Human Genome Research Institute. All about the Human Genome Project (HGP). 2014. Available from: http://www.genome.gov/10001772. Accessed 17 Jan 2018.
6. Pareek CS, Smoczynski R, Tretyn A. Sequencing technologies and genome sequencing. J Appl Genet. 2011;52:413–35.
7. Hayden EC. The $1,000 genome. Nature. 2014;507:295.
8. Goldfeder RL, Wall DP, Khoury MJ, JPA I. Human genome sequencing at population scale: a primer on high throughput DNA sequencing and analysis. Am J Epidemiol. 2017;186:1000–9.
9. Goodwin S, JD MP, WR MC. Coming of age: ten years of next-generation sequencing technologies. Nat Rev Genet. 2016;17:333–51.
10. Manolio TA, Collins FS, Cox NJ, Goldstein DB, Hindorff LA, Hunter DJ, et al. Finding the missing heritability of complex diseases. Nature. 2009;461:747–53.
11. Theodoratou E, Timofeeva M, Li X, Meng X, JPA I. Nature, nurture, and cancer risks: genetic and nutritional contributions to cancer. Annu Rev Nutr. 2017;21:293–320.
12. Willett WC. Balancing life-style and genomics research for disease prevention. Science (80-). 2002;296:695–8.
13. Rappaport SM, Smith MT. Environment and disease risks. Science (80-.). 2010;330:460–1.
14. Wild CP. Complementing the genome with an "Exposome": the outstanding challenge of environmental exposure measurement in molecular epidemiology. Cancer Epidemiol Biomarkers. 2005;14:1847. LP-1850
15. Patel CJ, Ioannidis JPA. Studying the elusive environment in large scale. JAMA. 2014;311:2173–4.
16. Romero R, Espinoza J, Gotsch F, Kusanovic JP, Friel LA, Erez O, et al. The use of high-dimensional biology (genomics, transcriptomics, proteomics, and metabolomics) to understand the preterm parturition syndrome. BJOG. 2006;113:118–35.
17. Wilson SH. Disease-first: a new paradigm for environmental health science research. Environ Health Perspect. 2006;114:2006.
18. Rattray NJW, Charkoftaki G, Rattray Z, Hansen JE, Vasiliou V, Johnson CH. Environmental influences in the etiology of colorectal cancer: the premise of metabolomics. Curr Pharmacol Reports. 2017;3:114–25.
19. Patti GJ, Yanes O, Siuzdak G. Innovation: Metabolomics: the apogee of the omics trilogy. Nat Rev Mol Cell Biol. 2012;13:263–9.
20. Ellis JK, Athersuch TJ, Thomas LD, Teichert F, Perez-Trujillo M, Svendsen C, et al. Metabolic profiling detects early effects of environmental and lifestyle exposure to cadmium in a human population. BMC Med. 2012;10:61.
21. Andra SS, Austin C, Wright RO, Arora M. Reconstructing pre-natal and early childhood exposure to multi-class organic chemicals using teeth: towards a retrospective temporal exposome. Environ Int. 2015;83:137–45.
22. Maitre L, Villanueva CM, Lewis MR, Ibarluzea J, Santa-Marina L, Vrijheid M, et al. Maternal urinary metabolic signatures of fetal growth and associated clinical and environmental factors in the INMA study. BMC Med. 2016;14:1–12.
23. Baker MG, Simpson CD, Lin YS, Shireman LM, Seixas N. Original article the use of metabolomics to identify biological signatures of manganese exposure. Ann Work Expo Heal. 2017;61:406–15.
24. Walker DI, Uppal K, Zhang L, Vermeulen R, Smith M, Hu W, et al. High-resolution metabolomics of occupational exposure to trichloroethylene. Int J Epidemiol. 2016;45:1517–27.
25. van Veldhoven K, Keski-Rahkonen P, Barupal DK, Villanueva CM, Font-Ribera L, Scalbert A, et al. Effects of exposure to water disinfection by-products in a swimming pool: a metabolome-wide association study. Environ Int Elsevier. 2018;111:60–70.
26. Patel CJ. Analytic complexity and challenges in identifying mixtures of exposures associated with phenotypes in the Exposome era. Curr Epidemiol Reports. 2017;4:22–30.
27. Patel CJ, Kerr J, Thomas DC, Mukherjee B, Ritz B, Chatterjee N, et al. Opportunities and challenges for environmental exposure assessment in population-based studies. Cancer Epidemiol Biomarkers Prev. 2017;26:cebp.0459.2017.
28. Wishart D, Arndt D, Pon A, Sajed T, Guo AC, Djoumbou Y, et al. T3DB: the toxic exposome database. Nucleic Acids Res. 2015;43:D928–34.
29. Lim E, Pon A, Djoumbou Y, Knox C, Shrivastava S, Guo AC, et al. T3DB: a comprehensively annotated database of common toxins and their targets. Nucleic Acids Res. 2009;38:781–6.
30. Warth B, Spangler S, Fang M, Johnson CH, Forsberg EM, Granados A, et al. Exposome-scale investigations guided by global metabolomics, pathway analysis, and cognitive computing. Anal Chem. 2017; In-Press
31. Richard AM, Williams CR. Distributed structure-searchable toxicity (DSSTox) public database network: a proposal. Mutat Res Fundam Mol Mech Mutagen. 2002;499:27–52.
32. Neveu V, Moussy A, Rouaix H, Wedekind R, Pon A, Knox C, et al. Exposome-explorer: a manually-curated database on biomarkers of exposure to dietary and environmental factors. Nucleic Acids Res. 2017;45:D979–84.
33. Wishart DS, Tzur D, Knox C, Eisner R, Guo AC, Young N, et al. HMDB: the human metabolome database. Nucleic Acids Res. 2007;35:521–6.
34. Smith CA, O'Maille G, Want EJ, Qin C, Trauger SA, Brandon TR, et al. A metabolite mass spectral database. Ther Drug Monit. 2005;27:747–51.
35. Cui Q, Lewis IA, Hegeman AD, Anderson ME, Li J, Schulte CF, et al. Metabolite identification via the Madison Metabolomics Consortium Database [3]. Nat Biotechnol. 2008;26:162–4.
36. Wishart DS, Knox C, Guo AC, Shrivastava S, Hassanali M, Stothard P, et al. DrugBank: a comprehensive resource for in silico drug discovery and exploration. Nucleic Acids Res. 2006;34:D668–72.
37. Kaiser J. Chemists want NIH to curtail database. Science (80-.). 2005;308:774.
38. Williams AJ, Grulke CM, Edwards J, AD ME, Mansouri K, Baker NC, et al. The CompTox chemistry dashboard: a community data resource for environmental chemistry. J Cheminform. 2017;9:61.
39. Beger RD, Dunn W, Schmidt MA, Gross SS, Kirwan JA, Cascante M, et al. Metabolomics enables precision medicine: "a white paper, community perspective". Metabolomics. 2016;12:149.
40. Chen C, Krausz KW, Idle JR, Gonzalez FJ. Identification of novel toxicity-associated metabolites by metabolomics and mass isotopomer analysis of acetaminophen metabolism in wild-type and Cyp2e1-null mice. J Biol Chem. 2008;283:4543–59.

41. Johnson CH, Krausz KW, Kang DW, Patterson AD, Kim J, Luecke H, et al. Novel metabolites and roles for a-tocopherol in humans and mice discovered by mass spectrometry-based metabolomics 1–5. Am J Clin Nutr. 2012;96:818–30.

42. Chen C, Ma X, Malfatti MA, Krausz KW, Kimura S, Felton JS, et al. A comprehensive investigation of 2-amino-1-methyl-6-phenylimidazo[4,5-b]pyridine (PhIP) metabolism in the mouse using a multivariate data analysis approach. Chem Res Toxicol. 2007;20:531–42.

43. Sharma AK, Jaiswal SK, Chaudhary N, Sharma VK. A novel approach for the prediction of species-specific biotransformation of xenobiotic/drug molecules by the human gut microbiota. Sci Rep. 2017;7:1–13.

44. Gavaghan CL, Holmes E, Lenz E, Wilson ID, Nicholson JK. An NMR-based metabonomic approach to investigate the biochemical consequences of genetic strain differences: application to the C57BL10J and Alpk:ApfCD mouse. FEBS Lett 2000;484:169–174.

45. Johnson CH, Ivanisevic J, Siuzdak G. Metabolomics: beyond biomarkers and towards mechanisms. Nat Rev Mol Cell Biol. 2016;17:451–9.

46. Mahieu NG, Patti GJ. Systems-level annotation of a metabolomics data set reduces 25 000 features to fewer than 1000 unique metabolites. Anal Chem. 2017;89:10397–406.

47. Kuhl C, Tautenhahn R, Böttcher C, Larson TR, Neumann S. CAMERA: an integrated strategy for compound spectra extraction and annotation of liquid chromatography/mass spectrometry data sets. Anal Chem. 2012;84:283–9.

48. Broeckling CD, Afsar FA, Neumann S, Ben-Hur A, Prenni JE. RAMClust: a novel feature clustering method enables spectral-matching-based annotation for metabolomics data. Anal Chem. 2014;86:6812–7.

49. Mahieu NG, Huang X, Chen YJ, Patti GJ. Credentialing features: a platform to benchmark and optimize untargeted metabolomic methods. Anal Chem. 2014;86:9583–9.

50. da Silva RR, Dorrestein PC, Quinn RA. Illuminating the dark matter in metabolomics. Proc Natl Acad Sci. 2015;112:12549–50.

51. Huan T, Forsberg EM, Rinehart D, Johnson CH, Ivanisevic J, Benton HP, et al. Systems biology guided by XCMS online metabolomics. Nat Methods. 2017;14:461–2.

52. Evans AM, DeHaven CD, Barrett T, Mitchell M, Milgram E. Integrated, nontargeted ultrahigh performance liquid chromatography/electrospray ionization tandem mass spectrometry platform for the identification and relative quantification of the small-molecule complement of biological systems. Anal Chem. 2009;81:6656–67.

53. Lankadurai BP, Nagato EG, Simpson MJ. Environmental metabolomics: an emerging approach to study organism responses to environmental stressors. Environ Rev. 2013;21:180–205.

54. Johnson CH, Ivanisevic J, Benton HP, Siuzdak G. Bioinformatics: the next frontier of metabolomics. Anal Chem. 2015;87:147–56.

55. Misra BB, van der Hooft JJJ. Updates in metabolomics tools and resources: 2014-2015. Electrophoresis. 2016;37:86–110.

56. Misra BB, Fahrmann JF, Grapov D. Review of emerging metabolomic tools and resources: 2015–2016. Electrophoresis. 2017;38:2257–74.

57. Chan Q, Loo R, Ebbels T, Van Horn L, Daviglus M, Stamler J, et al. Metabolic phenotyping for discovery of urinary biomarkers of diet, xenobiotics and blood pressure in the INTERMAP study: an overview. Hypertens Res. 2016;40:1–10.

58. Karaman I, Ferreira DLS, Boulangé CL, Kaluarachchi MR, Herrington D, Dona AC, et al. Workflow for integrated processing of multicohort untargeted ^{1}H NMR metabolomics data in large-scale metabolic epidemiology. J Proteome Res. 2016;15:4188–94.

59. Ioannidis J, Allison D, Ball C, Coulibaly I, Cui X, Culhane A, et al. Repeatability of published microarray gene expression analyses. Nat Genet. 2009;41:149–204.

60. Kraft P, Zeggini E, Ioannidis J. Replication in genome-wide association studies. Stat Sci. 2010;24:561–73.

61. Ioannidis JPA. Why most published research findings are false. PLoS Med. 2005;2:0696–701.

62. Chadeau-Hyam M, Athersuch TJ, Keun HC, De Iorio M, TMD E, Jenab M, et al. Meeting-in-the-middle using metabolic profiling—a strategy for the identification of intermediate biomarkers in cohort studies. Biomarkers. 2011;16:83–8.

63. Vineis P, Perera F. Molecular epidemiology and biomarkers in etiologic cancer research: the new in light of the old. Cancer Epidemiol Biomark Prev. 2007;16:1954–65.

64. Munafò MR, Nosek BA, Bishop DVM, Button KS, Chambers CD, Percie du Sert N, et al. A manifesto for reproducible science. Nat Hum Behav. 2017;1:1–9.

65. Xu Y, Goodacre R. Multiblock principal component analysis: an efficient tool for analyzing metabolomics data which contain two influential factors. Metabolomics. 2012;8:37–51.

66. Abdi H, Williams LJ, Valentin D. Multiple factor analysis: principal component analysis for multitable and multiblock data sets. Wiley Interdiscip Rev Comput Stat. 2013;5:149–79.

67. Vinaixa M, Samino S, Saez I, Duran J, Guinovart JJ, Yanes O. A guideline to univariate statistical analysis for LC/MS-based untargeted metabolomics-derived data. Meta. 2012;2:775–95.

68. Sanderson S, Tatt ID, Higgins JPT. Tools for assessing quality and susceptibility to bias in observational studies in epidemiology: a systematic review and annotated bibliography. Int J Epidemiol. 2007;36:666–76.

69. Szklo M, Nieto FJ. Epidemiology: beyond the basics. 3rd Ed. Aspen: Jones & Bartlett Learning; 2000.

70. van den Berg RA, Hoefsloot HCJ, Westerhuis JA, Smilde AK, van der Werf MJ. Centering, scaling, and transformations: improving the biological information content of metabolomics data. BMC Genomics. 2006;7:142.

71. Chin L, Hahn WC, Getz G, Meyerson M. Making sense of cancer genomic data. Genes Dev. 2011;25:534–55.

72. Dohoo IR, Ducrot C, Fourichon C, Donald A, Hurnik D. An overview of techniques for dealing with large numbers of independent variables in epidemiologic studies. Prev Vet Med. 1997;29:221–39.

73. Eriksson L, Antti H, Gottfries J, Holmes E, Johansson E, Lindgren F, et al. Using chemometrics for navigating in the large data sets of genomics, proteomics, and metabonomics (gpm). Anal Bioanal Chem. 2004;380:419–29.

74. DiBello JR, Kraft P, ST MG, Goldberg R, Campos H, Baylin A. Comparison of 3 methods for identifying dietary patterns associated with risk of disease. Am J Epidemiol. 2008;168:1433–43.

75. Westerhuis JA, Kourti T, MacGregor JF. Analysis of multiblock and hierarchical PCA and PLS models. J Chemom. 1998;12:301–21.

76. Zwanenburg G, Huub CJ, Westerhuis JA, Jansen JJ, Smilde AK. ANOVA-principal component analysis and ANOVA-simultaneous component analysis: a comparison. J Chemom. 2011;25:561–7.

77. Gromski PS, Muhamadali H, Ellis DI, Xu Y, Correa E, Turner ML, et al. A tutorial review: metabolomics and partial least squares-discriminant analysis—a marriage of convenience or a shotgun wedding. Anal Chim Acta. 2015;879:10–23.

78. Jombart T, Devillard S, Balloux F, Falush D, Stephens M, Pritchard J, et al. Discriminant analysis of principal components: a new method for the analysis of genetically structured populations. BMC Genet. 2010;11:94.

79. Ogutu JO, Schulz-Streeck T, Piepho H-P. Genomic selection using regularized linear regression models: ridge regression, lasso, elastic net and their extensions. BMC Proc. 2012;6:S10.

80. Acharjee A, Finkers R, Visser RG, Maliepaard C. Comparison of regularized regression methods for ~omics data. Metabolomics. 2013;3:126.

81. Tzoulaki I, Ebbels TMD, Valdes A, Elliott P, JPA I. Design and analysis of metabolomics studies in epidemiologic research: a primer on-omic technologies. Am J Epidemiol. 2014;180:129–39.

82. Abdi H. Partial least squares (PLS) regression. Encycl Res Methods Soc Sci. 2003;2003:792–5.

83. Bylesjo M, Rantalainen M, Cloarec O, Nicholson JK, Holmes E, Trygg J. OPLS discriminant analysis: combining the strengths of PLS-DA and SIMCA classification. J Chemom. 2006;20:3541–351.

84. Wold S, Sjöström M, Eriksson L. PLS-regression: a basic tool of chemometrics. Chemom Intell Lab Syst. 2001;58:109–30.

85. Waldron L, Pintilie M, Tsao MS, Shepherd FA, Huttenhower C, Jurisica I. Optimized application of penalized regression methods to diverse genomic data. Bioinformatics. 2011;27:3399–406.

86. Tibshirani R. Regression selection and shrinkage via the lasso. J R Stat Soc B. 1996;58:267–88.

87. Vaarhorst AAM, Verhoeven A, Weller CM, Böhringer S, Göraler S, Meissner A, et al. A metabolomic profile is associated with the risk of incident coronary heart disease. Am Heart J. 2014;168:45–52. e7

88. Baker M. Statisticians issue warning over misuse of P values. Nature. 2016;531:151.

89. Greenland S, Senn SJ, Rothman KJ, Carlin JB, Poole C, Goodman SN, et al. Statistical tests, P values, confidence intervals, and power: a guide to misinterpretations. Eur J Epidemiol Springer Netherlands. 2016;31:337–50.

90. Broadhurst DI, Kell DB. Statistical strategies for avoiding false discoveries in metabolomics and related experiments. Metabolomics. 2006;2:171–96.

91. Chavalarias D, Wallach JD, Li AHT, Ioannidis JPA, Gigerenzer G, Berlin J, et al. Evolution of reporting P values in the biomedical literature, 1990–2015. JAMA. 2016;315:1141.

92. The American Statistical Association. Statement on statistical significance and P-values. 2016;

93. Wasserstein RL, Lazar NA. The ASA's statement on p-values: context, process, and purpose. Am Stat. 2016;70:129–33.

94. Chong EY, Huang Y, Wu H, Ghasemzadeh N, Uppal K, Quyyumi AA, et al. Local false discovery rate estimation using feature reliability in LC/MS metabolomics data. Sci Rep. 2015;5:17221.

95. Sugimoto M, Kawakami M, Robert M, Soga T, Tomita M. Bioinformatics tools for mass spectroscopy-based metabolomic data processing and analysis. Curr Bioinforma. 2012;7:96–108.

96. Benjamini Y, Hochberg Y. Controlling the false discovery rate: a practical and powerful approach to multiple testing. J R Stat Soc B. 1995:289–300.

97. McDonald JH. Handbook of biological statistics. Baltimore: Sparky House Publishing; 2015.

98. Efron B. Size, power and false discovery rates. Ann Stat. 2007;35:1351–77.

99. Efron B. Microarrays, empirical Bayes and the two-groups model. Stat Sci. 2008;23:1–22.

100. Bartel J, Krumsiek J, Theis FJ. Statistical methods for the analysis of high-throughput metabolomics data. Comput Struct Biotechnol J. 2013;4: e201301009

101. Lewis FI, Ward MP. Improving epidemiologic data analyses through multivariate regression modelling. Emerg Themes Epidemiol. 2013;10:2–11.

102. Zapala MA, Schork NJ. Multivariate regression analysis of distance matrices for testing associations between gene expression patterns and related variables. Proc Natl Acad Sci U S A. 2006;103:19430–5.

103. Saccenti E, Hoefsloot HCJ, Smilde AK, Westerhuis JA, MMWB H. Reflections on univariate and multivariate analysis of metabolomics data. Metabolomics. 2014;10:361–74.

104. Li S, Park Y, Duraisingham S, Strobel FH, Khan N, Soltow QA, et al. Predicting network activity from high throughput metabolomics. PLoS Comput Biol. 2013;9

105. Kaever A, Landesfeind M, Feussner K, Mosblech A, Heilmann I, Morgenstern B, et al. MarVis-pathway: integrative and exploratory pathway analysis of non-targeted metabolomics data. Metabolomics. 2015;11:764–77.

106. Wrzodek C, Eichner J, Büchel F, Zell A. InCroMAP: integrated analysis of cross-platform microarray and pathway data. Bioinformatics. 2013;29:506–8.

107. Sergushichev AA, Loboda AA, Jha AK, Vincent EE, Driggers EM, Jones RG, et al. GAM: a web-service for integrated transcriptional and metabolic network analysis. Nucleic Acids Res. 2016;44:W194–200.

108. Zhang L, CM MH, Rothman N, Li G, Ji Z, Vermeulen R, et al. Systems biology of human benzene exposure. Chem Biol Interact. 2010;184:86–93.

109. Maitre L, Fthenou E, Athersuch T, Coen M, Toledano MB, Holmes E, et al. Urinary metabolic profiles in early pregnancy are associated with preterm birth and fetal growth restriction in the Rhea mother-child cohort study. BMC Med. 2014;12:1–14.

110. Roede JR, Uppal K, Park Y, Tran VL, Jones DP. Transcriptome-metabolome wide association study (TMWAS) of maneb and paraquat neurotoxicity reveals network level interactions in toxicologic mechanism. Toxicol Rep. 2014;1:435–44.

111. Cribbs SK, Uppal K, Li S, Jones DP, Huang L, Tipton L, et al. Correlation of the lung microbiota with metabolic profiles in bronchoalveolar lavage fluid in HIV infection. Microbiome. 2016;4:1–11.

112. Chandler JD, Hu X, Ko E-J, Park S, Lee Y-T, Orr ML, et al. Metabolic pathways of lung inflammation revealed by high-resolution metabolomics (HRM) of H1N1 influenza virus infection in mice. Am J Physiol Regul Integr Comp Physiol. [Internet]. 2016;ajpregu.00298.2016.

113. Johnson CH, Athersuch TJ, Collman GW, Dhungana S, Grant DF, Jones DP, et al. Yale school of public health symposium on lifetime exposures and human health: the exposome; summary and future reflections. Hum Genomics. 2017;11:32.

114. Wang SY, Kuo CH, Tseng YJ. Batch normalizer: a fast total abundance regression calibration method to simultaneously adjust batch and injection order effects in liquid chromatography/time-of-flight mass spectrometry-based metabolomics data and comparison with current calibration met. Anal Chem. 2013;85:1037–46.

115. Reisetter AC, Muehlbauer MJ, Bain JR, Nodzenski M, Stevens RD, Ilkayeva O, et al. Mixture model normalization for non-targeted gas chromatography/mass spectrometry metabolomics data. BMC Bioinformatics. 2017;18:84.

116. NCBI dbSNP Database - www.ncbi.nlm.nih.gov/projects/SNP. Accessed 6 Nov 2017.

117. Sims D, Sudbery I, Ilott NE, Heger A, Ponting CP. Sequencing depth and coverage: key considerations in genomic analyses. Nat Rev Genet. 2014;15:121–32.

118. Ramasamy A, Mondry A, Holmes CC, Altman DG. Key issues in conducting a meta-analysis of gene expression microarray datasets. PLoS Med. 2008;5:1320–32.

119. Stefano GB. Comparing bioinformatic gene expression profiling methods: microarray and RNA-Seq. Med Sci Monit Basic Res. 2014;20:138–42.

120. Siddiqui AS, Delaney AD, Schnerch A, Griffith OL, Jones SJM, Marra MA. Sequence biases in large scale gene expression profiling data. Nucleic Acids Res. 2006;34:e84.

121. Büscher JM, Czernik D, Ewald JC, Sauer U, Zamboni N. Cross-platform comparison of methods for quantitative metabolomics of primary metabolism. Anal Chem. 2009;81:2135–43.

122. Ivanisevic J, Zhu ZJ, Plate L, Tautenhahn R, Chen S, O'Brien PJ, et al. Toward 'omic scale metabolite profiling: a dual separation-mass spectrometry approach for coverage of lipid and central carbon metabolism. Anal Chem. 2013;85:6876–84.

123. Klein RJ, Zeiss C, Chew EY, Tsai JY, Sackler RS, Haynes C, et al. Complement factor H polymorphism in age-related macular degeneration. Science (80-.). 2005;308:385–9.

124. Panagiotou OA, Willer CJ, Hirschhorn JN, Ioannidis JPA. The power of meta-analysis in genome-wide association studies. Annu Rev Genomics Hum Genet. 2013;14:441–65.

125. Hong EP, Park JW. Sample size and statistical power calculation in genetic association studies. Genomics Inform. 2012;10:117–22.

126. Blaise BJ. Data-driven sample size determination for metabolic phenotyping studies. Anal Chem. 2013;85:8943–50.

127. Van Iterson M, 't Hoen PAC, Pedotti P, Hooiveld GJ, Den Dunnen JT, van Ommen GJ, et al. Relative power and sample size analysis on gene expression profiling data. BMC Genomics. 2009;10:439.

128. Ferreira JA, Zwinderman A. Approximate power and sample size calculations with the Benjamini-Hochberg method. Int J Biostat. 2006;2:1–36.

129. Langaas M, Lindqvist BH, Ferkingstad E. Estimating the proportion of true null hypotheses, with application to DNA microarray data. J R Stat Soc Ser B Stat Methodol. 2005;67:555–72.

130. Blaise BJ, Correia G, Tin A, Young JH, Vergnaud AC, Lewis M, et al. Power analysis and sample size determination in metabolic phenotyping. Anal Chem. 2016;88:5179–88.

131. Jaremek M, Yu Z, Mangino M, Mittelstrass K, Prehn C, Singmann P, et al. Alcohol-induced metabolomic differences in humans. Transl Psychiatry. 2013;3:e276.

132. Homuth G, Teumer A, Völker U, Nauck M. A description of large-scale metabolomics studies: increasing value by combining metabolomics with genome-wide SNP genotyping and transcriptional profiling. J Endocrinol. 2012;215:17–28.

133. Mäkinen V-P, Ala-Korpela M. Metabolomics of aging requires large-scale longitudinal studies with replication. Proc Natl Acad Sci 2016; 113:E3470–E3470.

134. Nicholson JK, Holmes E, Elliott P. The metabolome-wide association study: a new look at human disease risk factors. J Proteome Res. 2008;7:3637–8.

135. Chadeau-Hyam M, Ebbels TM, Brown IJ, Chan Q, Stamler J, et al. Metabolic profiling and the metabolome-wide association study: significance level for biomarker identification. J Proteome Res. 2010;9:4620-7.

136. Castagné R, Boulangé CL, Karaman I, Campanella G, Santos Ferreira DL, Kaluarachchi MR, et al. Improving visualization and interpretation of metabolome-wide association studies: an application in a population-based cohort using untargeted 1 H NMR metabolic profiling. J Proteome Res. 2017;16:3623–33.

137. Walker DI, Pennell KD, Uppal K, Xia X, Hopke PK, Utell MJ, et al. Pilot Metabolome-Wide Association Study of Benzo(a)pyrene in Serum From Military Personnel. J Occup Environ Med. 2016;58:S44-52.

138. Bictash M, Ebbels TM, Chan Q, Loo RL, Yap IKS, Brown IJ, et al. Opening up the "black box": metabolic phenotyping and metabolome-wide association studies in epidemiology. J Clin Epidemiol Elsevier Inc. 2010;63:970–9.

139. Toledo E, Wang DD, Ruiz-Canela opez M, Clish CB, Razquin C, Zheng Y, et al. Plasma lipidomic profiles and cardiovascular events in a randomized intervention trial with the Mediterranean diet. Am J Clin Nutr. 2017;106:973–83.

140. Li XS, Obeid S, Klingenberg R, Gencer B, Mach F, Räber L, et al. Gut microbiota-dependent trimethylamine N-oxide in acute coronary syndromes: a prognostic marker for incident cardiovascular events beyond traditional risk factors. Eur Heart J. 2017;14:814–24.

141. Wang Z, Klipfell E, Bennett BJ, Koeth R, Levison BS, Dugar B, et al. Gut flora metabolism of phosphatidylcholine promotes cardiovascular disease. Nature. 2011;472:57–63.
142. Igari M, Alexander JC, Ji Y, Qi XL, Papke RL, Bruijnzeel AW. Varenicline and cytisine diminish the dysphoric-like state associated with spontaneous nicotine withdrawal in rats. Neuropsychopharmacology. 2014;39:445–55.
143. Renier N, Adams EL, Kirst C, Wu Z, Azevedo R, Kohl J, et al. Mapping of Brain Activity by Automated Volume Analysis of Immediate Early Genes. Cell. 2016;165:1789–802.
144. Gupta M, Neavin D, Liu D, Biernacka J, Hall-Flavin D, Bobo WV, et al. TSPAN5, ERICH3 and selective serotonin reuptake inhibitors in major depressive disorder: pharmacometabolomics-informed pharmacogenomics. Mol Psychiatry. 2016;21:1717–25.
145. Ji Y, Hebbring S, Zhu H, Jenkins GD, Biernacka J, Snyder K, et al. Glycine and a glycine dehydrogenase (GLDC) SNP as citalopram/escitalopram response biomarkers in depression: pharmacometabolomics-informed pharmacogenomics. Clin Pharmacol Ther. 2011;89:97–104.
146. Kepp O, Loos F, Liu P, Kroemer G. Extracellular nucleosides and nucleotides as immunomodulators. Immunol Rev. 2017;280:83–92.
147. Johnson CH, Spilker ME, Goetz L, Peterson SN, Siuzdak G. Metabolite and microbiome interplay in cancer immunotherapy. Cancer Res. 2016;76:6146–52.
148. Ioannidis JPA, Greenland S, Hlatky MA, Khoury MJ, Macleod MR, Moher D, et al. Increasing value and reducing waste in research design, conduct, and analysis. Lancet. 2014;383:166–75.
149. Iqbal SA, Wallach JD, Khoury MJ, Schully SD, JPA I. Reproducible research practices and transparency across the biomedical literature. PLoS Biol. 2016;14:1–13.
150. Siskos AP, Jain P, Römisch-Margl W, Bennett M, Achaintre D, Asad Y, et al. Interlaboratory reproducibility of a targeted Metabolomics platform for analysis of human serum and plasma. Anal Chem. 2017;89:656–65.
151. Buesen R, Chorley BN, da Silva Lima B, Daston G, Deferme L, Ebbels T, et al. Applying 'omics technologies in chemicals risk assessment: report of an ECETOC workshop. Regul Toxicol Pharmacol. 2017:1–11.
152. Kauffmann HM, Kamp H, Fuchs R, Chorley BN, Deferme L, Ebbels T, et al. Framework for the quality assurance of 'omics technologies considering GLP requirements. Regul Toxicol Pharmacol. 2017;91:1–9.
153. Sud M, Fahy E, Cotter D, Azam K, Vadivelu I, Burant C, et al. Metabolomics workbench: an international repository for metabolomics data and metadata, metabolite standards, protocols, tutorials and training, and analysis tools. Nucleic Acids Res. 2016;44:D463–70.
154. Griffin JL, Nicholls AW, Daykin CA, Heald S, Keun HC, Schuppe-Koistinen I, et al. Standard reporting requirements for biological samples in metabolomics experiments: mammalian/in vivo experiments. Metabolomics. 2007;3:179–88.
155. Sumner LW, Amberg A, Barrett D, Beale MH, Beger R, Daykin CA, et al. Proposed minimum reporting standards for chemical analysis: chemical analysis working group (CAWG) metabolomics standards initiative (MSI). Metabolomics. 2007;3:211–21.
156. Goodacre R, Broadhurst D, Smilde AK, Kristal BS, Baker JD, Beger R, et al. Proposed minimum reporting standards for data analysis in metabolomics. Metabolomics. 2007;3:231–41.
157. Metabolomics workbench - www.metabolomicsworkbench.org/. Accessed 17 Jan 2018.
158. Schober D, Jacob D, Wilson M, Cruz JA, Marcu A, Grant JR, et al. nmrML: a community supported open data standard for the description, storage, and exchange of NMR data. Anal Chem. 2017. In-Press
159. Rocca-Serra P, Salek RM, Arita M, Correa E, Dayalan S, Gonzalez-Beltran A, et al. Data standards can boost metabolomics research, and if there is a will, there is a way. Metabolomics. 2016;12:1–13.
160. Weber RJM, Lawson TN, Salek RM, Ebbels TMD, Glen RC, Goodacre R, et al. Computational tools and workflows in metabolomics: an international survey highlights the opportunity for harmonisation through galaxy. Metabolomics. 2017;13:1–5.
161. Salek RM, Steinbeck C, Viant MR, Goodacre R, Dunn WB. The role of reporting standards for metabolite annotation and identification in metabolomic studies. Gigascience. 2013;2:13.
162. van Rijswijk M, Beirnaert C, Caron C, Cascante M, Dominguez V, Dunn WB, et al. The future of metabolomics in ELIXIR. F1000Research. 2017;6:1649.
163. WHO. 7 million premature deaths annually linked to air pollution - http://www.who.int/mediacentre/news/releases/2014/air-pollution/en/. Accessed 17 Jan 2018.
164. Shin SH, Bode AM, Dong Z. Precision medicine: the foundation of future cancer therapeutics. Precis Oncol. 2017;1:12.
165. FDA approves first cancer treatment for any solid tumor with a specific genetic feature - https://www.fda.gov/newsevents/newsroom/pressannouncements/ucm560167.htm. Accessed 17 Jan 2018.
166. Gilbert JA, Quinn RA, Debelius J, Xu ZZ, Morton J, Garg N, et al. Microbiome-wide association studies link dynamic microbial consortia to disease. Nature. 2016;535:94–103.
167. Zou H, Hastie T. Regularization and variable selection via the elastic-net. J R Stat Soc. 2005;67:301–20.

14

Caution needs to be taken when assigning transcription start sites to ends of protein-coding genes: a rebuttal

Niv Sabath[1], Anna Vilborg[2], Joan A. Steitz[2] and Reut Shalgi[1*] iD

Abstract

Naturally occurring stress-induced transcriptional readthrough is a recently discovered phenomenon, in which stress conditions lead to dramatic induction of long transcripts as a result of transcription termination failure. In 2015, we reported the induction of such *downstream of gene* (DoG) containing transcripts upon osmotic stress in human cells, while others observed similar transcripts in virus-infected and cancer cells. Using the rigorous methodology Cap-Seq, we demonstrated that DoGs result from transcriptional readthrough, not de novo initiation. More recently, we presented a genome-wide comparison of NIH3T3 mouse cells subjected to osmotic, heat, and oxidative stress and concluded that massive induction of transcriptional readthrough is a hallmark of the mammalian stress response. In their recent letter, Huang and Liu in contrast claim that DoG transcripts result from novel transcription initiation near the ends of genes. Their conclusions rest on analyses of a publicly available transcription start site (TSS-Seq) dataset from unstressed NIH3T3 cells. Here, we present evidence that this dataset identifies not only true transcription start sites, TSSs, but also 5'-ends of numerous snoRNAs, which are generally processed from introns in mammalian cells. We show that failure to recognize these erroneous assignments in the TSS-Seq dataset, as well as ignoring published Cap-Seq data on TSS mapping during osmotic stress, have led to misinterpretation by Huang and Liu. We conclude that, contrary to the claims made by Huang and Liu, TSS-Seq reads near gene ends cannot explain the existence of DoGs, nor their stress-mediated induction. Rather it is, as we originally demonstrated, transcriptional readthrough that leads to the formation of DoGs.

Keywords: TSS-Seq, Cap-Seq, snoRNAs

Background

In 2015, we reported the induction of *downstream of gene* (DoG) containing transcripts upon osmotic stress in human cells [1]. Other labs have observed similar transcripts following viral infection, in renal cancer, and more [2–4]. We used the rigorous methodology Cap-Seq [5], before and after subjecting human cells to osmotic stress, to capture transcription start sites (TSSs) in a genome-wide manner, in order to ask whether stress-induced DoGs are independent transcripts or rather continuous with their upstream gene, i.e., a product of transcriptional readthrough. Our data demonstrated that DoGs result from transcriptional readthrough, not de novo initiation. More recently, we presented a genome-wide comparison of NIH3T3 mouse cells subjected to osmotic, heat, and oxidative stress and concluded that massive induction of transcriptional readthrough is a hallmark of the mammalian stress response [6].

In their recent letter, Huang and Liu [7], in contrast, claim that DoG transcripts result from novel transcription initiation near the ends of genes. Their conclusions rest on the analysis of a publicly available transcription start site (TSS-Seq) dataset from unstressed NIH3T3 cells [8]. Here, we present evidence that this dataset identifies not only true TSSs but also 5'-ends of numerous snoRNAs, usually processed from introns. Neglecting to discard these highly abundant contaminants, as well as failure to carry out additional necessary controls, have led Huang and Liu to draw erroneous conclusions. We demonstrate here that, contrary to Huang and Liu's assertion, TSS-Seq peaks near gene ends cannot explain

* Correspondence: reutshalgi@technion.ac.il
[1]Department of Biochemistry, Rappaport Faculty of Medicine, Technion—Israel Institute of Technology, 31096 Haifa, Israel
Full list of author information is available at the end of the article

the existence of DoGs, nor can it explain their stress-mediated induction.

TSS-Seq data efficiently capture snoRNAs, which originate from transcript introns

In their recent letter [7], Huang and Liu reported that one pan-stress DoG-producing gene out of more than 1800 found in our recent study [6], Hspa8, exhibits a high TSS-Seq peak near its 3′-end. They concluded that the Hspa8 DoG, doHspa8, is not a readthrough transcript but a lncRNA with an independent promoter. Close examination of the same data reveals that this TSS-Seq peak marks the exact 5′-end of the snoRNA Snord14e (Fig. 1a) within the last intron of Hspa8. Huang and Liu noted the association between the TSS-Seq peak and Snord14e, but rather interpreted it as evidence of a novel lncRNA. As it is well known that snoRNAs are processed from introns [9], it is highly unlikely that these TSS-Seq peaks correspond to the start of lncRNAs.

We then looked for other snoRNAs in the NIH3T3 TSS-Seq dataset and uncovered more than 1000 snoRNAs with TSS-Seq reads genome-wide, of which 249 had significant peaks (> 50). Thus, it is evident that the TSS-Seq method captures highly structured snoRNAs quite efficiently. We further examined additional TSS-Seq datasets from DBTSS [8] of human and mouse cell lines. We identified a significant TSS-Seq peak for Snord14e for all mouse samples examined (Fig. 1a) and a lower level peak in the human DLD1 samples within the Hspa8 gene (Fig. 1b).

The high abundance of snoRNA reads in TSS-Seq data might be the result of incomplete removal of 5′-phosphates by alkaline phosphatase, an early step in the preparation of RNA for analysis [8], which can be ascribed to the tight RNA secondary structure at the 5′-ends of snoRNAs. In contrast, the Cap-Seq protocol relies on

the presence of a 5′-m7G cap in addition to the removal of 5′-phosphates, which makes it significantly more rigorous in identifying true TSSs [5]; indeed, it does not capture snoRNA 5′-ends (Fig. 1b).

Therefore, these analyses demonstrate that caution is advisable when analyzing TSS-Seq data. Strict filtering of snoRNA-related, and perhaps also other small RNA-related, TSS-Seq peaks should be performed prior to analysis and is required in order to draw conclusions.

Transcription start sites at ends of protein-coding genes do not explain stress-induced DoGs

Next, Huang and Liu compared the TSS-Seq tag counts in the last 1kb of pan-stress DoG-generating genes versus non-DoG genes and reported significant differences. We replicated the same analysis, while using stringent inclusion criteria for non-DoGs—genes that show no evidence of transcriptional readthrough. As in our previous study [6], we defined pan-stress DoGs as DoGs that exist in all three stress conditions, heat shock, oxidative, and osmotic stress. We further defined non-DoGs as the group of genes whose maximum reads per kilobase per million mapped reads (RPKM) over the 4kb region downstream of the gene end was lower than the minimal RPKM of the 4kb region downstream of the gene end of pan-stress DoGs, in all three stress conditions in our NIH3T3 RNA-seq dataset [6]. We analyzed the same TSS-Seq data, while performing rigorous multiple sub-sampling and expression-matching procedures, to generate 1000 matching pan-stress DoG and non-DoG sets, in order to ensure similar distributions of expression levels of their upstream associated genes (explained in [6]). We then calculated the median TSS tag counts in the last 1kb of the gene for both

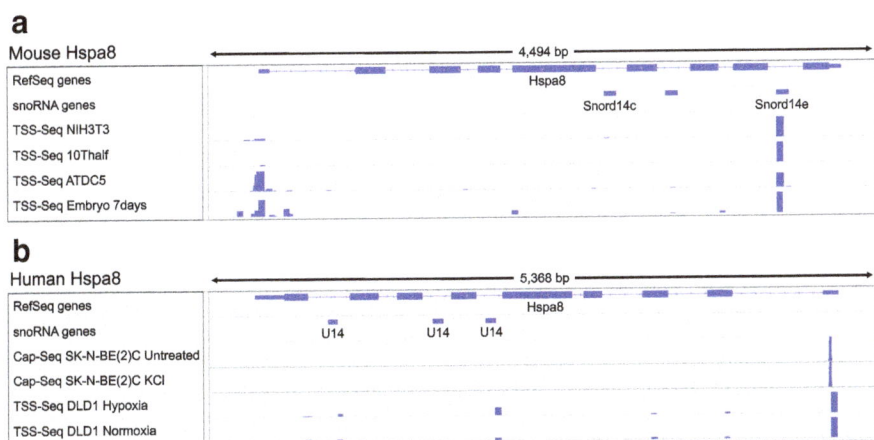

Fig. 1 The TSS-Seq peak near the 3′-end of Hspa8 originates from a snoRNA. **a** An IGV plot shows TSS-Seq data from four mouse cell lines [8] at the Hspa8 locus. snoRNA genes are indicated. **b** An IGV plot shows Cap-Seq data for unstressed and osmotic (KCl) stressed human SK-N-BE(2)C cells from [1] and TSS-Seq data from human DLD1 cells [8] at the Hspa8 locus. snoRNA genes are indicated

the pan-stress DoG and non-DoG groups. We found that the difference is significant ($p = 0.009$), but small: 11 for pan-stress DoGs and 8 for non-DoG genes (mean values are 65.8 and 62.9 for pan-stress DoGs and non-DoG genes, respectively. Fig. 2a, d). We then excluded 113 genes that harbor either a small RNA, e.g., a snoRNA or a TSS of another annotated transcript within their last 1kb, and repeated the same analysis. We found that although the median values remained identical, the difference now was only marginally significant ($p = 0.048$, mean values is now lower for pan-stress DoGs, 57.1, and 63.8 for non-DoG genes. Fig. 2b, e).

While examining the distributions of TSS-Seq tag counts in the last 1kb of pan-stress DoG- and non-DoG-associated genes (Fig. 2a, b), we noticed that the difference between them is mainly due to the number of genes with zero TSS-Seq tags. Indeed, when we excluded 545 additional genes that had zero TSS-Seq tag counts in their last 1kb, and repeated the analysis, the difference between pan-stress DoG- and non-DoG-associated genes completely disappeared, with median values of 20 and 21 for pan-stress DoG- and non-DoG-associated genes,

respectively ($p = 0.37$, mean values are 70.6 and 94.1 for pan-stress DoGs and non-DoG genes, respectively. Fig. 2c, f). Thus, the minute difference between the TSS-Seq tag counts in the last 1 kb of pan-stress DoG- and non-DoG-associated genes that was originally observed in the NIH3T3 cell TSS-Seq data was driven by snoRNAs and true TSSs that should have been filtered, in addition to genes with zero TSS-Seq tag counts.

Moreover, if DoGs are in fact generated by independent transcription, then TSS-Seq levels should correspond to DoG levels. However, we observe no correlation between TSS-Seq scores (TSS-Seq tag counts in the last 1kb of the gene) and DoG levels or DoG lengths (Pearson correlation of 0.03 and − 0.02 respectively, Fig. 3).

Finally, if independent transcription initiation were to produce DoGs, we should see significant increases in TSSs near the ends of all DoG-associated genes after stress. Huang and Liu analyzed data from unstressed cells only. Our previous Cap-Seq experiments, however, addressed this exact question by assaying stress-induced TSSs genome-wide, in human cells before and after osmotic stress [1]. Our results

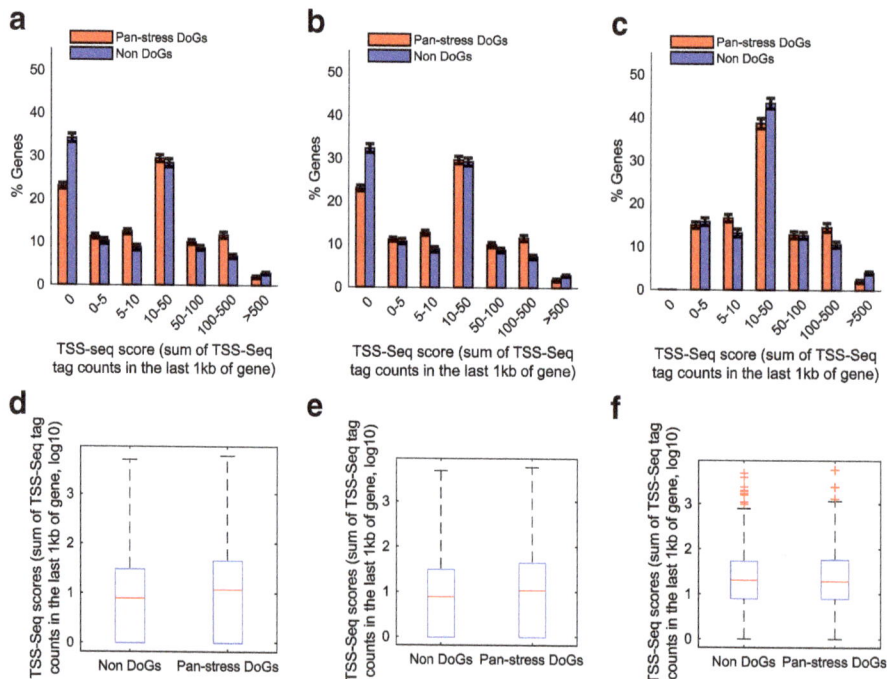

Fig. 2 TSS-Seq peaks do not explain DoGs. TSS-Seq data from NIH3T3 cells were downloaded from DBTSS [8], and TSS-Seq peaks in the last 1kb of all genes were extracted using bedtools. Bar graphs **a–c** show the mean (and standard deviation) percentage of genes in each bin of TSS-Seq scores (sum of tag counts in the last 1kb) as calculated from 1000 sub-samples of expression-matched pan-stress DoG-associated and non-DoG-associated genes. Boxplots **d–f** show the cumulative distribution of TSS-Seq peak scores in log10 scale according to three inclusion criteria: **a, d** No filter: median TSS-Seq scores are 11 for pan-stress DoGs and 8 for non-DoG genes, $p = 0.009$. **b, d** Excluding 113 genes that harbor either a snoRNA or a TSS of another transcript within their last 1kb: median TSS-Seq score is 11 for pan-stress DoGs and 8 for non-DoG genes, $p = 0.048$. **c, e** Excluding additional 545 genes with zero TSS-Seq tag count in their last 1kb: median TSS-Seq score is 20 for pan-stress DoG- and 21 for non-DoG-associated genes, $p = 0.37$

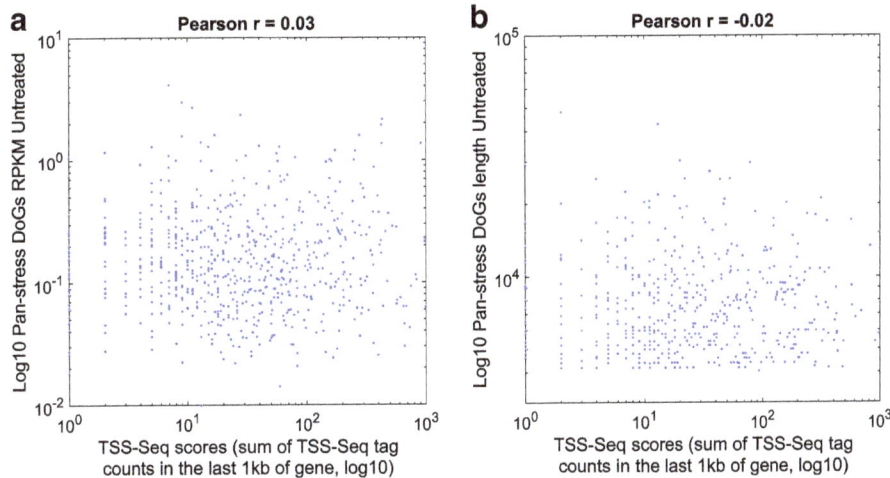

Fig. 3 TSS-Seq peaks do not explain DoG expression and length. **a** Scatter plot of the RPKMs of all pan-stress DoGs (in log10 scale, from untreated cell RNAseq data [6]) versus the TSS-Seq scores in log10 scale; they show no correlation. **b** DoG lengths in unstressed cells [6] (in log10 scale) are plotted versus the TSS-Seq scores for each pan-stress DoG; here too, the correlation is close to zero. Pearson correlation coefficients are indicated

detected a reduction, not an increase, in Cap-Seq peak induction near the 3'-ends of DoG-associated human genes in osmotic stress compared to untreated cells [1]. Thus, we ruled out that osmotic-stress DoGs result from independent transcription initiation [1]. While future experiments after heat shock and oxidative stress should generalize this conclusion, the great overlap in the identity of DoG-producing genes between stress conditions that we previously reported [6] argues that, during stress, DoGs are generated by transcriptional readthrough.

Abbreviations
DoG: Downstream of gene-containing transcript; lncRNA: Long non-coding RNA; RPKM: Reads per kilobase per million mapped reads; snoRNA: Small nucleolar RNA; TSS: Transcription start site

Acknowledgements
We thank Mingyi Xie for the useful advice on Cap-Seq and Ruth Hershberg for the critical reading of the manuscript.

Funding
N.S. and R.S. have received funding from the European Research Council under the European Union's Horizon 2020 program Grant 677776. J.A.S was supported by the NIH Common Fund Program, grant U01CA200147, as a Transformative Collaborative Project Award (TCPA-2017-NEUGEBAUER).

Authors' contributions
NS and RS designed the study. NS performed the data analysis. NS, AV, JAS, and RS wrote the manuscript. All authors read and approved the final manuscript.

Consent for publication
Not applicable.

Competing interests
The authors declare that they have no competing interests.

Author details
[1]Department of Biochemistry, Rappaport Faculty of Medicine, Technion—Israel Institute of Technology, 31096 Haifa, Israel. [2]Department of Molecular Biophysics and Biochemistry, Howard Hughes Medical Institute, Boyer Center for Molecular Medicine, Yale University School of Medicine, 295 Congress Avenue, New Haven, CT 06536, USA.

References
1. Vilborg A, Passarelli MC, Yario TA, Tycowski KT, Steitz JA. Widespread inducible transcription downstream of human genes. Mol Cell. 2015;59(3):449–61.
2. Grosso AR, Leite AP, Carvalho S, Matos MR, Martins FB, Vitor AC, Desterro JM, Carmo-Fonseca M, de Almeida SF. Pervasive transcription read-through promotes aberrant expression of oncogenes and RNA chimeras in renal carcinoma. Elife. 2015;4:1–16.
3. Rutkowski AJ, Erhard F, L'Hernault A, Bonfert T, Schilhabel M, Crump C, Rosenstiel P, Efstathiou S, Zimmer R, Friedel CC, et al. Widespread disruption of host transcription termination in HSV-1 infection. Nat Commun. 2015;6:7126.
4. Muniz L, Deb MK, Aguirrebengoa M, Lazorthes S, Trouche D, Nicolas E. Control of gene expression in senescence through transcriptional read-through of convergent protein-coding genes. Cell Rep. 2017;21(9):2433–46.
5. Xie M, Li M, Vilborg A, Lee N, Shu MD, Yartseva V, Sestan N, Steitz JA. Mammalian 5'-capped microRNA precursors that generate a single microRNA. Cell. 2013;155(7):1568–80.
6. Vilborg A, Sabath N, Wiesel Y, Nathans J, Levy-Adam F, Yario TA, Steitz JA, Shalgi R. Comparative analysis reveals genomic features of stress-induced transcriptional readthrough. Proc Natl Acad Sci U S A. 2017;114(40):E8362–71.
7. Huang MY, Liu JL. Transcription start sites at the end of protein-coding genes. Hum Genomics. 2018;12(1):15.
8. Tsuchihara K, Suzuki Y, Wakaguri H, Irie T, Tanimoto K, Hashimoto S, Matsushima K, Mizushima-Sugano J, Yamashita R, Nakai K, et al. Massive transcriptional start site analysis of human genes in hypoxia cells. Nucleic Acids Res. 2009;37(7):2249–63.
9. Kiss T, Fayet E, Jady BE, Richard P, Weber M. Biogenesis and intranuclear trafficking of human box C/D and H/ACA RNPs. Cold Spring Harb Symp Quant Biol. 2006;71:407–17.

15

Whole exome sequencing of a single osteosarcoma case—integrative analysis with whole transcriptome RNA-seq data

Ene Reimann[1,2*], Sulev Kõks[1,2], Xuan Dung Ho[3,4], Katre Maasalu[3,5] and Aare Märtson[3,5]

Abstract

Background: Osteosarcoma (OS) is a prevalent primary malignant bone tumour with unknown etiology. These highly metastasizing tumours are among the most frequent causes of cancer-related deaths. Thus, there is an urgent need for different markers, and with our study, we were aiming towards finding novel biomarkers for OS.

Methods: For that, we analysed the whole exome of the tumorous and non-tumour bone tissue from the same patient with OS applying next-generation sequencing. For data analysis, we used several softwares and combined the exome data with RNA-seq data from our previous study.

Results: In the tumour exome, we found wide genomic rearrangements, which should qualify as chromotripsis— we detected almost 3,000 somatic single nucleotide variants (SNVs) and small indels and more than 2,000 copy number variants (CNVs) in different chromosomes. Furthermore, the somatic changes seem to be associated to bone tumours, whereas germline mutations to cancer in general. We confirmed the previous findings that the most significant pathway involved in OS pathogenesis is probably the WNT/β-catenin signalling pathway. Also, the IGF1/IGF2 and IGF1R homodimer signalling and TP53 (including downstream tumour suppressor gene *EI24*) pathways may have a role. Additionally, the mucin family genes, especially *MUC4* and cell cycle controlling gene *CDC27* may be considered as potential biomarkers for OS.

Conclusions: The genes, in which the mutations were detected, may be considered as targets for finding biomarkers for OS. As the study is based on a single case and only DNA and RNA analysis, further confirmative studies are required.

Keywords: Osteosarcoma, Whole exome sequencing, Integrative analysis

Introduction

Osteosarcoma (OS) is a most prevalent primary malignant bone tumour and mostly occurs in children and adolescents—75% of patients with OS are 15 to 25 years old. The etiology is unknown; however, a genetic predisposition has been suggested [1,2]. Reviewed in [3], these tumours have high potential to metastasize and are one of the most frequent causes of cancer-related deaths. The survival rate increased up to 70% after chemotherapy became available [4]. However, no further improvements have been made in the last decades in terms of survival. Thus, the survival plateau forces scientists to look for new biomarkers (diagnostic, disease monitoring, response, resistance markers, drug targets), which could lead to, i.e. applying new therapeutic agents. While OS is rare and very heterogeneous (inter-patient, inter-tumour and intra-tumour heterogeneity), the clinical study progress is slow; thus, the preclinical studies are vital. Furthermore, finding the biomarkers and detecting the potential targets for new drugs are essential to improve the present situation.

There are several next-generation sequencing (NGS) and genome-wide association studies (GWAS) about OS, which associate different genes and pathways with pathogenesis of OS [5-7]. With whole exome sequencing

* Correspondence: ene.reimann@ut.ee
[1]Department of Pathophysiology, University of Tartu, 19 Ravila Street Tartu 50411, Estonia
[2]Department of Reproductive Biology, Estonian University of Life Sciences, 64 Kreutzwaldi Street, Tartu, Estonia
Full list of author information is available at the end of the article

(WES) and whole genome sequencing (WGS) studies, *TP53*, *PTEN* and *PRB2* are found to be mutated in significant frequency [5]. High mutation rate in *TP53* has also demonstrated in OS cell lines. Additionally, deletion of *CDKN2A/B* locus and amplification of *MDM2* were detected [8]. With GWAS studies, a single nucleotide variant (SNV) in *GRM4* was detected as potential biomarker for OS [7]. Gene expression studies reveal that, i. e. WNT inhibitory factor (*WIF1*) has a loss of expression in OS cell lines [9]; however, we found in our previous work that the expression has increased significantly [10]. Thus, as demonstrated, the expression pattern of WNT pathway genes in different OS cases may not be similar. When correlating the expression patterns of miRNA/ mRNA pairs, miRNAs regulating *TGFBR2*, *IRS1*, *PTEN* and *PI3K* have been detected [11]. In addition, several serine/threonine kinases (mechanistic target of rapamycin (mTOR)) or tyrosine kinases (SRC, IGF1R, PDGFR, KIT) are considered as targets in OS treatment [3,12,13].

When observing the related pathways, the WNT/β-catenin pathway is one of the most thoroughly studied among bone malignancies. For example, the tumour growth is regulated through this pathway and the overexpression of *BMP9* suppresses its activity [14,15]. Furthermore, PI3K/AKT/mTOR signalling pathway was brought forward as a potential target for therapy, and also, pathways associated to TP53 may be altered [5]. Hypoxia-HIF-1α-CXCR4 pathway plays a crucial role during the migration of human osteosarcoma cells [16]. These are just a few examples—the network of associated genes and pathways is complex.

OS has a very unstable genome—it may contain aberrant number of chromosomes, and in most cases, these chromosomes display major structural abnormalities including amplification, deletions and translocations. For example, several studies have demonstrated the gain of chromosomal arms 6p, 8q and 17p in the case of OS [17,18]. To be more precise, i.e. *VEGFA* amplification and *LSAMP* deletion have been detected in OS [19,20]. Thus, it is suggested that genomic instability is linked to the development of this tumour [18,21-23]. Furthermore, the genomic aberrations are more frequent in metastases than in primary tumours [24]. The genes responsible for cell cycle regulation are suggested to be associated to DNA breakage and genomic instability, i.e. *CDC5L* overexpression and mutations in *TP53* gene are correlated to the high genomic instability in OS [23,25]. Moreover, the chromothripsis event is characteristic to OS—it generates new fusion products. This may explain the sudden onset of OS and the complexity and heterogeneity of OS genome [26]. All these changes make it difficult to find biomarkers suitable for targeting OS, as there are so many different subtypes.

In the present work, we analysed the whole exome of the tumorous and non-tumour bone tissue from the same patient with osteosarcoma. We used next-generation sequencing to study how the coding region of the tumour genome has altered. Additionally, we analysed together the WES genotyping and RNA expression data (from our previous RNA-seq analysis).

Materials and methods
Subject

The protocols and informed consent form used in this study were approved by the Ethical Review Committee on Human Research of the University of Tartu. The patient signed a written informed consent, which also includes the acceptance of the report to be published. A 16-year-old Caucasian male patient with an OS diagnosis was studied. In more detail, the patient became ill with complaints of pain in the left knee area. History of trauma was missing, and GP administered painkillers and vitamins. After 6 months, the patient returned to GP with complaints of pain, swelling and dysfunction in the left distal femur and knee area. The swelling line was observed in the left femoral distal region, and the area was thicker and painful to touch. No changes in skin colour were detected. The X-ray investigation showed additional shading and structural change in the distal part of the left femur. For detailed investigation, the MRI was performed and as a result, malignant process was suspected. Patient was hospitalized, and bone biopsy was taken for histological investigation. The diagnosis of osteosarcoma was confirmed. Chemotherapy for osteosarcoma started by Scandinavian Sarcoma Group (SSG) XIV treatment protocol. The patient responded well to the therapy—the histological analysis confirmed the necrotic tissue in tumour. After 3 month of chemotherapy, surgical removal of tumour (distal part of femoral bone with knee joint) and replacement of the knee and the lower part of the femur with megaprosthesis was performed. Pathologist confirmed that resection line was without tumour cells and OS was referred as NAS (Not Further Specified). After the patient had recovered from surgery, the SSG XIV chemotherapy treatment protocol was followed. Materials for this study were collected from the surgically removed tissue.

Exome sequencing

The genomic DNA (gDNA) was extracted from two bone samples from different locations—one sample from tumour area and another sample from the uninvolved normal bone tissue as a control. For gDNA extraction, the tissue was homogenized applying liquid nitrogen and a mortar, and after that, the PureLink Genomic DNA kit (Life Technologies Corp., Carlsbad, CA, USA) was used according to manufacturer's protocol. The Target Seq Exome Enrichment System and SOLiD 5500 barcoded adaptors (Life Technologies Corp., Carlsbad, CA, USA)

were used to prepare the libraries. The SOLiD 5500xl platform and paired-end DNA sequencing chemistry (75 bp forward and 35 bp reverse direction) were applied to sequence the samples.

The data analysis

Offline cluster was used for data processing and analysis. For bioinformatic analysis, LifeScope version 2.5 was applied. LifeScope performed colour space mapping and pairing. Tertiary analysis consisted of SNV discovery (diBayes algorithm) and detection of small indels. Hg19 (GRCh37.p13) was used as a reference, and before mapping, the multifasta file was verified in order to increase the mapping quality.

The SNVs and small indel .gff3 files were used as input in ANNOVAR software (AS; www.openbioinformatics. org/annovar/) [27] and Ingenuity Variant Analysis (IVA; http://www.ingenuity.com) QIAGEN, Redwood City, MD, USA) software. Applying refGene hg19, dbSNP135 and dbCOSMIC67 databases, AS annotated and predicted the effects of SNVs and small indels we detected in our study samples. AS also provides other prediction tools in order to get prediction scores (PolyPhen-2, SIFT, ljb2 etc.) [28-30]. Comparative distribution of SNVs and small indels between different samples was performed with Galaxy software bundle [31,32]. IVA provided tools to annotate SNVs and small indels, which may be associated to cancer. The tumour and control samples were compared, and the lists for diseases, processes and pathways related to cancer were received as output.

The .bam and .bai files were used as input in CEQer software (CS) (www.ngsbicocca.org/html/ceqer.html), which is a tool for analysing copy number variants (CNVs) and loss of heterozygosity (LOH).

About the RNA-seq data analysis, please see our previous article, where we used the bone samples from the same patient [10].

Results

For comparing the tumour tissue and non-tumour tissue (control tissue) from the same individual, different approaches were applied. After mapping the data to a reference genome, we used several tools to perform the tertiary analysis.

Sequencing statistics from LifeScope software

In the case of the tumour tissue, over 130 million (58%) mappable reads were in target and the enrichment fold was 48%. Eighty-five percent of the detected targets were covered over 20 times, and the average coverage was 185.5. In the case of the control tissue, over 154 million (61%) mappable reads were in target and the enrichment fold was 51%. Eighty-three percent of the detected targets were covered over 20 times, and the average coverage was 157.

SNVs, small indels and CNVs

1) Results from ANNOVAR software

Using refGene hg19 database, AS was able to annotate 37,990 SNVs and 1,484 small indels. In the case of SNVs, we considered the data reliable, if the coverage was over 20; thus, 25,914 SNVs remained. In the case of SNVs, there were 23,767 germline mutations (9,067 in homozygous form and 14,700 in heterozygous form) and 2,147 somatic mutations (in the tumour tissue—116 in homozygous form and 2,031 in heterozygous form) (Table 1, Additional file 1). Furthermore, there were 896 germline small indels (278 in homozygous form and 618 in heterozygous form) and 588 somatic indels (in the tumour tissue—177 in homozygous form and 411 in heterozygous form).

Applying dbSNP135, we were able to annotate 5,281 SNVs and 239 small indels. With dbCOSMIC67, we annotated 2,569 SNVs and 59 small indels—none of these were noted to be associated to bone cancer. Applying ljb2 database, we found 469 SNVs to potentially cause a disease (average ljb2 score over 0.918), including 31 germline mutations and 4 somatic mutations (*ESX1*: c. A578G/p.K193R; *CDC27*: c.A17G/p.E6G; *TMEM120B*: c.G274A/p.D92N; *TMEM131*: c.C3947T/p.P1316L) in homozygous form in the tumour tissue.

2) Results from Ingenuity Variant Analysis software

Altogether, 207 cancer driver variants (CD-SNVs) were found in 123 genes according to IVA (Additional file 2). Fourteen CD-SNVs potentially gain and 186 lose the gene function. Only seven SNVs may have no drastic effect on gene function in the tumour tissue. Furthermore, according to IVA, none of these 207 SNVs affect the gene functionality in the control tissue. Thirteen of the CD-SNVs were homozygous in the tumour tissue (Table 2). There were no cancer-associated homozygous mutations present in the control tissue; thus, the homozygous CD-SNVs in the tumour tissue are all somatic.

According to IVA, six cancer-associated small indels were found (Table 2). Four of them are homozygous and two are heterozygous in the tumour tissue—the effect is most probably the loss of gene function. These indels are predicted to have no effect in the control tissue.

In most of the genes brought front by IVA, one CD-SNV was found in coding region in heterozygous form. However, some of the genes have more CD-SNVs in coding regions: *MUC4* had even 22, *ZNF717* had 8, *CTBP2* had 7 and *OR4C3* had 5 CD-SNVs, whereas these were not present in the control tissue (data not shown). When observing from a slightly different angle—the gene complexes, we can see that the mucin complex has the highest

Table 1 The numbers of SNV and small indel findings received from data analysis with ANNOVAR software

	Germline mutations		Somatic mutations			
	Homozygous: non-reference	Heterozygous	Homozygous in tumour / Heterozygous in control	Heterozygous in tumour / Homozygous in control	Homozygous in tumour / Homozygous (reference) in control	Heterozygous in tumour / Homozygous (reference) in control
SNVs						
Altogether	9,067	14,700	48	237	68	1,794
Exonic (includes ncRNA)	5,244	8,702	21	103	29	967
Nonsynounymous	2,435	4,035	15	52	18	500
Stopgain	6	50	0	2	0	7
Stoploss	2	5	0	0	0	1
Splicing (includes exonic)	11	20	0	0	0	2
Intronic (includes ncRNA)	3,091	4,846	22	111	35	681
5' UTR and 3' UTR	515	797	3	19	1	91
Downstream and upstream	51	76	1	2	1	13
Intergenic	155	259	1	2	2	40
Small indels						
Altogether	278	618	89	75	88	336
Exonic (includes ncRNA)	33	99	14	11	2	29
Frameshift	9	30	4	4	1	12
Stopgain	0	1	0	0	0	0
Splicing (includes exonic)	4	16	0	4	1	3
Intronic (includes ncRNA)	212	419	64	51	75	270
5' UTR and 3' UTR	25	73	10	7	8	24
Downstream and upstream	1	2	0	1	1	6
Intergenic	3	9	1	1	1	4

significance—three genes and 27 CD-SNVs are considered (Table 3). There are also other gene complexes, which are potentially associated to cancer processes, and in different complexes, the CD-SNVs are either somatic or germline (Table 3).

In the case of cancer-associated small indels, the statistically most significant results were with complexes related to *RELA* gene—NFKB1-RELA and RELA-REL complexes both had *p* value 7.56E-4.

IVA provided the first 100 cancer-associated processes and diseases related to CD-SNVs and small indels. Seventy-three genes and 135 CD-SNVs were found associated to process named as "disorder of genitourinary system" (Table 4). These findings were present in both the tumour and control tissues. There were also two processes associated to bone "myelopoiesis of bone marrow" (associated genes *NPM1, RARA*) and "quantity of trabecular bone" (associated genes *CREBBP, SMO*)— these findings were present only in the tumour tissue. In the case of small indels, all the findings were somatic and *ALK* and *RELA* genes were associated to "outgrowth

of bone marrow cells" and "inflammatory response of bone marrow-derived macrophages", respectively.

IVA found 111 genes with 202 germline CD-SNVs associated to cancer (Table 5). Fifteen genes, which had 43 somatic CD-SNVs were associated to "bone marrow cancer and tumours". In the case of small indel, all six genes, with a finding, are associated to cancer and the found small indels are all somatic. The disease named as "tumourigenesis of bone tumour" was associated to small indel in *ALK* gene and was present only in the tumour tissue.

With the osteosarcoma patient's tumour and control tissue, WES data IVA found six pathways associated to CD-SNVs and six to cancer driver small indels (Table 6). All the mutations considered here were somatic. In the case of CD-SNVs, the statistically most significant association was between tumour and WNT/β-catenin signalling pathway. In the case of small indels, associations with different cytokine pathways were found. Also, a pathway directly linked to the bone tissue—"RANK signalling in osteoclasts" was brought front.

Table 2 The somatic cancer driver SNVs and small indels found in data analysis with Ingenuity Variant Analysis software

Gene symbol	Chr number	Position	REF/ALT	Tumour zygosity	Effect on function	Control zygosity	Effect on function	dbSNP	SIFT function	Polyphen function	Transcript ID	Nucleotide change	Amino acid change	Gene region	Translation impact
SNVs															
RGPD3 (includes others)	2	110585652	A/G	1/1	Loss	0/0	Normal		Damaging	Benign	NM_001037866.1, NM_001123363.3, NM_005054.2, NM_032260.2	c.2393A > G	p.E798G	Exonic	Missense
PRDM9	5	23527251	C/T	1/1	Loss	0/0	Normal		Tolerated	Probably damaging	NM_020227.2	c.2054C > T	p.T685I	Exonic	Missense
FOXK1	7	4722436	A/G	1/1	Loss	0/0	Normal		Damaging	Benign	NM_001037165.1	c.497A > G	p.N166S	Exonic	Missense
CCZ1/CCZ1B	7	6841033	T/A	1/1	Loss	0/1	Normal		Tolerated		NM_198097.3	c.1228A > T	p.M410L	Exonic	Missense
PLAT[a]	8	42044965	G/A	1/1	Normal	0/0	Normal	2020921	Tolerated	Benign	NM_033011.2/ NM_000930.3	c.352C > T/ c.490C > T	p.R118W/ p.R164W	Exonic	Missense
AGTPBP1[a]	9	88292495	C/T	1/1	Loss	0/1	Normal		Tolerated	Benign	NM_015239.2	c.292G > A	p.G98R	Exonic	Missense
SARDH	9	136597592	T/C	1/1	Loss	0/0	Normal	149002589	Tolerated	Benign	NM_001134707.1, NM_007101.3	c.463A > G	p.I155V	Exonic	Missense
FAH	15	80472526	C/T	1/1	Normal	0/1	Normal	11555096	Damaging	Probably damaging	NM_000137.2	c.1021C > T	p.R341W	Exonic	Missense
CDC27	17	45266522	T/C	1/1	Loss	0/0	Normal	62077279	Damaging	Probably damaging	NM_001114091.1, NM_001256.3	c.17A > G	p.E6G	Exonic	Missense
SBF1[a]	22	50893287	T/C	1/1	Loss	0/1	Normal	200488568	Tolerated	Benign	NM_002972.2	c.4768A > G	p.T1590A	Exonic	Missense
LRRC37A3[a] (includes others)	17	44632540	T/C	1/1	Gain	0/0	Normal	144051917	Activating	Benign	NM_001006607.2	c.4882 T > C	p.W1628R	Exonic	Missense
ARL17A	17	44632540	T/C	1/1	Gain	0/0	Normal	144051917	Activating	Benign	NM_001113738.1/ NM_016632.2	c.*2182A > G/ c.259 + 15585A > G	-/ -	3'UTR/ Intronic	
LILRB3	19	54725835	G/C	1/1	Gain	0/0	Normal	201948566	Activating	Benign	NM_001081450.1, NM_006864.2	c.523C > G	p.R175G	Exonic	Missense
Small indels															
CTCFL	20	56073500	(N)103/T	1/1	Loss	0/0	Normal				NM_001269041.1/ NM_001269043.1/ NM_001269040.1/ NM_001269042.1/ NM_080618.3/ NM_001269046.1	c.*4_*105del(N)103/ c.1988 + 8_1988 + 109del(N)103/ c.*4_*105del(N)103/ c.*4_*105del(N)103/ c.*4_*105del(N)103/ c.*4_*105del(N)103		3' UTR/ Intronic/ 3' UTR/ 3' UTR/ 3' UTR/ 3' UTR	

Table 2 The somatic cancer driver SNVs and small indels found in data analysis with Ingenuity Variant Analysis software (*Continued*)

Gene	Chr	Position	Ref/Alt	GT	CNV	GT	Status		Transcript	Variant	Protein	Location	Effect
PRR23C	3	13876627	GTGC/G	1/1	Loss	0/1	Normal	63140560	NM_001134657.1	c.-168_-166delGCA		5' UTR	
CDCA7L	7	21941867	CTTAG/C	1/1	Loss	0/0	Normal		NM_001127371.2/	c.*69_*72delCTAA/		3' UTR/	
									NM_001127370.2/	c.*69_*72delCTAA/		3' UTR/	
									NM_018719.4	c.*69_*72delCTAA		3' UTR	
ALK	2	29416029	G/GATTG	1/1	Loss	0/0	Normal		NM_004304.4	c.*60_*61insCAAT		3' UTR	
DSPP	4	88537081	CAGCAGCAAT/C	0/1	Loss	0/0	Normal		NM_014208.3	c.3268_3276delAGCAGCAAT	p.S1090_N1092del	Exonic	In-frame
RELA	11	65422086	CTC/CTGTAGT	0/1	Loss	0/0	Normal		NM_001145138.1/	c.1408delGinsACTAC/	p.E470fs*19	Exonic/	Frameshift/
									NM_021975.3/	c.1417delGinsACTAC/		Exonic/	Frameshift/
									NM_001243984.1/	c.1210delGinsACTAC/		Exonic/	Frameshift/
									NM_001243985.1	c.1216-108delGinsACTAC		Intronic	-

[a] The expression pattern of these genes has changed in the tumour tissue compared to that in the control tissue.

Table 3 The gene complexes which are potentially associated to cancer processes

Complex name	p value	Number of genes associated	Number of variances found	Tumour tissue	Control tissue
Mucin	9.54E-05	3: *MUC2, MUC4, MUC6*	27	1	0
Bcl9-Cbp/p300-Ctnnb1-Lef/Tcf	2.46E-03	2: *CREBBP, TCF3*	2	1	0
Sox	4.55E-03	2: *SOX7, SOX10*	2	1	0
Cholesterol monooxygenase (side-chain-cleaving)	1.06E-02	1	1	1	0
CYP11A	1.06E-02	1	1	1	0
Sarcosine dehydrogenase	1.06E-02	1	1	1	0
Ctbp	1.59E-02	1	7	1	0
Cbp/p300	1.59E-02	1	1	1	0
Dimethylglycine dehydrogenase	1.59E-02	1	1	1	0
DRD1/5	1.59E-02	1	1	1	0
MAGI	2.64E-02	1	2	1	1
Magi-Pten	3.68E-02	1	2	1	1
Fumarylacetoacetase	1.06E-02	1	1	1	1

There are both somatic and germline cancer driver SNVs found in the tumour and control tissues.

3) Results from CEQer software

We applied CS to analyse CNVs in tumour and non-tumour tissue exomes. Compared to the control tissue, in the tumour tissue, the loss of coding sequences was found in 6 chromosomes and 183 genes and gain of coding sequences in 4 chromosomes and 65 genes (Figure 1). The loss or gain of coding sequences was altogether in 8 chromosomes, and the most altered were chromosomes 2 and 19 (193,701 bp and 115,358 bp, respectively; Figure 2). The loss of heterozygosity was detected altogether in 68 regions in 37 genes, located in 15 different chromosomes (Additional file 3).

Integrative analysis
The integrative analysis narrows down the large list of findings from NGS data. When combining the results from WES data (AS, IVA, CS) and RNA-seq data [10], we found some interesting and rather logical associations, which we would like to emphasize.

SNVs, small indels and RNA expression
To reduce down the complexity of data we received from AS, we decided to perform as follows. In the case of SNV data, we observed both somatic and germline SNVs, which are homozygous in the tumour tissue and should have an effect on translation (nonsynonymous, stopgain, stoploss findings). Thus, we got 527 homozygous germline SNVs (in 392 genes) and 8 homozygous somatic SNVs (in 7 genes), which are located in genes with altered expression in the tumour tissue compared to that in the control tissue. If also considering the ljb2 database scores, seven homozygous SNVs with high disease-causing probability remained (Table 7).

In the case of small indels detected with AS, we observed the somatic and germline indels, which were homozygous in the tumour tissue. There was 52 germline and 26 somatic indels in introns of the genes, which expression pattern has also changed (data not shown). Furthermore, there was five germline and three somatic indels in exons of the genes with altered expression. Thus, we found altogether three frameshift small indels, which possibly have an effect on translation (frameshift insertions and deletion in exons) (Table 7).

In the case of homozygous cancer driver SNVs and small indels found with IVA (Table 2), only four genes have altered expression pattern in the tumour tissue compared to that in the control tissue. The mRNA expression was increased in the case of *PLAT* (log fold change (logFC) = 3.65, false discovery rate (FDR; corrected statistical significance) = 8.27E-27), *AGTPBP1* (logFC = 0.91, FDR = 0.039) and *LRRC37A3* (logFC = 1.14, FDR = 0.0072) and decreased in the case of *SFB1* (logFC = −1.33, FDR = 0.0037).

CNVs, LOHs and RNA expression
When analysing the CNV results together with RNA expression results, we found that with gained copy numbers, there were altogether 22 genes, with altered expression profile—20 genes with increased and 2 genes with decreased mRNA expression. In the case of loss copy of number, 74 genes' expression profile had changed—11 genes with increased and 63 genes with decreased mRNA expression. In Table 8, the genes with the lowest FDR values for gene expression results are presented. Here, we would emphasize that the *INSR*, which has copy number loss in area covering 174,552 bp has also a remarkable decrease in mRNA expression

Table 4 The cancer-associated processes detected by IVA

Process name	p value	Number of genes associated	Number of variances found	Tumour tissue	Control tissue
CD-SNVs					
Disorder of genitourinary system	9.05E-14	73	135	1	1
Cell biology	4.08E-04	69	132	1	1
Cell signalling	3.83E-03	25	31	1	1
Morphology of body region	2.55E-03	23	24	1	1
Abnormal morphology of cells	1.73E-03	18	19	1	1
Abnormal morphology of body cavity	6.17E-04	17	18	1	1
Morphology of body cavity	1.36E-03	17	18	1	1
Morphology of cardiovascular system	5.80E-04	13	14	1	1
Abnormal morphology of cardiovascular system	7.22E-04	12	13	1	1
Abnormal morphology of thoracic cavity	1.22E-03	11	12	1	1
Myelopoiesis of bone marrow	**3.64E-03**	**2: NPM1, RARA**	**2**	**1**	**0**
Quantity of trabecular bone	**4.08E-03**	**2: CREBBP, SMO**	**2**	**1**	**0**
Small indels					
Tissue development	1.19E-03	5	5	1	0
Developmental process of tissue	1.35E-03	5	5	1	0
Development of organ	5.05E-03	4	4	1	0
Organogenesis	5.32E-03	4	4	1	0
Colony formation of tumour cell lines	6.25E-05	3	3	1	0
Colony formation of cells	4.44E-04	3	3	1	0
Colony formation	5.46E-04	3	3	1	0
Developmental process of tumour cells	3.81E-03	3	3	1	0
Colony formation of carcinoma cell lines	5.94E-05	2	2	1	0
Apoptosis of nervous tissue cell lines	2.49E-04	2	2	1	0
Outgrowth of bone marrow cells	**7.56E-04**	**1: ALK**	**1**	**1**	**0**
Inflammatory response of bone marrow-derived macrophages	**1.26E-03**	**1: RELA**	**1**	**1**	**0**

The sorting is performed by number of genes.
The bold data reflects the processes directly associated to bone.

(3.36 times; FDS = 9.67E-31). However, there are also several genes with CNVs, which could be associated to cancer.

Combining the LOH and mRNA expression data, we found that in the tumour tissue, the expression of four genes with LOH has increased significantly and expression of five genes with LOH has decreased significantly (Table 9). The rest of the genes with LOHs had no significant changes in mRNA expression level, and two genes were not detected with RNA-seq (*FLJ20518, MANSC4*) [10].

For additional information, please see the supplementary material as separate files for AS, IVA and CS combined with RNA-seq data.

Discussion

In this study, the exome profiles of the osteosarcoma patient's tumour and normal bone tissue were compared.

Additionally, the RNA-seq data from our previous work was used [10]. For WES data analysis, several softwares were applied and possibly some of them are better in detecting some mutations and not so effective in detecting others. Still, we think it is more beneficial to use different approaches and we believe it is easier to follow, if we discuss separately the results gained from each software.

The ANNOVAR software annotated a large amount of genes with SNVs and small indels, applying refGene hg19 database. Over 2,700 somatic SNVs and small indels were detected specifically in the tumour tissue, from which almost 300 are homozygous. These findings are located all over the exome. This demonstrates that the changes in OS genome are not concentrated into a single or few areas but are rather distributed.

When using ljb2 database, AS detected four homozygous somatic mutations in the tumour tissue, which could potentially cause a disease. These nonsynonymous

Table 5 The diseases associated to CD-SNVs and small indels

Disease name	p value	Number of genes associated	Number of variances found	Tumour tissue	Control tissue
CD-SNVs					
Cancer	7.04E-23	111	202	1	1
Tumourigenesis	8.21E-16	111	202	1	1
Cancers and tumours	3.37E-15	111	202	1	1
Organismal injury and abnormalities	9.45E-17	105	194	1	1
Carcinoma	3.46E-25	99	186	1	1
Solid tumour	2.64E-24	99	186	1	1
Epithelial neoplasia	3.34E-23	99	186	1	1
Epithelioma	3.34E-23	99	186	1	1
Breast or colorectal cancer	5.45E-23	83	164	1	1
Malignant neoplasm of abdomen	6.93E-20	83	169	1	1
Bone marrow cancer	**1.69E-03**	**15: CREBBP, EPHA2, FGFR2, KCNJ12, KMT2C, LILRB3, MUC17, MUC4, MYBPC3, NPM1, RARA, SMO, TCF3, TTN, TUBG1**	**43**	**1**	**0**
Bone marrow cancer and tumours	**1.69E-03**		**43**	**1**	**0**
Small indels					
Cancer	9.07E-03	6	6	1	1[a]
Hematologic cancer	2.36E-04	4	4	1	1[a]
Hematologic cancer and tumours	2.36E-04	4	4	1	1[a]
Hematological neoplasia	8.01E-04	4	4	1	1[a]
Lymphohematopoietic cancer	9.12E-04	4	4	1	1[a]
Disease of colon	7.88E-03	4	4	1	0
Hematological disease	8.15E-03	4	4	1	1[a]
Immunological disease	1.28E-02	4	4	1	1[a]
Gastrointestinal tract cancer	2.00E-02	4	4	1	0
Gastrointestinal tract cancer and tumours	2.02E-02	4	4	1	0
Tumourigenesis of bone tumour	**7.04E-03**	**1: ALK**	**1**	**1**	**0**

[a]Here, only one gene PRR23C has a small indel in heterozygous form, which most likely does not affect the gene function. See Table 2.
The bold data reflects the diseases directly associated to bone.

mutations were located in *ESX1*, *CDC27*, *TMEM120B* and *TMEM131*. Additionally, in the case of *TMEM120B* and *TMEM131*, the mRNA expression has decreased substantially in the tumour tissue compared to that in the control tissue [10]; however, further studies are needed to confirm the possible associations between found mutations and gene expression level. Available data about the possible associations between OS and these genes is very limited. In *TMEM120B*, a gene with an unclear function, the mutation COSM1599921 has been previously detected in glioma [33]. The *CDC27* is a gene possibly controlling the timing of mitosis and may have an important role in tumour cell division [34]. In addition to the somatic mutation, the *CDC27* had 33 heterozygous germline disease-causing mutations (non-synonymous) (data not shown). In the case of breast cancer, the *CDC27* has been demonstrated to be a

promising biomarker in predicting the disease progression and prognostication [35]. Thus, these somatic mutations may have some effect on OS pathogenesis. Especially the abundant changes in *CDC27* may be important in terms of regulating OS tumour cell division.

In the tumour tissue, we detected homozygous somatic small indels causing the frameshift in five genes—*EI24*, *ALG1L2*, *TIGD6*, *GPATCH4* and *SSPO*. None of these genes have previously been associated to OS, and according to our RNA-seq data, only *EI24* of these five genes has altered mRNA expression—it has decreased in the tumour tissue [10], which could be due to the insertion in exon 9. The *EI24* encodes a tumour suppressor and is an immediate-early induction target of TP53-mediated apoptosis—it binds to antiapoptotic BLC2. Furthermore, the *EI24* has found to be highly mutated in the case of aggressive breast cancer and is rather

Table 6 The pathways associated to cancer

Pathway name	p value	Number of genes	Genes	Number of variants	Tumour tissue	Control tissue
CD-SNVs						
Wnt/β-catenin signalling	7.07E-04	6	CREBBP, RARA, SMO, SOX10, SOX7, TCF3	6	1	0
Epithelial adherens junction Ssignalling	1.26E-02	4	IQGAP1, KEAP1, TCF3, TUBG1	4	1	0
Germ cell-sertoli cell junction signalling	2.10E-02	4	GSN, IQGAP1, KEAP1, TUBG1	5	1	0
Mouse embryonic stem cell pluripotency	2.59E-02	3	CREBBP, SMO, TCF3	3	1	0
Regulation of the epithelial-mesenchymal transition pathway	3.40E-02	4	FGFR2, SMO, TCF3, ZEB2	4	1	0
Hereditary breast cancer signalling	4.95E-02	3	CREBBP, NPM1, TUBG1	3	1	0
Small indels						
IL-17A signalling in gastric cells	8.79E-03	1	RELA	1	1	0
Role of JAK1, JAK2 and TYK2 in interferon signalling	9.54E-03	1	RELA	1	1	0
Interferon signalling	9.79E-03	1	RELA	1	1	0
IL-15 production	1.00E-02	1	RELA	1	1	0
TNFR2 signalling	1.05E-02	1	RELA	1	1	0
RANK signalling in osteoclasts	2.86E-02	1	RELA	1	1	0

associated to tumour invasiveness than development of the primary tumour [36-38]. In the present case, we found no mutations in *TP53* nor was the expression altered [10]; thus, according to this data, we may suggest that the TP53 is functional in the tumour tissue. However, the TP53 pathway may still be suppressed due to mutated and downregulated *EI24*. Moreover, the aggressive nature of OS is correlated to this finding.

Appling Ingenuity Variant Analysis software, we found over 200 cancer driver variants and 93% of these possibly

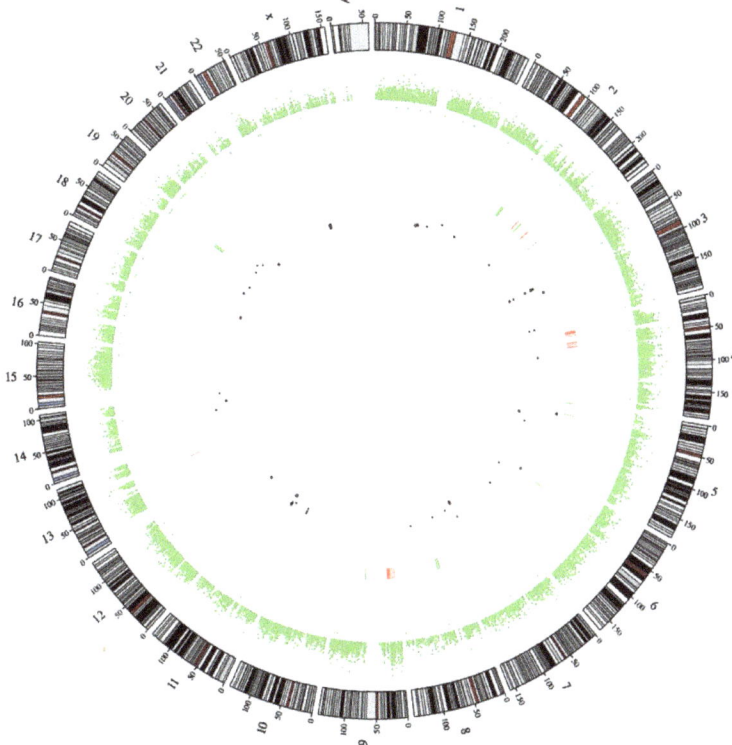

Figure 1 Circos plot illustrating the CNVs and LOHs in the OS tissue compared to that in the control tissue. CNVs are marked as lines in the centre: red—gain and green—loss. LOHs are marked as dots in the centre: black—copy neutral, green—copy gain and red—copy loss.

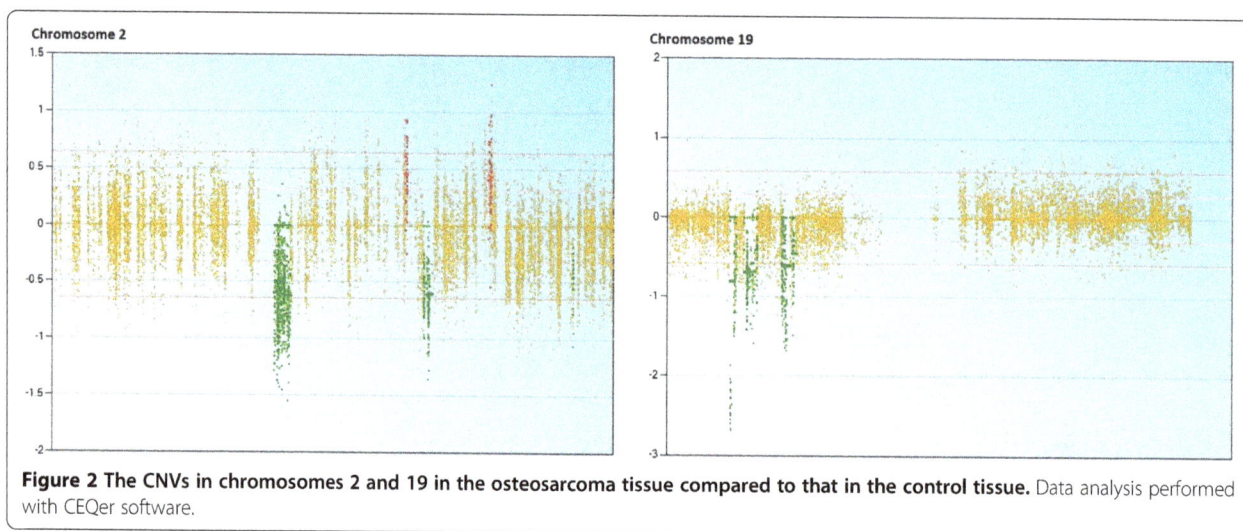

Figure 2 The CNVs in chromosomes 2 and 19 in the osteosarcoma tissue compared to that in the control tissue. Data analysis performed with CEQer software.

cause the loss of gene function. Thirteen homozygous somatic CD-SNVs were detected in different genes—*RGPD3*, *PRDM9*, *FOXK1*, *CCZ1*, *PLAT*, *AGTPBP1*, *SARDH*, *FAH*, *CDC27*, *SBF1*, *LRRC37A3*, *ARL17A* and *LILRB3*. The mRNA expression of *PLAT*, *AGTPBP1* and *LRRC37A3* has increased and of *SFB1* has decreased significantly [10]. We found no previous data about the associations between OS and these genes, except *SBF1*. With previous OS studies, another missense mutation (p.E1539K) has detected in *SBF1* [39]. SBF1 is a SET (a nuclear oncogene) binding

Table 7 The integrative analysis—genes with altered expression pattern [10] and SNVs annotated with ANNOVAR software

Gene name	Transcript name—exon number: nucleotide change/amino acid change	Ijb2 score/ indel	Chr number	Start	End	REF/ALT	logFC	FDR
Germline mutations homozygous in tumour tissue								
STEAP4	NM_024636—exon2: c.G364A/p.A122T	0.647	Chr7	87913221	87913221	C/T	3.015	1.44E-19
	NM_001205316—exon2: c.G364A/p.A122T							
	NM_001205315—exon3: c.G364A/p.A122T							
DDX60L	NM_001012967—exon18: c.T2491C/p.C831R	0.711	Chr4	169341435	169341435	A/G	2.349	2.67E-14
MT1A	NM_005946—exon3: c.A152G/p.K51R	0.785	Chr16	56673828	56673828	A/G	−3.094	0.00795
ACOX1	NM_004035—exon7: c.C936G/p.I312M	0.872	Chr17	73949540	73949540	G/C	−0.809	0.01538
	NM_007292—exon7: c.C936G/p.I312M							
	NM_001185039—exon7: c.C822G/p.I274M							
TMC7	NM_001160364—exon6: c.G431A/p.G144E	0.695	Chr16	19041595	19041595	G/A	1.266	0.01726
	NM_024847—exon6: c.G761A/p.G254E							
MYO7A	NM_001127179—exon27: c.3514_3535del/ p.1172_1179del	Frameshift deletion	Chr11	76895771	76895792	GGAGGC GGGGAC ACCAGG GCCT/-	1.541	0.03810
ATRNL1	NM_001276282—exon8: c.1399_1400insTT/p.L467fs	Frameshift insertion	Chr10	116931101	116931101	-/TT	2.321	0.04535
Somatic mutations homozygous in the tumour tissue								
TMEM120B	NM_001080825—exon3: c.G274A/p.D92N → X → COSM1599921	0.981	Chr12	122186317	122186317	G/A	−1.548	0.00064
TMEM131	NM_015348—exon31: c.C3947T/p.P1316L	0.945	Chr2	98409046	98409046	G/A	−0.799	0.01371
EI24	NM_001007277—exon9: c.733dupC/p.R244fs	Frameshift insertion	Chr11	125452300	125452300	-/C	−0.815	0.01569

These germline or somatic SNVs are all nonsynonymous and homozygous in the tumour tissue and according to Ijb2 database have a disease-causing effect.

Table 8 The integrative analysis—CNVs and RNA expression data [10] is observed together

Gene name	CNVs						RNA expression	
	Chr number	Start	End	Area length	CNV p value	Copy number fold change	logFC	FDR
Loss								
INSR	Chr19	7119459	7294011	174,552	3.18E-11	−6.64	−3.36	9.67E-31
NFIX	Chr19	13106583	13201204	94,621	0	−10.82	−2.45	1.63E-17
FARSA	Chr19	13034964	13044558	9,594	0	−10.82	−2.62	1.96E-16
RAD23A	Chr19	13056627	13063667	7,040	0	−10.82	−2.40	4.46E-16
GINS4	Chr8	41386724	41399418	12,694	8.28E-05	−3.94	−2.79	1.31E-15
GADD45GIP1	Chr19	13064971	13068050	3,079	0	−10.82	−2.82	3.67E-15
IFIH1	Chr2	163123588	163175218	51,630	0	−10.11	2.20	3.69E-14
RPL31	Chr2	101618690	101622885	4,195	0	−9.39	−2.05	5.08E-13
PLEKHG4B	Chr5	156185	181790	25,605	1.08E-05	−4.40	−2.17	1.58E-12
ZNF358	Chr19	7581003	7581135	132	3.18E-11	−6.64	−2.59	2.56E-11
ARHGEF18	Chr19	7459998	7532004	72,006	3.18E-11	−6.64	−1.96	1.44E-10
STX10	Chr19	13255223	13260987	57,64	0	−10.82	−2.55	1.96E-10
COL5A3	Chr19	10102679	10121147	18,468	4.14E-04	−3.53	−1.92	5.45E-10
MGAT4A	Chr2	99242185	99347589	105,404	0	−10.22	1.87	9.95E-10
Gain								
SLC40A1	Chr2	190428309	190428951	642	1.83E-05	4.28	2.22	1.05E-14
KIT	Chr4	55524094	55603446	79,352	1.69E-06	4.79	2.54	1.17E-13
PTPLAD2	Chr9	21008019	21031635	23,616	7.73E-14	7.48	3.02	4.49E-13
ATP8A1	Chr4	42571177	42629126	57,949	1.83E-07	5.22	2.65	4.54E-10
FOCAD	Chr9	20658308	20993327	335,019	7.73E-14	7.48	1.94	9.12E-08
FAM200B	Chr4	15683351	15692070	8,719	3.54E-05	4.14	1.83	8.59E-07
SLIT2	Chr4	20255234	20512189	256,955	4.16E-05	4.10	1.35	1.73E-05
MLLT3	Chr9	20353522	20622514	268,992	7.73E-14	7.48	1.94	2.19E-05
LCORL	Chr4	17887690	18023483	135,793	4.16E-05	4.10	1.40	5.04E-05

Only the genes with lowest FDR value are presented.

Table 9 The integrative analysis - loss of heterozygosity and RNA expression data observed together

Gene name	Chr number	LOHs				RNA expression	
		LOH position	Alleles	LOH	LOH p value	logFC	FDR
MS4A14	Chr11	60165358–60165379	G/C	CopyNeutralLOH	0.025	2.46	3.20E-08
DSC2	Chr18	28666554–28666556	A/C	CopyNeutralLOH	0.025	1.87	3.82E-07
RPS4X	ChrX	71495409–71495414	G/C	CopyNeutralLOH	0.01	−1.44	7.25E-07
RPS23	Chr5	81571874	A/C	CopyNeutralLOH	0.005	−1.43	1.04E06
IL7R	Chr5	35874575	C/T	1AlleleGain	0.025	1.59	6.69E-06
PCNXL2	Chr1	233398713	C/T	CopyNeutralLOH	0.01	1.20	0.00027
HILPDA (C7orf68)	Chr7	128098270	T/G	CopyNeutralLOH	0.0001	−1.12	0.00094
HRNR	Chr1	152188041	C/T	Allele(s)Loss	0.025	−3.09	0.00796
MUC4	Chr3	195515594, 195516630	C/G	CopyNeutralLOH	0.025	−2.22	0.01230

Only the genes with significant mRNA expression changes in the tumour tissue compared to that in the control tissue are presented.

factor 1 and may inhibit the cell division [40]. The decreased expression in the tumour tissue may be responsible for the increased cell proliferation. Some other associations, which might be interesting—*PLAT* gene is important for cell migration and tissue remodelling and the overexpression might cause hyperfibrinolysis [41], which has not previously described in the case of OS. Two mutations in *ARL17A* have detected in chondrosarcoma cells [42]. In the case of *CDC27*, the same mutation (p.E6G) was also brought front by AS as potentially disease causing, which is discussed above. Thus, it is highly likely that at least some of these genes participate in some level of OS pathogenesis.

Additionally, with IVA four homozygous somatic small indels were detected in the tumour tissue. These were in noncoding regions of genes *CTCFL*, *PRR23C*, *CDCA7L* and *ALK*; thus, the effect might be post-transcriptional. *CTCFL* is a genetic paralog of *CTCF*; latter is an important methylation pattern regulator. In the case of *CTCF*, it has previously demonstrated that in the OS tissue, the changes in its methylation pattern may also cause loss of imprinting of *IGF2* and *H19* genes, which further alters their expression pattern [43]. In our OS patient's tumour tissue, the mRNA expression of both *IGF2* and *H19* has increased significantly (FDR = 3.46E-15 and FDR = 0.0015, respectively) [10]. Thus, the association may be valid here also. In *PRR23C*, one missense mutation (p.R190W) has detected previously in the OS tissue [42]. *ALK* encodes a receptor tyrosine kinase and is rearranged, mutated or amplified in several tumours. However, in the case of OS, there are only few reports about ALK [44,45]. In addition, two heterozygous somatic small indels were detected in *DSPP* and *RELA* exons; however, we found no previous data about these findings and associations to OS. The small indels might have an effect on the expression of these genes both pre- and post-transcriptional level; however, these suggestions need to be further studied.

According to IVA, there were several genes with more than one mutation—in *MUC4*, there were even 22 somatic mutations in exons and 44 in introns, although they all were heterozygous. Thus, we found *MUC4* locus to be the most altered in the tumour tissue compared to that in the control tissue. This might explain why its mRNA expression in the tumour tissue has decreased (FDR = 0.012) [10]. Mucin 4 is among major constituents of mucus, and it has demonstrated that primary bone tumours rarely express MUC4 protein [46], which correlates to our finding. Furthermore, with IVA, we found mucin complex (*MUC2*, *MUC4*, *MUC6*) to have a highest significance in OS among others. However, there are also other mucin genes (MUC16, MUC17, MUC20) with somatic heterozygous CD-SNVs. The expression pattern of all other detected mucins has not changed significantly. Thus, mucins may have a role in OS pathogenesis, but we dear not to make any further conclusions.

With IVA, there was four bone-related processes brought front only in the case of the tumour tissue—"myelopoiesis of bone marrow" (*NPM1*, *RARA*), "quantity of trabecular bone" (*CREBBP*, *SMO*), "outgrowth of bone marrow cells" (*ALK*) and "inflammatory response of bone marrow-derived macrophages" (*RELA*). Furthermore, in disease list, 16 genes with over 40 somatic variations were associated to "bone marrow cancer" and "bone tumour"; however, there were also over 200 germline CD-SNVs associated to cancer. Thus, here, we would like to emphasize that in the case of both cancer-associated processes and diseases, the ones associated with bone are somatic mutations; however, the findings possibly promoting cancer are germline mutations. This is one of the phenomena, which we would like to observe in our future studies.

The most significant pathway found with IVA was "WNT/β-catenin signalling pathway" (altered genes: *CREBBP*, *RARA*, *SMO*, *SOX10*, *SOX7*, *TCF3*). Reviewed in [15], the pathway is required for bone development and has demonstrated to be altered in pathogenesis of OS—overexpression of numerous WNT pathway components including WNT ligands, FZDs and LRP receptors and epigenetic silencing of the pathway inhibiting genes, i.e. *WIF1*. However, in our previous study, we found *WNT7B* and *WNT11* to be downregulated and *WNT2B* and *WNT5B* upregulated; *FZD4* and *FZD8* upregulated and *LRP8* and *LRP12* downregulated and *DVL3* downregulated and *WIF1* and *SOST* upregulated. Additionally, genes with CD-SNVs—*RARA*, *SMO* and *SOX7* were upregulated [10]. Thus, our results are rather controversial to several previous studies demonstrating the WNT/β-catenin pathway to be upregulated [47-49]. However, there are also studies correlating to our findings [50,51]. As our study is based on a single case, we dear not to conclude, why the WNT/β-catenin pathway is rather downregulated here, but we suggest the controversial results may occur due to major heterogeneity of OS. Nevertheless, the present study demonstrates that in addition to altered expression patter, the genes involved in WNT/β-catenin signalling pathway carry the CD-SNVs.

In the case of small indels, the IVA brought front the pathways associated to RELA and these are mostly cytokine signalling pathways (Table 6). Previously, it has demonstrated that interaction of IL17A and IL17AR promotes metastasis in OS cells. Furthermore, IL17 stimulates osteoclast resorption [52]. In our previous study, we found IL17AR to be significantly upregulated [10]. Osteoclasts are important in pathogenesis of OS—the more active they are, the more aggressive the tumour is [53]. RELA is demonstrated to enhance the osteoclast differentiation [54]. As IVA predicts the loss of RELA functionality (at least partially, as the small indel is heterozygous), in the present case, the OS might not have been as aggressive as it usually would.

Previously, it has demonstrated that chromothripsis event is common to early stage of OS—hundreds of genomic rearrangements will appear in a single instability event [26]. In the present case, the CEQer software detected nearly 2,400 gain and loss events in 8 chromosomes involved, which should qualify as the chromothripsis. However, the initiating cause of this massive rearrangement is unknown, as there were no traumas or other environmental causes we are aware of.

In general, the gain of copy number should increase the mRNA expression and loss of copy number should decrease the expression [6]. In present work, this pattern was valid in the case of 86.5% of the genes with CNVs and altered expression. One of the strongest findings here was the amount of CNVs in *INSR*, which expression has decreased remarkably (Table 8). The main physiological role of the insulin receptor appears to be metabolic regulation [55]. However, together with IGF1R it forms a hybrid receptor for IGF1, latter together with IGF2 is thought to have a key role in driving the proliferation and survival of sarcoma cells [56]. Furthermore, the growth hormone and IGF1 axis controls the growth and bone modelling/remodelling [57]. Additionally, the IRS1, which is phosphorylated by the INSR, is important for both metabolic and mitogenic pathways [58]. In the present case, the mRNA expression of both *IGF1* and *IGF2* has increased (FDR = 4.65E-35 and FDR = 3.46E-15, respectively); however, the expression of *IGF1R* remained the same in the tumour tissue compared to that in the control tissue [10]. Furthermore, in *IGF1R* we found a heterozygous germline nonsynonymous mutation (p.G1117R) with AS, which according to ljb2 database is a disease causing (data not shown). Similarly to *INSR*, the mRNA expression of *IRS1* is decreased in the tumour tissue compared to that in the control tissue (FDR = 2.62E-10) [10]. Thus, in the present case it seems, the proliferation of tumour cells might be rather supported by increased effect of IGF1, IGF2 and IGF1R homodimer associations, than IGF1, IGF2 and INSR-IGF1R heterodimer associations or INSR effects on IRS1.

The loss of heterozygosity has been reported to be extensive in OS exomes [39]. In the present case, we did not detect whole chromosome or gene region loss; however, we did detect the loss of heterozygosity in smaller regions. The genes with LOH findings and increased mRNA expression—*MS4A14*, *DSC2*, *IL7R* and *PCNXL2* have not associated to OS previously. However, in the case of *DSC2*, the overexpression has demonstrated to be inversely correlated to bone metastasis-free survival [59]. The mutations in *IL7R* exon 6 have been demonstrated to be present in leukaemia patients' bone marrow samples but not associated to other solid tumours [60]. The five genes with LOHs and decreased mRNA expression—*RPS4X*, *RPS23*, *HILPDA* (*C7orf68*), *HRNR* and

MUC4 also do not have previous information associated to OS. Nonetheless, also the LOH analysis brought forward different genes in mucin family. In addition to *MUC4*, there were also other genes with LOHs but with insignificant mRNA expression changes in the tumour tissues—*MUC2*, *MUC6* and *MUC17*. Thus, these results also support the idea that mucins might have a role in pathogenesis of osteosarcoma.

In summary, the present case has several characteristics previously demonstrated in OS. The wide genomic arrangements have appeared—SNVs and small indels all over the genome and CNVs in some chromosomes; and in several cases, these rearrangements may have an effect on gene expression. Furthermore, the germline mutations seem to be associated to cancer in general and somatic mutations to bone tumours. The most significant pathway was the one probably most thoroughly studied in the case of OS—the WNT/β-catenin signalling pathway. We found several genes in this pathway carrying the cancer driver variances. Additionally, the IGF1/IGF2 and IGF1R homodimer signalling might have an essential effect on OS pathogenesis. Which also needs to be emphasized is that according to our data (based on DNA and RNA studies), there is no evidence of a nonfunctional *TP53*; however, the TP53 pathway might be suppressed in further levels—the downregulation of *EI24*. In addition, with this study, we found associations between different genes and OS pathogenesis, which have not demonstrated before in earlier studies. We found the *MUC4* locus to be the most altered in the tumour tissue compared to that in the control tissue; furthermore, several other mucin genes are also possibly associated to OS. The somatic mutation in *CDC27* was brought front by two different data analysis softwares and might have a role in OS pathogenesis.

Conclusions

All genes, in which the mutations were detected, may be considered as potential targets for additional studies (i.e. functional, histopathological, clinical studies) for finding OS biomarkers. The present study brought front the WNT pathway genes, IGF1/IGF2 and IGF1R homodimer signalling pathway genes, *TP53* together with *EI24*, *MUC4* together with other mucin genes and *CDC27* as potential biomarkers for OS. Finally, as this study is based on a single case and only DNA and RNA analysis, these data may not be taken as conclusive evidence and further studies are needed to confirm the present findings.

Additional files

Additional file 1: ANNOVAR software. The file contains the list of SNVs (coverage at least 20 times) and small indels detected from WES study.

Additionally, the dbSNP135, dbCOSMIC67, ljb2 scores and RNA-seq i nformation is added if available.

Additional file 2: Ingenuity Variant Analysis software. The file contains the list of SNVs, which according to IVA are associated to cancer. Additionally, the dbSNP135, SIFT and POLYPHEN functions and RNA-seq information is added if available.

Additional file 3: CEQer software. The file contains the list of genes where CNVs and LOHs were detected. Additionally, the RNA-seq information is added if available.

Competing interests
The authors declare that they have no competing interests.

Authors' contributions
ER participated in the research concept and design, data analysis and interpretation, writing of the article and critical revision of the article and performed the statistical analysis. SK involved in the research concept and design, data analysis and interpretation and critical revision of the article. XDH contributed in the research concept and design and carried out the collection and/or assembly of data. KM participated in the research concept and design, collection and/or assembly of data, data analysis and interpretation and critical revision of the article. AM involved in the research concept and design, collection and/or assembly of data, data analysis and interpretation and critical revision of the article. All authors read and approved the final manuscript.

Acknowledgements
This work was supported by the institutional research funding IUT (IUT20-46) of the Estonian Ministry of Education and Research, by the Centre of Translational Genomics of University of Tartu (SP1GVARENG) and by the European Regional Development Fund (Centre of Translational Medicine, University of Tartu).

Author details
[1]Department of Pathophysiology, University of Tartu, 19 Ravila Street Tartu 50411, Estonia. [2]Department of Reproductive Biology, Estonian University of Life Sciences, 64 Kreutzwaldi Street, Tartu, Estonia. [3]Department of Traumatology and Orthopaedics, University of Tartu, 8 Puusepa Street, Tartu, Estonia. [4]Department of Oncology, Hue University of Medicine and Pharmacy, 6 Ngo Quyen Street, Hue, Vietnam. [5]Traumatology and Orthopaedics Clinic, Tartu University Hospital, 8 Puusepa Street, Tartu, Estonia.

References
1. Picci P: Osteosarcoma (osteogenic sarcoma). *Orphanet J Rare Dis* 2007, **2**:6.
2. Ottaviani G, Jaffe N: The etiology of osteosarcoma. *Cancer Treat Res* 2009, **152**:15–32.
3. Botter SM, Neri D, Fuchs B: Recent advances in osteosarcoma. *Curr Opin Pharmacol* 2014, **16C**:15–23.
4. Allison DC, Carney SC, Ahlmann ER, Hendifar A, Chawla S, Fedenko A, Angeles C, Menendez LR: A meta-analysis of osteosarcoma outcomes in the modern medical era. *Sarcoma* 2012, **2012**:704872. http://www.ashg.org/2013meeting/programguide/files/assets/basic-html/page258.html.
5. Kiezun A, Janeway K, Tonzi P, Mora J, Aguiar S, Mercado G, Melendez J, Garraway L, Rodriguez-Galindo C, Orkin S, Golub T, Getz G, Yunes JA: Next generation sequencing of osteosarcoma identifies the PI3K/mTOR pathway as a unifying vulnerability to be exploited for targeted therapy. In.: ASHG meeting 2013; 2013.
6. Kuijjer ML, Hogendoorn PC, Cleton-Jansen AM: Genome-wide analyses on high-grade osteosarcoma: making sense of a genomically most unstable tumor. *Int J Cancer* 2013, **133**(11):2512–2521.
7. Savage SA, Mirabello L, Wang Z, Gastier-Foster JM, Gorlick R, Khanna C, Flanagan AM, Tirabosco R, Andrulis IL, Wunder JS, Gokgoz N, Patiño-Garcia A, Sierrasesúmaga L, Lecanda F, Kurucu N, Ilhan IE, Sari N, Serra M, Hattinger C, Picci P, Spector LG, Barkauskas DA, Marina N, de Toledo SR, Petrilli AS, Amary MF, Halai D, Thomas DM, Douglass C, Meltzer PS, et al: Genome-wide association study identifies two susceptibility loci for osteosarcoma. *Nat Genet* 2013, **45**(7):799–803.
8. Ottaviano L, Schaefer KL, Gajewski M, Huckenbeck W, Baldus S, Rogel U, Mackintosh C, de Alava E, Myklebost O, Kresse SH, Meza-Zepeda LA, Serra M, Cleton-Jansen AM, Hogendoorn PC, Buerger H, Aigner T, Gabbert HE, Poremba C: Molecular characterization of commonly used cell lines for bone tumor research: a trans-European EuroBoNet effort. *Genes Chromosomes Cancer* 2010, **49**(1):40–51.
9. Kansara M, Tsang M, Kodjabachian L, Sims NA, Trivett MK, Ehrich M, Dobrovic A, Slavin J, Choong PF, Simmons PJ, Dawid IB, Thomas DM: Wnt inhibitory factor 1 is epigenetically silenced in human osteosarcoma, and targeted disruption accelerates osteosarcomagenesis in mice. *J Clin Invest* 2009, **119**(4):837–851.
10. Märtson A, Kõks S, Reimann E, Prans E, Erm T, Maasalu K: Transcriptome analysis of osteosarcoma identifies suppression of wnt pathway and up-regulation of adiponectin as potential biomarker. *Genomic Discovery* 2013, **1**:1–9.
11. Namlos HM, Meza-Zepeda LA, Baroy T, Ostensen IH, Kresse SH, Kuijjer ML, Serra M, Burger H, Cleton-Jansen AM, Myklebost O: Modulation of the osteosarcoma expression phenotype by microRNAs. *PLoS One* 2012, **7**(10):e48086.
12. Hingorani P, Zhang W, Gorlick R, Kolb EA: Inhibition of Src phosphorylation alters metastatic potential of osteosarcoma in vitro but not in vivo. *Clin Cancer Res* 2009, **15**(10):3416–3422.
13. Yap TA, Arkenau HT, Camidge DR, George S, Serkova NJ, Gwyther SJ, Spratlin JL, Lal R, Spicer J, Desouza NM, Leach MO, Chick J, Poondru S, Boinpally R, Gedrich R, Brock K, Stephens A, Eckhardt SG, Kaye SB, Demetri G, Scurr M: First-in-human phase I trial of two schedules of OSI-930, a novel multikinase inhibitor, incorporating translational proof-of-mechanism studies. *Clin Cancer Res* 2013, **19**(4):909–919.
14. Lv Z, Wang C, Yuan T, Liu Y, Song T, Liu Y, Chen C, Yang M, Tang Z, Shi Q, Weng Y: Bone morphogenetic protein 9 regulates tumor growth of osteosarcoma cells through the Wnt/beta-catenin pathway. *Oncol Rep* 2014, **31**(2):989–994.
15. Cai Y, Cai T, Chen Y: Wnt pathway in osteosarcoma, from oncogenic to therapeutic. *J Cell Biochem* 2014, **115**(4):625–631.
16. Guo M, Cai C, Zhao G, Qiu X, Zhao H, Ma Q, Tian L, Li X, Hu Y, Liao B, Ma B, Fan Q: Hypoxia promotes migration and induces CXCR4 expression via HIF-1alpha activation in human osteosarcoma. *PLoS One* 2014, **9**(3):e90518.
17. Raymond A, Ayala A, Knuutila S: *Conventional osteosarcoma, genetics.* Lyon: IARC Press; 2002.
18. Lau CC, Harris CP, Lu XY, Perlaky L, Gogineni S, Chintagumpala M, Hicks J, Johnson ME, Davino NA, Huvos AG, Meyers PA, Healy JH, Gorlick R, Rao PH: Frequent amplification and rearrangement of chromosomal bands 6p12-p21 and 17p11.2 in osteosarcoma. *Genes Chromosomes Cancer* 2004, **39**(1):11–21.
19. Kresse SH, Ohnstad HO, Paulsen EB, Bjerkehagen B, Szuhai K, Serra M, Schaefer KL, Myklebost O, Meza-Zepeda LA: LSAMP, a novel candidate tumor suppressor gene in human osteosarcomas, identified by array comparative genomic hybridization. *Genes Chromosomes Cancer* 2009, **48**(8):679–693.
20. Pasic I, Shlien A, Durbin AD, Stavropoulos DJ, Baskin B, Ray PN, Novokmet A, Malkin D: Recurrent focal copy-number changes and loss of heterozygosity implicate two noncoding RNAs and one tumor suppressor gene at chromosome 3q13.31 in osteosarcoma. *Cancer Res* 2010, **70**(1):160–171.
21. Tarkkanen M, Karhu R, Kallioniemi A, Elomaa I, Kivioja AH, Nevalainen J, Bohling T, Karaharju E, Hyytinen E, Knuutila S: Gains and losses of DNA sequences in osteosarcomas by comparative genomic hybridization. *Cancer Res* 1995, **55**(6):1334–1338.
22. Tarkkanen M, Elomaa I, Blomqvist C, Kivioja AH, Kellokumpu-Lehtinen P, Bohling T, Valle J, Knuutila S: DNA sequence copy number increase at 8q: a potential new prognostic marker in high-grade osteosarcoma. *Int J Cancer* 1999, **84**(2):114–121.
23. Overholtzer M, Rao PH, Favis R, Lu XY, Elowitz MB, Barany F, Ladanyi M, Gorlick R, Levine AJ: The presence of p53 mutations in human osteosarcomas correlates with high levels of genomic instability. *Proc Natl Acad Sci U S A* 2003, **100**(20):11547–11552.
24. Yen CC, Chen WM, Chen TH, Chen WY, Chen PC, Chiou HJ, Hung GY, Wu HT, Wei CJ, Shiau CY, Wu YC, Chao TC, Tzeng CH, Chen PM, Lin CH, Chen YJ, Fletcher JA: Identification of chromosomal aberrations associated with disease progression and a novel 3q13.31 deletion involving LSAMP gene in osteosarcoma. *Int J Oncol* 2009, **35**(4):775–788.

Whole exome sequencing of a single osteosarcoma case—integrative analysis with whole transcriptome...

175

25. Lu XY, Lu Y, Zhao YJ, Jaeweon K, Kang J, Xiao-Nan L, Ge G, Meyer R, Perlaky L, Hicks J, Chintagumpala M, Cai WW, Ladanyi M, Gorlick R, Lau CC, Pati D, Sheldon M, Rao PH: **Cell cycle regulator gene CDC5L, a potential target for 6p12-p21 amplicon in osteosarcoma.** *MCR* 2008, **6**(6):937–946.

26. Stephens PJ, Greenman CD, Fu B, Yang F, Bignell GR, Mudie LJ, Pleasance ED, Lau KW, Beare D, Stebbings LA, McLaren S, Lin ML, McBride DJ, Varela I, Nik-Zainal S, Leroy C, Jia M, Menzies A, Butler AP, Teague JW, Quail MA, Burton J, Swerdlow H, Carter NP, Morsberger LA, Iacobuzio-Donahue C, Follows GA, Green AR, Flanagan AM, Stratton MR, et al: **Massive genomic rearrangement acquired in a single catastrophic event during cancer development.** *Cell* 2011, **144**(1):27–40.

27. Wang K, Li M, Hakonarson H: **ANNOVAR: functional annotation of genetic variants from high-throughput sequencing data.** *Nucleic Acids Res* 2010, **38**(16):e164.

28. Adzhubei IA, Schmidt S, Peshkin L, Ramensky VE, Gerasimova A, Bork P, Kondrashov AS, Sunyaev SR: **A method and server for predicting damaging missense mutations.** *Nat Methods* 2010, **7**(4):248–249.

29. Gonzalez-Perez A, Lopez-Bigas N: **Improving the assessment of the outcome of nonsynonymous SNVs with a consensus deleteriousness score.** *Condel Am J Human Genetics* 2011, **88**(4):440–449.

30. Kumar P, Henikoff S, Ng PC: **Predicting the effects of coding non-synonymous variants on protein function using the SIFT algorithm.** *Nat Protoc* 2009, **4**(7):1073–1081.

31. Blankenberg D, Von Kuster G, Coraor N, Ananda G, Lazarus R, Mangan M, Nekrutenko A, Taylor J: **Galaxy: a web-based genome analysis tool for experimentalists.** *Current protocols in molecular biology/edited by Frederick M Ausubel [et al.]* 2010, **Chapter 19**:Unit 19.10.1:1–21.

32. Giardine B, Riemer C, Hardison RC, Burhans R, Elnitski L, Shah P, Zhang Y, Blankenberg D, Albert I, Taylor J, Miller W, Kent WJ, Nekrutenko A: **Galaxy: a platform for interactive large-scale genome analysis.** *Genome Res* 2005, **15**(10):1451–1455.

33. Yost SE, Pastorino S, Rozenzhak S, Smith EN, Chao YS, Jiang P, Kesari S, Frazer KA, Harismendy O: **High-resolution mutational profiling suggests the genetic validity of glioblastoma patient-derived pre-clinical models.** *PLoS One* 2013, **8**(2):e56185.

34. Topper LM, Campbell MS, Tugendreich S, Daum JR, Burke DJ, Hieter P, Gorbsky GJ: **The dephosphorylated form of the anaphase-promoting complex protein Cdc27/Apc3 concentrates at kinetochores and chromosome arms in mitosis.** *Cell Cycle* 2002, **1**(4):282–292.

35. Talvinen K, Karra H, Pitkanen R, Ahonen I, Nykanen M, Lintunen M, Soderstrom M, Kuopio T, Kronqvist P: **Low cdc27 and high securin expression predict short survival for breast cancer patients.** *APMIS: acta pathologica, microbiologica, et immunologica Scandinavica* 2013, **121**(10):945–953.

36. Zhao YG, Zhao H, Miao L, Wang L, Sun F, Zhang H: **The p53-induced gene Ei24 is an essential component of the basal autophagy pathway.** *J Biol Chem* 2012, **287**(50):42053–42063.

37. Gu Z, Flemington C, Chittenden T, Zambetti GP: **ei24, a p53 response gene involved in growth suppression and apoptosis.** *Mol Cell Biol* 2000, **20**(1):233–241.

38. Zhao X, Ayer RE, Davis SL, Ames SJ, Florence B, Torchinsky C, Liou JS, Shen L, Spanjaard RA: **Apoptosis factor EI24/PIG8 is a novel endoplasmic reticulum-localized Bcl-2-binding protein which is associated with suppression of breast cancer invasiveness.** *Cancer Res* 2005, **65**(6):2125–2129.

39. Joseph CG, Hwang H, Jiao Y, Wood LD, Kinde I, Wu J, Mandahl N, Luo J, Hruban RH, Diaz LA Jr, He TC, Vogelstein B, Kinzler KW, Mertens F, Papadopoulos N: **Exomic analysis of myxoid liposarcomas, synovial sarcomas, and osteosarcomas.** *Genes, chromosomes & cancer* 2014, **53**(1):15–24.

40. Firestein R, Cleary ML: **Pseudo-phosphatase Sbf1 contains an N-terminal GEF homology domain that modulates its growth regulatory properties.** *J Cell Sci* 2001, **114**(Pt 16):2921–2927.

41. Booth NA, Bennett B, Wijngaards G, Grieve JH: **A new life-long hemorrhagic disorder due to excess plasminogen activator.** *Blood* 1983, **61**(2):267–275.

42. Tarpey PS, Behjati S, Cooke SL, Van Loo P, Wedge DC, Pillay N, Marshall J, O'Meara S, Davies H, Nik-Zainal S, Beare D, Butler A, Gamble J, Hardy C, Hinton J, Jia MM, Jayakumar A, Jones D, Latimer C, Maddison M, Martin S, McLaren S, Menzies A, Mudie L, Raine K, Teague JW, Tubio JM, Halai D, Tirabosco R, Amary F, et al: **Frequent mutation of the major cartilage collagen gene COL2A1 in chondrosarcoma.** *Nat Genet* 2013, **45**(8):923–926.

43. Ulaner GA, Vu TH, Li T, Hu JF, Yao XM, Yang Y, Gorlick R, Meyers P, Healey J, Ladanyi M, Hoffman AR: **Loss of imprinting of IGF2 and H19 in osteosarcoma is accompanied by reciprocal methylation changes of a CTCF-binding site.** *Hum Mol Genet* 2003, **12**(5):535–549.

44. Pant V, Jambhekar NA, Madur B, Shet TM, Agarwal M, Puri A, Gujral S, Banavali M, Arora B: **Anaplastic large cell lymphoma (ALCL) presenting as primary bone and soft tissue sarcoma—a study of 12 cases.** *Ind J Pathology & Microbiology* 2007, **50**(2):303–307.

45. Choy E, Hornicek F, MacConaill L, Harmon D, Tariq Z, Garraway L, Duan Z: **High-throughput genotyping in osteosarcoma identifies multiple mutations in phosphoinositide-3-kinase and other oncogenes.** *Cancer* 2012, **118**(11):2905–2914.

46. Tirabosco R, Berisha F, Ye H, Halai D, Amary MF, Flanagan AM: **Assessment of MUC4 expression in primary bone tumours.** *Histopathology* 2013, **63**(1):142–143.

47. Flores RJ, Li Y, Yu A, Shen J, Rao PH, Lau SS, Vannucci M, Lau CC, Man TK: **A systems biology approach reveals common metastatic pathways in osteosarcoma.** *BMC Syst Biol* 2012, **6**:50.

48. Ma Y, Ren Y, Han EQ, Li H, Chen D, Jacobs JJ, Gitelis S, O'Keefe RJ, Konttinen YT, Yin G, Li TF: **Inhibition of the Wnt-beta-catenin and Notch signalling pathways sensitizes osteosarcoma cells to chemotherapy.** *Biochem Biophys Res Commun* 2013, **431**(2):274–279.

49. Leow PC, Tian Q, Ong ZY, Yang Z, Ee PL: **Antitumor activity of natural compounds, curcumin and p KF118–310, as Wnt/beta-catenin antagonists against human osteosarcoma cells.** *Investig New Drugs* 2010, **28**(6):766–782.

50. Cleton-Jansen AM, Anninga JK, Briaire-de Bruijn IH, Romeo S, Oosting J, Egeler RM, Gelderblom H, Taminiau AH, Hogendoorn PC: **Profiling of high-grade central osteosarcoma and its putative progenitor cells identifies tumourigenic pathways.** *Br J Cancer* 2009, **101**(12):2064.

51. Cai Y, Mohseny AB, Karperien M, Hogendoorn PC, Zhou G, Cleton-Jansen AM: **Inactive Wnt/beta-catenin pathway in conventional high-grade osteosarcoma.** *J Pathol* 2010, **220**(1):24–33.

52. Van Bezooijen RL, Papapoulos SE, Lowik CW: **Effect of interleukin-17 on nitric oxide production and osteoclastic bone resorption: is there dependency on nuclear factor-kappaB and receptor activator of nuclear factor kappaB (RANK)/RANK ligand signalling?** *Bone* 2001, **28**(4):378–386.

53. Avnet S, Longhi A, Salerno M, Halleen JM, Perut F, Granchi D, Ferrari S, Bertoni F, Giunti A, Baldini N: **Increased osteoclast activity is associated with aggressiveness of osteosarcoma.** *Int J Oncol* 2008, **33**(6):1231–1238.

54. Vaira S, Alhawagri M, Anwisye I, Kitaura H, Faccio R, Novack DV: **RelA/p65 promotes osteoclast differentiation by blocking a RANKL-induced apoptotic JNK pathway in mice.** *J Clin Invest* 2008, **118**(6):2088–2097.

55. Lee J, Pilch PF: **The insulin receptor: structure, function, and signalling.** *Am J Physiol* 1994, **266**(2 Pt 1):C319–C334.

56. Kim SY, Toretsky JA, Scher D, Helman LJ: **The role of IGF-1R in pediatric malignancies.** *Oncologist* 2009, **14**(1):83–91.

57. Canalis E, McCarthy T, Centrella M: **Growth factors and the regulation of bone remodeling.** *J Clin Invest* 1988, **81**(2):277–281.

58. Dearth RK, Cui X, Kim HJ, Kuiatse I, Lawrence NA, Zhang X, Divisova J, Britton OL, Mohsin S, Allred DC, Hadsell DL, Lee AV: **Mammary tumorigenesis and metastasis caused by overexpression of insulin receptor substrate 1 (IRS-1) or IRS-2.** *Mol Cell Biol* 2006, **26**(24):9302–9314.

59. Sanz-Pamplona R, Garcia-Garcia J, Franco S, Messeguer X, Driouch K, Oliva B, Sierra A: **A taxonomy of organ-specific breast cancer metastases based on a protein-protein interaction network.** *Mol BioSyst* 2012, **8**(8):2085–2096.

60. Kim MS, Chung NG, Kim MS, Yoo NJ, Lee SH: **Somatic mutation of IL7R exon 6 in acute leukemias and solid cancers.** *Hum Pathol* 2013, **44**(4):551–555.

16

Exome sequencing identifies novel and recurrent mutations in GJA8 and CRYGD associated with inherited cataract

Donna S Mackay, Thomas M Bennett, Susan M Culican and Alan Shiels*

Abstract

Background: Inherited cataract is a clinically important and genetically heterogeneous cause of visual impairment. Typically, it presents at an early age with or without other ocular/systemic signs and lacks clear phenotype-genotype correlation rendering both clinical classification and molecular diagnosis challenging. Here we have utilized trio-based whole exome sequencing to discover mutations in candidate genes underlying autosomal dominant cataract segregating in three nuclear families.

Results: In family A, we identified a recurrent heterozygous mutation in exon-2 of the gene encoding γD-crystallin (*CRYGD*; c.70C > A, p.Pro24Thr) that co-segregated with 'coralliform' lens opacities. Families B and C were found to harbor different novel variants in exon-2 of the gene coding for gap-junction protein α8 (*GJA8*; c.20T > C, p.Leu7Pro and c.293A > C, p.His98Pro). Each novel variant co-segregated with disease and was predicted *in silico* to have damaging effects on protein function.

Conclusions: Exome sequencing facilitates concurrent mutation-profiling of the burgeoning list of candidate genes for inherited cataract, and the results can provide enhanced clinical diagnosis and genetic counseling for affected families.

Keywords: Cataract, Exome sequencing, *CRYGD*, *GJA8*

Background

Hereditary forms of cataract constitute a clinically and genetically heterogeneous condition affecting the ocular lens [1-3]. Typically, inherited cataract has an early onset (<40 years) and most cases are diagnosed at birth (congenital), during infancy, or during childhood accounting for 10%–25% of all pediatric cataract cases [2]. Congenital and infantile forms of cataract are a clinically important cause of impaired visual development that accounts for 3%–39% of childhood blindness, worldwide [4]. Despite advances in surgical treatment, pediatric cataract poses a long-term risk of postoperative complications including secondary glaucoma, nystagmus, and retinal detachment [5-9].

Cataract can be inherited, either, as an isolated lens phenotype—usually with autosomal dominant transmission and full penetrance—or as part of a genetic/metabolic disorder (http://www.omim.org) involving additional ocular defects (e.g., anterior segment dysgenesis MIM107250) and/or systemic abnormalities (e.g., galactosemia MIM230400). Under slit-lamp examination, inherited cataract exhibits considerable inter- and intrafamilial phenotypic variation in location, size, shape, density, progression rate, and even color of the lens opacities [10]. Currently, genetic studies have identified over 39 genes and loci for inherited cataract, with or without other ocular signs [1,3]. These include gene coding for α-, β-, and γ-crystallins (e.g., *CRYAA*, *CRYBB2*, *CRYGD*), α-connexins (*GJA3*, *GJA8*) and other lens membrane or cytoskeleton proteins (e.g., *MIP*, *BFSP2*), several transcription factors (e.g., *HSF4*, *PITX3*), and an expanding group of functionally divergent genes (e.g., *EPHA2*, *TDRD7*, *FYCO1*). Since mutations in the same gene can cause morphologically different lens opacities and mutations in different genes can cause similar opacities, there is little genotype-phenotype correlation for inherited cataract

_____
* Correspondence: shiels@vision.wustl.edu
Department of Ophthalmology and Visual Sciences, Washington University School of Medicine, 660 S. Euclid Ave., Box 8096, St. Louis, Missouri 63110, USA

rendering both clinical classification and molecular diagnosis challenging.

Traditionally, linkage analysis in extended pedigrees has been used to map cataract disease loci to specific chromosome regions and thereby limit the number of positional candidate genes that need to be conventionally sequenced in order to discover underlying mutations. However, the advent of next-generation (massively parallel) sequencing has facilitated the concurrent screening of multiple candidate genes in nuclear families and cases without a family history. Here, we have undertaken affected child-parent-trio-based whole-exome next-generation sequencing in order to identify mutations underlying autosomal dominant cataract in three nuclear families.

Results

Cataract families

We investigated three Caucasian-American pedigrees segregating cataract with autosomal dominant transmission in the absence of other ocular and/or systemic abnormalities (Figures 1A and 2A,D). A review of ophthalmic records indicated that bilateral cataract was diagnosed at birth (congenital) or during infancy in all three families with age-at-surgery ranging from 3 months to 1 year. In family A, the lens opacities appeared similar to those first described by Gunn in 1895 as resembling a piece of coral or coralliform [11]. No clinical images of lens opacities were available for family B or C, and none of the families had a sufficient number of meiotic events (≥10) to support

independent linkage analysis. Instead, an affected child–parent plus spouse trio from each family was selected for whole-exome sequencing.

Candidate genes and exome sequences

We pre-selected 39 candidate genes for inherited cataract (Additional file 1) cited in the OMIM (http://www.omim. org), Cat-Map (http://cat-map.wustl.edu/), and iSyTE (http:// bioinformatics.udel.edu/Research/iSyTE) databases [3,12]. The candidate list comprises genes known to be highly expressed in the lens including those coding for cystallins, connexins, and other lens membrane/cytoskeletal proteins, along with several more widely expressed genes that are associated with cataract and other limited eye/systemic conditions. Collectively, these candidate genes span over 111,000 bps of the genome and contain 300 exons located on chromosomes 1–13, 16, 17, 19–22, and X.

For all nine exome samples, over 98% of total paired-end reads were mapped to the reference genome (Additional file 2). Approximately 72%–84% of mapped reads were present in the captured exomes, and the average mean-mapped read-depth was 149.2X. With the exception of one sample in family C (C-I:1), >97% of each exome achieved a read-depth of ≥10X coverage, yielding a total of >38,900 single nucleotide polymorphisms (SNPs), of which >8,400 were non-synonymous and >1,400 were novel. For exome C-I:1, 80.61% reached ≥10X coverage yielding a total of 34,435 SNPs (7,639 non-synonymous and 1,331 novel). In addition, exome C-I:1 contained several more unexpected

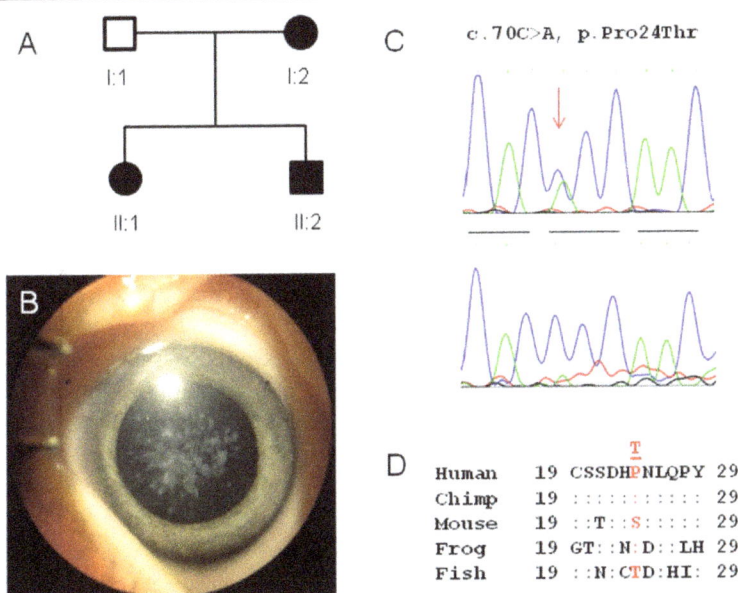

Figure 1 Mutation analysis of inherited cataract in family A. (A) Pedigree of family A. Squares denote males, circles denote females, and filled symbols denote affected status. The trio of individuals I:1, I:2, and II:1 was subject to exome sequencing. **(B)** Photograph of coralliform lens opacities in the left eye of individual II:2 just prior to surgery at 3 months of age. **(C)** Sanger sequence of *CRYGD* showing the heterozygous c.70 C > A and p.Pro24Thr mutation found in affected individuals I:2, II:1, and II:2 (upper trace) but not in the unaffected spouse I:1 (lower trace). Horizontal bars indicate the codon reading frame. **(D)** Amino acid alignment of *CRYGD* showing low cross-species conservation of Pro24.

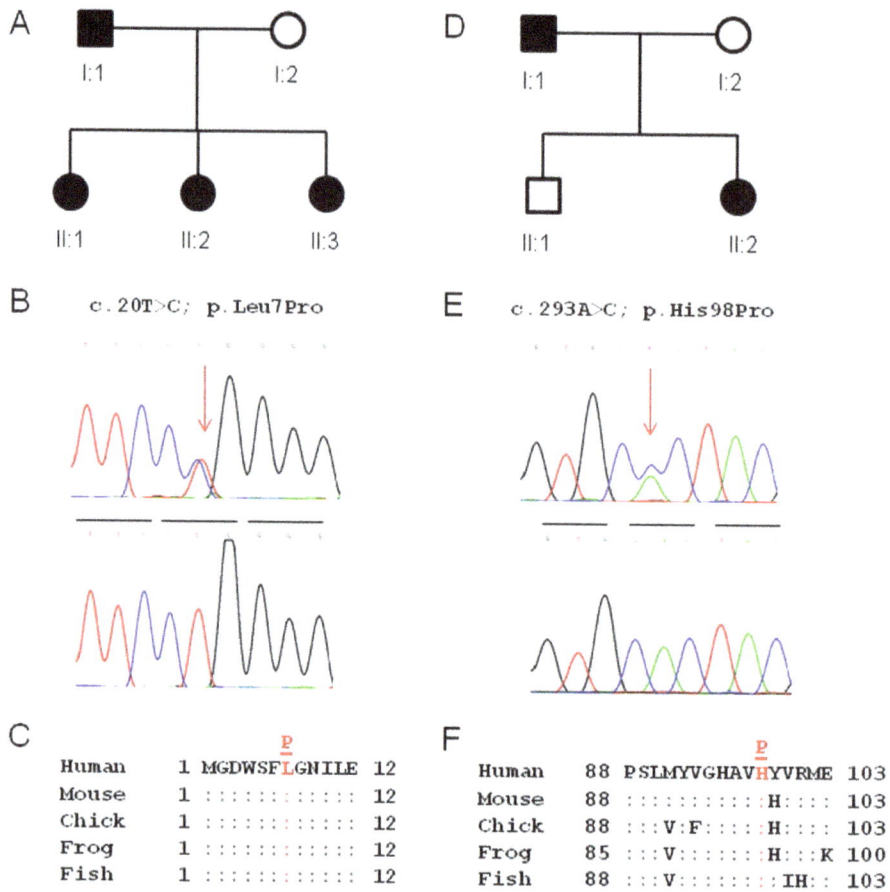

Figure 2 Mutation analysis of inherited cataract in family B and family C. (**A**) Pedigree of family B. The trio of individuals I:1, I:2, and II:1 was subject to exome sequencing. (**B**) Sanger sequence of *GJA8* showing the heterozygous c.20 T > C, and p.Leu7Pro mutation found in affected individuals I:1, II:1, II:2, and II:3 but not in the affected spouse I:2 (lower trace). Horizontal bars indicate the codon reading frame. (**C**) Amino acid alignment of *GJA8* showing high cross-species conservation of Leu7. (**D**) Pedigree showing family C. The trio of individuals I:1, I:2, and II:2 was subject to exome sequencing. (**E**) Sanger sequence of *GJA8* showing the heterozygous c. 293A > C and p.His98Pro mutation found in affected individuals I:1 and II:2 (upper trace) but not in the unaffected individuals I:2 and II:1 (lower trace). Horizontal bars indicate the codon reading-frame. (**F**) Amino acid alignment of *GJA8* showing high cross-species conservation of His98.

regions of low coverage (gaps) than those detected in the other exomes (Additional file 2). However, the reduced coverage of exome C-I:1 did not compromise analysis of variants in the candidate genes of interest. Coverage of the 39 candidate genes exceeded a read-depth of >10X with three exceptions. The iron response element (IRE) of *FTL* is located in the 5′-UTR (untranslated region) and was not covered by capture probes. In addition, coverage of the single exons that code for *FOXE3* and *MAF* was incomplete as previously reported [13]. We excluded mutations in all three missing gene regions by Sanger sequencing of an affected member of each family essentially as described [13,14]. Collectively, from the nine exomes sequenced, 112 variants were identified in 32 of the 39 candidate genes (Additional file 3). Of these variants, only five did not have genome reference sequence (rs) numbers and were potentially novel variants.

Family A variants

A review of the exome SNPs in family A with the list of candidate genes for cataract identified a total of 76 variants in 28 of 39 genes (Additional file 1 and Additional file 3). Of these, six variants (two coding/missense and four non-coding/synonymous) in five candidate genes were found in both affected relatives and not in the unaffected spouse. However, five of these variants associated with four candidate genes (*SLC16A12*, *PAX6*, *CRYAB*, *GALK1*) were excluded as disease-causing mutations as they have minor allele frequencies (MAFs) >0.01% (range 8.5%–52.4%) in Caucasians (Additional file 3). We note that the variant rs3740030 in *SLC16A12* on chromosome 10 (chr10:91,222,287) has previously been associated with age-related cataract [15]. As rs3740030 was first thought to be a non-coding variant located in the 5′-UTR, the authors proposed a complex functional mechanism that

involved modulation of translational efficiency. However, rs3740030 is now known to be located in exon-3 of *SLC16A12* (c.49T > G) and was predicted to result in a non-conservative tryptophan-to-glycine substitution at codon 17 (p.Trp17Gly). While this variant was also predicted *in silico* to have a damaging effect on protein function (PolyPhen-2 score =0.997), it had a MAF value of 8.5% in Caucasians, suggesting that it is unlikely to be disease causing in family A. The remaining variant, rs28931605, occurred in exon-2 of *CRYGD* (c.70C > A) on chromosome 2 (chr2:208,989,018) and was predicted to result in the non-conservative substitution of proline-to-threonine at codon 24 (p.Pro24Thr) (Table 1). While this variant was predicted *in silico* to be tolerated, benign, or neutral with respect to protein function (Table 2), it has been previously associated with autosomal dominant cataract in multiple families (Additional file 4). The p.Pro24Thr variant has also been documented as p.Pro23Thr based on N-terminal processing of the of the *CRYGD* protein that removes the initiator methionine residue. Here, we have adopted the recommended nomenclature in order to avoid confusion and re-numbering of other mutations in *CRYGD* associated with inherited cataract [16]. Sanger sequencing of all four members of family A (Figure 1A,C) confirmed that the p.Pro24Thr variant co-segregated with disease providing further support for its role as a causal mutation.

Family B variants

A review of the exome SNPs in family B with the candidate gene list, revealed a total of 73 variants in 22 of 39 genes (Additional file 1 and Additional file 3). Only 13 of these variants (12 non-coding or synonymous) associated with 7 of the candidate genes were found in both affected relatives and not in the unaffected spouse. All 12 non-coding or synonymous variants had MAF values >0.01% (range 0.4%–45.80%) and were effectively excluded as disease-causing mutations. The remaining variant was located in exon-2 of *GJA8* (c.20T > C) on chromosome 1 (chr1:147,380,102) and was predicted to result in the substitution of leucine-to-proline at codon 7 (p.Leu7Pro) (Table 1). The p.Leu7Pro substitution represented a relatively conservative change with the non-polar, side-chain of leucine ($[CH_3]_2$-CH-CH2-) replaced by the unusual non-polar, side-ring of proline (–CH_2-CH_2-CH_2-). However, Leu7 is phylogenetically conserved

in *GJA8* (Figure 2C), and the Pro7 substitution was predicted *in silico* to have probably damaging effects on protein function (Table 2). Sanger sequencing of all five members of family B (Figure 2A,B) confirmed that the novel p.Leu7Pro variant in *GJA8* co-segregated with cataract further suggesting that it was the disease-causing mutation.

Family C variants

A review of the exome SNPs in family C using the candidate gene list yielded a total of 82 variants in 23 of the 39 genes (Additional file 1 and Additional file 3). However, only three of these variants associated with the candidate genes, *WFS1*, *BFSP1*, and *GJA8*, were present in both affected relatives but not in the unaffected spouse. The variants associated with *WFS1* (rs734312) and *BFSP1* (rs2281207) had MAF values of 54.69% and 25.74%, respectively, and were excluded as causative mutations. The remaining variant occurred in exon-2 of *GJA8* (c.293A > C) on chromosome 1 (chr1:147,380,375) and was predicted to result in a non-conservative substitution of histidine-to-proline at codon 98 (p.His98Pro) (Table 1). Histidine 98 is phylogenetically conserved across species (Figure 2F), and this variant was also predicted to have damaging effects on protein function using six mutation prediction programs (Table 2). Sanger sequencing of all four members of family C (Figure 2D,E) confirmed that the novel p.His98Pro variant in *GJA8* co-segregated with cataract, consistent with it being the disease-causing mutation.

Discussion

Several recent studies have employed exome sequencing of index patients or probands in multiple families in order to discover mutations in candidate genes underlying autosomal dominant and recessive forms of cataract [13,17-19]. In this study, we have used trio-based exome sequencing to uncover a recurrent missense mutation in *CRYGD* (p.Pro24Thr) and two novel missense mutations in *GJA8* (p.Leu7Pro, p.His98Pro) associated with autosomal dominant cataract in three nuclear families. Child–parent trios offer the initial benefit of co-segregation testing during exome variant analysis, but this advantage may be offset in larger cohorts of families by the additional sequencing costs. The p.Pro24Thr

Table 1 Summary of mutations detected by exome sequencing of trios from families A, B, and C

Pedigree (numbers affected)	Physical location of variant	Candidate gene (ID)	Exon	cDNA variant	Protein variant	Allele frequency (EVS)*	Status
A (3)	chr2:208,989,018	*CRYGD* (1412)	2	c.70C > A	p.Pro24Thr	0/8,600	Recurrent (Additional file 5)
B (4)	chr1:147,380,102	*GJA8* (2703)	2	c.20T > C	p.Leu7Pro	0/8,600	Novel
C (2)	chr1:147,380,375	*GJA8* (2703)	2	c.293A > C	p.His98Pro	0/8,600	Novel

*Allele frequencies for European Americans listed on the Exome Variant Server.

Table 2 *In silico* predictions of functional effects for the three mutations identified in this study

Prediction program		CRYGD c.70C > A p.Pro24Thr	GJA8 c.20T > C p.Leu7Pro	GJA8 c.293A > C p.His98Pro
SIFT	Value (<0.05)	0.10	0.00	0.00
	Prediction	Tolerated	Not tolerated	Not tolerated
Polyphen-2	Score	0.084	0.991	1.000
	Prediction	Benign	Probably damaging	Probably damaging
PMUT	NN output	0.0936	0.8749	0.7822
	Reliability	8	7	5
	Prediction	Neutral	Pathological	Pathological
PANTHER	subPSEC (<−3)	−2.35974	−4.3388	−3.98807
	$P_{deleterious}$ (>0.5)	0.34519	0.79148	0.72871
PON-P2	Probability for pathogenicity	0.479	0.947	0.444
	Standard error	0.332	0.050	0.090
	Prediction	Unknown	Pathogenic	Unknown
MutPred	Probability of deleterious mutation	0.840	0.918	0.889
	Molecular mechanisms disrupted		Loss of stability (p =0.0067) Gain of disorder (p =0.0099)	

substitution in *CRYGD* has now been identified in some 14 different families, mostly segregating coralliform cataract that affects more than 133 individuals with varied ethnic backgrounds and constitutes the most recurrent missense mutation in a crystallin gene to be associated with inherited cataract (Additional file 4). The novel mutations found in *GJA8* increase the mutation spectrum of this connexin gene to at least 32 different mutations segregating in 38 families making it one of the most common non-crystallin genes to be associated with inherited cataract in humans (Additional file 5).

CRYGD (MIM: 123690) consists of three exons and encodes γD-crystallin—a hydrophylic protein of 174 amino acids that is characterized by two βγ-crystallin domains each formed by two repeat Greek-key motifs of approximately 40 residues. *CRYGD* is expressed at high concentrations in fiber cells of mammalian lenses and plays an important structural role in establishing lens transparency and gradient refractive index [20]. Proline at position 24 is located within the first Greek-key motif of human *CRYGD* but is not well conserved across species (replaced by serine in the mouse and threonine in the zebrafish). Consequently, *in silico* analysis predicted that the Pro24Thr substitution was benign (Table 2). Further, NMR-spectroscopy and X-ray crystallography have indicated that the Pro24 and Thr24 proteins are structurally similar overall [21,22]. However, the Thr24 mutant exhibits local conformational and dynamic differences that may initiate aggregation or polymerization and *in vitro* experiments have shown that the Thr24

protein exhibits reduced solubility—a property that is likely to trigger cataract formation [23-25].

GJA8 (MIM: 600897) comprises two exons with exon-2 coding for the entire 433 amino acid residues of gap-junction protein α8 or connexin 50. *GJA8* contains four transmembrane domains that are joined by two extracellular loops and one cytoplasmic loop and flanked by cytoplasmic N- and C-termini. By forming hexamers, or hemi-channels, that can dock between adjacent cells to create gap-junction channels, *GJA8* plays an important role in lens intercellular communication [26]. Of the 32 known coding mutations in *GJA8*, 30 result in missense substitutions that, with one exception, are associated with autosomal dominant cataract, and the remaining two are frameshift mutations associated with autosomal recessive cataract (Additional file 5). Most of the missense substitutions are located within the N-terminal half of the protein, which also contains the conserved connexin domain (pfam00029; amino acids 3–109). The novel p.Leu7Pro substitution found in family B is the first to be located at the cytosolic N-terminal end of human *GJA8*. Support for its pathogenicity in humans is provided by the SHR-Dca rat strain, which inherits semi-dominant cataract [27]. Heterozygous (+/Dca) mutants develop nuclear pulverulent opacities and smaller eyes than wild-type, while homozygotes (Dca/Dca) present with severe microphthalmia and a hypoplastic lens. The underlying mutation has been identified as a missense mutation in *GJA8* (c.20T > A) that is predicted to result in a non-conservative p.Leu7Gln substitution. Both the rat p.Leu7Gln and

human p.Leu7Pro mutations result in the substitution of a highly conserved leucine residue with uncharged amino acids, suggesting that they may exert similar deleterious effects on *GJA8* function.

The novel p.His98Pro mutation identified in family C, is located near the junction of the second transmembrane domain with the cytoplasmic loop of *GJA8*. Four other mutations, p.Val79Leu, p.Pro88Ser, p.Pro88Gln, and p.Pro88Thr, have previously been localized to the second transmembrane domain (Additional file 5). Functional expression studies of the relatively conservative p.Val79Leu substitution results in functional gap-junction channels with altered voltage-gating and a reduction in the single-channel open probability [28]. By contrast, neither of the non-conservative p.Pro88Gln and p.Pro88Ser substitutions was targeted to the plasma membrane, with the former accumulating in the endoplasmic-reticulum (ER)-Golgi-complex and the latter forming discrete cytoplasmic inclusions [26]. Based on the non-conservative nature of the p.His98Pro substitution, we speculate that this mutant will also fail to reach the plasma membrane and form functional gap-junction channels

Conclusions

Exome sequencing provides a rational approach to concurrently screen over 39 candidate genes for inherited cataract in nuclear families or even sporadic cases. In addition, exome sequencing may enable the discovery of novel genes underlying inherited cataract and, potentially, genes associated with age-related cataract. However, considerable supporting evidence (e.g., additional mutations, functional expression *in vitro*, and/or an animal model) will be required to verify disease causation. In a clinical setting, results from exome sequencing are unlikely to be 'clinically actionable' with respect to surgical treatment and subsequent management of inherited cataract. However, such data can contribute to a gene-centric clinical classification of inherited cataract and provide enhanced diagnosis and genetic counseling for affected families.

Methods

Ethics statement

Ethical approval for this study was obtained from the Washington University Human Research Protection Office (HRPO), and written informed consent was provided by all participants prior to enrollment in accordance with the tenets of the Declaration of Helsinki and Health Insurance Portability and Accountability Act (HIPAA) regulations.

Family participants

Three Caucasian-American pedigrees segregating autosomal dominant cataract were ascertained through ophthalmic records in the Department of Ophthalmology and

Visual Sciences at Washington University School of Medicine. Blood samples were obtained from available family members including a spouse (Figures 1 and 2). Leukocyte genomic DNA was purified using the Gentra Puregene Blood kit (Qiagen, Valencia, CA) and quantified by absorbance at 260 nm (NanoDrop 2000, Wilmington, DE).

Exome sequencing

Whole exome capture was achieved using the SureSelect Human All Exon V5 (50.4 Mb) Kit, according to manufacturer's instructions (Agilent Technologies). Briefly, genomic DNA (3 µg) was fragmented (150–200 bp) by acoustic shearing, ligated to adapter primers, and PCR-amplified. Following denaturation (95°C, 5 min), amplified DNA-fragment libraries (~500 ng) were hybridized in a solution under high stringency (65°C, 24 h) with biotinylated RNA capture probes (~120 bp). Resulting DNA/RNA hybrids were recovered by streptavidin-coated magnetic bead separation (Dynal, Invitrogen, Carlsbad, CA). Captured DNA was eluted (NaOH) and then subject to solid phase (flow-cell) next-generation (massively parallel) sequencing on a HiSeq2000 System (Illumina, San Diego, CA) using the Illumina Multiplexing Sample Preparation Oligo-nucleotide Kit and the HiSeq 2000 Paired-End Cluster Generation Kit according to the manufacturer's instructions. Briefly, hybrid-capture libraries were amplified to add indexing (identifying) tags and sequencing primers then subjected to paired-end (2 × 101 bp read length), multiplex sequencing-by-synthesis using fluorescent, cyclic reversible (3′-blocked) terminators. A pool of three exome samples (representing a family trio) was sequenced in a single lane of the sequencer's flow-cell.

Exome variant analysis

Raw sequence data was aligned to the human reference genome (build hg19) by NovoalignMPI (www.novocraft.com), and sequence variants called using the Sequence Alignment/Map format (SAMtools) and Picard programs (http://samtools.sourceforge.net/) and further annotated using SeattleSeq (http://snp.gs.washington.edu/SeattleSeq Annotation138/). Target coverage and read-depth were reviewed by the Integrated Genomics Viewer (IGV; http://www.broadinstitute.org/igv/). Variants were filtered using the Ingenuity variant analysis website (IVA http://ingenuity.com) or the gNOME project pipeline (http://gnome.tchlab.org/) [29]. Identified variants in the pre-selected candidate genes (Additional file 1) were then reviewed for presence/absence and frequency in various websites including dbSNP (http://www.ncbi.nlm.nih.gov/snp/), 1000 genomes (http://www.1000genomes.org/), and the Exome Variant Server database (http://evs.gs.washington.edu/EVS/). The predicted effect on protein function was analyzed using the SIFT (http://sift.jcvi.org), PolyPhen-2 (http://genetics.bwh.harvard.edu/pph2/), PMUT (http://mmb2.pcb.ub.es:8080/PMut/),

PON-P2 (http://structure.bmc.lu.se/PON-P2/), PANTHER (http://www.pantherdb.org/tools/csnpScoreForm.jsp), and MutPred (http://mutpred.mutdb.org/) *in silico* mutation prediction programs [30-34].

Sanger sequencing

Genomic DNA (2.5 ng/µl, 10 µl reactions) was amplified (35 cycles) in a GeneAmp 9700 thermal cycler using Top Taq mastermix kit (Qiagen) and 20 pmol of gene-specific primers (Additional file 6). Resulting PCR amplicons were enzyme-purified with ExoSAP-IT (USB Corporation, Cleveland, OH). The purified amplicons were direct cycle-sequenced in both directions with BigDye Terminator Ready Reaction Mix (v3.1)(Applied Biosystems, Grand Island, NY) containing M13 forward or reverse sequencing primers, then ethanol precipitated and detected by capillary electrophoresis on a 3130xl Genetic Analyzer running Sequence Analysis (v.6.0) software (Applied Biosystems) and Chromas (v2.23) software (Technelysium, Tewantin, Queensland, Australia).

Additional files

Additional file 1: Table S1. Candidate genes for inherited cataract evaluated in this study.

Additional file 2: Table S2. Sample metrics for exome sequencing of trios from families A, B, and C.

Additional file 3: Table S3. Exome sequencing variants found in candidate genes for inherited cataract (Additional file 1, Table S1) in family trios A, B, and C.

Additional file 4: Table S4. Recurrence of the p.Pro24Thr mutation in *CRYGD* associated with autosomal dominant cataract.

Additional file 5: Table S5. Mutation profile of *GJA8*.

Additional file 6: Table S6. PCR primers used for Sanger sequencing of mutations found in *CRYGD* and *GJA8*. Each primer was tailed with M13 sequences to aid in Sanger sequencing. The forward primers were tagged with 'tgtaaaacgacggccagt' and the reverse primers with 'caggaaacagctatgacc'.

Abbreviations
OMIM: Online Mendelian Inheritance in Man; MAF: Minor allele frequency; SHR-Dca: Spontaneously hypertensive rat-Dominant cataract; NMR: Nuclear magnetic resonance.

Competing interests
The authors declare that they have no competing interests.

Authors' contributions
DSM and TMB were involved in acquisition and analysis of exome sequencing data and bioinformatics analyses. SMC coordinated ascertainment and recruitment of patients and was involved in acquisition and analysis of clinical data. DSM and AS conceived the study, participated in its design and coordination, and drafted the manuscript. All authors read and approved the final manuscript.

Acknowledgements
We thank the families for their participation in this study and the Genome Technology Access Center (GTAC) at Washington University School of Medicine for help with genomic analysis. GTAC is partially supported by National Institutes of Health (NIH) grants P30 CA91842 and UL1 TR000448. This work was supported by NIH grants EY012284 (to A.S.) and EY02687 (Core Grant for Vision Research) and by an unrestricted grant to the Department of Ophthalmology and Visual Sciences from Research to Prevent Blindness (RPB).

References
1. Shiels A, Bennett T, Hejtmancik J: Cat-Map: putting cataract on the map. *Mol Vis* 2010, 16:2007–2015.
2. Trumler A: Evaluation of pediatric cataracts and systemic disorders. *Curr Opin Ophthalmol* 2011, 22:365–379.
3. Shiels A, Hejtmancik JF: Genetics of human cataract. *Clin Genet* 2013, 84:120–127.
4. Mickler C, Boden J, Trivedi RH, Wilson ME: Pediatric cataract. *Pediatr Ann* 2011, 40:83–87.
5. Mataftsi A, Haidich AB, Kokkali S, Rabiah PK, Birch E, Stager DRJ, Cheong-Leen R, Singh V, Egbert JE, Astle WF, Lambert SR, Amitabh P, Khan AO, Grigg J, Arvanitidou M, Dimitrakos SA, Nischal KK: Postoperative glaucoma following infantile cataract surgery: an individual patient data meta-analysis. *JAMA Ophthalmol* 2014, doi:10.1001/jamaophthalmol.2014.1042. Published online June 12, 2014.
6. Lambert SR, Purohit A, Superak HM, Lynn MJ, Beck AD: Long-term risk of glaucoma after congenital cataract surgery. *Am J Ophthalmol* 2013, 156:355–361. e352.
7. Ruddle JB, Staffieri SE, Crowston JG, Sherwin JC, Mackey DA: Incidence and predictors of glaucoma following surgery for congenital cataract in the first year of life in Victoria, Australia. *Clin Experiment Ophthalmol* 2013, 41:653–661.
8. Young MP, Heidary G, VanderVeen DK: Relationship between the timing of cataract surgery and development of nystagmus in patients with bilateral infantile cataracts. *J AAPOS* 2012, 16:554–557.
9. Haargaard B, Andersen EW, Oudin A, Poulsen G, Wohlfahrt J, la Cour M, Melbye M: Risk of retinal detachment after pediatric cataract surgery. *Invest Ophthalmol Vis Sci* 2014, 55:2947–2951.
10. Amaya L, Taylor D, Russell-Eggitt I, Nischal KK, Lengyel D: The morphology and natural history of childhood cataracts. *Surv Ophthalmol* 2003, 48:125–144.
11. Gunn RM: Peculiar coralliform cataract with crystals in the lens. *Trans Ophthalmol Soc UK* 1895, XV:119.
12. Lachke SA, Ho JW, Kryukov GV, O'Connell DJ, Aboukhalil A, Bulyk ML, Park PJ, Maas RL: iSyTE: integrated Systems Tool for Eye gene discovery. *Invest Ophthalmol Vis Sci* 2012, 53:1617–1627.
13. Reis LM, Tyler RC, Muheisen S, Raggio V, Salviati L, Han DP, Costakos D, Yonath H, Hall S, Power P, Semina EV: Whole exome sequencing in dominant cataract identifies a new causative factor, CRYBA2, and a variety of novel alleles in known genes. *Hum Genet* 2013, 132:761–770.
14. Bennett TM, Maraini G, Jin C, Sun W, Hejtmancik JF, Shiels A: Noncoding variation of the gene for ferritin light chain in hereditary and age-related cataract. *Mol Vis* 2013, 19:835–844.
15. Zuercher J, Neidhardt J, Magyar I, Labs S, Moore AT, Tanner FC, Waseem N, Schorderet DF, Munier FL, Bhattacharya S, Berger W, Kloeckener-Gruissem B: Alterations of the 5'untranslated region of SLC16A12 lead to age-related cataract. *Invest Ophthalmol Vis Sci* 2010, 51:3354–3361.
16. den Dunnen JT, Antonarakis SE: Nomenclature for the description of human sequence variations. *Hum Genet* 2001, 109:121–124.
17. Aldahmesh MA, Khan AO, Mohamed JY, Hijazi H, Al-Owain M, Alswaid A, Alkuraya FS: Genomic analysis of pediatric cataract in Saudi Arabia reveals novel candidate disease genes. *Genet Med* 2012, 14:955–962.
18. Sun W, Xiao X, Li S, Guo X, Zhang Q: Exome sequencing of 18 Chinese families with congenital cataracts: a new sight of the NHS gene. *PLoS One* 2014, 9:e100455.
19. Prokudin I, Simons C, Grigg JR, Storen R, Kumar V, Phua ZY, Smith J, Flaherty M, Davila S, Jamieson RV: Exome sequencing in developmental eye disease leads to identification of causal variants in GJA8, CRYGC, PAX6 and CYP1B1. *Eur J Hum Genet* 2014, 22:907–915.
20. Slingsby C, Wistow GJ, Clark AR: Evolution of crystallins for a role in the vertebrate eye lens. *Protein Sci* 2013, 22:367–380.
21. Jung J, Byeon IJ, Wang Y, King J, Gronenborn AM: The structure of the cataract-causing P23T mutant of human gammaD-crystallin exhibits distinctive local conformational and dynamic changes. *Biochemistry* 2009, 48:2597–2609.
22. Ji F, Koharudin LM, Jung J, Gronenborn AM: Crystal structure of the cataract-causing P23T gammaD-crystallin mutant. *Proteins* 2013, 81:1493–1498.

23. Mackay D, Andley U, Shiels A: **A missense mutation in the gammaD crystallin gene (CRYGD) associated with autosomal dominant "coral-like" cataract linked to chromosome 2q.** *Mol Vis* 2004, **10**:155–162.

24. Evans P, Wyatt K, Wistow GJ, Bateman OA, Wallace BA, Slingsby C: **The P23T cataract mutation causes loss of solubility of folded γD-crystallin.** *J Mol Biol* 2004, **343**:435–444.

25. Pande A, Ghosh KS, Banerjee PR, Pande J: **Increase in surface hydrophobicity of the cataract-associated P23T mutant of human gammaD-crystallin is responsible for its dramatically lower, retrograde solubility.** *Biochemistry* 2010, **49**:6122–6129.

26. Beyer E, Ebihara L, Berthoud V: **Connexin mutants and cataracts.** *Front Pharmacol* 2013, **4**:43.

27. Liska F, Chylíková B, Martínek J, Kren V: **Microphthalmia and cataract in rats with a novel point mutation in connexin 50–L7Q.** *Mol Vis* 2008, **7**:828–828.

28. Rubinos C, Villone K, Mhaske P, White TW, Srinivas M: **Functional effects of Cx50 mutations associated with congenital cataracts.** *Am J Physiol Cell Physiol* 2014, **306**:C212–C220.

29. Lee I-H, Lee K, Hsing M, Choe Y, Park J-H, Kim SH, Bohn JM, Neu MB, Hwang KB, Green RC, Kohane IS, Kong SW: **Prioritizing disease-linked variants, genes, and pathways with an interactive whole-genome analysis pipeline.** *Hum Mut* 2014, **35**:537–547.

30. Ferrer-Costa C, Gelpi JL, Zamakola L, Parraga I, de la Cruz X, Orozco M: **PMUT: a web-based tool for the annotation of pathological mutations on proteins.** *Bioinformatics* 2005, **21**:3176–3178.

31. Thomas PD, Campbell MJ, Kejariwal A, Mi H, Karlak B, Daverman R, Diemer K, Muruganujan A, Narechania A: **PANTHER: a library of protein families and subfamilies indexed by function.** *Genome Res* 2003, **13**:2129–2141.

32. Adzhubei IA, Schmidt S, Peshkin L, Ramensky VE, Gerasimova A, Bork P, Kondrashov AS, Sunyaev SR: **A method and server for predicting damaging missense mutations.** *Nat Methods* 2010, **7**:248–249.

33. Li B, Krishnan VG, Mort ME, Xin F, Kamati KK, Cooper DN, Mooney SD, Radivojac P: **Automated inference of molecular mechanisms of disease from amino acid substitutions.** *Bioinformatics* 2009, **25**:2744–2750.

34. Sim N-L, Kumar P, Hu J, Henikoff S, Schneider G, Ng PC: **SIFT web server: predicting effects of amino acid substitutions on proteins.** *Nucl Acids Res* 2012, **40**:W452–W457.

17

Changing genetic paradigms: creating next-generation genetic databases as tools to understand the emerging complexities of genotype/phenotype relationships

Bruce Gottlieb[1,2,3*], Lenore K Beitel[1,3,4] and Mark Trifiro[1,2,3,4]

Abstract

Understanding genotype/phenotype relationships has become more complicated as increasing amounts of inter- and intra-tissue genetic heterogeneity have been revealed through next-generation sequencing and evidence showing that factors such as epigenetic modifications, non-coding RNAs and RNA editing can play an important role in determining phenotype. Such findings have challenged a number of classic genetic assumptions including (i) analysis of genomic sequence obtained from blood is an accurate reflection of the genotype responsible for phenotype expression in an individual; (ii) that significant genetic alterations will be found only in diseased individuals, in germline tissues in inherited diseases, or in specific diseased tissues in somatic diseases such as cancer; and (iii) that mutation rates in putative disease-associated genes solely determine disease phenotypes. With the breakdown of our traditional understanding of genotype to phenotype relationships, it is becoming increasingly apparent that new analytical tools will be required to determine the relationship between genotype and phenotypic expression. To this end, we are proposing that next-generation genetic database (NGDB) platforms be created that include new bioinformatics tools based on algorithms that can evaluate genetic heterogeneity, as well as powerful systems biology analysis tools to actively process and evaluate the vast amounts of both genomic and genomic-modifying information required to reveal the true relationships between genotype and phenotype.

Keywords: Human genetic variation, Next-generation sequencing, Genotype to phenotype relationships, Next-generation genetic databases

Introduction

The problem of understanding the relationships between genotype and phenotype has become very much more complicated with the explosion of genetic information produced by next-generation sequencing (NGS). This information has greatly complicated not only our ability to understand complex traits, but also our understanding of monogenic traits is no longer quite so straight forward. Indeed, recent articles have suggested the need to develop new approaches to come to grips with the

ever-expanding complexity of genotype/phenotype relationships, such as 'systems genetics' [1] and 'particle genetics' [2].

However, perhaps the most confusing from a 'traditional' genetics standpoint has been the revelation of unexpected amounts of genetic variation in normal individuals, e.g., through the 1000 Genomes Project Consortium [3,4] (www.1000genomes.org), and The Cancer Genome Atlas (www.cancergenome.nih.gov) projects. Further, multiple sequence comparisons both between and within an individual's tissues have revealed extensive inter- and intra-tissue genetic heterogeneity [5-7]. These discoveries have raised some fundamental questions about our most basic genetics assumptions, among which are the following: (i) Can genetic studies still rely on a *unique* DNA or RNA sequence derived from blood or diseased tissue to determine

* Correspondence: bruce.gottlieb@mcgill.ca
[1]Lady Davis Institute for Medical Research, 3755 Côte Ste Catherine Road, Montreal, QC H3T 1E2, Canada
[2]Segal Cancer Centre, Jewish General Hospital, 3755 Côte Ste Catherine Road, Montreal, QC H3T 1E2, Canada
Full list of author information is available at the end of the article

phenotype?; (ii) Does a definitive and practical human genome reference sequence really exist, or at least can the reference sequence adopted by the NCBI (RefSeqGen) be practically useful in determining genotype/phenotype relationships?; and (iii) Does genetic heterogeneity in normal and diseased tissues imply that in certain tissues an individual's genome will naturally undergo somatic changes from conception to death as suggested in Figure 1. In particular, newly revealed genetic heterogeneity data could help explain the long observed, but poorly understood concepts of variable expressivity and reduced penetrance. Traditionally, their effects on phenotypic differences have been considered to be relatively insignificant, particularly so for variable expressivity. To further complicate matters, phenotypic variations have been found, where identical gene alterations have been associated with (i) considerably different disease phenotypes, e.g., in phenylalanine hydroxylase deficiency (PAH) [8], or (ii) in a more extreme manner in the androgen receptor (AR) gene, with both androgen insensitivity syndrome (AIS) and prostate cancer [9].

In addition, there has also been an increase in the discovery of significant phenotype-modifying events, including epigenetic modifications, RNA editing, and protein interactions that can clearly influence transcriptional and non-transcriptional events involved in determining the phenotype. Thus, these complex influences are also likely to render our traditional understanding of the relationship between genotype and phenotype problematical. Further, a recent review of genotype/phenotype dissociation that discussed the possible molecular basis of reduced penetrance in human inherited disease, highlighted 12 molecular events that can influence reduced penetrance [10], some of which are also likely involved in situations of variable expressivity. In Figure 2, we have suggested a model that incorporates some of

these processes, and how they might influence phenotype, with special emphasis on the influence of intra-organismal and intra-tissue genetic heterogeneity. Traditionally, genetic databases have been the tools of choice in determining genotype/phenotype relationships; however, in their present form, they are totally inadequate to deal with these issues. Therefore, we are suggesting that it is time to create next-generation genetic databases (NGDB) that will be able to incorporate and analyze all of the factors that can contribute to the dissociation of genotype from phenotype, including those that may contribute to reduced penetrance and variable expressivity.

Factors that have been shown to influence phenotype
Somatic mutations that result in intra-organismal and intra-tissue genetic heterogeneity

Until recently, it has been assumed that somatic mutations are almost exclusively associated with cancers and are uniform within an individual neoplasm. However, different sets of somatic mutations have been found within a single individual's cancer tissues, as in a recent study of primary high-grade serous ovarian cancers that revealed a considerable amount of intra-tumor genetic heterogeneity [11].

Somatic sequence variants in normal tissues have also been examined in relation to oncogenesis. One study concluded that somatic sequence variants in normal cell populations could be the earliest stage of oncogenesis [12]. Evidence that altered mammary gland development and predisposition to breast cancer is due to *in utero* exposure to endocrine disruptors has suggested that selection of cells with different phenotypic properties, presumably as a result of very early somatic mutations, may take place at the very earliest stages of breast tissue development [13]. Thus, we may need to reconsider whether accumulation of a critical number of oncogenic

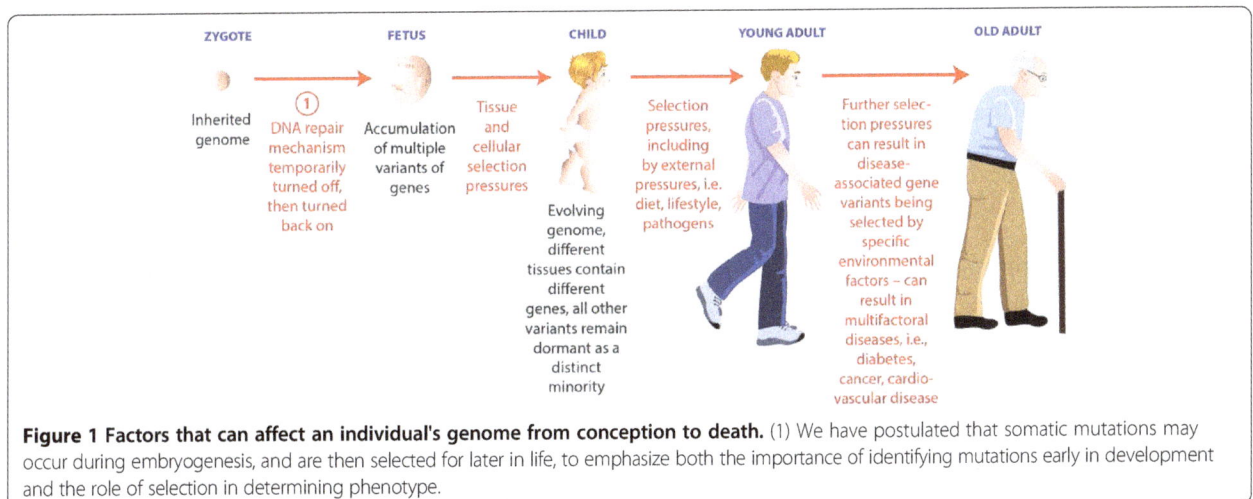

Figure 1 Factors that can affect an individual's genome from conception to death. (1) We have postulated that somatic mutations may occur during embryogenesis, and are then selected for later in life, to emphasize both the importance of identifying mutations early in development and the role of selection in determining phenotype.

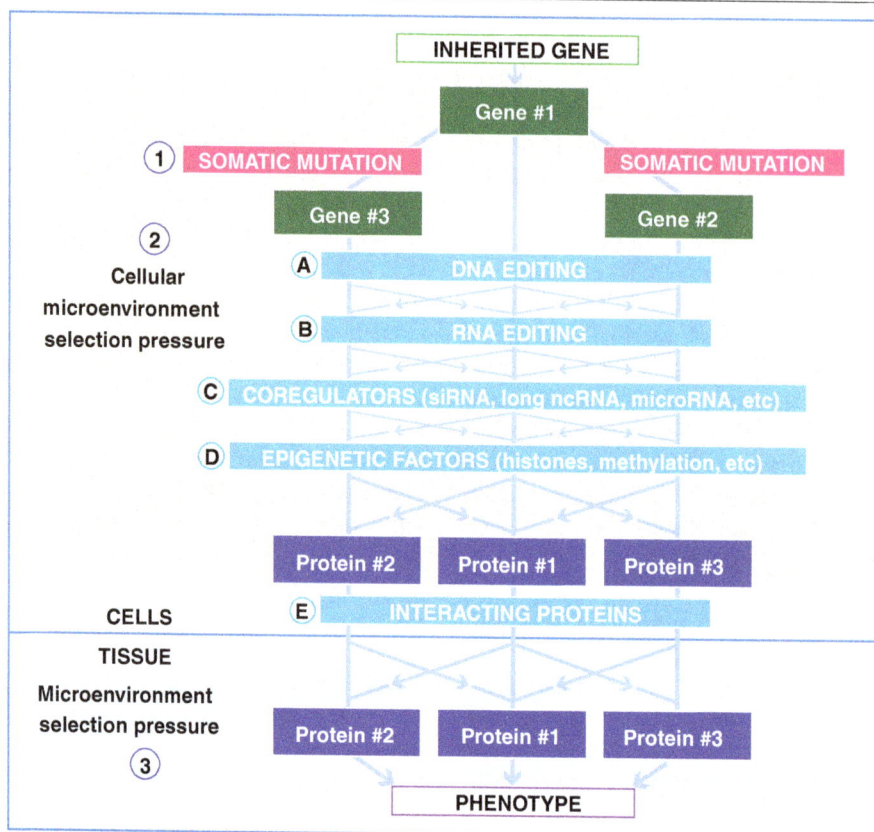

Figure 2 Phenotypic modifying factors. (1) Somatic mutations can include both single nucleotide variants and structural alterations such as copy number variations that can then result in somatic and clonal mosaicism. (2) Cellular microenvironment selection pressure can work at the (i) DNA level, i.e., due to somatic mutations or (A) DNA editing; (ii) RNA level, i.e., due to (B) RNA editing, (C) interacting RNAs, or (D) epigenetic factors, etc.; or (iii) protein level, i.e., due to (E) protein-protein interactions. (3) Tissue microenvironment selection pressure can select a different protein product. *Crossing arrows* reflect the fact that selection can go in either direction.

mutations, e.g., the buildup of driver somatic mutations, is the reason that many cancers occur later in life. Rather, it has been proposed that while the genetic origins of cancer may occur early in fetal development, *later selection pressure* could explain the relationship between aging and cancer [14]. Interestingly, a possible mechanism to produce very early somatic mutations, namely the temporarily deferring of the repair of DNA lesions encountered during tissue replication, that has been termed damage bypass, has been identified as responsible for somatic hypermutation of the immunoglobin gene [15]. Regardless of which oncogenesis hypothesis is eventually proven, the implications for construction of NGDB for cancers is likely to be profound, as NGDBs will need to consider incorporating sequence data from much earlier stages in a tissue development, particularly from tissues that have the potential to become cancerous. Obviously, the ability to do so at the moment is not practical, but it is possible to envision that in the future, new micro-sampling techniques, together with the continued dramatic decline in

the cost of NGS, will make such an approach much more realistic.

In addition, as specific tissues are being sequenced routinely, the number of other diseased tissues in which somatic mutations have been found has increased considerably [16]. More detailed studies have also reported somatic mosaicism in a number of other conditions, including the Proteus syndrome [17] and hemimegalencephaly [18].

Further, a study of copy-number variants (CNVs) in somatic human tissues revealed a significant number of intra-individual genomic changes between tissues [19]. Other studies of chromosomal abnormalities, including CNVs have revealed clonal mosaicism associated with aging and cancer [14], as well as related it to a higher risk of hematological cancer [20].

DNA editing

At the present state of our knowledge, this process is still considered to be extremely rare and of little phenotypic significance [21].

RNA editing

Recent, though controversial, evidence has suggested that RNA editing occurs more frequently than previously thought [22,23], although questions of how common it actually is in normal tissues and the validity of the original report have arisen [24-26]. However, there do appear to be cases where modifications of disease phenotypes are related to RNA editing [27,28].

Coregulators: non-coding RNAs

In recent years, non-coding RNAs (ncRNA) have been found to play an important role in the phenotypic expression of the transcribed genomic output. This family of untranslated RNAs includes small nucleolar RNAs (snoRNAs), which facilitate mRNA splicing, regulate transcription factors, and repress gene expression [via microRNAs (miRNAs)]. Small nuclear RNAs (snRNAs) that alter cellular proliferation and apoptosis by means of small interfering RNAs (siRNAs) have also been identified [29]. Long non-coding RNAs (lncRNAs) have also been identified as possible regulators of gene transcription and expression. Thus, the use of NGS to infer transcript expression levels in general, specifically *via* ncRNAs, is becoming increasingly common in molecular and clinical laboratories [30]. Therefore, it is not surprising that ncRNAs have been implicated as being responsible for a number of disease phenotypes [31].

Epigenetic factors

Epigenetics describes chromatin-based events that regulate DNA-templated processes and result in stable reprogramming of gene expression in response to transient external stimuli. Primary epigenetic factors include modifications to DNA and histones that are dynamically added and removed by chromatin-modifying enzymes in a highly regulated manner. Epigenetic mechanisms identified include DNA methylation, phosphorylation, ubiquitylation, sumoylation, RNA interference, and histone variance. Further, such epigenetic modifications play a critical role in the regulation of DNA-based processes such as transcription, DNA repair and replication, which can affect phenotype expression. Thus, abnormal expression patterns or genomic changes in chromatin regulators can have profound effects on human disease processes [32]. Indeed, epigenetics is considered a unifying factor in the etiology of some complex traits [33].

Regulators and other types of interacting proteins

Over the past few years, phenotypic expression has also found to be influenced by interacting proteins. Alterations in the interacting surfaces of a specific molecule [34] or the interacting proteins themselves can result in faulty protein-protein interactions and contribute to a disease phenotype [35].

Selection pressure by cellular and tissue microenvironments

It has been proposed that tumor morphology and phenotype are driven by selective pressure from the tissue microenvironment [36,37]. This hypothesis has been expanded to include other genetically determined diseased and non-diseased phenotypes [38]. The ability to perform ultra-deep sequencing using next-generation sequencers has revealed many more variants of a gene within tissues and thus the possibility that evolution at the tissue level contributes to disease phenotypes such as cancer [37,38].

Genotype/phenotype disconnects and possible mechanisms

In light of all the potential phenotype-modifying factors (Figure 2), which are generally not documented in traditional genetic databases, it is easy to understand why such databases, in their attempt to link a defined genotype with a specific phenotype, tend to avoid commenting on genotype/phenotype disconnects, due to the lack of information regarding the mechanisms that could produce such effects. However, a recent review highlighted the importance of understanding these disconnects, with over 650 references cited in proposing 12 molecular mechanisms to explain reduced penetrance [10]. Similarly, a number of possible mechanisms have been suggested to explain variable expressivity, e.g., somatic mosaicism [39], modifier genes [40], microRNA [41], epigenetic processes [42], and allelic heterogeneity [43]. Originally, the concept of reduced penetrance was based on studies of well-known genetic conditions in which a family tree predicted a disease phenotype, but this phenotype was not observed. While in most cases, the likelihood of reduced penetrance was small, it did serve a useful purpose in calculating the possibility of an individual having a diseased phenotype. The concept was further expanded when large-scale studies started to record the presence of mutations in specific genes associated with multifactorial diseases, such as cancer, a prime example being the breast cancer BRCA genes. In these cases, predicting penetrance was considered important in assessing the risk of disease. What has further complicated the issue, as we have noted, has been recent data from the 1000 Genomes Project and other large scale sequencing projects, which have reported that normal individuals can contain tens of potentially severe disease-associated alleles [10]. Thus, rather than talk about reduced penetrance of a pathogenic variant in a cohort that is known to express the disease phenotype, we now have to consider why these pathogenic variants are non-penetrant in a significant number of normal healthy individuals.

Redefining the human genome reference sequence

Clearly, the arrival of relatively inexpensive whole genome sequencing, and the subsequent sequencing of large

numbers of non-diseased individuals, has revealed the increasing presence of known disease-associated gene variants within non-diseased individuals. This was initially shown when the first Korean genome sequence was compared to other Asian genomes [44]. More detailed studies found sequence variants in genes associated with specific genetic disorders, in individuals with normal phenotypes. Such examples were recently discovered in a genomic analysis of 10 healthy individuals, where each individual had what was said to be 'healthy variance' in 19 to 31 OMIM genes, as they did not exhibit any of the signs, symptoms, or phenotypes of the associated genetic disorders [45]. However, it should be noted that not all sequence variants in OMIM genes are always pathogenic, as has recently been comprehensively reported [10]. Nevertheless, a systematic survey of loss-of-function (LoF) variants identified 26 known and 21 predicted severe disease-causing variants in analysis of 2,951 putative LoF variants obtained from 185 human genomes [46]. What is even more problematic is that our own work has identified specific pathogenic sequence variants in the AR gene in individuals with completely normal phenotypes, i.e., exactly the same AR variants as found in diseased individuals [9].

We believe this data calls into question the validity of our present methods of defining the so-called normal human genome. In particular, normal tissue genotype/phenotype disconnects have clearly created questions regarding the practicality of relying on a single unique reference sequence as the definitive predictor of phenotype. The Human Genome Variation Society (HGVS) nomenclature committee has studied this issue (www.hgvs.org/mutnomen/refseq.html) and recommended that the NCBI RefSeqGen be used and that the reference sequence guidelines should follow the Locus Reference Genomic (LRG) sequence format [47], which suggests using a single-file record containing a unique stable reference sequence. These recommendations were appropriate at the start of NGS, when the extent of variance in normal individuals, was relatively unknown. Naturally, we understand that a definitive reference sequence is important in defining exonic, intronic, and other structural parameters of genes. However, the issue of correlating phenotype with a specific sequence has clearly become much more complex.

To deal with this issue, the increasing amount of sequence variability in normal individuals has been incorporated into the latest version of the NCBI RefSeqGen (GRC37p13) (www.ncbi.nlm.nih.gov/projects/genome/assembly/grc/human), with the idea that these variants could be used as a contextual filter to determine the relationship between genotype and phenotype. Furthermore, additional tools have been set up to deal with the issue of normal variance, such as considering population-specific references where the major alleles are included at every location, or generating a reference sequence where all the alleles have been identified as part of the common ancestral lineage of modern humans. However, we would argue that just integrating normal human variance, however nuanced, into an overall version of the RefSeqGen fails to deal with the increasing problem of the association of the same gene variant with both normal and diseased phenotypes. Thus, relying solely on a DNA-based reference sequence, however sophisticated, will make it very difficult to distinguish between benign and disease-causing gene alterations, at least in traditional genetic databases, where the phenotypic classification of specific gene variants is based on having a unique reference sequence that is exclusively associated with a normal phenotype.

Possible organization of next-generation genetic databases
As an overlying principle, NGDBs need to be organized to take into consideration, particularly for multifactorial diseases, the overall genetic *context* of any identified mutation. However, context involves both intra-organismal genetic heterogeneity as well as other phenotype-modifying factors (Figure 2). These modifying factors also need to be considered in the context of 'pathway analysis' [48]. In light of the many contextual factors that can affect the genotype/phenotype expression, it seems reasonable that future of locus-specific databases (LSDBs) should be organized to take into account as much specific phenotype information as possible, including genotype-modifying factors, as opposed to most present LSDBs that are primarily genotype centered.

The issue of how to deal with the increasing identification of somatic mutations and intra-organismal genetic heterogeneity also needs to be investigated. Traditionally, somatic mutations have not been associated with databases unless a cancer phenotype was involved. At present, most disease-based databases associated with common multifactorial diseases such as cancer, diabetes and cardiovascular diseases often lack tissue and individual specific data. Indeed, only the COSMIC database [49] lists a comprehensive spectrum of somatic mutations associated with specific tissues and individual samples. Furthermore, currently, there is no description of the germline susceptibility variants found in matching control tissues, therefore making it difficult to draw definitive conclusions as to the significance of many somatic mutations. The situation will become even more complex when inter- and intra-tumor genetic heterogeneity data is added. Clearly, traditional flat-file databases will be unable to deal with such data and what are needed are radically different database structures that

include much more powerful analysis tools. In particular, it will be necessary to incorporate complex 'system analysis tools' that can analyze the intricate relationships between genotypic and phenotypic ontology [50]. Such analysis tools will need to incorporate extremely powerful knowledge analysis engines, possibly similar in design and organization to those developed by Google and other search engine companies.

These knowledge engines, for 'systems genetics analysis', will require the creation of powerful new bioinformatics tools and tremendously expanded database resources, particularly for disease-based databases. In particular, they will be required to analyze integrated *genetic* and *non-genetic* variation across many datasets, from different ethnic sub-groups or geographic populations, with the ultimate goal of integrating all genetic and non-genetic databases for a particular condition, especially if an initial population-based analysis fails to generate any significant insights into genotype/phenotype relationships. At the moment, such a task is clearly far beyond our capabilities; however, initial studies using mice have started to generate the bioinformatics tools and database resources required to create such NGDBs [51]. As NGDBs will include inter- and intra-tissue genetic heterogeneity, one factor that needs to be considered is the importance of quantifying variants that result in genetic heterogeneity, particularly if they are present within individual genes, rather than simply recording their presence. Indeed, we recently analyzed intra-tissue genetic heterogeneity in the AR gene in both cancer and non-cancer tissues taken from breast tumors and quantified AR variants in individual tissue samples using a new NGS technique [52]. Another approach has been to consider what has been termed 'particle genetics', where every cell is considered to be genetically unique, using probabilistic trait loci (PTL) to link genomic regions to probabilities of cellular characteristics [2].

Taking all of these factors into consideration, we would propose a NGDB model that integrates separate databases for each of the potential genome-modifying factors, together with a genotype database that incorporates genetic heterogeneity, with all of the individual databases linked to an associated phenotype database, and the data is then processed and analyzed through a very sophisticated knowledge engine (Figure 3).

Summary of possible actions required to create NGDBs
The following are some of the most significant actions that need to be undertaken in creating NGDBs:

1. Work in conjunction with the 1000 Genomes Project consortium and the Human Variome Project (HVP) to define the limits and significance of normal genome variation.

2. Incorporate individual inter- and intra-individual genetic heterogeneity into NGDBs.
3. Establish guidelines as to the significance of the number of reads needed to confirm a particular variant. Note, that initial NGS sequencing depth started at 4× to 10× coverage and rapidly rose to where 30× to 50× coverage is considered normal. However, recent studies show that increased coverage is likely to result in increased detection of variants [53,54], which in the case of tumor diagnostics coverage has now reached up to 20,000 reads.
4. Determine how the different frequency of occurrence of multiple gene variants within individuals should be incorporated into NGDBs. It should be noted that, at the moment, such frequencies are generally not incorporated into databases, particularly not into LSDBs. It would also clearly help to integrate structural variant data such as CNVs into LSDBs.
5. Incorporate expression data effectively into phenotype data parameters in NGDBs. Note that examples of tissue-specific variations in gene expression have now been reported [55]. In addition, data from the Genotype-Tissue Expression project [56] could be invaluable in determining relationships between tissue gene expression and disease phenotype.
6. Finally, research the bioinformatics and data parameters required to construct NGDBs that can incorporate and analyze all of the above data. To be truly effective, we believe that this effort should involve experts in genetics, bioinformatics, and systems biology-based search and knowledge engines, as well as a worldwide effort to collect genetic variation as for instance, proposed by the HVP.

Suggestions for future actions to be taken by the HVP
We believe that HVP is an organization that could play a leading role in developing NGDBs first by creating a special committee to look into future genetic database designs to deal with some of the issues raised in this article. Such a committee might include not only nomenclature experts, but also experts in creating both the algorithms required to design the databases, as well as the search and analytical engines. Based on the recommendations of this committee, the HVP could then set up an Institute for Genetic Database Research, which in addition to being responsible for NGDB design, could create a working model of the infrastructure required to run such databases on a worldwide scale. In particular, it will be important to establish a universal design structure so that all NGDBs will have a high degree of compatibility, and we believe that if such a design is coordinated through HVP, which already plays such a role in genetic

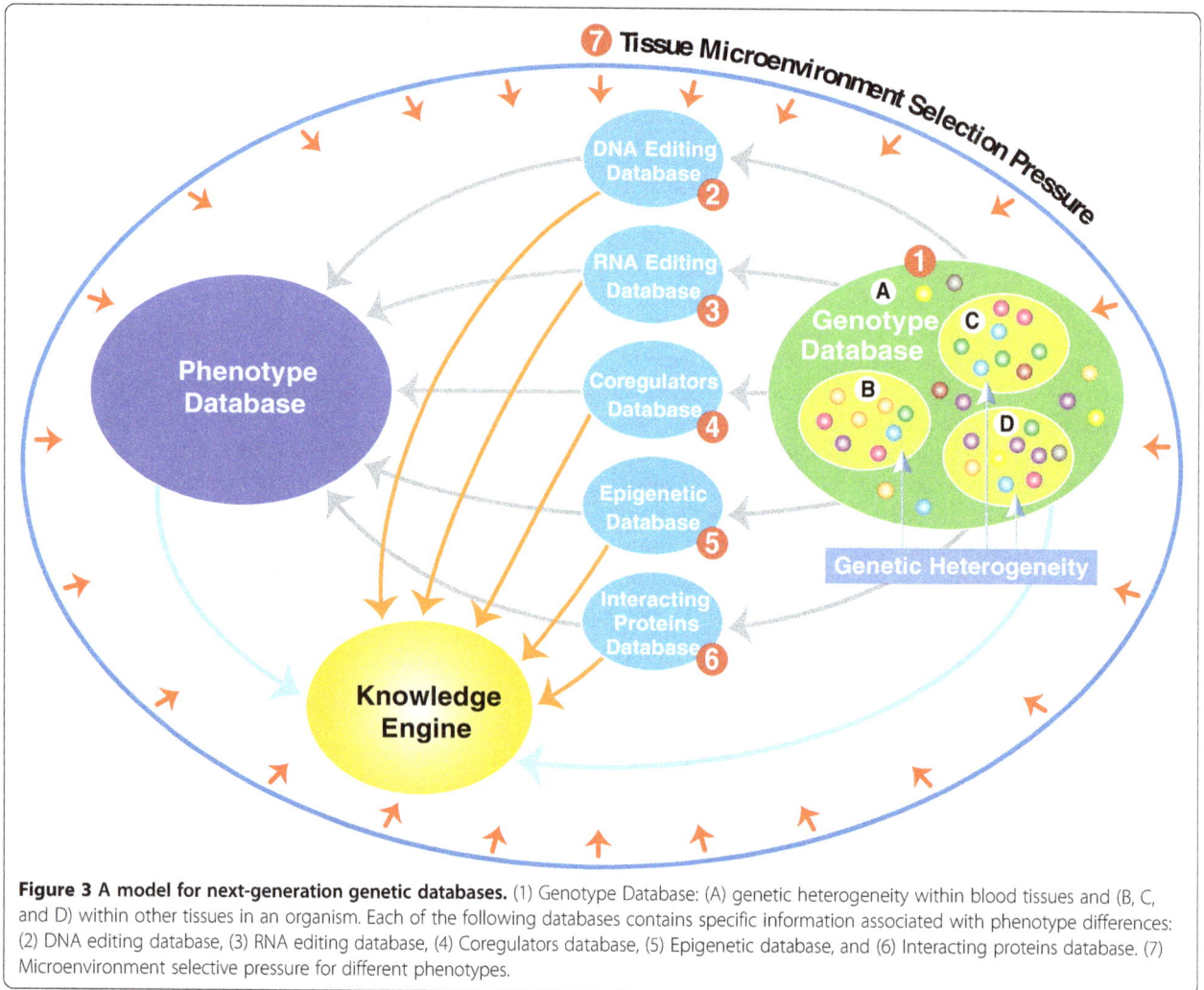

Figure 3 A model for next-generation genetic databases. (1) Genotype Database: (A) genetic heterogeneity within blood tissues and (B, C, and D) within other tissues in an organism. Each of the following databases contains specific information associated with phenotype differences: (2) DNA editing database, (3) RNA editing database, (4) Coregulators database, (5) Epigenetic database, and (6) Interacting proteins database. (7) Microenvironment selective pressure for different phenotypes.

nomenclature, it is much more likely to be accepted. Finally, in the age of data clouds and sophisticated communication platforms, such an institution need not have a physical structure, but rather could be a virtual institute, that would then allow experts from all over the world to participate.

Conclusion

For many years, genetics and related medical research have been based on the concept that genetic diseases are the result of alterations to a basically stable human genome that has limited natural variation within individuals, so that single or, in the case of multifactorial diseases, a number of very rare alterations to the human genome are directly responsible for specific diseases. Our initial response to the discovery of increased genetic complexity, particularly in multifactorial diseases, has been to use statistical-based approaches, such as GWAS to try to identify significant rare variants. However, most of these studies have yet to produce the breakthroughs

initially predicted, perhaps because they are still analyzing 'silos of genetic information' and ignoring the fact that the genomic makeup and phenotypic modifications of every individual are both complex and dynamic. Indeed, the increasing use of NGS, together with more accurate expression and pathway analysis tools, is further broadening our understanding of genotype/phenotype relationships, by revealing that the new genetic landscape is infinitely more complex, not only between individuals, but also within individuals. In such a genetic scenario, multifaceted worldwide NGDBs are likely to be essential tools in our fight to treat genetic-based disease.

Competing interests
The authors declare that they have no competing interests.

Authors' contributions
BG conceived and drafted the article, LKB and MT contributed to the discussion of the concepts and ideas presented and helped edit the text. All authors read and approved the final manuscript.

Acknowledgements

The authors acknowledge the support to BG of an operating grant from the Weekend to End Breast Cancer Fund of the Segal Cancer Centre of the Jewish General Hospital, Montreal, Quebec, Canada.

Author details

[1]Lady Davis Institute for Medical Research, 3755 Côte Ste Catherine Road, Montreal, QC H3T 1E2, Canada. [2]Segal Cancer Centre, Jewish General Hospital, 3755 Côte Ste Catherine Road, Montreal, QC H3T 1E2, Canada. [3]Department of Human Genetics, McGill University, Montreal, QC, Canada. [4]Department of Medicine, McGill University, Montreal, QC, Canada.

References

1. Civelek M, Lusis A: Systems genetics approaches to understand complex traits. Nat Rev Genet 2014, 15:34–48.
2. Yvert G: 'Particle genetics': treating every cell as unique. Trends Genet 2014, 30:49–56.
3. The Thousand Genomes Project Consortium: A map of human genome variation from population-scale sequencing. Nature 2010, 467:1061–1073.
4. The Thousand Genomes Project Consortium: An integrated map of genetic variation from 1092 human genomes. Nature 2012, 491:56–65.
5. Bertos NR, Park M: Breast cancer – one term, many entities. J Clin Invest 2011, 121:3789–3796.
6. Russnes HG, Navin N, Hicks J, Borresen-Dale A-L: Insight into the heterogeneity of breast cancer through next-generation sequencing. J Clin Invest 2011, 121:3810–3818.
7. Gerlinger M, Rowan AJ, Horswell S, Larkin J, Endesfelder D, Gronroos E, Martinez P, Matthews N, Stewart A, Tarpey P, Varela I, Phillimore B, Begum S, McDonald NQ, Butler A, Jones D, Raine K, Latimer C, Santos CR, Nohadani M, Eklund AC, Spencer-Dene B, Clark G, Pickering L, Stamp G, Gore M, Szallasi Z, Downward J, Futreal PA, Swanton C: Intratumor heterogeneity and branched evolution revealed by multiregion sequencing. N Engl J Med 2012, 366:883–892.
8. Zhu T, Ye J, Han L, Qiu W, Zhang H, Liang L, Gu X: Variations in genotype-phenotype correlations in phenylalanine hydroxylase deficiency in Chinese Han population. Gene 2013, 529:80–87.
9. Gottlieb B, Beitel LK, Nadarajah A, Paliouras M, Trifiro M: The androgen receptor gene mutations database: 2012 update. Hum Mutat 2012, 33:887–894.
10. Cooper DN, Krawczak M, Polychronakos C, Tyler-Smith C, Kehrer-Sawatzki H: Where genotype is not predictive of phenotype: towards an understanding of the molecular basis of reduced penetrance in human inherited disease. Hum Genet 2013, 132:1077–1130.
11. Bashashati A, Ha G, Tone A, Ding J, Prentice LM, Roth A, Rosner J, Shumansky K, Kalloger S, Senz J, Yang W, McConechy M, Melnyk N, Anglesio M, Luk MT, Tse K, Zeng T, Moore R, Zhao Y, Marra MA, Gilks B, Yip S, Huntsman DG, McAlpine JN, Shah SP: Distinct evolutionary trajectories of primary high-grade serous ovarian cancers revealed through spatial mutational profiling. J Pathol 2013, 231:21–34.
12. Howk CL, Voller Z, Beck BB, Dai D: Genetic diversity in normal cell populations is the earliest stage of oncogenesis leading to intra-tumor heterogeneity. Front Oncol 2013, 3:61. doi:10.3389/fonc.2013.00061.
13. Soto AM, Brisken C, Schaeberle C, Sonnenschein C: Does cancer start in the womb? Altered mammary gland development and predisposition to breast cancer due to in utero exposure to endocrine disruptors. J Mammary Gland Biol Neoplasia 2013, 18:199–208.
14. Jacobs KB, Yeager M, Zhou W, Wacholder S, Wang Z, Rodriguez-Santiago B, Hutchinson A, Deng X, Liu C, Horner M-J, Cullen M, Epstein CG, Burdett L, Dean MC, Chatterjee N, Sampson J, Chung CC, Kovaks J, Gapstur SM, Stevens VL, Teras LT, Gaudet MM, Albanes D, Weinstein SJ, Virtamo J, Taylor PR, Freedman ND, Abnet CC, Goldstein AM, Hu N, et al: Detectable clonal mosaicism and its relationship to aging and cancer. Nat Genet 2012, 44:651–658.
15. Sale JE, Batters C, Edmunds CE, Philips LG, Simpson LJ, Szuts D: Timing matters: error–prone gap filling and translation synthesis in immunoglobin gene hypermutation. Philos Trans R Soc Lond B Biol Sci 2009, 364:595–603.
16. Erickson RP: Somatic gene mutation and human disease other than cancer: An update. Mutat Res 2010, 705:96–106.
17. Lindhurst MJ, Sapp JC, Teer JK, Johnston JJ, Finn EM, Peters K, Turner J, Cannons JL, Bick B, Blackemore L, Blumhorst C, Brockman K, Calder P, Cherman N, Deardorff MA, Everman DB, Golas G, Greenstein RM, Kato BM, Keppler-Noreuil KM, Kuznetsov SA, Miyamoto RT, Newman K, Ng D, O'Brien K, Rothenberg S, Schwartzentruber DJ, Singhal V, Tirabosco R, Upton J, et al: A mosaic activating mutation is associated with the Proteus syndrome. N Engl J Med 2011, 365:611–619.
18. Evrony GD, Cai X, Lee E, Hills LB, Elhosary PC, Lehmann HS, Parker JJ, Atabay KD, Gilmore EC, Poduri A, Park PJ, Walsh CA: Single-neuron sequencing analysis of L1 retrotransposition and somatic mutation in the human brain. Cell 2012, 151:483–496.
19. O'Huallachain M, Karczewski KJ, Weissman SM, Urban AE, Snyder MP: Extensive genetic variation in somatic human tissues. Proc Natl Acad Sci USA 2012, 109:18018–18023.
20. Laurie CC, Laurie CA, Rice K, Dohey KF, Zelnick LR, McHugh CP, Ling H, Hetrick KN, Pugh EW, Amos C, Wei Q, Wang L-E, Lee JE, Barnes KC, Hansel NN, Mathias R, Daley D, Beaty TH, Scott AF, Ruczinski I, Scharpf RB, Bierut LJ, Hartz SM, Landi MT, Freedman ND, Goldin LR, Ginsburg D, Li J, Desch KC, Strom SS, et al: Detectable clonal mosaicism from birth to old age and its relationship to cancer. Nat Genet 2012, 44:642–650.
21. Zaranek AW, Levanon EY, Zecharia T, Clegg T, Church GM: A survey of genomic traces reveals a common sequencing error, RNA editing, and DNA editing. PLoS Genet 2010, 6:e1000954. doi:10.1371/journal.pgen.1000954.
22. Li M, Wang IX, Li Y, Bruzel A, Richards AL, Toung JM, Cheung VG: Widespread RNA and DNA sequence differences in the human transcriptome. Science 2011, 333:53–58.
23. Li M, Cheung VG: Response to comment on "Widespread RNA and DNA sequence differences in the human transcriptome". Science 2012, 335:335–1302F.
24. Kleinman CL, Majewski J: Comment on "Widespread RNA and DNA sequence differences in the human transcriptome". Science 2012, 335:335–302c.
25. Pickrell JK, Gilad Y, Pritchard JK: Comment on "Widespread RNA and DNA sequence differences in the human transcriptome". Science 2012, 335:335–1302e.
26. Lin W, Piskol R, Tan MH, Li JB: Comment on "Widespread RNA and DNA sequence differences in the human transcriptome". Science 2012, 335:335–1302e.
27. Gottlieb B, Chalifour LE, Mitmaker B, Sheiner N, Obrand D, Abraham C, Meilleur M, Sugahara T, Bkaily G, Schweitzer M: BAK1 gene variation and abdominal aortic aneurysms. Hum Mutat 2009, 30:1043–1047.
28. Costa V, Aprile M, Esposito R, Ciccodicola A: RNA-Seq and human complex diseases: recent accomplishments and future perspectives. Eur J Hum Genet 2013, 12:134–142.
29. Hauptman N, Glavac D: MicroRNAs and long non-coding RNAs: prospects in diagnostics and therapy of cancer. Radiol Oncol 2013, 47:311–318.
30. Isakov O, Roy Ronen R, Kovarsky J, Gabay A, Gan I, Modai S, Shomron N: Novel insight into the non-coding repertoire through deep sequencing analysis. Nucleic Acids Res 2012, 40:e86.
31. Taft RJ, Pang KC, Mercer TR, Dinger M, Mattick JS: Non-coding RNAs: regulators of disease. J Pathol 2010, 220:126–139.
32. Relton C, Smith DG: Epigenetic epidemiology of common complex disease: prospects for prediction, prevention, and treatment. PLoS Med 2010, 7:e1000356. doi:10.1371/journal.pmed.1000356.
33. Petronis A: Epigenetics as a unifying principle in the aetiology of complex traits. Nature 2010, 465:712–727.
34. Schuster-Böckler B, Bateman A: Protein interactions in human genetic diseases. Genome Biol 2008, 9:R9.
35. Al-Khoury R, Coulombe B: Defining protein interactions that regulate disease progression. Expert Opin Ther Targets 2009, 13:13–17.
36. Anderson AR, Weaver AM, Cummings PT, Quaranta V: Tumor morphology and phenotypic evolution driven by selective pressure from the microenvironment. Cell 2006, 127:905–915.
37. Greaves M, Maley CC: Clonal evolution in cancer. Nature 2012, 481:306–313.
38. Gottlieb B, Beitel LK, Trifiro MA: Selection and mutation in the "new" genetics: an emerging hypothesis. Hum Genet 2010, 31:491–501.
39. Gottlieb B, Beitel LK, Trifiro MA: Post-zygotic mutations and somatic mosaicism in androgen insensitivity syndrome. Trends Genet 2001, 17:628–632.

40. Nadeau JH: **Modifier genes in mice and humans.** *Nat Rev Genet* 2001,
     2:165–174.

41. Ahluwalia JK, Hariharan M, Bargaje R, Pillai B, Brahmachan V: **Incomplete
     penetrance and variable expressivity: is there a microRNA connection?**
     Bioessays 2009, **31:**981–992.

42. Blewitt M, Whitelaw E: **The use of mouse models to study epigenetics.**
     Cold Spring Harb Perspect Biol 2013, **5:**a017939.

43. Kuo DS, Labelle-Dumais C, Mao M, Jeanne M, Kauffman WB, Allen J, Favor J,
     Gould DB: **Allelic heterogeneity contributes to variability in ocular
     dysgenesis, myopathy and brain malformations caused by** *Col4a1* **and**
     Col4a2. *Hum Mol Genet* 2014, **23:**1709–1722.

44. Ahn SM, Kim TH, Lee S, Kim D, Ghang H, Kim DS, Kim BC, Kim SY, Kim WY,
     Kim C, Park D, Lee YS, Kim S, Reja R, Jho S, Kim CG, Cha JY, Kim KH, Lee B,
     Bhak J, Kim SJ: **The first Korean genome sequence and analysis: full
     genome sequencing for a socio-ethnic group.** *Genome Res* 2009,
     19:1622–1629.

45. Moore B, Hu H, Singleton M, De La Vega FM, Reese MG, Yandell M: **Global
     analysis of disease-related DNA sequence variation in 10 healthy individuals:
     implications for whole genome-based clinical diagnostics.** *Genet Med* 2011,
     13:210–217.

46. MacArthur DG, Balasubramanian S, Frankish A, Huang N, Morris J, Walter K,
     Habegger L, Pickrell JK, Montgomery SB, Albers CA, Zhang ZD, Conrad DF,
     Lunter G, Zheng H, Ayub Q, DePristo MA, Banks E, Hu M, Handsaker RE,
     Rosenfeld JA, Fromer M, Jin M, Mu XJ, Khurana E, Ye K, Kay M, Saunders GI,
     Suner MM, Hunt T: **A systematic survey of loss-of-function variants in
     human protein-coding genes.** *Science* 2012, **335:**823–828.

47. Daglish R, Fileck P, Cunningham F, Atashyn A, Tully RE, Proctor G, Chen Y,
     McLaren WM, Larsson P, Vaughan BW, Beroud C, Dobson G, Lehväslaiho H,
     Taschner PE, den Dunnen JT, Devereau A, Birney E, Brookes AJ, Maglott DR:
     **Locus Reference Genomic sequences: an improved basis for describing
     human DNA variants.** *Genome Med* 2010, **2:**24.

48. Khatri P, Sirota M, Butte AT: **Ten years of pathway analysis: current
     approaches and outstanding challenges.** *PLoS Comput Biol* 2012, **8:**e1002375.

49. Forbes SA, Bindal N, Bamford S, Cole C, Kok CY, Beare D, Jia M, Shepherd R,
     Leung K, Menzies A, Teague JW, Campbell PJ, Stratton MR, Futreal PA:
     **COSMIC: mining complete cancer genomes in the Catalogue of Somatic
     Mutations in Cancer.** *Nucl Acids Res* 2011, **39:**D945–D950.

50. Robinson PN, Mundlos S: **The human phenotype ontology.** *Clin Genet*
     2010, **77:**525–534.

51. Durrant C, Swertz Alberts R, Arends D, Moller S, Mott R, Primns P, van der
     Velde KJ, Jansen RC, Schughart K: **Bioinformatics tools and database
     resources for systems genetics analysis in mice – a short review and an
     evaluation of future needs.** *Brief Bioinform* 2012, **13:**135–142.

52. Gottlieb B, Alvarado C, Wang C, Gharizadeh B, Babrzadeh F, Richards B,
     Batist G, Basik M, Beitel LK, Trifiro M: **Making sense of intra-tumor genetic
     heterogeneity: altered frequency of androgen receptor CAG repeat
     length variants in breast cancer tissues.** *Hum Mutat* 2013, **34:**610–618.

53. Li M, Stoneking M: **A new approach for detecting low-level mutations in
     next-generation sequencing data.** *Genome Biol* 2012, **13:**R34.

54. Schmitt MW, Kennedy SR, Salk JJ, Fox EJ, Hiatt JB, Loeb LA: **Detection of
     ultra-rare mutations by next-generation sequencing.** *Proc Natl Acad Sci
     U S A* 2012, **109:**14508–14513.

55. Fu J, Wolfs MGM, Deelen P, Westra H-J, Fehrmann RSN, te Meerman GJ,
     Buurman WA, Rensen SSM, Groen HJM, Weersma RK, van den Berg LH,
     Veldink J, Ophoff RA, Snieder H, van Heel D, Jansen RC, Hofker MH,
     Wijmenga C, Franke L: **Unraveling the regulatory mechanisms underlying
     tissue-dependent genetic variation of gene expression.** *PLoS Genet* 2012,
     8:e1002431. doi:10.1371/journal.pgen.

56. The GTEx Consortium: **The Genotype-Tissue Expression (GTEx) project.**
     Nat Genet 2013, **45:**580–585.

18

Role of conserved cis-regulatory elements in the post-transcriptional regulation of the human MECP2 gene involved in autism

Joetsaroop S Bagga[1,3] and Lawrence A D'Antonio[2*]

Abstract

Background: The MECP2 gene codes for methyl CpG binding protein 2 which regulates activities of other genes in the early development of the brain. Mutations in this gene have been associated with Rett syndrome, a form of autism. The purpose of this study was to investigate the role of evolutionarily conserved *cis*-elements in regulating the post-transcriptional expression of the MECP2 gene and to explore their possible correlations with a mutation that is known to cause mental retardation.

Results: A bioinformatics approach was used to map evolutionarily conserved *cis*-regulatory elements in the transcribed regions of the human MECP2 gene and its mammalian orthologs. *Cis*-regulatory motifs including G-quadruplexes, microRNA target sites, and AU-rich elements have gained significant importance because of their role in key biological processes and as therapeutic targets. We discovered in the 5'-UTR (untranslated region) of MECP2 mRNA a highly conserved G-quadruplex which overlapped a known deletion in Rett syndrome patients with decreased levels of MeCP2 protein. We believe that this 5'-UTR G-quadruplex could be involved in regulating MECP2 translation. We mapped additional evolutionarily conserved G-quadruplexes, microRNA target sites, and AU-rich elements in the key sections of both untranslated regions. Our studies suggest the regulation of translation, mRNA turnover, and development-related alternative MECP2 polyadenylation, putatively involving interactions of conserved *cis*-regulatory elements with their respective *trans* factors and complex interactions among the *trans* factors themselves. We discovered highly conserved G-quadruplex motifs that were more prevalent near alternative splice sites as compared to the constitutive sites of the MECP2 gene. We also identified a pair of overlapping G-quadruplexes at an alternative 5' splice site that could potentially regulate alternative splicing in a negative as well as a positive way in the MECP2 pre-mRNAs.

Conclusions: A Rett syndrome mutation with decreased protein expression was found to be associated with a conserved G-quadruplex. Our studies suggest that MECP2 post-transcriptional gene expression could be regulated by several evolutionarily conserved *cis*-elements like G-quadruplex motifs, microRNA target sites, and AU-rich elements. This phylogenetic analysis has provided some interesting and valuable insights into the regulation of the MECP2 gene involved in autism.

Keywords: G-quadruplex, Post-transcriptional regulation, MECP2, MicroRNAs, AU-rich elements, Autism

Background

The methyl CpG binding protein 2 gene codes for the protein MeCP2, which is essential for normal brain development [1]. This protein is responsible for regulated transcription of neuron-specific genes and is vital for connecting nerve cells, where cell–cell communication takes place. Mutations in the MECP2 gene can cause a form of autism called Rett syndrome. Victims of this syndrome are typically females between the ages of 6 and 18 months. Additionally, Rett syndrome patients experience a loss of acquired skills, impaired speech, and abnormal stereotypical movements. In some cases, young patients have experienced frequent seizures and mental retardation [2]. Rett syndrome is in fact one of the most common causes of mental retardation in females.

* Correspondence: ldant@ramapo.edu
[2]Ramapo College of New Jersey, 505 Ramapo Valley Rd., Mahwah, NJ 07430, USA

Several types of mutations have been mapped to the MECP2 gene from affected patients [3,4]. Many of the mutations affect the coding region and either result in a MeCP2 protein with altered function or a non-functional protein. Mutations that lead to altered gene expression have been mapped to the 5′- and 3′-untranslated regions (UTRs) [3,5,6]. Several mutations in the genomic MECP2 sequence lead to altered splicing of the gene [3].

Cis-regulatory motifs located in the untranslated regions and in the vicinity of splice junctions are known to interact with RNA binding proteins for regulating post-transcriptional gene expression. Studying cis-element regulation of MECP2 gene expression can help provide better insights into the molecular mechanism of MECP2 regulation and deeper understanding of the genetic disorders caused by alteration of its expression.

Guanine-rich sequences can form highly stable structures. Instead of the Watson and Crick DNA duplex, four consecutive tetrads of G-rich sequences in a nucleic acid can form G-quadruplexes [7]. The G-quadruplexes are known to have important roles in biological processes and human disease and as therapeutic targets [8-11]. These structures have been found in telomeres, promoter regions, and other biologically important regions in the DNA influencing DNA replication, transcription, and epigenetic mechanisms [12,13]. Computationally predicted G-quadruplex structures have been reported in the MECP2 gene [14]. However, the biological role of these motifs in the MECP2 DNA remains to be determined. Recently, it became possible to quantitatively visualize the formation of genomic G-quadruplexes in living mammalian cells [15]. RNA G-quadruplexes are more likely to be formed in vivo [16] and are more stable than the DNA G-quadruplexes [17]. There is ample evidence for cis-regulatory roles of G-quadruplexes in the post-transcriptional gene expression [18]. RNA G-quadruplexes located in the 5′-UTR have been known to be involved in regulated translational initiation [19,20] as well as translation repression [21-23]. G-quadruplex motifs found in the translated regions have been shown to affect folding and proteolysis of hERα protein [24]. G-rich sequences in the 3′-UTR have been shown to influence polyadenylation [25], RNA turnover [26], and subcellular mRNA localization [27]. A 3′-UTR polymorphism that affects G-quadruplex structure has been shown to modulate gene expression of the KiSS1 mRNA [28]. There is evidence for direct G-quadruplex role in regulated alternative splicing of fragile X mental retardation 1 (FMR1) transcripts [29] and of beta-site amyloid precursor protein (APP) cleaving enzyme 1 (BACE1) involved in Alzheimer disease [30].

Development of bioinformatics techniques has made it possible to study the prevalence and distribution of G-quadruplex forming sequence motifs at genomic levels [31-34]. Consequently, there has been a tremendous increase in published literature and reviews on this subject [34-36]. Large scale computational studies have identified an association of G-quadruplex forming sequences in both 5′- as well as 3′-UTRs [37]. However, computational predictions have difficulty in distinguishing between a G-quadruplex sequence motif which occurs by chance and the one that forms a structure with a biological role in the cell.

In this study, we have used a bioinformatics approach to map evolutionarily conserved G-quadruplex motifs, microRNA target sites, and AU-rich elements (AREs) in the transcribed regions of the human MECP2 gene and its mammalian orthologs. Identifying evolutionarily conserved motifs helps validate computational predictions, improving accuracy, and providing evidence for their biological relevance. The goal of this project was to study the role of conserved cis-regulatory motifs in regulating the post-transcriptional expression of the MECP2 gene and explore their possible correlations with a mutation that is known to cause mental retardation.

The translation and destabilization of large number of eukaryotic mRNAs are known to be regulated via microRNA-mediated pathways, which have received significant attention [38]. MeCP2 protein expression has been shown to be influenced by microRNA targeting [39]. Similarly, AU-rich elements in the 3′-UTRs of developmentally expressed mRNAs have been associated with regulated stability [40]. Therefore, in addition to the G-quadruplexes, the roles of microRNA targeting and AREs as post-transcriptional regulators and their interrelationships were also investigated in this project.

Results and discussion

A total of four MECP2 mammalian orthologs, Homo sapiens, Canis lupus familiaris, Mus musculus, and Rattus norvegicus were chosen for the current studies (Table 1). Although the MeCP2 protein orthologs were quite similar, the nucleotide sequence similarities among the mRNAs were relatively lower due to variation in the 5′- and 3′- untranslated regions (human, dog, and mouse MECP2 genes are known to have multiple isoforms. Orthologous isoforms with comparable exon/intron structures were chosen for sequence alignments.).

A conserved G-quadruplex in the 5′-UTR of MECP2 orthologs

A G-quadruplex highly conserved in relative location to the translation start site was discovered in the 5′-UTR of human, dog, and mouse MECP2 mRNAs (Figure 1). Existence of a conserved motif within an otherwise highly variable region signifies its functional role. This conserved G-quadruplex motif, which we named 'CG', is located 110 bases upstream of the translation initiation site in the human MECP2 mRNA and is likely to play a

Table 1 A total of four MECP2 mammalian orthologs were chosen for the current studies

MECP2 ortholog	Nucleotide sequence (mRNA) identity to human	Protein sequence identity/similarity to human	Protein length (amino acids)
Homo sapiens [RefSeq:NM_004992.3]	100%	100%	486
Canis lupus familiaris [RefSeq:XM_848395.1]	92.50%	96%/97%	486
Mus musculus [RefSeq:NM_001081979.1]	85.00%	95%/96%	501
Rattus norvegicus [RefSeq:NM_022673.1]	78.20%	94%/95%	492

RNAs were aligned pairwise by semi-global method which does not penalize end-gaps. Therefore, the percentage similarity calculations do not include differences in the lengths of untranslated regions. Values in parentheses represent NCBI accession numbers of the corresponding mRNAs.

role in the regulation of translation. There have been several reports of 5'-UTR G-quadruplexes that are involved in translation regulation. A G-quadruplex structure located in the 5'-UTR of human fibroblast growth factor 2 (FGF2) acts as an internal ribosomal entry site (IRES) for translation initiation [19]. On the other hand, formation of G-quadruplexes can also play inhibitory roles for translation of NRAS oncogene [21], Ying Yang 1 involved in tumorigenesis [41], and ADAM10 responsible for anti-amyloidogenic processing of the APP [22]. The CG G-quadruplex conserved in the 5'-UTR of human, dog, and mouse MECP2 mRNA orthologs (Figures 1 and 2) is of particular interest because it maps to a known mutation in the MECP2 gene leading to Rett syndrome [42]. An 11-bp deletion (GCGAGGAGGAG) (Figure 2) in the 5'-UTR results in the lack of MeCP2 protein in about 25% of the tested cells even though the mRNA is

detectable and the coding sequence (CDS) of the mRNA is apparently intact [42].

We believe that the MECP2 5'-UTR G-quadruplex CG is in fact the translation regulatory motif which gets affected due to the 11-bp deletion in some Rett syndrome patients. Nucleotide sequence mutations and polymorphisms that destroy G-quadruplex folding or change the G-quadruplex conformation are known to affect gene expression [22,28]. Two possible mechanisms may lead to G-quadruplex-mediated regulation of translation in the MECP2 mRNA. Interaction of RNA binding proteins with the G-quadruplexes in the 5'-UTR is known to modulate translation. For example, nucleolin protein binds to G-rich sequences to positively influence protein translation [43]. We have tested several nucleolin targets [43] with the quadruplex forming G-rich sequences (QGRS) Mapper software [31] and found them to be capable of forming

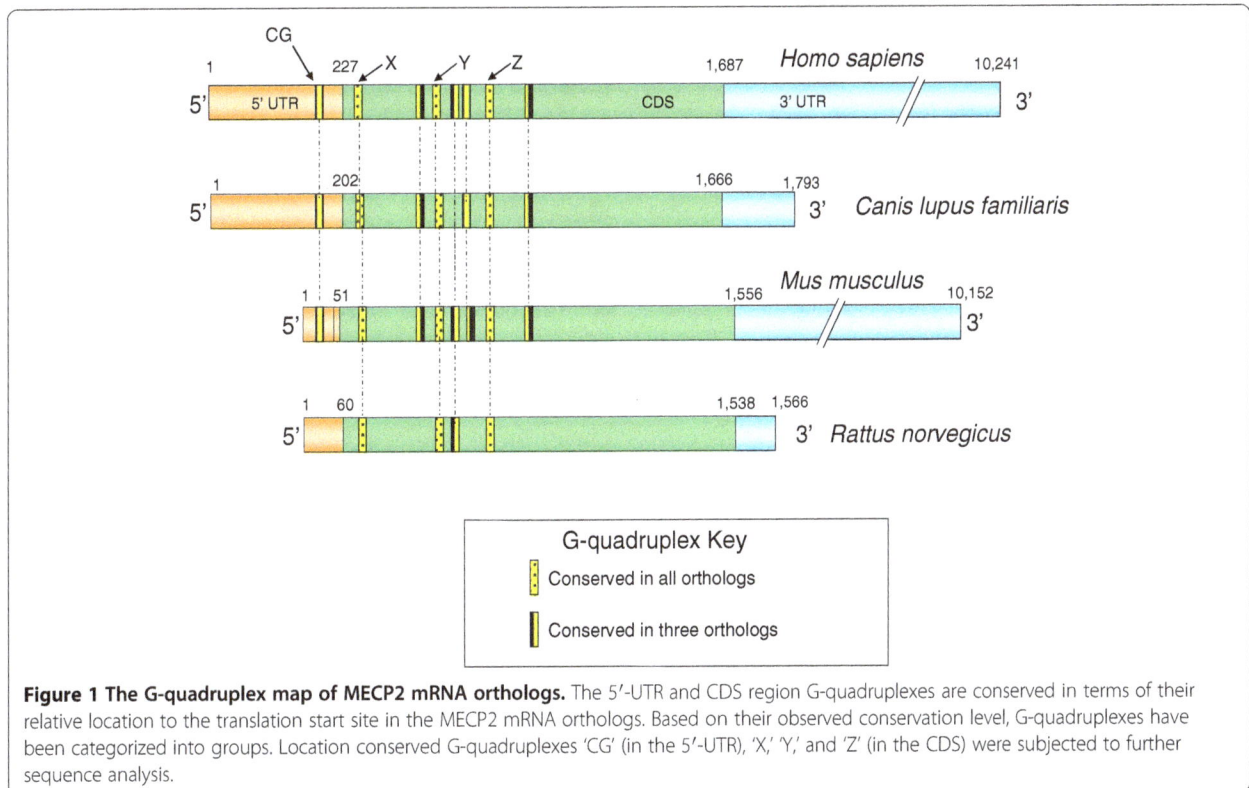

Figure 1 The G-quadruplex map of MECP2 mRNA orthologs. The 5'-UTR and CDS region G-quadruplexes are conserved in terms of their relative location to the translation start site in the MECP2 mRNA orthologs. Based on their observed conservation level, G-quadruplexes have been categorized into groups. Location conserved G-quadruplexes 'CG' (in the 5'-UTR), 'X,' 'Y,' and 'Z' (in the CDS) were subjected to further sequence analysis.

```
                              -120        -110        -100
Homo sapiens:                 GGAGGAGGAGGCGAGGAGGAGAGACTGCTC
                              -120        -110        -100
Canis lupus familiaris:       GGAGGAGGAGGCGAGGAGGAGAGACTGCTC
                              -30         -20         -10
Mus musculus:                 GGAGAGAGGGCTGTGGTAAAACCCGTCCGG

Consensus:                    ****---*-*--*-**---*-----*-*--
```

Figure 2 The conserved G-quadruplex motif (CG). This motif in the 5′-UTR of *Homo sapiens*, *Canis lupus familiaris*, and the *Mus musculus* MECP2 mRNAs maps to a known deletion in the human MECP2 gene leading to Rett syndrome. The predicted tetrad forming G-tracts of the quadruplex are underlined. This conserved G-quadruplex is likely to get disrupted due to the 11-bp deletion which is known to affect MECP2 translation in some Rett syndrome patients. The deletion is marked in the box with bold characters. The numbers represent upstream distances from the CDS start sites of the respective mRNAs.

G-quadruplexes (data not shown). A disruption in the 5′-UTR G-quadruplex of the MECP2 mRNA could consequently lead to lower protein translation. The fragile X mental retardation protein (FMRP) is also known to regulate translation by binding to G-quadruplexes on its target mRNAs [44]. Altered function of FMRP could lead to atypical synapse development in the brain and impaired learning resulting in mental retardation [45]. Several other genes implicated in autism have been shown to form G-quadruplexes [44,46]. A change in the 5′-UTR G-quadruplex region is likely to affect FMRP binding and hence translation of MECP2 mRNA, possibly leading to genetic defects like Rett syndrome.

Alternatively, the 5′-UTR G-quadruplex may be an important component of IRES [19,20] which is responsible for translation of the Mecp2 mRNA. The 11-bp deletion in the G-quadruplex motif, and therefore disruption of IRES, may affect the translation of the Mecp2 mRNA.

Conserved G-quadruplexes in the coding region of MECP2 orthologs

We mapped several conserved G-quadruplexes within the CDS region of the MECP2 mRNA orthologs. Three G-quadruplexes ('X', 'Y', and 'Z', Figure 1) were highly conserved within the MECP2 CDS region of all four species. The G-quadruplex 'Y' showed a high level of sequence conservation across the four mammalian species (Figure 3). Regardless of the modest variation in sequence conservation, all of the three CDS G-quadruplexes exhibited high conservation at a position relative to the translation start site and at the predicted structure level. G-quadruplexes within the coding regions of mRNAs are known to be involved in regulating the RNA stability [47], translation [43], and protein folding [24].

```
                          G-quadruplex Y
Homo sapiens:             GGAACTGGCAGAGGCCGGGG
Canis lupus familiaris:   GGAACTGGCAGAGGTCGGGG
Mus musculus:             GGAACTGGCAGGGGTCGGGG
Rattus norvegicus:        GGAACTGGCAGGGGTCGGGG
Consensus:                ***********-********
```

Figure 3 Location-conserved G-quadruplex 'Y' in the CDS of MECP2 orthologs. This motif showed a high level of sequence conservation across the four mammalian species. G-quadruplex 'Y' refers to the corresponding marked map position in the Figure 1 above.

Conserved *cis*-regulatory elements in the 3′-UTR of MECP2 orthologs

The MECP2 mRNAs analyzed in this work included two alternatively spliced isoforms each for human, dog, and mouse orthologs and one MECP2 transcript of rat. Both MECP2 isoforms of mouse and human isoform 1, each have long 3′-UTRs (>8.5 kb). Both of the dog MECP2 isoforms, isoform 2 of human MECP2 and the rat mRNA each have short 3′-UTRs (<0.5 kb). The longer MECP2 isoforms contain at least two polyadenylation signals and their corresponding cleavage/polyadenylation sites. Alternative polyadenylation in MECP2 can lead to transcript isoforms with the longer or shorter version of the 3′-UTRs [48]. The longer human isoform has been found to be in higher abundance in the fetal neuronal tissues and involved in the development of the brain while shorter transcripts are prevalent within the adult brain [48]. Long 3′-UTRs are likely to play pivotal roles in post-transcriptional regulation of MECP2 mRNA, especially during the early developmental process when gene expression needs to be tightly regulated. Therefore, this part of our project explored the capability of 3′-UTRs of MECP2 mammalian orthologs and isoforms to form evolutionarily conserved G-quadruplexes, especially in the vicinity of other conserved *cis*-regulatory elements: AREs, microRNA target sites, and alternative polyadenylation signals.

First, we studied the overall phylogenic conservation of the MECP2 gene particularly in the 3′-UTR regions. Based on sequence alignments among mammalian orthologs of MECP2 mRNAs, we found that most of the MECP2 3′-UTR sequence is highly variable. However, regions surrounding polyadenylation signals/sites showed much better conservation (data not presented). This suggests important biological roles of the conserved regions in

the regulation of alternative polyadenylation involved in the developmental regulation of MECP2.

The 3′-UTR of MECP2 is highly variable; however, the majority of the conserved *cis*-regulatory elements that we analyzed (microRNA target sites, AU-rich elements, and G-quadruplexes) mapped to evolutionarily conserved regions in the 3′-UTR of the long MECP2 isoform, which is involved in early brain development (Figure 4) (all four mammalian orthologs of MECP2 were analyzed. Only data from human and mouse isoforms is presented. Human MECP2 alignments with its dog and rat orthologs were very similar to the alignments between human and mouse orthologs). The short human MECP2 mRNA isoform 2, expressed mostly in the adult brain, lacked conserved microRNA targets, ARE, or G-quadruplexes. Our results suggest that these conserved *cis*-elements could have important regulatory roles in post-transcriptional MECP2 expression during early development stages of the brain.

There is sufficient evidence to indicate a role for 3′-UTR G-quadruplex in post-transcriptional regulation of gene expression [28,43,49-51]. G-quadruplexes in the 3′-UTR are known to regulate translation [43]. Interactions between RNA binding proteins like hnRNP F/H and quadruplex forming G-rich sequences are known to regulate splicing and 3′-end processing [49-51]. In our studies, a highly conserved G-quadruplex was found to be associated with one alternative poly(A) site but not the second site (Figure 4). The conserved G-quadruplex was present 17 bases downstream of the poly(A) site 1 (Figure 5), well within the range of the cleavage/polyadenylation complex formation associated with G-quadruplex-mediated regulation of 3′-end formation [49]. Mutations of G-rich sequences in this region of MECP2 RNA have been shown to reduce polyadenylation efficiency *in vivo* [52]. We did not find any evidence of G-quadruplex forming sequences within 200 bases downstream of the alternative poly(A) site 2 responsible for the long isoform of the human MECP2 gene (Figure 6 and data not shown). These results suggest a G-quadruplex role in alternative cleavage/polyadenylation associated with brain development-specific MECP2 gene expression. The mechanism of alternative 3′-end processing regulation may involve dynamic formation or resolution of the RNA G-quadruplex near poly(A) Site 1 via specific helicases such as RHAU [53]. The role of

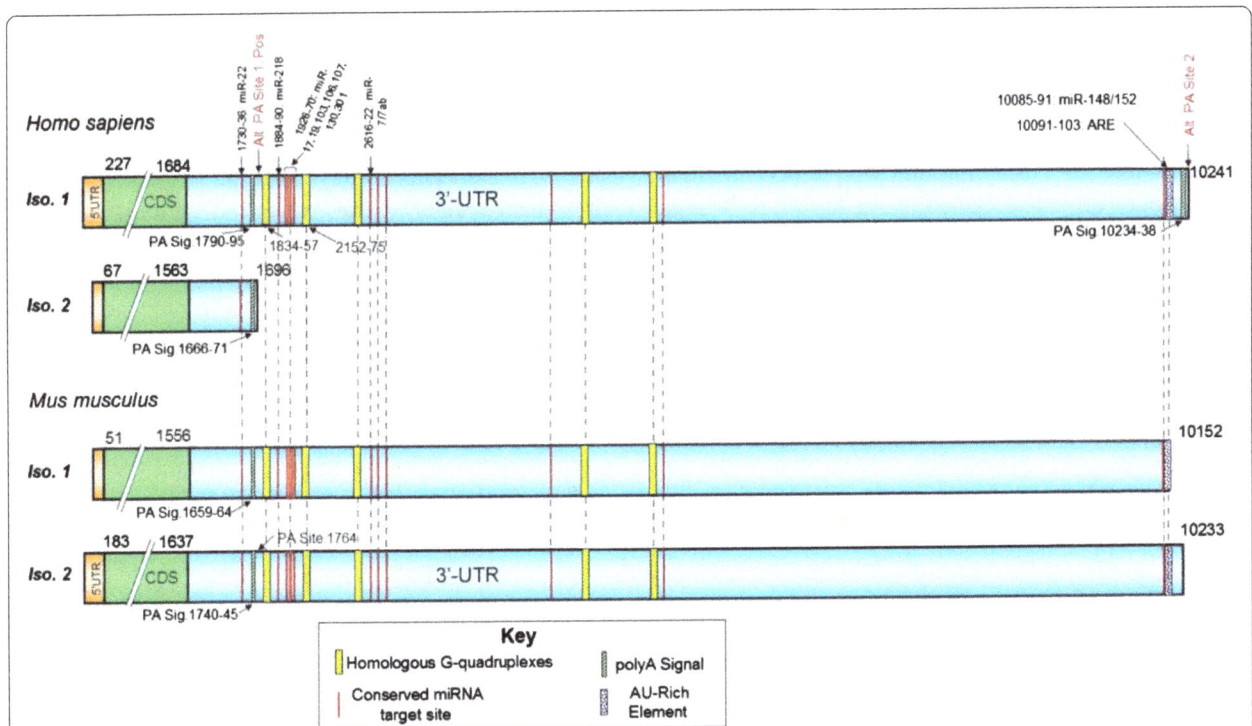

Figure 4 Conserved 3′-UTR *cis*-regulatory elements map of MECP2 mRNA orthologs and isoforms. Majority of the *cis*-regulatory elements mapped to the evolutionarily conserved regions of long MECP2 isoform 3′-UTR which is involved in early brain development. All four mammalian orthologs of MECP2 mRNAs from human, dog, mouse, and rat were analyzed. Only human and mouse mRNA alignments are displayed. Human MECP2 alignments with its dog and rat orthologs were very similar to the alignment shown. The short human MECP2 isoform 2 lacked conserved microRNA targets, ARE, or G-quadruplexes. A highly conserved G-quadruplex is present selectively near one alternative polyadenylation signal/site. Most evolutionarily conserved G-quadruplexes were preferentially associated with microRNA target sites. Evolutionarily conserved AU-rich element (ARE) and mi-R148/152 target sites were associated with the second alternative poly(A) site which results in the expression of longer isoform during the early development of the human brain.

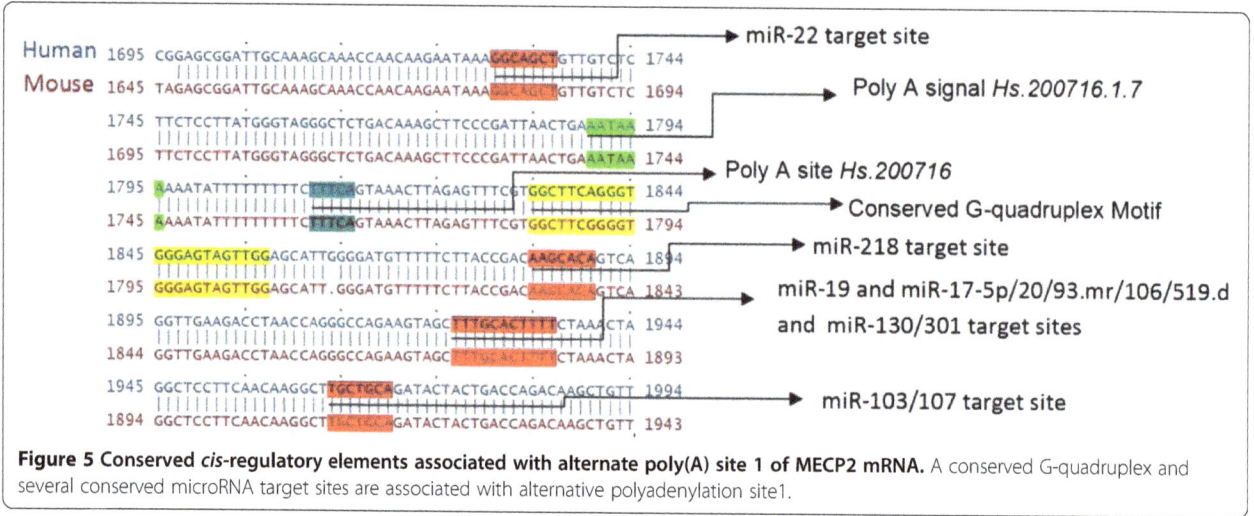

Figure 5 Conserved *cis*-regulatory elements associated with alternate poly(A) site 1 of MECP2 mRNA. A conserved G-quadruplex and several conserved microRNA target sites are associated with alternative polyadenylation site1.

G-quadruplexes in polyadenylation can be modulated by interactions with different proteins. For example, while binding of hnRNP H/H′ to quadruplex forming G-rich sequences can enhance polyadenylation [49,54], hnRNP F (which also has affinity for G-rich tracts) has been shown to interfere with polyadenylation [55].

Most of the evolutionarily conserved microRNA target sites were located in 3′-UTR of the long isoform; many of them are approximately 100 bp downstream of the poly(A) site 1 which is closer to the MECP2 coding region (Figure 4). The translation and destabilization of a large number of eukaryotic mRNAs, especially those under strict expression regulation, are known to be regulated via microRNA-mediated pathways [38]. Therefore, it was not surprising to discover microRNA target sites in the 3′-UTR of developmentally regulated long MECP2 isoform. MicroRNA targeting the long 3′-UTR MECP2 isoform has been previously shown to modulate MeCP2 protein levels in the developing human brain [56].

We noticed that most evolutionarily conserved G-quadruplexes were preferentially associated with conserved microRNA target sites in the 3′-UTR (Figure 4),

suggesting a potential interplay between microRNAs/microRNP (microribonucleoprotein) and G-quadruplex binding proteins. G-quadruplex binding proteins like FXR1 (fragile X retardation 1, a paralog of FMRP and involved in mental retardation) are known to be part of microRNP complexes [57]. FXR1 is also involved in directing microRNAs to the ARE for regulation of translation [57]. Therefore, a regulatory role for some G-quadruplexes in 3′-UTR of MECP2 may also have to do with mRNA translation.

Evolutionarily conserved ARE and mi-R148/152 target sites were associated with the second alternative poly(A) site which results in the expression of longer isoform (Figures 5 and 6). AU-rich elements in the 3′-UTRs of developmentally expressed mRNAs have been associated with regulated stability via the 3′-5′ exosome pathway following deadenylation [40]. The *cis*-acting AREs can interact with a variety of proteins to promote [58] or delay [59] ARE-mediated mRNA degradation (AMD). Recent studies and reviews have suggested that microRNAs can regulate post-transcriptional gene expression by targeting AMD as well as translation [60,61]. Association of evolutionarily

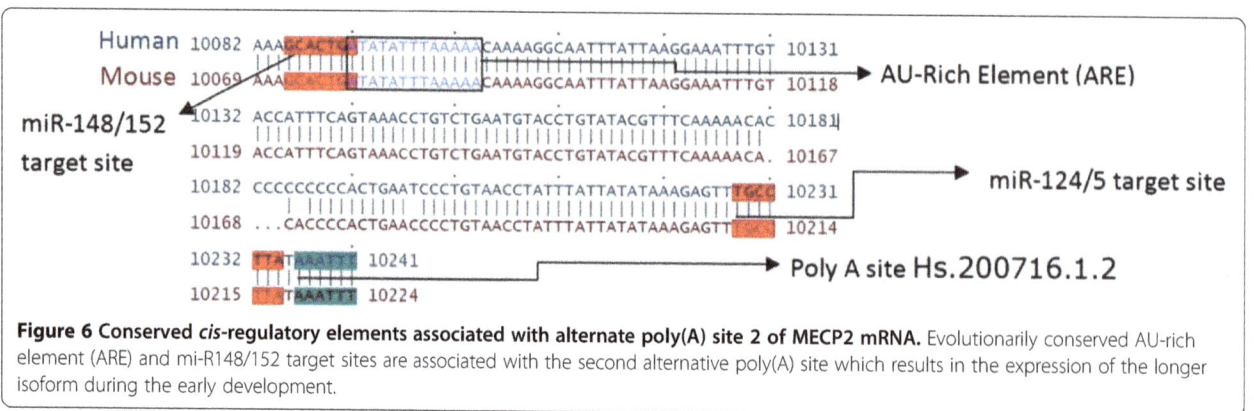

Figure 6 Conserved *cis*-regulatory elements associated with alternate poly(A) site 2 of MECP2 mRNA. Evolutionarily conserved AU-rich element (ARE) and mi-R148/152 target sites are associated with the second alternative poly(A) site which results in the expression of the longer isoform during the early development.

conserved mi-R148/152 target sites along with ARE in the long isoform suggests a potential cooperation between microRNAs/microRNP and ARE-binding proteins (ARE-BPs) for ARE-mediated post-transcriptional regulation of MECP2 transcripts.

Conserved G-quadruplex motifs near splice sites of the MECP2 pre-mRNA orthologs

We focused our attention to the conserved G-quadruplex motifs located in the vicinity of splice sites, especially those that are alternatively regulated. Human, dog, and mouse MECP2 orthologs are known to have two alternatively spliced isoforms each. The human isoform 1 (also known as MECP2A) of MECP2 has an extra exon. This isoform is predominantly expressed in the neurons during early development while the human isoform 2 is prevalent in adults in a variety of tissues including the brain.

Many G-quadruplexes were mapped in the isoforms of four mammalian pre-mRNA orthologs. A total of 33 G-quadruplexes, which were conserved in all the four mammalian orthologs, were mapped to the vicinity of 18 constitutive and 6 alternative splice sites. A bias in the overall distribution of conserved G-quadruplexes was noticed (Figure 7). Conserved G-quadruplexes were more likely to be associated with alternative splice sites of the mammalian MECP2 orthologs, suggesting a prospective biological role for them in regulated splicing. Almost all the alternatively spliced sites of MECP2 mammalian orthologs were associated with at least one conserved G-quadruplex (Figure 8). Alternative splice site G-quadruplexes were more or less equally distributed among exons and introns.

G-quadruplex forming sequences have the potential to affect alternative tissue-specific splicing through their interactions with hnRNP H family of proteins [62]. For example, the hnRNP F protein, with an affinity for quadruplex forming G-rich sequences, is needed for nervous tissue-specific alternative splicing [10]. A G-quadruplex in FMR1 RNA can act as an alternative exonic enhancer by binding to its own FMRP protein involved in mental retardation [29]. An intronic G-quadruplex in the tumor suppressor TP53 gene is also responsible for alternative splicing [63]. A G-quadruplex in the third exon of beta-site APP cleaving enzyme 1 (BACE1) involved in Alzheimer disease has been shown to regulate splice site selection [30]. Alternative splicing in the human and mouse MECP2 pre-mRNAs involve the second exon which gets skipped. Conserved G-quadruplexes were located near both splice sites of this skippable exon in the human and mouse MECP2 orthologs. While one of the G-quadruplexes (A) was near the 3′ splice site in the intron, there were two conserved overlapping G-quadruplexes (B/B′) near the 5′ splice site in this exon. The locations of these conserved G-quadruplexes seem optimal for direct

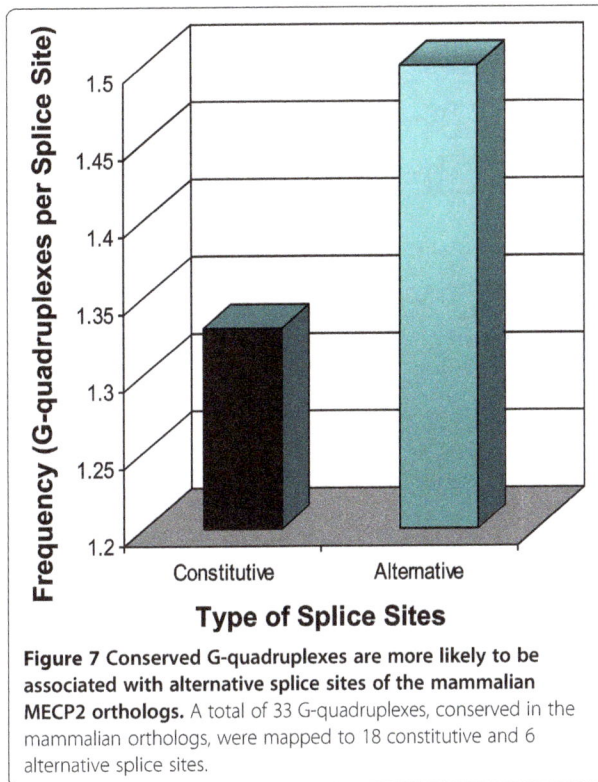

Figure 7 Conserved G-quadruplexes are more likely to be associated with alternative splice sites of the mammalian MECP2 orthologs. A total of 33 G-quadruplexes, conserved in the mammalian orthologs, were mapped to 18 constitutive and 6 alternative splice sites.

involvement in the regulated, development-related alternative splicing via interactions with splice regulatory proteins. In one of the dog MECP2 isoforms, the last exon gets interrupted by a short intron resulting in a total of five rather than four exons due to this alternative splicing (Figure 8). A conserved G-quadruplex was also discovered near the alternative 5′ splice site of the alternative intron. Our findings from this experiment suggest a good possibility that G-quadruplexes are involved in regulated alternative splicing in the MECP2 gene.

Multiple sequence alignments revealed that three location-conserved G-quadruplexes (A, B/B′, and D, Figure 8) near the alternative splice sites of all mammalian MECP2 orthologs have highly conserved motifs as well. A highly stable G-quadruplex (C) not found near an alternative splice site is relatively less well conserved at the sequence level (Figure 9). This data demonstrates a difference in the nature of G-quadruplexes found near alternatively spliced sites and other G-quadruplexes conserved in the same gene.

Location-conserved G-quadruplex B′ is also highly conserved at the sequence level in all four mammalian MECP2 orthologs (Figure 10). G-quadruplex B′ partially overlaps with G-quadruplex B (Figure 8). Additionally, the B′ G-quadruplex was found to overlap the second 5′ splice site of MECP2 pre-mRNA (Figure 8). This particular site is known to be alternatively spliced in human and mouse MECP2 orthologs. The highly conserved

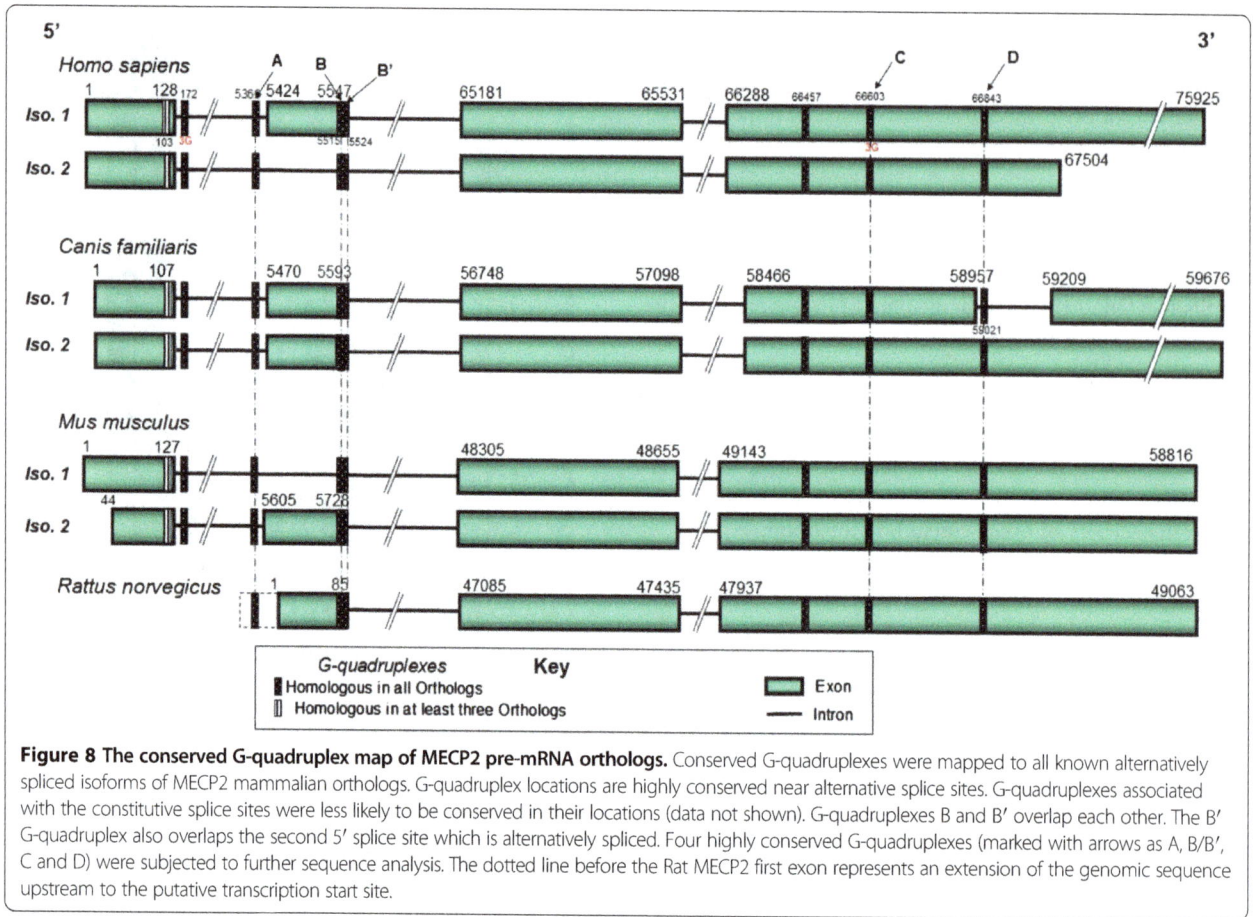

Figure 8 The conserved G-quadruplex map of MECP2 pre-mRNA orthologs. Conserved G-quadruplexes were mapped to all known alternatively spliced isoforms of MECP2 mammalian orthologs. G-quadruplex locations are highly conserved near alternative splice sites. G-quadruplexes associated with the constitutive splice sites were less likely to be conserved in their locations (data not shown). G-quadruplexes B and B′ overlap each other. The B′ G-quadruplex also overlaps the second 5′ splice site which is alternatively spliced. Four highly conserved G-quadruplexes (marked with arrows as A, B/B′, C and D) were subjected to further sequence analysis. The dotted line before the Rat MECP2 first exon represents an extension of the genomic sequence upstream to the putative transcription start site.

G-quadruplex B is found 5 bases upstream of the alternative 5′ splice site in the human MECP2 pre-mRNA sequence (Figure 11). This is a convenient location for a G-quadruplex to function as an exonic splicing enhancer (ESE) regulatory motif. Previous studies have demonstrated that G-quadruplex structures found near the splice sites in the exons of genes expressed in the brain can act as ESEs by interacting with FMRP protein [29]. The B′ G-quadruplex, which is also highly conserved across the mammalian species, overlaps the B G-quadruplex motif as well as the alternative 5′ splice site. At a given time, only one of these G-quadruplexes is likely to be formed in the cell. Therefore, quadruplexes B and B′ are likely to be

mutually exclusive. While G-quadruplex B can perform as an ESE, B′, when formed, may act as an inhibitor of alternative splicing since formation of this structure is likely to make the 5′ splice site unavailable. This data suggests that the B/B′ G-quadruplex pair can regulate alternative splicing in a negative as well as a positive way in the MECP2 pre-mRNAs.

Regulated alternative pre-mRNA splicing is an essential component of post-transcriptional gene expression and is important for biological processes. MECP2 produces multiple isoforms and its expression is highly regulated among different tissues, especially in the brain during different developmental stages. Our study has identified

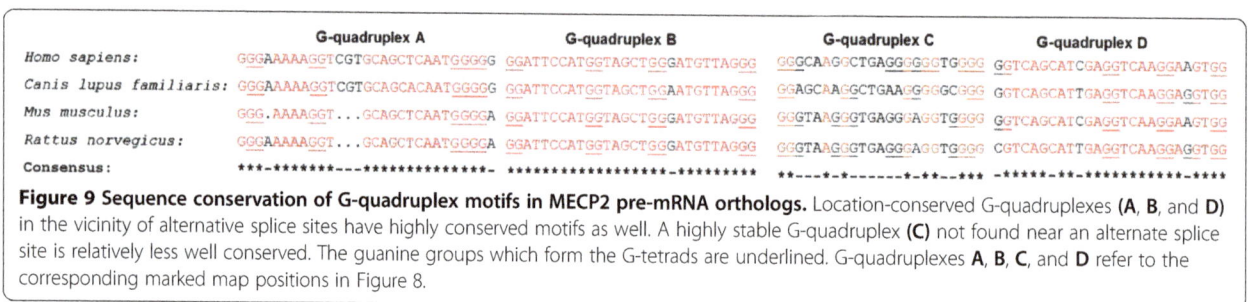

Figure 9 Sequence conservation of G-quadruplex motifs in MECP2 pre-mRNA orthologs. Location-conserved G-quadruplexes (**A**, **B**, and **D**) in the vicinity of alternative splice sites have highly conserved motifs as well. A highly stable G-quadruplex (**C**) not found near an alternate splice site is relatively less well conserved. The guanine groups which form the G-tetrads are underlined. G-quadruplexes **A**, **B**, **C**, and **D** refer to the corresponding marked map positions in Figure 8.

Homo sapiens:	GGTAGCTGGGATGTTAGGGCTCAGG
Canis lupus familiaris:	GGTAGCTGGAATGTTAGGGCTCAGG
Mus musculus:	GGTAGCTGGGATGTTAGGGCTCAGG
Rattus norvegicus:	GGTAGCTGGGATGTTAGGGCTCAGG
Consensus:	*********-***************

Figure 10 Sequence conservation of G-quadruplex B′ motif that overlaps alternatively spliced 5′ splice site of MECP2 pre-mRNA. Location-conserved G-quadruplex B′ (which partially overlaps with G-quadruplex B) is also highly conserved at the sequence level in all the four mammalian MECP2 orthologs. Additionally, the B′ G-quadruplex was found to overlap the second 5′ splice site of MECP2 pre-mRNA. This particular site is known to be alternatively spliced in human and mouse MECP2 orthologs. The guanine groups which form the G-tetrads are underlined. G-quadruplex B′ refers to the corresponding marked map position in Figure 8.

evolutionarily conserved G-quadruplexes associated with alternative splicing of MECP2 mammalian orthologs.

Conclusions

The goal of this project was to perform evolutionary analysis of four MECP2 mammalian orthologs in order to identify conserved *cis*-regulatory elements that may regulate post-transcriptional expression of this gene which is known to be associated with mental retardation syndromes. Our bioinformatics based studies focused on G-quadruplexes, microRNA target sites, and AU-Rich elements which we mapped to the transcribed regions of MECP2 orthologs.

We identified a highly conserved G-quadruplex in the 5′-UTR of three mammalian MECP2 orthologs which overlapped with a known 11-bp deletion in Rett syndrome patients with decreased levels of MeCP2 protein but normal transcripts [42]. We believe that this 5′-UTR G-quadruplex could be involved in regulating MECP2 post-transcriptional expression either as an IRES [19,20], or by interacting with specific proteins such as nucleolin [43], or FMRP [44]. Altered levels of MeCP2 protein during the early brain development can interfere with neuronal connections, leading to autism.

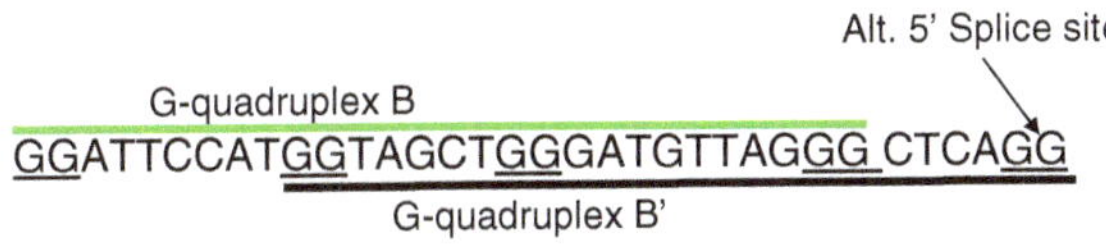

Figure 11 The B/B′ G-quadruplex pair may regulate alternative splicing at the 5′ splice site of human and msouse MECP2 orthologs. The highly conserved G-quadruplex B is found 5 bases upstream of the alternative 5′ splice site in the human MECP2 pre-mRNA sequence and may function as an exonic splicing enhancer (ESE) regulatory motif. The B′ G-quadruplex, which is also highly conserved across the mammalian species, overlaps the B motif as well as the alternative 5′ splice site. At a given time, only one of the G-quadruplex is likely to be formed in the cell. Therefore, B and B ′ are likely to be mutually exclusive. G-quadruplex B′ when formed may act as an inhibitor of alternative splicing since formation of this structure is likely to make the 5′ splice site unavailable. Underlined Gs represent the bases involved in the G-tetrad formation in the G-quadruplex. G-quadruplexes B and B′ refer to the corresponding marked map positions in Figure 8. Human and mouse B/B′ G-quadruplex sequence motifs are identical.

The majority of the conserved *cis*-regulatory elements analyzed (G-quadruplexes, microRNA target sites, and AREs) mapped to the evolutionarily conserved regions of the otherwise variable 3′-UTR of the long MECP2 isoform which requires tight regulation during the early brain development. The short isoform which has a more stable adult expression primarily lacks most of the conserved 3′-UTR *cis*-regulatory elements analyzed. Most evolutionarily conserved G-quadruplexes were preferentially associated with microRNA target sites, suggesting an interplay between microRNAs/microRNA ribonucleoprotein (miRNP) and G-quadruplex binding proteins. A highly conserved G-quadruplex present selectively near alternative polyadenylation site 1 could be responsible for alternative polyadenylation which is the primary mechanism of differential MECP2 expression in the early brain development.

Evolutionarily conserved ARE and mi-R148/152 target sites were associated with the second alternative poly(A) site which results in the expression of longer isoform. Our data suggests that the stability and/or translation of the long MECP2 isoform, which is expected to be under strict post-transcriptional control, is potentially regulated via a cooperation between microRNAs/miRNPs and ARE-BPs.

G-quadruplex locations were found to be highly conserved near alternative splice sites of the MECP2 gene. Location-conserved G-quadruplexes in the vicinity of alternative splice sites are also highly conserved at sequence levels as compared to the G-quadruplexes found elsewhere in the MECP2 gene. We also discovered a bias in the overall distribution of conserved G-quadruplexes which were more likely to be associated with alternative splice sites of the mammalian MECP2 orthologs. Our data suggests a prospective biological role for G-quadruplexes in regulated alternative splicing of the MECP2 pre-mRNAs. We identified a pair of overlapping G-quadruplexes at an alternative 5′ splice site that could regulate alternative splicing in a negative as well as a positive way in the MECP2 pre-mRNAs.

This phylogenic analysis has provided some interesting and valuable insights into the post-transcriptional regulation of MECP2 gene by conserved *cis*-regulatory elements. The findings can help us further our understanding of mental retardation associated with this gene.

Methods

Several freely available public databases and bioinformatics sequence analysis tools were used for this project.

Sources of MECP2 Gene related information

The majority of the gene and sequence-related information was obtained from the database resources of National Center for Biotechnology Information (NCBI) [64]. Nucleotide and amino acid sequences of the human MECP2 gene and its orthologs were obtained from the RefSeq database [65]. The Entrez Gene database [66] was useful for obtaining alternative MECP2 isoforms and gene-related information. Exon/intron patterns were compared between the mRNA isoforms of the respective MECP2 orthologs to identify alternative and constitutive splice sites. MECP2 orthologs were identified with the help of Homologene database [64]. Several allelic variations and mutations were mapped to the human MECP2 gene with the help of OMIM database [4]. RettBASE [3] was also found to be a comprehensive collection of a wide variety of MECP2 mutations and phenotypes.

Sequence alignments

Pairwise sequence alignments were performed with a commercial program based on the Needleman and Wunsch algorithm [67]. Unless otherwise specified, all pairwise alignments used the semi-global method rather than the full global alignment because of the variation between the lengths of untranslated regions across orthologous mRNAs. ClustalW program [68] was used for multiple sequence alignments.

Mapping G-quadruplex sequence motifs

The QGRS Mapper [31] software program and the G-rich sequence database (GRSDB) [32] database were used to map QGRS (predicted G-quadruplexes) in the mRNA and pre-mRNA sequences of human MECP2 orthologs and generate information about the composition and distribution of QGRS in the nucleotide sequence entries. QGRS Mapper and GRSDB identify QGRS based on established algorithms which we have previously described in detail [31,69]. Briefly, the putative G-quadruplexes are identified using the motif $G_xN_{y1}G_xN_{y2}G_xN_{y3}G_x$. The motif consists of four guanine (G) tracts of equal size interspersed by three loops. The size of each G-tract corresponds to the number of stacked G-tetrads forming the quadruplex structure.

While quadruplexes with at least three G-tetrads have been accepted as stable structures, two G-tetrad quadruplexes are not uncommon [70,71]. In fact, stable two G-tetrad RNA G-quadruplexes capable of significantly influencing gene expression *in vivo* have been reported [16]. Lower stability, in fact, may allow more sensitive control of gene expression [16]. Two G-tetrads are expected to be far more prevalent in the genomes as compared to the three G-tetrads. We have employed two approaches to carefully weed out potential false positive predictions. All

predicted G-quadruplexes below a G-score [69] threshold of 13, representing the bottom 25% of all the G-quadruplexes in the entire human transcriptome predicted in our lab (data not presented), were discarded. Secondly, only the predicted G-quadruplexes which are phylogenetically conserved across a minimum of three mammalian MECP2 orthologs were analyzed, thereby validating our predictions.

It is widely accepted that the biological roles of G-quadruplexes depend primarily on their structure and location within the gene, rather than their sequence. The determinants of G-quadruplex homology are expected to be similarities in their specific locations on the aligned transcripts, number of tetrads, loop lengths, and overall lengths. Therefore, these criteria were adopted to identify evolutionarily conserved G-quadruplexes.

Polyadenylation signal and site mapping

Poly(A) signals and sites information was obtained either from the NCBI nucleotide database records [65] or polyA_DB database [72] which reports evolutionarily conserved sites.

AU-rich element mapping

AREs were mapped on the MECP2 mRNA orthologs with the help of the ARED database [73,74].

Mapping microRNA target sites

MicroRNA target sites were mapped to the 3'-UTRs of MECP2 mRNA orthologs with the help of TargetScan [75,76] which reports target sites conserved across multiple species.

Abbreviations

AMD: ARE-mediated mRNA degradation; APP: Amyloid precursor protein; ARE: AU-rich element; ARE-BPs: ARE-binding proteins; CDS: Coding sequence; DNA: Deoxyribose nucleic acid; ESE: Exonic splicing enhancer; FMR1: Fragile X mental retardation 1; FMRP: The fragile X mental retardation protein; GRSDB: G-rich sequence database; hnRNP: Heterogeneous nuclear ribonucleoprotein; IRES: Internal ribosomal entry site; MeCP2: Methyl CpG binding protein-2; miRNA: microRNA; miRNP: microRNA ribonucleoprotein; NCBI: National Center for Biotechnology Information; OMIM: Online mendelian inheritance in man; QGRS: Quadruplex forming G-rich sequences; RNA: Ribonucleic acid; UTR: Untranslated region; YY1: Ying Yang 1.

Competing interests

The authors declare that they have no competing interests.

Authors' contributions

JB initiated the project and performed the data collection and analysis. LD helped with the design and coordination of the project and with the draft of the manuscript. Both authors have read and approved the final manuscript.

Authors' information

JB was a high school student when the project began. He is now studying at Carnegie-Mellon University. LD is a Professor of Mathematics and Computer Science at Ramapo College of New Jersey.

Author details
[1]John P. Stevens High School, 855 Grove Ave., Edison, NJ 08820, USA. [2]Ramapo College of New Jersey, 505 Ramapo Valley Rd., Mahwah, NJ 07430, USA. [3]Carnegie Mellon University, 5000 Forbes Ave., Pittsburgh, PA 15213, USA.

References

1. Chadwick LH, Wade PA: MeCP2 in Rett syndrome: transcriptional repressor or chromatin architectural protein? *Curr Opin Genet Dev* 2007, 17:121–125.

2. Ben Zeev Ghidoni B: Rett syndrome. *Child Adolesc Psychiatr Clin N Am* 2007, 16:723–743.

3. Christodoulou J: RettBASE: IRSF MECP2 Variation Database. http://mecp2.chw.edu.au/.

4. Amberger J, Bocchini CA, Scott AF, Hamosh A: McKusick's Online Mendelian Inheritance in Man (OMIM). *Nucleic Acids Res* 2009, 37:D793–D796.

5. Hoffbuhr K, Devaney JM, LaFleur B, Sirianni N, Scacheri C, Giron J, Schuette J, Innis J, Marino M, Philippart M, Narayanan V, Umansky R, Kronn D, Hoffman EP, Naidu S: MeCP2 mutations in children with and without the phenotype of Rett syndrome. *Neurology* 2001, 56:1486–1495.

6. Coutinho AM, Oliveira G, Katz C, Feng J, Yan J, Yang C, Marques C, Ataíde A, Miguel TS, Borges L, Almeida J, Correia C, Currais A, Bento C, Mota-Vieira L, Temudo T, Santos M, Maciel P, Sommer SS, Vicente AM: MECP2 coding sequence and 3'UTR variation in 172 unrelated autistic patients. *Am J Med Genet B Neuropsychiatr Genet* 2007, 144B:475–483.

7. Gellert M, Lipsett MN, Davies DR: Helix formation by guanylic acid. *Proc Natl Acad Sci U S A* 1962, 48:2013–2018.

8. Balasubramanian S, Neidle S: G-quadruplex nucleic acids as therapeutic targets. *Curr Opin Chem Biol* 2009, 13:345–353.

9. Patel DJ, Phan AT, Kuryavyi V: Human telomere, oncogenic promoter and 5'-UTR G-quadruplexes: diverse higher order DNA and RNA targets for cancer therapeutics. *Nucleic Acids Res* 2007, 35:7429–7455.

10. Wu Y, Brosh RM Jr: G-quadruplex nucleic acids and human disease. *Febs J* 2010, 277:3470–3488.

11. Faudale M, Cogoi S, Xodo LE: Photoactivated cationic alkyl-substituted porphyrin binding to g4-RNA in the 5'-UTR of KRAS oncogene represses translation. *Chem Commun (Camb)* 2012, 48:874–876.

12. Baral A, Kumar P, Pathak R, Chowdhury S: Emerging trends in G-quadruplex biology - role in epigenetic and evolutionary events. *Mol Biosyst* 2013, 9(7):1568–1575.

13. Kumar P, Yadav VK, Baral A, Kumar P, Saha D, Chowdhury S: Zinc-finger transcription factors are associated with guanine quadruplex motifs in human, chimpanzee, mouse and rat promoters genome-wide. *Nucleic Acids Res* 2011, 39:8005–8016.

14. Saunders CJ, Friez MJ, Patterson M, Nzabi M, Zhao W, Bi C: Allele drop-out in the MECP2 gene due to G-quadruplex and i-motif sequences when using polymerase chain reaction-based diagnosis for Rett syndrome. *Genet Test Mol Biomarkers* 2010, 14:241–247.

15. Biffi G, Tannahill D, McCafferty J, Balasubramanian S: Quantitative visualization of DNA G-quadruplex structures in human cells. *Nat Chem* 2013, 5:182–186.

16. Wieland M, Hartig JS: RNA quadruplex-based modulation of gene expression. *Chem Biol* 2007, 14:757–763.

17. Mergny JL, De Cian A, Ghelab A, Sacca B, Lacroix L: Kinetics of tetramolecular quadruplexes. *Nucleic Acids Res* 2005, 33:81–94.

18. Bugaut A, Balasubramanian S: 5'-UTR RNA G-quadruplexes: translation regulation and targeting. *Nucleic Acids Res* 2012, 40:4727–4741.

19. Bonnal S, Schaeffer C, Créancier L, Clamens S, Moine H, Prats AC, Vagner S: A single internal ribosome entry site containing a G quartet RNA structure drives fibroblast growth factor 2 gene expression at four alternative translation initiation codons. *J Biol Chem* 2003, 278:39330–39336.

20. Morris MJ, Negishi Y, Pazsint C, Schonhoft JD, Basu S: An RNA G-quadruplex is essential for cap-independent translation initiation in human VEGF IRES. *J Am Chem Soc* 2010, 132:17831–17839.

21. Kumari S, Bugaut A, Huppert JL, Balasubramanian S: An RNA G-quadruplex in the 5' UTR of the NRAS proto-oncogene modulates translation. *Nat Chem Biol* 2007, 3:218–221.

22. Lammich S, Kamp F, Wagner J, Nuscher B, Zilow S, Ludwig AK, Willem M, Haass C: Translational repression of the disintegrin and metalloprotease ADAM10 by a stable G-quadruplex secondary structure in its 5'-untranslated region. *J Biol Chem* 2011, 286:45063–45072.

23. Halder K, Wieland M, Hartig JS: Predictable suppression of gene expression by 5'-UTR-based RNA quadruplexes. *Nucleic Acids Res* 2009, 37:6811–6817.

24. Endoh T, Kawasaki Y, Sugimoto N: Stability of RNA quadruplex in open reading frame determines proteolysis of human estrogen receptor alpha. *Nucleic Acids Res* 2013, 41(12):6222–6231.

25. Arhin GK, Boots M, Bagga PS, Milcarek C, Wilusz J: Downstream sequence elements with different affinities for the hnRNP H/H' protein influence the processing efficiency of mammalian polyadenylation signals. *Nucleic Acids Res* 2002, 30:1842–1850.

26. Millevoi S, Moine H, Vagner S: G-quadruplexes in RNA biology. *Wiley Interdiscip Rev RNA* 2012, 3:495–507.

27. Subramanian M, Rage F, Tabet R, Flatter E, Mandel JL, Moine H: G-quadruplex RNA structure as a signal for neurite mRNA targeting. *EMBO Rep* 2011, 12:697–704.

28. Huijbregts L, Roze C, Bonafe G, Houang M, Le Bouc Y, Carel JC, Leger J, Alberti P, Roux N: DNA polymorphisms of the KiSS1 3' untranslated region interfere with the folding of a G-rich sequence into G-quadruplex. *Mol Cell Endocrinol* 2012, 351:239–248.

29. Didiot MC, Tian Z, Schaeffer C, Subramanian M, Mandel JL, Moine H: The G-quartet containing FMRP binding site in FMR1 mRNA is a potent exonic splicing enhancer. *Nucleic Acids Res* 2008, 36:4902–4912.

30. Fisette JF, Montagna DR, Mihailescu MR, Wolfe MS: A G-rich element forms a G-quadruplex and regulates BACE1 mRNA alternative splicing. *J Neurochem* 2012, 121:763–773.

31. Kikin O, D'Antonio L, Bagga PS: QGRS Mapper: a web-based server for predicting G-quadruplexes in nucleotide sequences. *Nucleic Acids Res* 2006, 34:W676–W682.

32. Kikin O, Zappala Z, D'Antonio L, Bagga PS: GRSDB2 and GRS_UTRdb: databases of quadruplex forming G-rich sequences in pre-mRNAs and mRNAs. *Nucleic Acids Res* 2008, 36:D141–D148.

33. Huppert JL, Balasubramanian S: Prevalence of quadruplexes in the human genome. *Nucleic Acids Res* 2005, 33:2908–2916.

34. Todd AK: Bioinformatics approaches to quadruplex sequence location. *Methods* 2007, 43:246–251.

35. Huppert JL: Hunting G-quadruplexes. *Biochimie* 2008, 90:1140–1148.

36. Huppert JL: Structure, location and interactions of G-quadruplexes. *FEBS J* 2010, 277:3452–3458.

37. Huppert JL, Bugaut A, Kumari S, Balasubramanian S: G-quadruplexes: the beginning and end of UTRs. *Nucleic Acids Res* 2008, 36:6260–6268.

38. Zhang R, Su B: Small but influential: the role of microRNAs on gene regulatory network and 3'UTR evolution. *J Genet Genomics* 2009, 36:1–6.

39. Wada R, Akiyama Y, Hashimoto Y, Fukamachi H, Yuasa Y: miR-212 is downregulated and suppresses methyl-CpG-binding protein MeCP2 in human gastric cancer. *Int J Cancer* 2010, 127:1106–1114.

40. Khabar KS: The AU-rich transcriptome: more than interferons and cytokines, and its role in disease. *J Interferon Cytokine Res* 2005, 25:1–10.

41. Huang W, Smaldino PJ, Zhang Q, Miller LD, Cao P, Stadelman K, Wan M, Giri B, Lei M, Nagamine Y, Vaughn JP, Akman SA, Sui G: Yin Yang 1 contains G-quadruplex structures in its promoter and 5'-UTR and its expression is modulated by G4 resolvase 1. *Nucleic Acids Res* 2011, 40(3):1033–1049.

42. Saxena A, de Lagarde D, Leonard H, Williamson SL, Vasudevan V, Christodoulou J, Thompson E, MacLeod P, Ravine D: Lost in translation: translational interference from a recurrent mutation in exon 1 of MECP2. *J Med Genet* 2006, 43:470–477.

43. Abdelmohsen K, Tominaga K, Lee EK, Srikantan S, Kang MJ, Kim MM, Selimyan R, Martindale JL, Yang X, Carrier F, Zhan M, Becker KG, Gorospe M: Enhanced translation by nucleolin via G-rich elements in coding and non-coding regions of target mRNAs. *Nucleic Acids Res* 2011, 39:8513–8530.

44. Darnell JC, Jensen KB, Jin P, Brown V, Warren ST, Darnell RB: Fragile X mental retardation protein targets G quartet mRNAs important for neuronal function. *Cell* 2001, 107:489–499.

45. Wang H, Ku L, Osterhout DJ, Li W, Ahmadian A, Liang Z, Feng Y: Developmentally-programmed FMRP expression in oligodendrocytes: a potential role of FMRP in regulating translation in oligodendroglia progenitors. *Hum Mol Genet* 2004, 13:79–89.

46. Nishimura Y, Martin CL, Vazquez-Lopez A, Spence SJ, Alvarez-Retuerto AI, Sigman M, Steindler C, Pellegrini S, Schanen NC, Warren ST, Geschwind DH: Genome-wide expression profiling of lymphoblastoid cell lines distinguishes different forms of autism and reveals shared pathways. *Hum Mol Genet* 2007, 16:1682–1698.

47. Simonsson T: G-quadruplex DNA structures–variations on a theme. *Biol Chem* 2001, 382:621–628.

48. Coy JF, Sedlacek Z, Bachner D, Delius H, Poustka A: A complex pattern of evolutionary conservation and alternative polyadenylation within the long 3'-untranslated region of the methyl-CpG-binding protein 2 gene (MeCP2) suggests a regulatory role in gene expression. *Hum Mol Genet* 1999, **8**:1253–1262.

49. Bagga PS, Arhin GK, Wilusz J: DSEF-1 is a member of the hnRNP H family of RNA-binding proteins and stimulates pre-mRNA cleavage and polyadenylation in vitro. *Nucleic Acids Res* 1998, **26**:5343–5350.

50. Millevoi S, Decorsière A, Loulergue C, Iacovoni J, Bernat S, Antoniou M, Vagner S: A physical and functional link between splicing factors promotes pre-mRNA 3' end processing. *Nucleic Acids Res* 2009, **37**:4672–4683.

51. Decorsière A, Cayrel A, Vagner S, Millevoi S: Essential role for the interaction between hnRNP H/F and a G quadruplex in maintaining p53 pre-mRNA 3'-end processing and function during DNA damage. *Genes Dev* 2011, **25**:220–225.

52. Newnham CM, Hall-Pogar T, Liang S, Wu J, Tian B, Hu J, Lutz CS: Alternative polyadenylation of MeCP2: influence of cis-acting elements and trans-acting factors. *RNA Biol* 2010, **7**:361–372.

53. Lattmann S, Giri B, Vaughn JP, Akman SA, Nagamine Y: Role of the amino terminal RHAU-specific motif in the recognition and resolution of guanine quadruplex-RNA by the DEAH-box RNA helicase RHAU. *Nucleic Acids Res* 2010, **38**:6219–6233.

54. Bagga PS, Ford LP, Chen F, Wilusz J: The G-rich auxiliary downstream element has distinct sequence and position requirements and mediates efficient 3' end pre-mRNA processing through a trans-acting factor. *Nucleic Acids Res* 1995, **23**:1625–1631.

55. Veraldi KL, Arhin GK, Martincic K, Chung-Ganster LH, Wilusz J, Milcarek C: hnRNP F influences binding of a 64-kilodalton subunit of cleavage stimulation factor to mRNA precursors in mouse B cells. *Mol Cell Biol* 2001, **21**:1228–1238.

56. Han K, Gennarino VA, Lee Y, Pang K, Hashimoto-Torii K, Choufani S, Raju CS, Oldham MC, Weksberg R, Rakic P, Liu Z, Zoghbi HY: Human-specific regulation of MeCP2 levels in fetal brains by microRNA miR-483-5p. *Genes Dev* 2013, **27**:485–490.

57. Steitz JA, Vasudevan S: miRNPs: versatile regulators of gene expression in vertebrate cells. *Biochem Soc Trans* 2009, **37**:931–935.

58. Stoecklin G, Colombi M, Raineri I, Leuenberger S, Mallaun M, Schmidlin M, Gross B, Lu M, Kitamura T, Moroni C: Functional cloning of BRF1, a regulator of ARE-dependent mRNA turnover. *Embo J* 2002, **21**:4709–4718.

59. Peng SS, Chen CY, Xu N, Shyu AB: RNA stabilization by the AU-rich element binding protein, HuR, an ELAV protein. *Embo J* 1998, **17**:3461–3470.

60. Bindra RS, Wang JTL, Bagga PS: Bioinformatics methods for studying microRNA and ARE mediated regulation of post-transcriptional gene expression. *Int J Knowl Discov Bioinform* 2010, **1**:97–112.

61. von Roretz C, Gallouzi IE: Decoding ARE-mediated decay: is microRNA part of the equation? *J Cell Biol* 2008, **181**:189–194.

62. Chou MY, Rooke N, Turck CW, Black DL: hnRNP H is a component of a splicing enhancer complex that activates a c-src alternative exon in neuronal cells. *Mol Cell Biol* 1999, **19**:69–77.

63. Marcel V, Tran PL, Sagne C, Martel-Planche G, Vaslin L, Teulade-Fichou MP, Hall J, Mergny JL, Hainaut P, Van Dyck E: G-quadruplex structures in TP53 intron 3: role in alternative splicing and in production of p53 mRNA isoforms. *Carcinogenesis* 2011, **32**:271–278.

64. Acland AAR, Barrett T, Beck J, Benson DA, Bollin C, Bolton E, Bryant SH, Canese K, Church DM, Clark K, DiCuccio M, Dondoshansky I, Federhen S, Feolo M, Geer LY, Gorelenkov V, Hoeppner M, Johnson M, Kelly C, Khotomlianski V, Kimchi A, Kimelman M, Kitts P, Krasnov S, Kuznetsov A, Landsman D, Lipman DJ, Lu Z, Madden TL, Madej T, et al: Database resources of the national center for biotechnology information. *Nucleic Acids Res* 2013, **41**:D8–D20.

65. Pruitt KD, Tatusova T, Brown GR, Maglott DR: NCBI Reference Sequences (RefSeq): current status, new features and genome annotation policy. *Nucleic Acids Res* 2012, **40**:D130–D135.

66. Maglott D, Ostell J, Pruitt KD, Tatusova T: Entrez Gene: gene-centered information at NCBI. *Nucleic Acids Res* 2011, **39**:D52–D57.

67. Gotoh O: An improved algorithm for matching biological sequences. *J Mol Biol* 1982, **162**:705–708.

68. Thompson JD, Gibson TJ, Higgins DG: Multiple sequence alignment using ClustalW and ClustalX. *Curr Protoc Bioinformatics* 2002, **2**:Unit 2.3.

69. D'Antonio L, Bagga PS: Computational methods for predicting intramolecular G-quadruplexes in nucleotide sequences. *Comput Syst Bioinform, IEEE: CSB* 2004, **2004**:561–562.

70. Kankia BI, Barany G, Musier-Forsyth K: Unfolding of DNA quadruplexes induced by HIV-1 nucleocapsid protein. *Nucleic Acids Res* 2005, **33**:4395–4403.

71. Zarudnaya MI, Kolomiets IM, Potyahaylo AL, Hovorun DM: Downstream elements of mammalian pre-mRNA polyadenylation signals: primary, secondary and higher-order structures. *Nucleic Acids Res* 2003, **31**:1375–1386.

72. Lee JY, Yeh I, Park JY, Tian B: PolyA_DB 2: mRNA polyadenylation sites in vertebrate genes. *Nucleic Acids Res* 2007, **35**:D165–D168.

73. Halees AS, El-Badrawi R, Khabar KS: ARED Organism: expansion of ARED reveals AU-rich element cluster variations between human and mouse. *Nucleic Acids Res* 2008, **36**:D137–D140.

74. Bakheet T, Williams BR, Khabar KS: ARED 3.0: the large and diverse AU-rich transcriptome. *Nucleic Acids Res* 2006, **34**:D111–D114.

75. Lewis BP, Burge CB, Bartel DP: Conserved seed pairing, often flanked by adenosines, indicates that thousands of human genes are microRNA targets. *Cell* 2005, **120**:15–20.

76. Grimson A, Farh KK, Johnston WK, Garrett-Engele P, Lim LP, Bartel DP: MicroRNA targeting specificity in mammals: determinants beyond seed pairing. *Mol Cell* 2007, **27**:91–105.

Permissions

List of Contributors

James J Galligan
Department of Pharmacology, University of Colorado Anschutz Medical Campus, Aurora, CO 80045, USA

Dennis R Petersen
Molecular Toxicology and Environmental Health Sciences Program, Department of Pharmacy and Pharmaceutical Sciences, University of Colorado Anschutz Medical Campus, Aurora, CO 80045, USA

Suqin Liu, Hongjiang Wang, Lizhi Zhang, Liying Ban, Aman Wang and Tao Zhou
The First Affiliated Hospital of Dalian Medical University, Dalian, Liaoning, China

Zhiyuan Liu, Feng Lou, Dandan Zhang, Hong Sun, Haichao Dong, Guangchun Zhang, Zhishou Dong, Baishuai Guo, He Yan, Chaowei Yan, Lu Wang, Ziyi Su, Yangyang Li, Chuanning Tang and Hua Ye
San Valley Biotechnology Incorporated, Beijing, China

Xue F Huang, Si-Yi Chen and Lindsey Jones
Norris Liu et al. Human Genomics, Comprehensive Cancer Center, Department of Molecular Microbiology and Immunology, Keck School of Medicine, University of Southern California, Los Angeles, CA, USA

Abijeet Singh Mehta, Kirti Snigdha and Panagiotis A Tsonis
Department of Biology, University of Dayton, Dayton, OH, USA

M Sharada Potukuchi
School of Biotechnology, Shri Mata Vaishno Devi University, Katra, J&K, India

N. V. Punina and A. F. Topunov
Bach Institute of Biochemistry, Russian Academy of Science, Moscow 119071, Russia

N. M. Makridakis
Tulane University School of Public Health and Tropical Medicine, New Orleans, LA 70112, USA

M. A. Remnev
The Federal State Unitary Enterprise All-Russia Research Institute of Automatics, Moscow 127055, Russia

Mohammad Shabbir Hasan and Liqing Zhang
Department of Computer Science, Virginia Tech, Blacksburg, VA 24061, USA

Xiaowei Wu
Department of Statistics, Virginia Tech, Blacksburg, VA 24061, USA

Matahi Moarii and Jean-Philippe Vert
CBIO-Centre for Computational Biology, Mines Paristech, PSL-Research University, 35 Rue Saint-Honore, F-77300 Fontainebleau, France
Department of Bioinformatics, Biostatistics and System Biology, Institut Curie, 11-13 Rue Pierre et Marie Curie, F-75248 Paris, France
U900, INSERM, 11-13 Rue Pierre et Marie Curie, F-75248 Paris, France

Fabien Reyal
UMR932, Immunity and Cancer Team, Institut Curie, 26 Rue d'Ulm, 75006 Paris, France
Department of Translational Research, Residual Tumor and Response to Treatment Team, Institut Curie, 26 Rue d'Ulm, 75006 Paris, France
Department of Surgery, Institut Curie, 26 Rue d'Ulm, 75006 Paris, France

Jianxin Wang
School of Information Science and Engineering, Central South University, Changsha 410083, China

Xueyong Li
School of Information Science and Engineering, Central South University, Changsha 410083, China
Department of Information and Computing Science, Changsha University, Changsha 410003, China

Bihai Zhao
Department of Information and Computing Science, Changsha University, Changsha 410003, China

Fang-Xiang Wu
Department of Mechanical Engineering and Division of Biomedical Engineering, University of Saskatchewan, Saskatoon, SK S7N 5A9, Canada

Yi Pan
Department of Computer Science, Georgia State University, Atlanta, GA 30302-4110, USA

Songtao Ben, Tracey M. Ferrara and Richard A. Spritz
Human Medical Genetics Program, University of Colorado, Aurora, CO 80045, USA

Rhonda M. Cooper-DeHoff
Center for Pharmacogenomics, College of Pharmacy, University of Florida, Gainesville, FL 32610, USA

Hanna K. Flaten and Oghenero Evero
Department of Emergency Medicine, School of Medicine, University of Colorado, 12401 E. 17th Ave, B215, Aurora, CO 80045, USA

Andrew A. Monte
Department of Emergency Medicine, School of Medicine, University of Colorado, 12401 E. 17th Ave, B215, Aurora, CO 80045, USA
Rocky Mountain Poison & Drug Center, Denver Health and Hospital Authority, Denver, CO 80203, USA

Dan Xi, Jinzhen Zhao and Zhigang Guo
Division of Cardiology, Huiqiao Medical Center, Nanfang Hospital, Southern Medical University, 1838 North Guangzhou Avenue, Guangzhou 510515, Guangdong, People's Republic of China

Wenyan Lai
Laboratory of Department of Cardiology, Nanfang Hospital, Southern Medical University, Guangzhou 510515, Guangdong, People's Republic of China

Jin-Woo Park, Jae-Mok Lee and Jo-Young Suh
Department of Periodontology, School of Dentistry, Kyungpook National University, Daegu 41940, Korea

Yong-Gun Kim
Department of Periodontology, School of Dentistry, Kyungpook National University, Daegu 41940, Korea
Institute for Hard Tissue and Bone Regeneration, Kyungpook National University, Daegu 41940, Korea

Jae-Young Kim and Youngkyun Lee
Institute for Hard Tissue and Bone Regeneration, Kyungpook National University, Daegu 41940, Korea
Department of Biochemistry, School of Dentistry, Kyungpook National University, 2177 Dalgubeol-daero, Joong-gu, Daegu 41940, Korea

Minjung Kim
Department of Life and Nanopharmaceutical Sciences, Kyung Hee University, Seoul 02447, Korea

Jae-Hyung Lee
Department of Life and Nanopharmaceutical Sciences, Kyung Hee University, Seoul 02447, Korea
Department of Maxillofacial Biomedical Engineering, School of Dentistry, Kyung Hee University, 26 Kyunghee-daero, Dongdaemun-gu, Seoul 02447, Korea

Ji Hyun Kang and Hyo Jeong Kim
Department of Biochemistry, School of Dentistry, Kyungpook National University, 2177 Dalgubeol-daero, Joong-gu, Daegu 41940, Korea

Bihai Zhao, Sai Hu, Xueyong Li, Fan Zhang, Qinglong Tian and Wenyin Ni
Department of Mathematics and Computing Science, Changsha University, Changsha, Hunan 410022, China

Benjamin Capps
Department of Bioethics, Faculty of Medicine, Dalhousie University, Halifax, Canada

Ruth Chadwick
School of Law, University of Manchester, Manchester, UK

Yann Joly
Department of Human Genetics, Centre of Genomics and Policy, McGill University, Québec, Canada

John J. Mulvihill
Department of Pediatrics, University of Oklahoma Health Sciences Center, Oklahoma, USA

Tamra Lysaght
Centre for Biomedical Ethics, Yong Loo Lin School of Medicine, National University of Singapore, Singapore, Singapore

Hub Zwart
Faculty of Science, Department of Philosophy and Science Studies, Radboud University Nijmegen, Nijmegen, The Netherlands

Nicholas J. W. Rattray and Nicole C. Deziel
Department of Environmental Health Sciences, Yale School of Public Health, Yale University, New Haven, CT, USA

Joshua D. Wallach
Collaboration for Research Integrity and Transparency (CRIT), Yale Law School, New Haven, CT, USA
Center for Outcomes Research and Evaluation (CORE), Yale-New Haven Health System, New Haven, CT, USA

Sajid A. Khan
Department of Surgery, Section of Surgical Oncology, Yale University School of Medicine, New Haven, CT, USA
Yale Cancer Center, Yale University School of Medicine, New Haven, CT, USA

Vasilis Vasiliou and Caroline H. Johnson
Department of Environmental Health Sciences, Yale School of Public Health, Yale University, New Haven, CT, USA
Yale Cancer Center, Yale University School of Medicine, New Haven, CT, USA

Niv Sabath and Reut Shalgi
Department of Biochemistry, Rappaport Faculty of Medicine, Technion — Israel Institute of Technology, 31096 Haifa, Israel

Anna Vilborg and Joan A. Steitz
Department of Molecular Biophysics and Biochemistry, Howard Hughes Medical Institute, Boyer Center for Molecular Medicine, Yale University School of Medicine, 295 Congress Avenue, New Haven, CT 06536, USA

Ene Reimann and Sulev Kõks
Department of Pathophysiology, University of Tartu, 19 Ravila Street Tartu 50411, Estonia
Department of Reproductive Biology, Estonian University of Life Sciences, 64 Kreutzwaldi Street, Tartu, Estonia

John P. A. Ioannidis
Stanford Prevention Research Center, Department of Medicine, Stanford University, Stanford, CA, USA
Department of Health Research and Policy, Stanford University, Stanford, CA, USA
Department of Biomedical Data Science, Stanford University, Stanford, CA, USA
Department of Statistics, Stanford University, Stanford, CA, USA
Meta-Research Innovation Center at Stanford, Stanford University, Stanford, CA, USA

Xuan Dung Ho
Department of Traumatology and Orthopaedics, University of Tartu, 8 Puusepa Street, Tartu, Estonia
Department of Oncology, Hue University of Medicine and Pharmacy, 6 Ngo Quyen Street, Hue, Vietnam

Katre Maasalu and Aare Märtson
Department of Traumatology and Orthopaedics, University of Tartu, 8 Puusepa Street, Tartu, Estonia
Traumatology and Orthopaedics Clinic, Tartu University Hospital, 8 Puusepa Street, Tartu, Estonia

Donna S Mackay, Thomas M Bennett, Susan M Culican and Alan Shiels
Department of Ophthalmology and Visual Sciences, Washington University School of Medicine, 660 S. Euclid Ave., St. Louis, Missouri 63110, USA

Bruce Gottlieb
Lady Davis Institute for Medical Research, 3755 Côte Ste Catherine Road, Montreal, QC H3T 1E2, Canada
Segal Cancer Centre, Jewish General Hospital, 3755 Côte Ste Catherine Road, Montreal, QC H3T 1E2, Canada
Department of Human Genetics, McGill University, Montreal, QC, Canada

Lenore K Beitel
Lady Davis Institute for Medical Research, 3755 Côte Ste Catherine Road, Montreal, QC H3T 1E2, Canada
Department of Human Genetics, McGill University, Montreal, QC, Canada
Department of Medicine, McGill University, Montreal, QC, Canada

Mark Trifiro
Lady Davis Institute for Medical Research, 3755 Côte Ste Catherine Road, Montreal, QC H3T 1E2, Canada
Segal Cancer Centre, Jewish General Hospital, 3755 Côte Ste Catherine Road, Montreal, QC H3T 1E2, Canada
Department of Human Genetics, McGill University, Montreal, QC, Canada
Department of Medicine, McGill University, Montreal, QC, Canada

Joetsaroop S Bagga
John P. Stevens High School, 855 Grove Ave., Edison, NJ 08820, USA
Carnegie Mellon University, 5000 Forbes Ave., Pittsburgh, PA 15213, USA

Lawrence A D'Antonio
Ramapo College of New Jersey, 505 Ramapo Valley Rd., Mahwah, NJ 07430, USA

Index

www.ingramcontent.com/pod-product-compliance
Lightning Source LLC
Chambersburg PA
CBHW082032190326
41458CB00010B/3342